Security, Design, and Architecture for Broadband and Wireless Network Technologies

Naveen Chilamkurti
La Trobe University, Australia

Managing Director: Lindsay Johnston
Editorial Director: Joel Gamon
Production Manager: Jennifer Yoder
Publishing Systems Analyst: Adrienne Freeland
Assistant Acquisitions Editor: Kayla Wolfe
Typesetter: Lisandro Gonzalez
Cover Design: Jason Mull

Published in the United States of America by
Information Science Reference (an imprint of IGI Global)
701 E. Chocolate Avenue
Hershey PA 17033
Tel: 717-533-8845
Fax: 717-533-8661
E-mail: cust@igi-global.com
Web site: http://www.igi-global.com

Library of Congress Cataloging-in-Publication Data

Security, design, and architecture for broadband and wireless network technologies / Naveen Chilamkurti, editor.
pages cm
Summary: "This book provides a discussion on the latest research achievements in wireless networks and broadband technology, highlighting new trends, applications, developments, and standards"--Provided by publisher.
Includes bibliographical references.
ISBN 978-1-4666-3902-7 (hardcover) -- ISBN 978-1-4666-3903-4 (ebook) -- ISBN 978-1-4666-3904-1 (print & perpetual access) 1. Broadband communication systems--Research. 2. Wireless communication systems--Research. 3. Computer networks--Research. 4. Computer security--Research. 5. Computer architecture--Research. I. Chilamkurti, Naveen, 1974-
TK5103.4.S42 2013
005.8--dc23
2012048659

British Cataloguing in Publication Data
A Cataloguing in Publication record for this book is available from the British Library.

Editorial Advisory Board

Associate Editors

Table of Contents

Preface xvii

Chapter 1
Joint Angular and Time Diversity of Multi-Antenna CDMA Systems in Wireless Fading Channels ... 1
Feng She, Alcatel-Lucent Shanghai Bell, China
Hsiao Hwa Chen, National Cheng Kung University, Taiwan
Hongyang Li, National Sun Yat-Sen University, Taiwan

Chapter 2
Improving WLAN Performance by Modifying an IEEE 802.11 Protocol 15
Nurul I. Sarkar, Auckland University of Technology, New Zealand

Chapter 3
Insights from Experimental Research on Distributed Channel Assignment in Wireless Testbeds 33
Felix Juraschek, Freie Universität Berlin, Germany
Mesut Günes, Freie Universität Berlin, Germany
Matthias Philipp, Freie Universität Berlin, Germany
Bastian Blywis, Freie Universität Berlin, Germany

Chapter 4
HTTP Traffic Model for Web2.0 and Future WebX.0 53
Vladimir Deart, Moscow Technical University of Communications and Informatics, Russia
Alexander Pilugin, Moscow Technical University of Communications and Informatics, Russia

Chapter 5
DMT Optimal Cooperative MAC Protocols in Wireless Mesh Networks with Minimized Signaling Overhead 60
Benoît Escrig, Universite de Toulouse, France

Chapter 6
Performance Evaluation of a Three Node Client Relay System 78
Sergey Andreev, Tampere University of Technology, Finland
Olga Galinina, Tampere University of Technology, Finland
Alexey Vinel, Tampere University of Technology, Finland

Chapter 7
BER Fairness and PAPR Study of Interleaved OFDMA System 91
Sabbir Ahmed, Ritsumeikan University, Japan
Makoto Kawai, Ritsumeikan University, Japan

Chapter 8
Lifetime Maximization in Wireless Sensor Networks 107
Vivek Katiyar, National Institute of Technology Hamirpur, India
Narottam Chand, National Institute of Technology Hamirpur, India
Surender Soni, National Institute of Technology Hamirpur, India

Chapter 9
Doubly Cognitive Architecture Based Cognitive Wireless Sensor Networks 121
Sumit Kumar, International Institute of Information Technology, Hyderabad, India
Deepti Singhal, International Institute of Information Technology, Hyderabad, India
Garimella Rama Murthy, International Institute of Information Technology, Hyderabad, India

Chapter 10
Synchronous Relaying in Vehicular Ad-hoc Networks 127
D. Moltchanov, Tampere University of Technology, Finland
A. Vinel, Tampere University of Technology, Finland
J. Jakubiak, Tampere University of Technology, Finland
Y. Koucheryavy, Tampere University of Technology, Finland

Chapter 11
Mobile WiMAX Bandwidth Reservation Thresholds: A Heuristic Approach 133
Sondes Khemiri, University of Paris 6, France
Khaled Boussetta, University of Paris 13, France
Nadjib Achir, University of Paris 13, France
Guy Pujolle, University of Paris 6, France

Chapter 12
IP Paging for Mobile Hosts in Distributed and Fixed Hierarchical Mobile IP 154
Paramesh C. Upadhyay, Sant Longowal Institute of Engineering & Technology, India
Sudarshan Tiwari, Motilal Nehru National Institute of Technology, India

Chapter 13
Cooperation Among Members of Online Communities: Profitable Mechanisms to Better
Distribute Near-Real-Time Services 170
M. L. Merani, University of Modena and Reggio Emilia, Italy
M. Capetta, University of Modena and Reggio Emilia, Italy
D. Saladino, University of Modena and Reggio Emilia, Italy

Detailed Table of Contents

In this paper, a multi-antenna based receiver structure for direct sequence code division multiple access (DS/CDMA) system is proposed. The proposed scheme exploits the excellent time resolution of a CDMA RAKE receiver and uses an antenna array beamforming structure to resolve multipath returns in both angular and time domains. A much higher diversity gain than that based only on the time domain diversity can be achieved. This work suggests a new space-time diversity paradigm, namely angular-time diversity, which differs from traditional Alamouti-type space-time coded schemes. The impairments caused by multipath and multiuser interference are analyzed. The performance of the proposed receiver in multipath fading channel is explicitly evaluated. An expression for uncoded system bit error probability is derived. Simulation results show the performance improvement in terms of BER due to the use of multi-antenna in the receiver, and the results illustrate that the multi-antenna based receiver works effectively in resolving multipaths in both angular and time domains to achieve performance improvement due to angular and time diversity gain provided by the multi-antenna system.

One of the limitations of the IEEE 802.11 distributed coordination function (DCF) protocol is its low bandwidth utilization under medium-to-high traffic loads resulting in low throughput and high packet delay. To overcome performance problems, traditional IEEE 802.11 DCF ("DCF") protocol is modified to the buffer unit multiple access (BUMA) protocol. The BUMA protocol achieves a better system performance by introducing a temporary buffer unit at the medium access control (MAC) layer to accumulate multiple packets and combine them into a single packet (with a header and a trailer) before transmission. This paper provides an in-depth performance evaluation (by simulation) of BUMA for multiuser ad hoc and infrastructure networks. Results obtained show that the BUMA is more efficient than that of DCF. The BUMA protocol is simple and its algorithm (software) can be upgraded to 802.11 networks requiring no hardware changes. The BUMA protocol is described and simulation results are presented to verify the performance.

This article presents the DES-Chan framework for experimental research on distributed channel assignment algorithms in wireless mesh testbeds. The implementation process of channel assignment algorithms is a difficult task for the researcher since common operating systems do not support channel assignment algorithms. DES-Chan provides a set of common services required by distributed channel assignment algorithms and eases the implementation effort. The results of experiments to measure the channel characteristics in terms of intra-path and inter-path interference according to the channel distance on the DES-Testbed are also presented. The DES-Testbed is a multi-radio WMN with more than 100 nodes located on the campus of the Freie Universität Berlin. These measurements are an important input to validate common assumptions of WMNs and derive more realistic, measurement-based interference models in contrast to simplified heuristics.

This article presents a method of estimation for HTTP traffic quality service parameters mean delay and lost packets percentage. This method, based on statistic measurements, includes simulation and analytical modeling. Statistical HTTP traffic models presented earlier take into account typical features of WEB2.0 Internet traffic, which were used for the simulation model. Developed universal simulation models make it possible to research service quality parameters under setting network conditions over a wide range considering Internet development. The presented analytical method based on batch packet arrival model allows an accuracy estimation of mean HTTP-packets delay in Core Router by simple calculations. Objective results of HTTP traffic quality service parameters can be used in QoS standard development for WEB traffic and model QoE standard development.

In this paper, a cooperative protocol is proposed for wireless mesh networks. Two features are implemented: on-demand cooperation and selection of the best relay. First, cooperation is activated by a destination terminal when it fails in decoding the message from a source terminal. Second, a selection of the best relay is performed when cooperation is needed. The robustness of wireless links are increased while the resource consumption is minimized. The selection of the best relay is performed by a splitting algorithm, ensuring a fast selection process, the duration of which is now fully characterized. Only terminals that improve the direct link participate in the relay selection and inefficient cooperation is avoided. The proposed protocol is demonstrated to achieve an optimal diversity-multiplexing trade-off. This study focuses on Nakagami- wireless channel models to encompass a variety of fading models in the context of wireless mesh networks.

In this paper, the authors examine a client relay system comprising three wireless nodes. Closed-form expressions for mean packet delay, as well as for throughput, energy expenditure, and energy efficiency of the source nodes are also obtained. The precision of the established parameters is verified by means of simulation.

For an OFDMA system, the role of interleavers is analyzed to ensure fairness of BER performance among the active users and investigate their respective PAPR properties. In this paper, the authors consider a generic system and show that for a slowly changing multipath channel, individual user's BER performance can vary, implying that the propagation channel effect is unfairly distributed on the users. Applying different types of frequency interleaving mechanisms, the authors demonstrate that random interleaving can ensure BER fairness on an individual user basis but the associated system overhead for de-interleaving is very high. In this context, the authors introduce the application of cyclically shifted random interleaver and demonstrate its effectiveness in achieving BER fairness (dispersion in individual users BER reduced by 94% compared to no interleaving at 20dB SNR) with little system overhead. The authors also explore the comparative performance of different interleavers for scenarios with varying number of total subcarriers and subcarriers per user. Based on the scenario specific results, the authors conclude that for a heavily loaded system, i.e., relatively low number of subcarriers per user, cyclically shifted random interleavers can effectively ensure uniform performance among active users with reduced system complexity and manageable PAPR.

One of the fundamental requirements in wireless sensor networks (WSNs) is to prolong the lifetime of sensor nodes by minimizing the energy consumption. The information about the energy status of sensor nodes can be used to notify the base station about energy depletion in any part of the network. An energy map of WSN can be constructed with available remaining energy at sensor nodes. The energy map can increase the lifetime of sensor networks by adaptive clustering, energy centric routing, data aggregation, and so forth. In this paper, the authors describe use of energy map techniques for WSNs and summarize the applications in routing, aggregation, clustering, data dissemination, and so forth. The authors also present an energy map construction algorithm that is based on prediction.

Chapter 9
Doubly Cognitive Architecture Based Cognitive Wireless Sensor Networks 121
Sumit Kumar, International Institute of Information Technology, Hyderabad, India
Deepti Singhal, International Institute of Information Technology, Hyderabad, India
Garimella Rama Murthy, International Institute of Information Technology, Hyderabad, India

Scarcity of spectrum is increasing not only in cellular communication but also in wireless sensor networks. Adding cognition to the existing wireless sensor network (WSN) infrastructure has helped. As sensor nodes in WSN are limited with constraints like power, efforts are required to increase the lifetime and other performance measures of the network. In this article, the authors propose Doubly Cognitive WSN, which works by progressively allocating the sensing resources only to the most promising areas of the spectrum and is based on pattern analysis and learning. As the load of sensing resource is reduced significantly, this approach saves the energy of the nodes and reduces the sensing time dramatically. The proposed method can be enhanced by periodic pattern analysis to review the strategy of sensing. Finally the ongoing research work and contribution on cognitive wireless sensor networks in Communication Research Centre (IIIT-H) is discussed.

Chapter 10
Synchronous Relaying in Vehicular Ad-Hoc Networks 127
D. Moltchanov, Tampere University of Technology, Finland
A. Vinel, Tampere University of Technology, Finland
J. Jakubiak, Tampere University of Technology, Finland
Y. Koucheryavy, Tampere University of Technology, Finland

In this paper, the authors propose a simple concept for emergency information dissemination in vehicular ad-hoc networks. Instead of competing for the shared wireless medium when transmitting the emergency information, the authors' proposed method requires nodes to cooperate by synchronizing their transmissions. The proposed scheme is backward compatible with IEEE 802.11p carrier sense multiple access with collision avoidance. The authors also briefly address some of the implementation issues of the proposed scheme.

Chapter 11
Mobile WiMAX Bandwidth Reservation Thresholds: A Heuristic Approach.................................. 133
Sondes Khemiri, University of Paris 6, France
Khaled Boussetta, University of Paris 13, France
Nadjib Achir, University of Paris 13, France
Guy Pujolle, University of Paris 6, France

This paper addresses the issue of wireless bandwidth partitioning of a Mobile WiMAX cell. The authors consider a Complete Partitioning strategy, where the wireless bandwidth capacity of a cell is divided into trunks. Each partition is strictly reserved to a particular type of connection. Four IEEE 802.16e 2005 service classes are distinguished: UGS, rtPS, nrtPS, and ErtPS. The authors consider mobility and differentiate new call request from handoffs. In addition, the authors take into consideration the Adaptive Modulation and Coding (AMC) scheme, through the partition of the cell into different areas associated to a particular modulation and coding scheme. The purpose of the paper is to determine, using an analytical model and a heuristic approach, the nearly optimal sizes of the partition sizes dedicated to each type of connection, which is characterized by its service class, type of request and modulation, and coding scheme.

The concept of Paging has been found useful in existing cellular networks for mobile users with low call-to-mobility ratio (CMR). It is necessary for fast mobility users to minimize the signaling burden on the network. Reduced signaling, also, conserves scarce wireless resources and provides power savings at user terminals. However, Mobile IP (MIP), a base protocol for IP mobility, does not support paging concept in its original form. Several paging schemes and micro-mobility protocols, centralized and distributed, have been proposed in literature to alleviate the inherent limitations of Mobile IP. In this paper, the authors propose three paging schemes for Distributed and Fixed Hierarchical Mobile IP (DFHMIP) and develop analytical models for them. Performance evaluations of these schemes have been carried out and results have been compared with DFHMIP without paging and with Dynamic Hierarchical Mobile IP (DHMIP) for low CMR values.

Today some of the most popular and successful applications over the Internet are based on Peer-to-Peer (P2P) solutions. Online Social Networks (OSN) represent a stunning phenomenon too, involving communities of unprecedented size, whose members organize their relationships on the basis of social or professional friendship. This work deals with a P2P video streaming platform and focuses on the performance improvements that can be granted to those P2P nodes that are also members of a social network. The underpinning idea is that OSN friends (and friends of friends) might be more willing to help their mates than complete strangers in fetching the desired content within the P2P overlay. Hence, an approach is devised to guarantee that P2P users belonging to an OSN are guaranteed a better service when critical conditions build up, i.e., when bandwidth availability is scarce. Different help strategies are proposed, and their improvements are numerically assessed, showing that the help of direct friends, two-hops away friends and, in the limit, of the entire OSN community brings in considerable advantages. The obtained results demonstrate that the amount of delivered video increases and the delay notably decreases, for those privileged peers that leverage their OSN membership within the P2P overlay.

The internet offers services for users which can be accessed in a collaborative shared manner. Users control these services, such as online gaming and social networking sites, with handheld devices. Wireless mesh networks (WMNs) are an emerging technology that can provide these services in an efficient manner. Because services are used by many users simultaneously, security is a paramount concern. Although many security solutions exist, they are not sufficient. None have considered the concept of load balancing with secure communication for online services. In this paper, a load aware multiparty secure group communication for online services in WMNs is proposed. During the registration process of a new client in the network, the Load Balancing Index (LBI) is checked by the router before issuing a certificate/key. The certificate is issued only if the value of LBI is less than a predefined threshold. The authors evaluate the proposed solution against the existing schemes with respect to metrics like storage and computation overhead, packet delivery fraction (PDF), and throughput. The results show that the proposed scheme is better with respect to these metrics.

Wireless Mesh Networks (WMNs) have emerged as a key technology for the next generation of wireless networking. Instead of being another type of ad-hoc networking, WMNs diversify the capabilities of ad-hoc networks. Several protocols that work over WMNs include IEEE 802.11a/b/g, 802.15, 802.16 and LTE-Advanced. To bring about a high throughput under varying conditions, these protocols have to adapt their transmission rate. This paper proposes a scheme to improve channel conditions by performing rate adaptation along with multiple packet transmission using packet loss and physical layer condition. Dynamic monitoring, multiple packet transmission and adaptation to changes in channel quality by adjusting the packet transmission rates according to certain optimization criteria provided greater throughput. The key feature of the proposed method is the combination of the following two factors: 1) detection of intrinsic channel conditions by measuring the fluctuation of noise to signal ratio via the standard deviation, and 2) the detection of packet loss induced through congestion. The authors show that the use of such techniques in a WMN can significantly improve performance in terms of the packet sending rate. The effectiveness of the proposed method was demonstrated in a simulated wireless network testbed via packet-level simulation.

Chapter 16
A Source Based On-Demand Data Forwarding Scheme for Wireless Sensor Networks 218
Martin Brandl, Danube University Krems, Austria
Andreas Kos, St. Pölten University of Applied Sciences, Austria
Karlheinz Kellner, Danube University Krems, Austria
Christian Mayerhofer, St. Pölten University of Applied Sciences, Austria
Thomas Posnicek, Danube University Krems, Austria
Christian Fabian, St. Pölten University of Applied Sciences, Austria

Wireless Sensor Networks (WSNs) are becoming more important in the medical and environmental field. The authors propose an on-demand routing protocol using sensor attractiveness-metric (Pa) gradients for data forwarding decisions within the network. Attractiveness-based routing provides an efficient concept for data-centric routing in wireless sensor networks. The protocol works on-demand, is source-initiated, has a flat hierarchy and has its origin in the idea of pheromone-based routing. The algorithm supports node-to-sink data traffic and is therefore a lightweight approach to generalized multihop routing algorithms in WSNs. The performance evaluation of the proposed protocol is done by extensive simulation using a multi-agent based simulation environment called NetLogo. The efficiency of the attractiveness-based routing algorithm is compared in simulations with the well known Dynamic Source Routing algorithm (DSR). The authors conclude that the Pa based routing algorithm is well suited for easy to set up WSNs because of its simplicity of implementation and its adaptability to different scenarios by adjustable weighting factors for the node's attractiveness metric.

Chapter 17
Wireless Transport Layer Congestion Control Evaluation 241
Sanjay P. Ahuja, University of North Florida, USA
W. Russell Shore, University of North Florida, USA

The performance of transport layer protocols can be affected differently due to wireless congestion, as opposed to network congestion. Using an active network evaluation strategy in a real world test-bed experiment, the Transport Control Protocol (TCP), Datagram Congestion Control Protocol (DCCP), and Stream Control Transport Protocol (SCTP) were evaluated to determine their effectiveness in terms of throughput, fairness, and smoothness. Though TCP's fairness was shown to suffer in wireless congestion, the results showed that it still outperforms the alternative protocols in both wireless congestion, and network congestion. In terms of smoothness, the TCP-like congestion control algorithm of DCCP did outperform TCP in wireless congestion, but at the expense of throughput and ensuing fairness. SCTP's congestion control algorithm was also found to provide better smoothness in wireless congestion. In fact, it provided smoother throughput performance than in the network congestion.

Recently data dissemination using Road Side Units (RSUs) in Vehicular Ad Hoc Networks (VANETs) received considerable attention for overcoming the vehicle to vehicle frequent disconnection problem. An RSU becomes overloaded due to its mounting location and/or during rush hour overload. As an RSU has short wireless transmission coverage range and vehicles are mobile, a heavily overloaded RSU may experience high deadline miss rate in effect of serving too many requests beyond its capacity. In this work, the authors propose a co-operative multiple-RSU model, which offers the opportunity to the RSUs with high volume workload to transfer some of its overloaded requests to other RSUs that have light workload and located in the direction in which the vehicle is heading. Moreover, for performing the load balancing, the authors propose three different heuristic load transfer approaches. By a series of simulation experiments, the authors demonstrate the proposed co-operative multiple-RSU based load balancing model significantly outperforms the non-load balancing multiple-RSU based VANETs model against a number of performance metrics.

Video streaming over vehicular networks is an attractive feature to many applications, such as emergency video transmission and inter-vehicle video transmission. Vehicles accessing road-side units have a few seconds to download information and experience a high packet loss rate. Hence, this paper proposes Cooperative Error Control (CEC) mechanism combining Cognitive Technology (CT) for video streaming over wireless vehicular networks. CEC mechanism combining CT uses a cooperative error recover scheme to recover lost packets not only from a road-side unit which uses the primary channel but also from the other vehicles using a free channel. Hence, CEC mechanism with CT can enhance error recovery performance and quality of video streaming over vehicular networks. Simulation results show the error recover performance of the CEC mechanism combining CT performs better than the other related mechanisms.

Chapter 20
A Framework for External Interference-Aware Distributed Channel Assignment 293
Felix Juraschek, Humboldt Universität zu Berlin, Germany
Mesut Günes, Freie Universität Berlin, Germany
Bastian Blywis, Freie Universität Berlin, Germany

DES-Chan is a framework for experimentally driven research on distributed channel assignment algorithms in wireless mesh networks. DES-Chan eases the development process by providing a set of common services required by distributed channel assignment algorithms. A new challenge for channel assignment algorithms are sources of external interferences. With the increasing number of wireless devices in the unlicensed radio spectrum, co-located devices that share the same radio channel may have a severe impact on the network performance. DES-Chan provides a sensing component to detect such external devices and predict their future activity. As a proof of concept, the authors present a reference implementation of a distributed greedy channel assignment algorithm. The authors evaluate its performance in the DES-Testbed, a multi-transceiver wireless mesh network with 128 nodes at the Freie Universität Berlin.

Chapter 21
An Efficient Data Dissemination Scheme for Warning Messages in Vehicular Ad Hoc Networks 308
Muhammad A. Javed, The University of Newcastle, Australia
Jamil Y. Khan, The University of Newcastle, Australia

Vehicular ad hoc networks (VANETs) are expected to be used for the dissemination of emergency warning messages on the roads. The emergency warning messages such as post crash warning notification would require an efficient multi hop broadcast scheme to notify all the vehicles within a particular area about the emergency. Such emergency warning applications have low delay and transmission overhead requirements to effectively transmit the emergency notification. In this paper, an adaptive distance based backoff scheme is presented for efficient dissemination of warning messages on the road. The proposed scheme adaptively selects the furthest vehicle as the next forwarder of the emergency message based on channel conditions. The detailed performance figures of the protocol are presented in the paper using simulations in the OPNET network simulator. The proposed protocol introduces lower packet delay and broadcast overhead as compared to standard packet broadcasting protocols for vehicular networks.

Compilation of References 328

About the Contributors 351

Index 355

Preface

A wide variety of technologies have been proposed in recent years, most of which are based on user applications and demand. New applications and technologies bring in more challenges for researchers, which need to be standardized for more interoperability. In recent years, high-speed cellular and next generation networks such as Long Term Evolution (LTE) also known as 4G, have migrated to a single modulation technology called Orthogonal Frequency Division Multiplexing (OFDM), which offers higher spectral efficiency and performance. Both LTE and WiMAX uses OFDM access with Frequency Division/Time Division multiplexing to achieve basic service bit rates in the range 10-20 Mb/s, but things changed dramatically when Multiple Input Multiple Output (MIMO) antennas were introduced. MIMO signal processing techniques elevated peak bit rate to the range of 100 Mb/s. Further, using a combination of OFDM and MIMO, 802.11n has achieved a peak bit rate of 300Mb/s. Looking ahead, using higher frequency unlicensed spectrum bands such as 60 GHz, network peak bit rate of 1 Gb/s is possible in 802.11ad wireless technology (Raychaudhuri & Mandayam, 2012).

Direct Sequence Code Division Multiple Access (DS/CDMA) is a modulation and multiple access technique currently being implemented in cellular mobile radio communication systems. In a wireless fading channel, transmitted signal may reach a receiver at different incoming rays, which may arrive at the same receiver at different times. This translates directly into induced time diversity. Another diversity approach to further improve the system performance is the use of spatial signal processing with an antenna array at a receiver in order to resolve signal rays arriving at different angles of arrival. The issues on performance analysis of a traditional DS/CDMA have been well addressed in the literature. However, very few works have discussed the issues on how to achieve a joint angular and time diversity gain in a CDMA-based system. In "Joint Angular and Time Diversity of Multi-Antenna CDMA Systems in Wireless Fading Channels" chapter, the authors proposed a framework to design a joint angular and time diversity scheme that can be implemented in a cost-effective way using existing multi-antenna and CDMA technologies. This chapter also analyses performance of this novel scheme in terms of bit error probability.

In the context of Channel access and assignment, a distributed channel assignment algorithm is more commonly used in Wireless Mesh Networks (WSN). Distributed channel approach can react faster to topology changes due to node failures, which makes it less protocol overhead compared to the centralized approach. In "Insights from Experimental Research on Distributed Channel Assignment in Wireless Testbeds," the authors designed a framework as DES-Chan, which distributes channel assignment in a multi-transceiver WMNs that provides a wide range of common services distributed channel assignment algorithms. On the other hand, the authors also presented the results of experiments in order to measure the channel characteristics of the DES-Testbed. The results enable a better estimate of the increase in performance of channel assignment algorithms since the interference estimation of two channels is closer to reality than simplified models.

In OFDM access, the task of subcarrier allocation amongst active users is a crucial resource allocation issue. There have been extensive studies conducted on resource allocation problem using various parameters such as type of data traffic, fairness, computational complexity, etc. High Peak to Average Ratio (PAPR) is another important issue for any OFDM-based system, and higher PAPR means the possibility of Bit Error Rate (BER) due to non-linearity in power amplifier is also high. In "BER Fairness and PAPR Study of Interleaved OFDMA System," the authors investigated the impact of subcarrier allocation on the physical layer performance of the system on an individual user basis. This chapter shows that for a slowly changing multipath propagation environment, i.e., channel co-efficient remain unchanged over many OFDM Access (OFDMA) symbols, if subcarriers are allocated on a contiguous basis, there may be signification amount of dispersion in the individual user's BER performance. This disparity or non-uniformity is also due to lack of fairness in BER. In order to improve fairness as required by individual users, the power control scheme needs to be strictly adjusted on an individual user basis. The authors of this chapter replaced this strict power control requirement by a central frequency interleaving mechanism and show that it can achieve BER fairness amongst the users. For achieving BER fairness with little system overhead, the authors introduced the application of a cyclically shifted random interleaver. Cyclically shifted random interleaver is a modified version of pure random interleaver that needs less memory space and low-side information transmission. In addition, cyclically shifted interleaver has an identical PAPR property to that of random interleavers, and the low complexity PAPR reduction scheme can be applied without increasing system complexity too much.

With the increase in number of devices using wireless communications, it is obvious that these devises will compete for the wireless medium and can interfere with each other, thus decreasing the achievable network performance in terms of throughput and latency. It is therefore important to consider external interference while performing channel allocation assignments. A distributed channel allocation approach is more common due to its fast adaptation to topology changes due to node failures. On top of this, a slow channel switching approach which usually switches the interfaces to a particular channel for a long period can be advantageous. In "A Framework for External Interference-Aware Distributed Channel Assignment," the authors analysed several algorithms for distributed channel assignment and derive the required services for their implementation. In this chapter, the authors also described the architecture of DES-Chan that provides these services as well as the implementation of the *Greedy Distributed Algorithm* (DGA) as a proof of concept.

Wireless Sensor Networks (WSN) is a class of wireless ad-hoc networks in which sensors gather data from the environment, process and transfer it to neighbour nodes and Base Station (BS). Due to recent advances in hardware technology, cheap, smart sensors are readily available and have been widely used for wide variety of applications. But, they have low memory, poor processing capabilities, and are limited (Okdem, Ozturk, & Karaboga, 2012). There are several constraints among which the constraint of resources (spectrum and power) is the most appealing one in a WSN. One of the methods proposed to make it more energy efficient is to cluster the networks where some nodes act as cluster head and will forward aggregated messages from a group of nodes. Several protocols have been proposed and show that the network lifetime can be improved using clustering method. Typically, Sensor nodes avoid direct communication with a distant destination since a high destination power is needed to achieve a reliable transmission. Instead, sensor networks communicate by forming a multi-hop network to forward messages to the collector node, which is also called the sink node. In this regard, routing in such multi-hop networks becomes crucial in achieving energy efficiency (Minal, & Ashour, 2013).

Although in WSN the nodes are constraint in resources mainly in terms of battery power, these days there is scarcity increasing in terms of spectrum availability. Traditionally, all the WSN works in ISM band (2.4 GHz), but in the same band, there are many competing technologies working simultaneously, like WLAN 802.11 a/b/g and ZigBee 802.15.4, Wi-Fi, Bluetooth. Hence, in such an environment where all these competing technologies are working simultaneously, it becomes difficult to find free spectrum and transmit without an error.

In "Lifetime Maximization in Wireless Sensor Networks," the authors presented applications of energy map in various fields of WSNs like data aggregation, routing, clustering, data dissemination, etc. The authors also proposed a prediction of energy consumption using energy map so that this information can be used by cluster head to prolong the lifetime of sensor node by distributing load. Simulation result proves the effectiveness of the proposed approach, which reduces message exchange rate up to 30% for energy map construction compared to a naive approach.

In "A Source Based on Demand Data Forwarding Scheme for Wireless Sensor Networks," the authors developed an attractiveness-metric gradient based routing strategy (*Pa*), which provides a new concept for a data-centric WSN routing protocol. The attractiveness-metric considers the actual energy states of neighbouring nodes for the routing decision and therefore ensures energy-aware operation. In detail, the routing costs consider the sensor nodes' energy status, as well as the received signal strength and current buffer (on board memory) fill level. It has a flat hierarchy, works on-demand, is source-initiated, and has its origin in the concept of ant-based routing. During development of the algorithm, the attractiveness-metric was turned into a special factor for link costs and is the sole determinant that remains similar to the ant-based commencements. The authors conclude that the *Pa*-based routing algorithm is well suited for easy to set up WSNs because of its simplicity of implementation and its adaptability to different scenarios by adjustable weighting factors for the node's attractiveness metric.

The authors of "Receiver Diversity for Distributed Detection in Wireless Sensor Networks" applied parallel distributed detection in Wireless Sensor Network (WSN). This is considered where the sensors process the observations to make local decisions and send these decisions to a central device called fusion centre. Receiver diversity technique is proposed here for the distributed detection system in order to enhance the system reliability by improving the detection performance. The fusion centre is assumed to be a multiple antenna device in order to imply the idea of receiver diversity. Different combining schemes at the fusion centre are used to reduce the fading effects in the case of receiver diversity. In this work, the authors considered optimal and sub-optimal fusion rules for each case study. Simulation results show the performance improvement obtained as compared to the conventional distributed detection system in which no diversity is used.

One of the biggest problems with licensed band spectrum is that the spectrum is almost free for 85% of the time. Hence, recent researchers are focusing on either finding a free channel in the unlicensed band to do wireless transmission or finding a free channel in the licensed band to use for wireless communication. In "Doubly Cognitive Architecture-Based Cognitive Wireless Sensor Networks," the authors proposed doubly Cognitive WSN, which works by progressively allocating the sensing resources only to the most promising areas of the spectrum and is based on pattern analysis and learning. As the load of sensing resource is reduced significantly, this approach saves the energy of the nodes and reduces the sensing time dramatically.

While increasing the usage of wireless sensor networks in different applications, effective data collecting and transmission of data towards Base Station is a crucial thing. Similarly, effective cost minimization of the total sensor field is required, and it is reduced by proper node deployment. In “Non-Uniform Grid-Based Cost Minimization and Routing in Wireless Sensor Networks,” the authors developed a new algorithm for sensor placement for target location with cost minimization and coverage in a non-uniform plane. Sensor placement for target location implies that they are given different types of sensors with different costs and ranges for given points on plane, which are to be covered with minimum cost. The proposed algorithm is verified using k-mean clustering and strongly connected graphs.

Road safety has been an important concern in the world as the number of vehicles on the road increases day by day. Automated highway systems and Intelligent Transportation Systems (ITS) were introduced to accelerate the development and use of intelligent integrated safety systems that use information and communication technologies as an intelligent solution in order to increase road safety and reduce the number of accidents on future roads. Vehicular Communication Networks (VCNs) are a cornerstone of the envisioned Intelligent Transportation Systems (ITS). By enabling vehicles to communicate with each other via Inter-Vehicle Communication (IVC) as well as with roadside base stations via Roadside-to-Vehicle Communication (RVC), vehicular networks could contribute to safer and more efficient roads (Jerbi, Senouci, Ghamri-Doudane, & Cherif, 2009).

Applications for ITS can be divided into broad categories, namely safety related, traffic management and transportation efficiency, user infotainment services, and Internet connectivity. Safety-related applications include lane change assistance, cooperative forward incident warning, intersection collision avoidance, emergency or incident warning. Traffic management applications form part of a greater Intelligent Transportation System (ITS) and include toll collection, intersection management, cooperative adaptive cruise control and detour or delay warning (Booysen, Zeadally, & van Rooyen, 2011).

Even though research on ITS is progressing at a reasonably fast pace, technologically, many aspects in ITS still need ideas and results from research. Some of the most critical ones include high performance and efficient physical layer transmission schemes, scalable MAC, efficient data dissemination protocols, energy efficient routing protocols, congestion and error control, and security. A fast and reliable communication between vehicle-to-vehicle (V2V) and/or vehicle-to-infrastructure (V2I) are essential for future safety message systems. Safety messages require fast and guaranteed access and a short transmission delay, while messages are relatively short. Even though V2V communication may be beneficial, wireless communication is typically unreliable. Many factors, for example, channel fading, packet collisions, and communication obstacles, can prevent messages from being correctly delivered in time.

In “Synchronous Relaying in Vehicular Ad-Hoc Networks,” the authors proposed a simple concept for emergency information dissemination in vehicular ad-hoc networks. Instead of competing for the shared wireless medium when transmitting the emergency information, the authors’ proposed method requires nodes to cooperate by synchronizing their transmissions. The proposed scheme is backward compatible with IEEE 802.11p carrier sense multiple access with collision avoidance. This proposed scheme is concerned with the dissemination speed and coverage of emergency messages, i.e., those, having the utmost importance. In order to protect the safety of information, for a protocol delivering such information, the authors’ stipulated two requirements. First of all, emergency information needs to be spread out to all vehicles in a certain neighbourhood, *h*, of a car detecting the hazard. This geographic area may include infrastructure nodes that may disseminate notifications further using wide area wireless networks such as cellular systems. Secondly, emergency messages need to be disseminated as quickly as possible. CSMA/CA protocol, which has been standardized in IEEE 802.11p, is far from the optimal solution for these problems due to the inherent properties of random access. The proposed scheme can prove to be particularly useful in dense traffic conditions, e.g., in city centers or highways during rush hours, where each node has a significant number of stations within its antenna range.

Dissemination of emergency/safety messages would require an efficient multi hop broadcast scheme to notify all vehicles in the vicinity. There has been some research focused on this issue, but there are still some challenging issues that need to be addressed. In "An Efficient Data Dissemination Scheme for Warning Messages in Vehicular Ad Hoc Networks," the authors proposed an adaptive distance-based backoff scheme for efficient dissemination of warning messages on the road. The proposed scheme adaptively selects the furthest vehicle as the next forwarder of the emergency message based on channel conditions. The detailed performance figures of the protocol are presented in the chapter using simulations using OPNET network simulator. The proposed protocol introduces lower packet delay and broadcast overhead as compared to standard packet broadcasting protocols for vehicular networks.

For Road-to-Vehicle communications, the system architecture assumes access points with IEEE 802.11p network interfaces to be set up at least in dedicated locations (such as road intersections), whereas the system is still able to deliver information even when no access point is available within the communication range of a vehicle. In Road-to-Vehicle communications, the roadside infrastructure and installation can be expensive. A complementary solution to the deployment of roadside access points consists of roadside wireless sensors. These devices represent a cost effective solution and allow creating Wireless Sensor Networks (WSN), but are subject to energy and processing constraints. WSN islands could be rolled out along the road, such as on the road surface or at road boundaries (curves, tunnels and bridges), and even on a much wider scale. They can be used to measure physical data, like temperature, humidity, light, or detect and track movements (Festag, Hessler, Baldessari, Le, Zhang, & Westhoff, 2008). However, these Roadside Units (RSU) can become overloaded due to their mounting location and/or during rush hour overload. As an RSU has short wireless transmission coverage range and vehicles are mobile, a heavily overloaded RSU may experience high deadline miss rate in effect of serving too many requests beyond its capacity. In "Cooperative Load Balancing in Vehicular Ad Hoc Networks (VANETs)," the authors propose a cooperative multiple-RSU model that offers the opportunity to the RSUs with high volume workload to transfer some overloaded requests to other RSUs that have light workload and are located in the direction in which the vehicle is heading. Moreover, for performing the load balancing, the authors propose three different heuristic load transfer approaches. By a series of simulation experiments, the authors also demonstrate the proposed cooperative multiple-RSU-based load balancing model significantly outperforms the non-load balancing multiple-RSU based VANETs model against a number of performance metrics.

Video streaming over vehicular networks is an attractive feature to many applications, such as emergency video transmission and inter-vehicle video transmission. Vehicles accessing roadside units have a few seconds to download information and experience a high packet loss rate. In "Cooperative Error Control Mechanism Combining Cognitive Technology for Video Streaming over Vehicular Networks," the authors proposed a Cooperative Error Control (CEC) mechanism combining Cognitive Technology (CT) for video streaming over wireless vehicular networks. CEC mechanism combining CT uses a cooperative error recover scheme to recover lost packets not only from a roadside unit that uses the primary channel but also from the other vehicles using a free channel. Hence, CEC mechanism with CT can enhance error recovery performance and quality of video streaming over vehicular networks. Simulation results show the error recover performance of the CEC mechanism combining CT performs better than the other related mechanisms.

One of the main aims of vehicular communications is to avoid collision between vehicles and thus reduce road accidents. Even though there has been active research in this area, still the fatalities are high on the road. One of the safety systems recently introduced is Forward Collision Warning System (FCWS). FCWS uses radar and image processing to detect imminent collision and provide a timely warning to vehicles nearby to avoid collision. In "Implementation of Dedicated Short Range Communications Combined with Radar Detection for Forward Collision Warning System," the authors proposed a Dedicated Short Range Communications (DSRC) system combined with radar detection for an FCWS mechanism. The mechanism proposed actively probes the emergency brakes of the vehicle in front and broadcast warning information with the Global Positioning System (GPS) position. Moreover, this mechanism uses warning information based on the GPS position to calculate the time of collision in order to alert the driver.

Wireless Mesh Network (WMN) is a promising wireless technology for several emerging and commercially interesting applications, e.g., broadband home networking, community and neighbourhood networks, coordinated network management, intelligent transportation systems. In WMN architecture, while static mesh routers form the wireless backbone, mesh clients access the network through mesh routers as well as directly meshing with each other. WMNs also have special features such as being dynamically self-organised and self-configured (Gungor, Natalizio, Pace, & Avallone, 2008). The advantages of WMN over traditional wireless networks include rapidly deployable, low transmission power, extended coverage, scalability, and self-healing. Thus, WMN is widely used in disaster relief, intelligent transport system, metropolitan area networks and building automation, etc. (Marwaha, Indulska, & Portmann, 2008). In addition, with the use of advanced radio technologies, e.g., multiple radio interfaces and smart antennas, network capacity in WMNs is increased significantly. Moreover, the gateway and bridge functionalities in mesh routers enable the integration of wireless mesh networks with various existing wireless networks, such as wireless sensor networks, Wireless-Fidelity (Wi-Fi), WiMAX, and LTE Advanced.

Even though WMNs can be relatively easily established, obtaining high data rates in the WMNs is still a big challenge since the bandwidth in wireless is limited. In recent research proposals, adaptation of different modulations schemes has been strongly suggested. However, when data rate is increased, the BER also increases. The performance of each modulation scheme is measured by its capability to precisely maintain the encoded data, which is represented by the low Bit Error Rate (BER). Variation in the BER is directly related to the received Signal-to-Noise Ratio (SNR). BER and SNR are inversely related: when the SNR is decreased the BER increases. In "Adaptive Sending Rate over Wireless Mesh Networks using SNR," the authors proposed a mechanism that does dynamic monitoring and adapts to changes in channel quality, thereby maximally taking advantage of the dynamic nature of WMNs. Greater throughput was achieved by the combination of dynamic monitoring, multiple packet transmission, and adaptation to changes in channel quality by adjusting the packet transmission rates according to certain optimization criteria. The effectiveness of the proposed method was demonstrated in a simulated wireless network testbed via packet-level simulation. The simulation results showed that regardless of the channel condition, there is an improved performance as shown by the increase in throughput.

Apart from increasing the signal quality of the links, the mesh architecture allows the cooperative forwarding of data packets through intermediate terminals in the network. Cooperative communication networks manage and share the available spectrum more efficiently. Such cooperative techniques can also help to improve power efficiency and hence are applicable to emerging "green" information technology initiatives (Raychaudhuri & Mandayam, 2012). Cooperative communications can also enable data transmission between two terminals through an alternate path when the direct wireless link is experiencing deep fade.

In "DMT Optimal Cooperative MAC Protocols in Wireless Mesh Networks with Minimized Signaling Overhead," the authors proposed a cooperative protocol for wireless mesh networks. Two features are implemented: on-demand cooperation and selection of the best relay. First, cooperation is activated by a destination terminal when it fails in decoding the message from a source terminal. Second, a selection of the best relay is performed when cooperation is needed. The robustness of wireless links is increased while the resource consumption is minimized. The selection of the best relay is performed by a splitting algorithm, ensuring a fast selection process, the duration of which is now fully characterized. Only terminals that improve the direct link participate in the relay selection, and inefficient cooperation is avoided. The proposed protocol is demonstrated to achieve an optimal diversity-multiplexing trade-off. This study focuses on Nakagami-m wireless channel models to encompass a variety of fading models in the context of wireless mesh networks. The rationale of this proposal is to provide a fast and opportunistic method in order to overcome the temporary failures of a wireless link between two terminals. The use of cooperative communications in this context avoids the need to re-establish a whole route when one link is temporarily dropped.

The recent expansion of wireless network technologies has facilitated emergence of novel applications, such as mobile TV, video on demand, Voice over IP, and vehicular communications. The next generation networks, such as Long Term Evolution (LTE), is based on TCP/IP (Transmission Control Protocol/ Internet Protocol) architectural suite, which is fully IP-based. However, the consequence of this transition is that an IP-based suite will pose more risk in terms of security and reliability. As the number of users using the network grows and more resource available to share, security at various levels will become paramount. Although security in wireless networks has been investigated before from various points of view, such as data confidentiality, integrity, trust management, etc., there has been no complete solution for WMNs. The existing solutions for the security are divided into two folds, namely for centralized and distributed scenarios. In the case of centralized scenarios, standard encryption/decryption mechanisms are applied, but these techniques have a single point of failure. Whereas in the distributed approach, users are divided into several subgroups with each group using a separate shared key for communication or multicasting the message. Both centralized and distributed strategies have their advantages and disadvantages. As the MCs are distributed in different regions, which are controlled by the respective mesh points, this may raise security challenges such as data confidentiality, integrity, and authentication. Hence, any solution for the secure group communication must be scalable and efficient with respect to the available resources of the network. In "Load Balancing Aware Multiparty Secure Group Communication for Online Services in Wireless Mesh Networks," the author proposed a load aware multiparty secure group communication for online services in WMNs. During the registration process of a new client in the network, the Load Balancing Index (LBI) is checked by the router before issuing a certificate/key. The certificate is issued only if the value of LBI is less than a predefined threshold. The author evaluated the proposed solution against the existing schemes with respect to metrics like storage and computation overhead, Packet Delivery Fraction (PDF), and throughput. The results show that the proposed scheme is better with respect to these metrics.

In the last decade, there has been a large increase in mobile devices such as smart phones, which are capable of connecting to wireless networks. A Mobile Ad Hoc Network (MANET) is a collection of mobile nodes like smart phones, PDAs and laptops, which form a temporary network, on ad-hoc basis using existing infrastructure (wireless access points or base stations) (Marwaha, Indulska, & Portmann, 2008). Similar to WMNs, MANETS have become increasingly popular due to their self-organising, self-configuration, self-adaptation characteristics. MANETs also have the unique ability to extend the range of

LANs and MANs without additional infrastructure cost. However, due to dynamic nature of the mobile nodes connected to a MANET, at present, the issue of dynamic changes that occur in the connectivity over certain period of time is one of the main concerns. Because MANET may not have fixed structure or density, there can be uneven distribution of nodes in an area where there can be limited connectivity. This can result in slow network performance. On the other hand, frequent solicitation of routing information via broadcasts performed by MANET nodes in densely populated areas exposes the network to a problem known as "broadcast storm." This event occurs when a high number of broadcast activities are performed simultaneously at a certain point in time and trigger torrents of redundant broadcast requests and replies that will eventually lead the contention-based link layer of MANETs to suffer a blackout. In "Effect of Node Mobility on Density-Based Probabilistic Routing Algorithm in Ad-Hoc Networks," the authors study the effect of inaccurate location information caused by node mobility under a rich set of scenarios. They identify three different environments: a high density, a variable density, and a sparse density. Simulation results show noticeable improvement under the three environments. Under the settings the authors examined, their proposed algorithm achieved up to 22% longer link lifetime than AODV (Ad Hoc On-Demand Distance Vector) and 45% longer link lifetime than OLSR (Optimized Link State Routing) at the three environments, on average, without incurring any additional routing overheads or intense computation.

WiMAX is short for Worldwide Interoperability for Microwave Access. It describes a 4G metro-area wireless technology defined in the IEEE 802.16 standards and promoted by the WiMAX Forum. A Mobile WiMAX uses OFDM as a multiple-access technique, whereby different users can be allocated different subsets of the OFDM tones.

In the WiMAX access protocol, outbound transmissions are broadcast to all stations in a cell or sector in a format that includes a device and a connection address; each station picks off and decrypts the frames addressed to it (WiMAX, n.d.). In WiMAX, a base station controls all inbound transmissions, it can eliminate inbound collisions, and, most importantly, it can implement QoS capabilities, where more time-sensitive inbound transmissions (e.g., voice and video) are given precedence over others. In order to guarantee the quality of service required by video and voice (VoIP)-based applications, the WiMAX standard defines five service classes, namely, UGS for VoIP without silence suppression, rtPS for video streaming, nrtPS for file downloading, ErtPS for voice with silence suppression, and finally, BE for Web and mailing applications. In order to guarantee these QoS constraints, the WiMAX provider must use an efficient Call Admission Control (CAC) mechanism and bandwidth allocation. In "Mobile WiMAX Bandwidth Reservation Thresholds: A Heuristic Approach," the authors investigate work on dimensioning a WiMAX network by introducing a method of how to configure parameter settings of such CAC for an optimal operating system that satisfies QoS constraints. The purpose of this work is to determine, using an analytical model and a heuristic approach, the nearly-optimal sizes of the partition sizes that are dedicated to each type of connection in an aim to fulfill two objectives: (1) statistically guarantee that the connection blocking probabilities remain under a given threshold, and (2) maximize the average gain of the wireless link. As BE (Best Effort) service does not requires any QoS guarantees, the authors distinguish in their model four IEEE 802.16e 2005 service classes: UGS, rtPS, nrtPS, and ErtPS. They also differentiate new call request from hand off ones and integrate into the system model the Adaptive Modulation and Coding (AMC) scheme.

Currently, Web-based online social communities is a growing phenomenon. Due to increase in Internet access, these types of social activities are becoming more popular. This has evolved into a different category of applications, which are based on Peer-to-Peer (P2P) communications. In the past, text-based data such as live chats used to be common, but with the advance in technology, live video streaming and multimedia applications are picking up. In "Cooperation Among Members of Online Communities: Profitable Mechanisms to Better Distribute Near-Real-Time Services," the authors proposed a P2P video streaming platform and focused on the performance improvements that can be granted to those P2P nodes that are also members of a social network. The underpinning idea is that OSN (Online Social Networks) friends (and friends of friends) might be more willing to help their mates than complete strangers in fetching the desired content within the P2P overlay. Hence, an approach is devised to guarantee that P2P users belonging to an OSN are guaranteed a better service when critical conditions build up, i.e., when bandwidth availability is scarce. Different help strategies are proposed, and their improvements are numerically assessed, showing that the help of direct friends, two-hops away friends, and, in the limit, of the entire OSN community brings considerable advantages. The obtained results demonstrate that the amount of delivered video increases and the delay notably decreases, for those privileged peers that leverage their OSN membership within the P2P overlay.

Naveen Chilamkurti
La Trobe University, Australia

REFERENCES

Booysen, Zeadally, & van Rooyen. (2011). Survey of media access control protocols for vehicular ad hoc networks. *IET Communicaitons, 5*(11), 1619-1631.

Festag, A., Hessler, A., Baldessari, R., Le, L., Zhang, W., & Westhoff, D. (2008). Vehicle-to-vehicle and road-side sensor communication for enhanced road safety. In *Proceedings of the ITS World Congress and Exhibition*. New York, NY: ACM.

Gungor, N. Pace, & Avallone. (2008). Challenges and issues in designing architectures and protocols for wireless mesh networks. In E. Hossain & K. K. Leung (Eds.), Wireless Mesh Networks: Architecture and Protocols. Berlin: Spinger.

Jerbi, M., Senouci, S., Ghamri-Doudane, Y., & Cherif, M. (2009). Vehicular communications networks: Current trends and challenges. In Pierre, S. (Ed.), *Handbook of research on next generation networks and ubiquitous computing*. Hershey, PA: IGI Global.

Marwaha, S., Indulska, J., & Portmann, M. (2008a). Challenges and recent advances in QoS provisioning in wireless mesh networks. In *Proceedings of the 8th International Conference on Computer and Information Technology*, (pp. 618-623). IEEE.

Marwaha, S., Indulska, J., & Portmann, M. (2008b). Challenges and recent advances in QoS provisioning, signaling, routing and MAC protocols for MANETs. In *Proceedings of the Australasian Telecommunication Networks and Applications Conference (ATNAC)*, (pp. 97–102). ATNAC.

Minal, E., & Ashour, M. (2013). Energy aware classification for wireless sensor networks routing. In *Proceedings of the 2013 15th International Conference on Advanced Communication Technology (ICACT)*, (pp. 66-71). ICACT.

Okdem, S., Ozturk, C., & Karaboga, D. (2012). A comparative study on differential evolution based routing implementations for wireless sensor networks. In *Proceedings of the 2012 International Symposium on Innovations in Intelligent Systems and Applications (INISTA)*, (pp. 1-5). INISTA.

Raychaudhuri, D., & Mandayam, N. (2012). Frontiers of wireless and mobile communications. *Proceedings of the IEEE, 100*(4), 824–840. doi:10.1109/JPROC.2011.2182095.

WiMAX Website. (n.d.). Retrieved from http://www.wimax.com/whiltepapers/sprint-mobile-wimax.pdf

Chapter 1
Joint Angular and Time Diversity of Multi-Antenna CDMA Systems in Wireless Fading Channels

Feng She
Alcatel-Lucent Shanghai Bell, China

Hsiao Hwa Chen
National Cheng Kung University, Taiwan

Hongyang Li
National Sun Yat-Sen University, Taiwan

ABSTRACT

In this paper, a multi-antenna based receiver structure for direct sequence code division multiple access (DS/CDMA) system is proposed. The proposed scheme exploits the excellent time resolution of a CDMA RAKE receiver and uses an antenna array beamforming structure to resolve multipath returns in both angular and time domains. A much higher diversity gain than that based only on the time domain diversity can be achieved. This work suggests a new space-time diversity paradigm, namely angular-time diversity, which differs from traditional Alamouti-type space-time coded schemes. The impairments caused by multipath and multiuser interference are analyzed. The performance of the proposed receiver in multipath fading channel is explicitly evaluated. An expression for uncoded system bit error probability is derived. Simulation results show the performance improvement in terms of BER due to the use of multi-antenna in the receiver, and the results illustrate that the multi-antenna based receiver works effectively in resolving multipaths in both angular and time domains to achieve performance improvement due to angular and time diversity gain provided by the multi-antenna system.

DOI: 10.4018/978-1-4666-3902-7.ch001

1. INTRODUCTION

Direct sequence code division multiple access (DS/CDMA) is a modulation and multiple access technique currently being implemented in cellular mobile radio communication systems (Schilling, Pickholtz, & Milstein, 1990). The robustness of the CDMA based technologies has been well demonstrated by the successful commercial operation in 2G and 3G wireless communications around the world. There are several features that make CDMA an attractive transmission technique. One of the most salient features of CDMA is its ability to distinguish different multipath returns in a very cost-effective way to achieve a multipath diversity gain. This is an advantage that the other multiple access technologies (such as orthogonal frequency division multiple access -- OFDMA) may not have (Viterbi & Viterbi, 1993).

In a wireless communication system, the underlined physical layer should work reliablly in an environment where multipath propagation exists. In a wireless fading channel, transmitted signal may reach a receiver at different incoming rays, which may arrive at the same receiver at different times. This translates directly into an induced time diversity. When the desired signal is propagated through different multipath rays, the signal itself can provide a very good time diversity with the help of a RAKE receiver, as long as we can distinguish them successfully. The resolution of multipath returns should be less than a chip duration in a CDMA system. If the time difference of the path delays is less than one chip duration, the multipath signals can not be resolved and no multipath diversity gain can be achieved.

Another diversity approach to further improve the system performance is the use of spatial signal processing with an antenna array at a receiver in order to resolve signal rays arriving at different angles of arrival (Simanapalli, 1994; Naguib, Paulraj, & Kailath, 1994; Swales, Beach, Edwards, & McGeehn, 1990; Alamouti, 1998; Yoo & Goldsmith, 2006; Tarokh, Seshadri, & Calderbank, 1998). Using spatial signal processing with antenna arrays, we can achieve an angular diversity gain on top of the time diversity gain in a CDMA based wireless system. The joint exploitation of angular and time diversity can significantly improve the overall performance of a CDMA system, based on the existing beamforming or smart antenna technologies. In addition, the application of joint angular and time diversity techniques can also contribute to a great reduction of multipath interference and multiple access interference.

It is noted that multi-antenna techniques alone have been well discussed for a long time in the literature. Alexiou and Haardt (2004) analyzed the performance of smart antenna systems. In a smart antenna system, an angular diversity gain can be obtained. Exploitation of the spatial domain signal processing can help improve the performance of a wireless system due to the improvement in its link quality through the mitigation of impairments in mobile communications, such as multipath interference and co-channel interference. It can also help increase the data rate through simultaneous transmission of multiple data streams by different antennas, as suggested in the multiple-input-multiple-output (MIMO) systems. However, mobile terminals usually never use a large number of antennas due to their size and power constraints (Yoo & Goldsmith, 2006). Thus, the smart antenna or antenna array are normally only used in base stations.

The issues on performance analysis of a traditional DS/CDMA has been well addressed in the literature. However, very few works have been reported to discuss the issues on how to achieve a joint angular and time diversity gain in a CDMA based system. In the text followed, we would like to briefly summarize the researches related to either RAKE based time domain multipath resolution schemes or spatial domain signal separation schemes.

Jalloul and Holtzman (1994) proposed a way to approximate the bit error probability of a CDMA system which works based on the resolution of multipath signal energies. It was shown in the work that the approximation is accurate for the resolution of multipath energies within a range of realistic coefficient variation. System performance was also evaluated in terms of the capacity which was defined as the number of users that can be supported at a given bit error probability. In Jalloul and Holtzman (1994) it was suggested that when Walsh Hadamard code is used as the spreading code, the BER performance of the system under multipath environment is not acceptable. This is caused mainly by rampant multiuser interference induced by the poor correlation functions of the Walsh Hadamard code in multpath channel (the same applies to both down link and up link channels). To solve this problem, there are two schemes proposed in the literature. The first approach was suggested by Naguib and Paulraj (1996) who proposed to use an antenna array in the base station receiver for a direct-sequence code-division multiple-access (DS/CDMA) system working with M -ary orthogonal modulation. The base station uses an antenna array beamformer receiver with non-coherent equal gain combining. The receiver consists of a "front end" beam-steering processor that is followed by a conventional non-coherent RAKE combiner (Naguib & Paulraj, 1996). The other way is to use orthogonal complementary sequences to implement a CDMA system. It is noted that Turyn first proposed complementary sequences theory in Turyn (1963) and Chen, Chu, and Guizani (2008) extended the theory to the generation of perfect complementary sequences. With the help of the perfect complementary sequences in a CDMA system, multipath interference and multiuser interference can be mitigated effectively, at least from the theoretical point of view. In Kim, Kwon, Hong, and Whang (2000) the performance of a RAKE receiver in a DS/CDMA system was studied. In this paper, it was concluded that when the duration of the multipath delay spread is shorter than a chip width, distinct multipath rays can not be resolved easily. In this case, as a consequence, a time diversity gain can not be achieved by a RAKE receiver alone. As a matter of fact, it can be shown in this paper that a diversity gain can still be achieved even if a RAKE receiver can not distinguish different incoming rays, as long as the signals can be further divided by different angle of arrivals (AOAs), with the help of a multi-antenna system. The joint angular and time diversity can be exploited in a relatively cost-effective way using the existing technologies. This is the issue we will address in this paper.

To the best of our knowledge, no previous works have thoroughly investigated the joint impact of angular spread and delay spread on a CDMA system which uses an antenna array based beamformer, and the exact analytical evaluation of the performance in such a CDMA system with the joint angular and time diversity scheme is still a widely open issue (Spagnolini, 2004). The work has its significance due to its timeliness when we are looking for some new multiple access technologies for futuristic wireless communication systems, such as 4G and beyond 4G wireless.

The contributions of this paper are two-fold, as summarized below.

1. We propose a framework to design a joint angular and time diversity scheme which can be implemented in a cost-effective way using existing multi-antenna and CDMA technologies.
2. We analyze the performance of a CDMA based system using a joint angular and time diversity signal reception scheme, through explicit derivation of numerical performance expression in terms of bit error probability.

The rest of this paper is outlined as follows. Section 2 gives the description of transmitter and receiver models used in the analysis followed. In Section 3, explicit analysis is conducted.

Performance analysis is given in Section 4. The simulation results are presented in Section 5, followed by the conclusion.

2. SYSTEM MODEL

In this section, we will give the mathematical modeling of the transmitted and received signals in a CDMA system working with a joint angular and time diversity scheme.

A. Transmitter Model

The environment of multi-user and multipath transmissions in a DS/CDMA system is shown in Figure 1, where an antenna array is used at the receiver to capture rays from different angles. Let us consider the uplink channel in which signals are transmitted form the k th user to the base station equipped with multiple receiving antennas. Let the transmitted signal of the k th user be

$$m_k(t) = c_k(t)s_k(t)\sqrt{\frac{2}{T_b}}\cos(2\pi f_c t), \tag{1}$$

where c_k , s_k , T_b and f_c are spreading code, transmitted data, bit duration and carrier frequency of user k , respectively. Thus, we can write the impulse response of the multipath fading channel as

$$\begin{aligned} &h(\varphi_{k,a,l}, t, \tau_{k,a,l}) \\ &= \sum_{k=1}^{K}\sum_{a=1}^{A}\sum_{l=1}^{L_{k,a}} \alpha_{k,a,l}(\varphi_{k,a,l}, t) e^{j\theta_{k,a,l}(\varphi_{k,a,l}, t)} \delta(t - \tau_{k,a,l}), \end{aligned} \tag{2}$$

where $\alpha_{k,a,l}$ is signal amplitude, $\varphi_{k,a,l}$ is signal phase, $\tau_{k,a,l}$ is delay. $\theta_{k,a,l}$ denotes the channel phase which corresponds to the multipath delay. The subscript indices k , a , and l denote the parameters for the a th angle and the l th path of user k signal. K is the total number of users, A denotes the total number of incidence angles, and $L_{k,a}$ is the number of multipath signals, respectively.

B. Receiver Model

The major part of this proposed joint angular and time diversity scheme is its receiver structure. The block diagram of the antenna array receiver is shown in Figure 2. The underlined multi-antenna system at the receiver enables multipath tracking. This means that the receiver has a "beamformer" structure where several multipath components can be tracked in both time and space domains. There are L_A RAKE modules, which share the same antenna array consisting of N_t antennas. Thus, each RAKE module will be able to track an individual multipath ray coming from a particular angle of arrival (AOA)

Figure 1. Environment of multiuser and multipath transmissions, where an antenna array is used at the receiver to capture rays from different angles

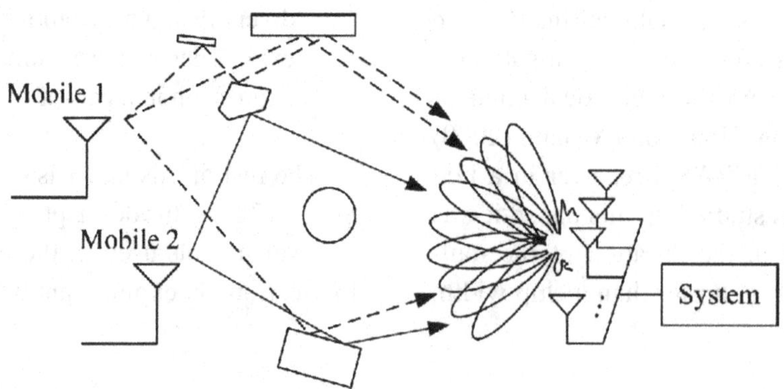

Figure 2. Block diagram of antenna array based CDMA receiver with angular-time diversity scheme

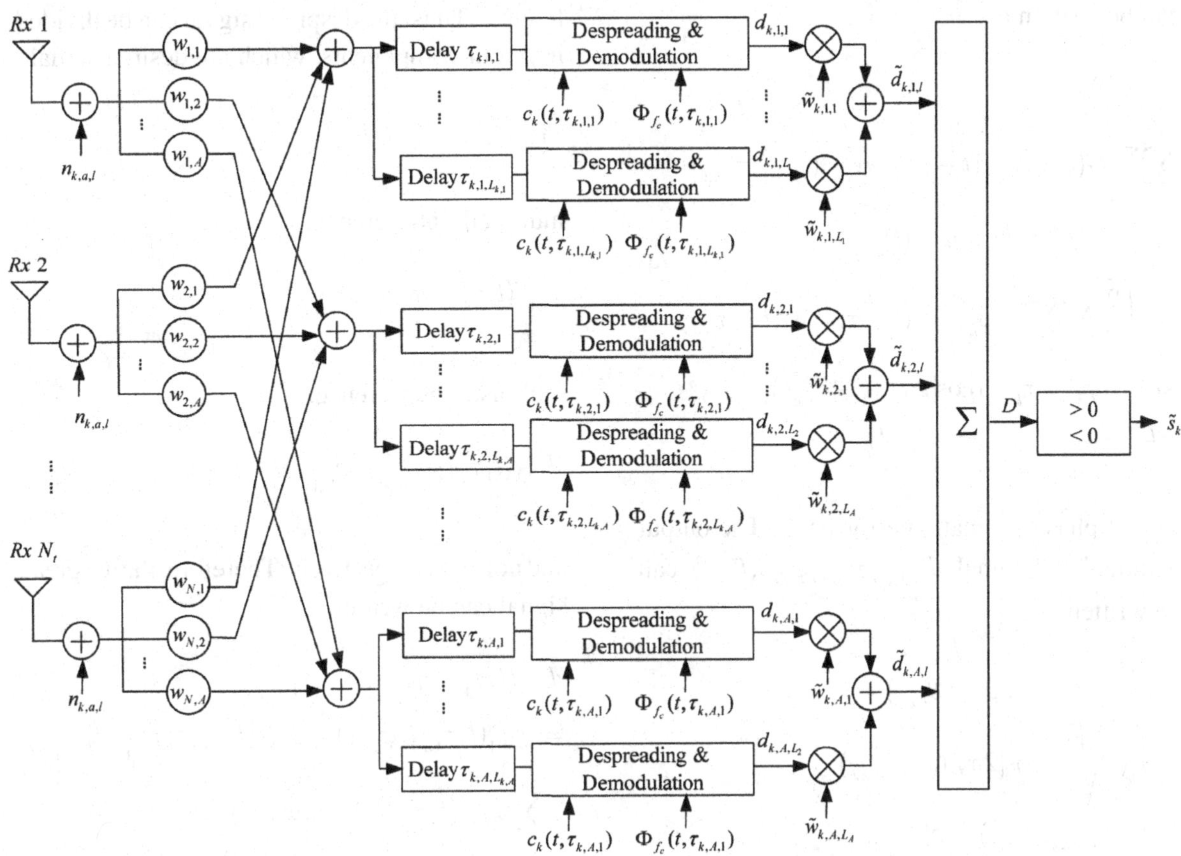

using a set of A beam-steering coefficients $(w_{n,1}, w_{n,2}, \ldots, w_{n,A})$, where

$$n = 1,2,\ldots,N_t$$

We assume that some sort of adaptive tracking algorithm has been used in the receiver so that each beam will follow a particular impinging multipath ray. The implementation of an adaptive tracking algorithm can be done with the help of pilot signal as well as numerous available adaptive algorithms, such as least mean square (LMS) and recursive least square (RLS) algorithms.

After being down converted to the baseband, the output signals from the low-pass filter (LPF) are fed into a bank of correlators as shown in Figure 2. The received signal is the sum of all the user signals plus noise, which is expressed by

$$r(t) = \sum_{k=1}^{K}\sum_{a=1}^{A}\sum_{l=1}^{L_{k,a}} G(\varphi_{k,a,l}) m(t-\tau_{k,a,l}) \alpha_{k,a,l}(\varphi_{k,a,l}, t) \times e^{j\theta_{k,a,l}(\varphi_{k,a,l},\tau_{k,a,l})} + n_{k,a,l}(t) \tag{3}$$

where $G(\varphi_{k,a,l})$ is antenna array gain, $\theta_{k,a,l}$ denotes the channel phase which corresponds to the multipath delay, and $n_{k,a,l}(t)$ is noise with parameters

k, a, and l. The received signal of the a th angle can be written as

$$r_a(t) = \sum_{k=1}^{K}\sum_{l=1}^{L_{k,a}} G(\varphi_{k,a,l}) m_k(t-\tau_{k,a,l}) \alpha_{k,a,l}(\varphi_{k,a,l},\tau_{k,a,l}) \times e^{j\theta_{k,a,l}(\varphi_{k,a,l},\tau_{k,a,l})} + n_{k,a,l}(t)$$
$$= \sqrt{\frac{2}{T_b}} \sum_{k=1}^{K}\sum_{l=1}^{L_{k,a}} G(\varphi_{k,a,l}) c_k(t-\tau_{k,a,l}) s_k(t-\tau_{k,a,l}) \times \alpha_{k,a,l}(\varphi_{k,a,l},\tau_{k,a,l}) \cos[2\pi f_c(t-\tau_{k,a,l})] + \theta_{k,a,l}(\varphi_{k,a,l},\tau_{k,a,l}) + n_{k,a,l}(t) \tag{4}$$

Let phase estimation error be ε. The output demodulated signal $\hat{d}_{k,a,l}(t,\tau_{k,a,l},\varphi_{k,a,l},\theta_{k,a,l})$ can be written as

$$\hat{d}_{k,a,l}(t,\tau_{k,a,l},\varphi_{k,a,l},\theta_{k,a,l}) = r_a(t)\sqrt{\frac{2}{T_b}} \cos[2\pi f_c(t-\tau_{k,a,l}) + \theta_{k,a,l}(\varphi_{k,a,l},\tau_{k,a,l}) + \varepsilon]$$
$$= \frac{2}{T_b} \sum_{k=1}^{K}\sum_{l=1}^{L_{k,a}} G(\varphi_{k,a,l}) c_k(t-\tau_{k,a,l}) s_Q^{(k)}(t-\tau_{k,a,l}) \times \alpha_{k,a,l}(\varphi_{k,a,l},\tau_{k,a,l}) \cos[2\pi f_c(t-\tau_{k,a,l}) + \theta_{k,a,l}(\varphi_{k,a,l},\tau_{k,a,l})] \times \cos[2\pi f_c(t-\tau_{k,a,l}) + \theta_{k,a,l}(\varphi_{k,a,l},\tau_{k,a,l}) + \varepsilon] + \cos[2\pi f_c(t-\tau_{k,a,l} + \theta_{k,a,l}(\varphi_{k,a,l},\tau_{k,a,l}) + \varepsilon] n_{k,a,l}(t) \tag{5}$$

3. INTERFERENCE ANALYSIS

If the spreading codes of different users are not synchronous, the interference always exists. The interference of despread signal includes two parts, namely multipath interference and multiuser interference. Thus, the despread signal can be divided into four components, which are desired signal

$$g_{1,1,1}(t,\tau_{1,1,1},\varphi_{1,1,1})$$

multipath interference,

$$I_{1,1,1}(t,\tau_{1,1,1},\tau_{k',1,l'},\varphi_{k',1,l'})$$

multiuser interference,

$$J_{1,1,1}(t,\tau_{1,1,1},\tau_{k',1,l'},\varphi_{k',1,l'})$$

and noise η, respectively. Therefore, the despread signal can be written as

$$d_{1,1,1}(t,\tau_{1,1,1},\tau_{k,1,l'},\varphi_{1,1,1},\varphi_{k',1,l'}) = g_{1,1,1}(t,\tau_{1,1,1},\varphi_{1,1,1}) + I_{1,1,1}(t,\tau_{1,1,1},\tau_{k',1,l'},\varphi_{k',1,l'}) + \sum_{k=2}^{K} J_{1,1,1}(t,\tau_{1,1,1},\tau_{k',1,l'},\varphi_{k',1,l'}) + \eta(t), \tag{6}$$

where

$$\eta(t) = \sum_{k=1}^{K}\sum_{l=1}^{L_a} \int_{\tau}^{\tau+T_b} c_1(\tau-\tau_{1,1,1}) n_{k,1,l}(\tau) d\tau$$

Next, we proceed to the analysis of multipath interference and multi-user interference in the subsections followed.

A. Multipath Interference (MI) Analysis

Assume that the parameters of desired signal are $k=1$, $a=1$, and $l=1$. Let $s_1(t)$ be transmitted binary signal, and T_b, N, and T_c be bit duration, spreading code length and chip width, respectively. Without loss of generality, let the value of the transmit signal be $s_1(t)$ is $+1$. Let the signal of user 1 be the desired signal. Then all the other signals will be considered as the interference signals. The conceptual diagram of multipath interference in a CDMA system is shown as Figure 3. When the spreading codes of different users are not synchronized, there will be relative delays. Let the relative delays of multipath to the first path be $\tilde{\tau}_l$, which can be written as $\tilde{\tau}_l = \tau - \tau_{1,1,l}$, and $b_{j-1}^{(1)}$, where $b_j^{(1)}$ denote two consecutive bits. Without loss of generality, let value of $b_{j-1}^{(1)}$ and $b_j^{(1)}$ be $+1$ for illustration simplicity.

Figure 3. Multipath interference

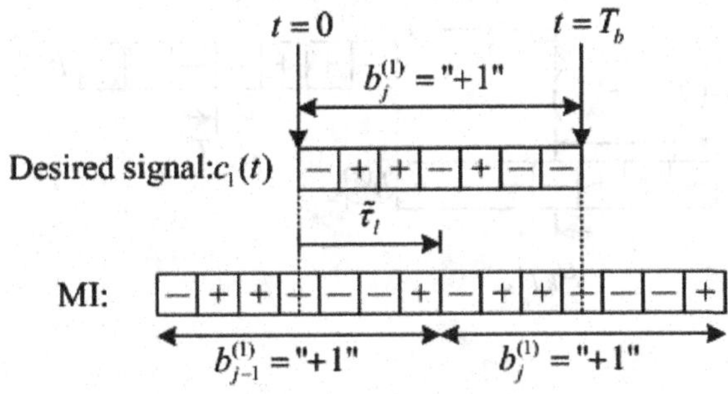

Figure 4. Multipath interference decomposed into two aperiodic cross correlation functions

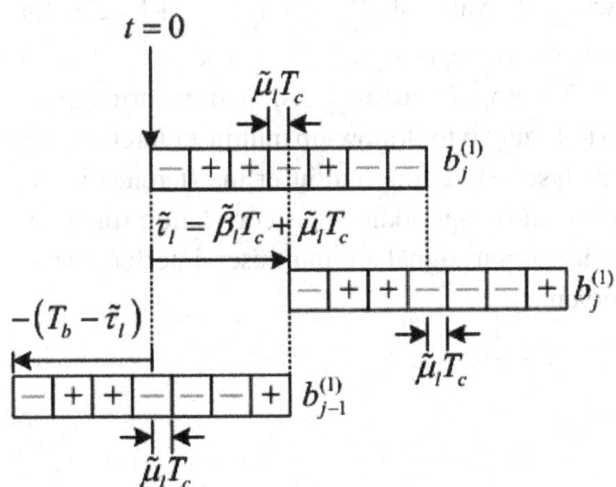

Next, we convert Figure 3 to Figure 4 for further analysis of multipath interference. Where the multipath interference is described as aperiodic cross correlation function. Let $\tilde{\mu}_l$ be uniformly distributed among 0 and 1, and let $\tilde{\beta}_l$ denote the number of shifted chips. Then, the relative delay of multipath returns to the first path can be written as

$$\tilde{\tau}_l = \tilde{\beta}_l T_c + \tilde{\mu}_l T_c, 0 \le \beta \le N-1. \tag{7}$$

If the relative delay $\tilde{\tau}_l$ takes a value from $[0, T_b]$, the output multipath interference signal is given by

$$\begin{aligned} & I'_{1,1,1}(t, \tau_{1,1,1}, \tau_{1,1,l'}, \varphi_{1,1,l'}) \\ & = \tilde{w}_{1,1,1}\sqrt{E_b}\cos(\varepsilon)\sum_{l'=2}^{L_1} G(\varphi_{1,1,l'})\alpha_{1,1,l'}(\varphi_{1,1,l'}, \tau_{1,1,l'}) \\ & \times \left[\begin{aligned} & b_{j-1}^{(1)} \int_{\tilde{\tau}_{l'}}^{T_b} c_1\left(t - \tilde{\tau}_{l'}\right) c_1(t)dt \\ & + b_j^{(1)} \int_0^{\tilde{\tau}_{l'}} c_1(t - \tilde{\tau}_{l'}) c_1(t)dt \end{aligned} \right] \end{aligned} \tag{8}$$

Let $R_1(q)$ be an aperiodic auto-correlation function, and let the two spreading codes be $c_1 = (a_0^{(1)}, a_1^{(1)}, a_2^{(1)}, \ldots, a_{N-1}^{(1)})$. Then we have

$$R_1(q) = \begin{cases} \sum_{p=0}^{N-1-q} a_p^{(1)} a_{p+q}^{(1)}, & 0 \le q \le N-1 \\ \sum_{p=0}^{N-1-q} a_{p-q}^{(1)} a_p^{(1)}, & -(N-1) \le q \le 0 \\ 0, & otherwise \end{cases} \tag{9}$$

Due to $0 \le \tilde{\mu}_l \le 0$, the auto-correlation is given by

$$\int_0^{\tilde{\tau}_l} c_1(t-\tilde{\tau}_l)c_1(t)dt$$
$$= R_1(-(N-\tilde{\beta}_l-1))\tilde{\mu}_l$$
$$+R_1(-(N-\tilde{\beta}_l))(1-\tilde{\mu}_l)$$
$$= R_1(\tilde{\beta}_l-N)$$
$$+\left[R_1(1+\tilde{\beta}_l-N)-R_1(\tilde{\beta}_l-N)\right]\tilde{\mu}_l$$

and we also have

$$\int_{\tilde{\tau}_l}^{T_b} c_1(t-\tilde{\tau}_l)c_1(t)dt$$
$$= R_1(\tilde{\beta}_l)(1-\tilde{\mu}_l)+R_1(\tilde{\beta}_l+1)\tilde{\mu}_l \quad (10)$$
$$= R_1(\tilde{\beta}_l)+\left[R_1(\tilde{\beta}_l+1)-R_1(\tilde{\beta}_l)\right]\tilde{\mu}_l.$$

Therefore, the multipath interference $I_{1,1,1}$ is expressed by

$$I'_{1,1,1}(t,\tau_{1,1,1},\tau_{1,1,l'},\varphi_{1,1,l'})$$
$$= \tilde{w}_{1,1,1}\sqrt{E_b}\cos(\varepsilon)\sum_{l'=2}^{L_1}G(\varphi_{1,1,l'})\alpha_{1,1,l'}(\varphi_{1,1,l'},\tau_{1,1,l'})$$
$$\times\{b_{j-1}^{(1)}\{R_1(\tilde{\beta}_{l'}-N)+[R_1(1+\tilde{\beta}_{l'}-N)$$
$$-R_1(\tilde{\beta}_{l'}-N)]\tilde{\mu}_{l'}\}$$
$$+b_j^{(1)}\{R_1(\tilde{\beta}_{l'})+[R_1(\tilde{\beta}_{l'}+1)-R_1(\tilde{\beta}_{l'})]\tilde{\mu}_{l'}\}\}. \quad (11)$$

B. Multi-User Interference

Let the transmit signal of user k $s_k(t)$ be binary signal, and T_b, N and T_c the bit width, spreading code length and chip width, respectively. Without loss of generality, we assume $s_k(t)=+1$ and $s_1(t)=+1$. Also assume that the first path signal of user 1 is desired signal. Then all the other signals will be treated as interference. As shown in Figure 5, when spreading codes of different users are not synchronous, the relative delay to the first user is $\tau'_{k,l}=\tau_{1,1,1}-\tau_{1,1,l}$. Let $b_{j-1}^{(k)}$ and $b_j^{(k)}$ be two consecutive bits of user k. The multiuser interference is shown as Figure 5, where the value of $b_{j-1}^{(1)}$ and $b_j^{(1)}$ are $+1$ again for illustration clarity.

To simplify our analysis, we transform Figure 5 to Figure 6 for follow-up multiuser interference analysis. Here, the multiuser interference is expressed by aperiodic cross correlation function. The output signal of multiuser interference is given by

Figure 5. multiuser interference

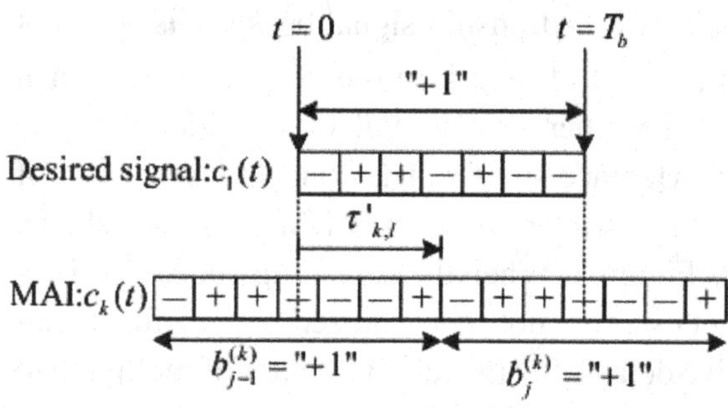

Figure 6. Multiuser interference decomposed into two aperiodic cross correlation functions

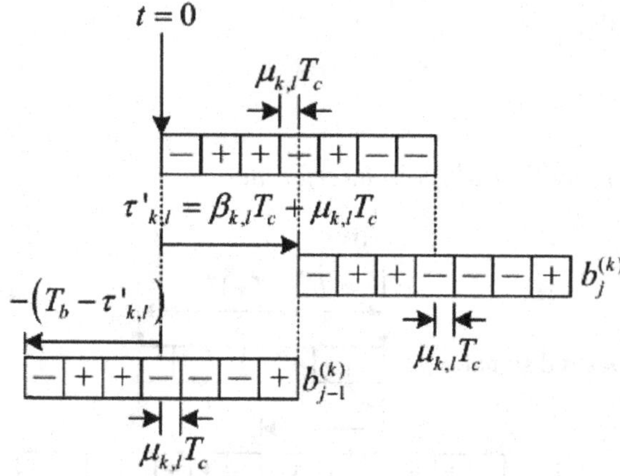

$$J'_{1,1,1}(t,\tau_{1,1,1},\tau_{k',1,l'},\varphi_{k',1,l'})$$
$$=\tilde{w}_{1,1,1}\sqrt{E_b}\cos(\varepsilon)\sum_{k=2}^{K}\sum_{l=1}^{L_{k,1}}G(\varphi_{1,1,l'})\alpha_{1,1,l'}(\varphi_{1,1,l'},\tau_{1,1,l'})$$
$$\times[b_{j-1}^{(k)}\int_0^{\tau'_{k,l}}c_k(t-\tau'_{k,l})c_1(t)dt+b_j^{(k)}\int_{\tau'_{k,l}}^{T_b}] \tag{12}$$

Let $R_{1,k}(q)$ denote the discrete aperiodic auto-correlation function of users 1 and k, and the corresponding spreading code be

$$c_1=(a_0^{(1)},a_1^{(1)},a_2^{(1)},\ldots,a_{N-1}^{(1)})$$

and

$$c_k=(a_0^{(1)},a_1^{(k)},a_2^{(k)},\ldots,a_{N-1}^{(k)})$$

respectively. Then $R_{1,k}(q)$ is given by

$$R_{1,k}(q)=\begin{cases}\frac{1}{N}\sum_{p=0}^{N-1-q}a_p^{(k)}a_{p+q}^{(1)}, & 0\le q\le N-1\\ \frac{1}{N}\sum_{p=0}^{N-1+q}a_{p-q}^{(k)}a_p^{(1)}, & -(N-1)\le q\le 0\\ 0, & otherwise\end{cases} \tag{13}$$

The multiuser interference $J'_{1,1,1}$ can be written as

$$J'_{1,1,1}(t,\tau_{1,1,1},\tau_{k',1,l'},\varphi_{1,1,1},\varphi_{k',1,l'})$$
$$=\tilde{w}_{1,1,1}\sqrt{E_b}\cos(\varepsilon)\sum_{k=2}^{K}\sum_{l=1}^{L_{k,1}}G(\varphi_{k,1,l})\alpha_{k,1,l}(\varphi_{k,1,l},\tau_{k,1,l})$$
$$\times\{b_{j-1}^{k}R_{1,k}(\beta_{k,l}-N)+[R_{1,k}(1+\beta_{k,l}-N)]\mu_{k,l}$$
$$+b_j^kR_{1,k}(\beta_{k,l})+[R_{1,k}(\beta_{k,l}+1)-R_{1,k}(\beta_{k,l})]\mu_{k,l}\}. \tag{14}$$

4. PERFORMANCE ANALYSIS

In this section, we derive the uncoded BER performance with hard decision. For the derivation of the other performance metrics, similar approach can be used. Let

$$D(t,\tau_{a,l},\tau_{a,l'},\varphi_{a,l},\varphi_{a,l'})$$

be the signal at the receiver before decision unit, and $\tilde{w}_{a,l}$ be the coefficients of the a th and l th paths. Then we have

$$D(t,\tau_{a,l},\tau_{a,l'},\varphi_{a,l},\varphi_{a,l'})$$
$$=\sum_{a=1}^{A}\sum_{l=1}^{L_a}\tilde{w}_{1,a,l}g_{1,a,l}(t,\tau_{1,a,l},\varphi_{1,a,l})$$
$$+\sum_{a=1}^{A}\sum_{l=1,l\neq l'}^{L_{1,a}}\tilde{w}_{1,a,l}I_{1,a,l}(t,\tau_{1,a,l},\tau_{1,a,l'},\varphi_{1,a,l})$$
$$+\sum_{a=1}^{A}\sum_{l=1,l\neq l'}^{L_{1,a}}\tilde{w}_{1,a,l}J_{1,a,l}(t,\tau_{1,a,l},\tau_{1,a,l'},\varphi_{1,a,l})$$
$$+\sum_{a=1}^{A}\sum_{l=1}^{L_a}\tilde{w}_{1,a,l}\int_t^{t+T_b}c_1(t-\tau_{1,1,1})\hat{n}_{k,1,l}(t)dt \qquad (15)$$
$$=\sum_{a=1}^{A}\sum_{l=1}^{L_a}g'_{1,a,l}(t,\tau_{1,a,l},\varphi_{1,a,l})$$
$$+\sum_{a=1}^{A}\sum_{l=1,l\neq l'}^{L_{1,a}}I'_{1,a,l}(t,\tau_{1,a,l},\tau_{1,a,l'},\varphi_{1,a,l}$$
$$+\sum_{a=1}^{A}\sum_{l=1,l\neq l'}^{L_{1,a}}J'_{1,a,l}(t,\tau_{1,a,l},\tau_{1,a,l'},\varphi_{1,a,l'})$$
$$+\sum_{a=1}^{A}\sum_{l=1}^{L_a}\tilde{w}_{1,a,l}\int_t^{t+T_b}c_1(t-\tau_{1,1,1})\hat{n}_{k,1,l}(t)dt$$

The signal to noise ratio (SNR) γ is given by

$$\gamma=\frac{E_b\cos^2(\varepsilon)\sum_{a=1}^{A}\sum_{l=1}^{L_{1,a}}|G(\varphi_{1,a,l})|^2\,|\alpha_{1,a,l}(\varphi_{1,a,l},\tau_{1,a,l})|^2}{\sigma_I^2+\sigma_J^2+N_0\sum_{a=1}^{A}L_{1,a}} \qquad (16)$$

The error probability can be expressed as the conditional expectation of SNR γ, and thus we get

$$P_b=E[P_b(x<0\mid\gamma)]$$
$$=\int_{-\infty}^{\infty}\frac{1}{\pi}\int_0^{\frac{\pi}{2}}e^{-\frac{\gamma}{2\sin^2u}}du f_\gamma(\gamma)d\gamma$$
$$=\frac{1}{\pi}\int_0^{\frac{\pi}{2}}\prod_{a=1}^{A}\prod_{l=1}^{L_a}[1 \qquad (17)$$
$$+\frac{E_b\cos^2(\varepsilon)\,|G(\varphi_{1,a,l})|^2}{\sin^2u(\sigma_I^2+\sigma_J^2+N_0\sum_{a=1}^{A}L_{1,a})}]^{-1}du$$

The bit error probability derived above is conditional probability and is a function of γ, which itself is a function of fading channel.

Figure 7. Data rate = 384 kps, $T_b=2.6$ μs, $T_c=40.6$ ns, spread code length N =64, user number K =1, number of multipaths L =1, number of receive antennas $N_t=12$, carrier phase error $e=0^o$, signal arrive angle and delay $(\varphi,\tau)=(0^o,0ns)$, antenna array gain $g=1$

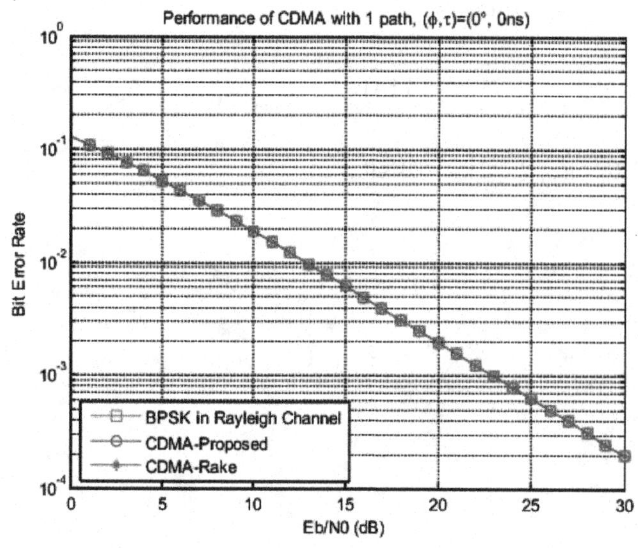

Figure 8. Data rate = 384 kps, $T_b = 2.6$ μs, $T_c = 40.6$ ns, user number $K = 1$, number of multipaths $L = 3$, number of receive antennas $N_t = 12$, carrier phase error $e = 0^o$, signal arrive angle and delay $(\varphi, \tau) = (0^o, 0ns), (0^o, 30ns), (0^o, 45ns)$, antenna array gain $g = 1$

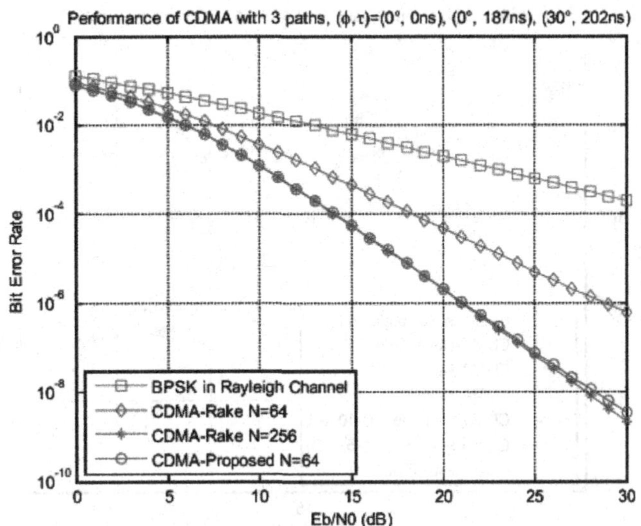

5. SIMULATION RESULTS

In this section, we provide some numerical examples to illustrate the performance of the underlined system.

Experiment 1: We first study the BER performance of a generic CDMA system as well as an antenna array based CDMA system. In the simulations, the channel is assumed to have only one path. As shown in Figure 7, the generic CDMA system has the same BER performance as that of the antenna array based CDMA system.

Experiment 2: In the next experiment, we study the BER performance with multipath channel. We assume that the number of multipath returns is three. The direction of arrivals (DOAs) and the delays of the three paths are $(\varphi, \tau) = (0^o, 0ns)$, $(0^o, 30ns)$, $(30^o, 45ns)$, respectively. From Figures 7 through 10 we can see that when code length is $N = 64$, the multipath signals can not be resolved in the generic CDMA system. When the code length is $N = 256$, the corresponding chip width would be $T_c = 10.2$ ns. Thus the chip width is shorter than the relative delay. As a result, the three paths will be resolved successfully. While in the proposed antenna array based CDMA system, the multipath signals will be resolved in all the scenarios of code lengths, such as $N = 64$ and $N = 256$. This proves that the antenna array based CDMA is more effective than the generic CDMA system in resolving multipath signals.

Experiment 3: This experiment is to study the impact of multiuser interference on the performance. As did in Experiment 2, let us assume that the DOA and delay of the three paths are:

$$(\varphi, \tau) = (0^o, 0ns), (0^o, 30ns), (30^o, 45ns)$$

respectively. When code length is $N = 64$, the generic CDMA system can only resolve two paths of arrived signal. As the increase of user number, the results in Figure 9 show that the BER performance is worse off.

Figure 9. Data rate = 384 kps, $T_b = 2.6$ μs, $T_c = 40.6$ *ns, code length* $N = 64$*, number of multipaths* $L = 3$*, number of receive antennas* $N_t = 12$*, carrier phase error* $e = 0^o$*, signal arrive angle and delay* $(\varphi, \tau) = (0^o, 0ns), (0^o, 30ns), (0^o, 45ns)$*, antenna array gain* $g = 1$

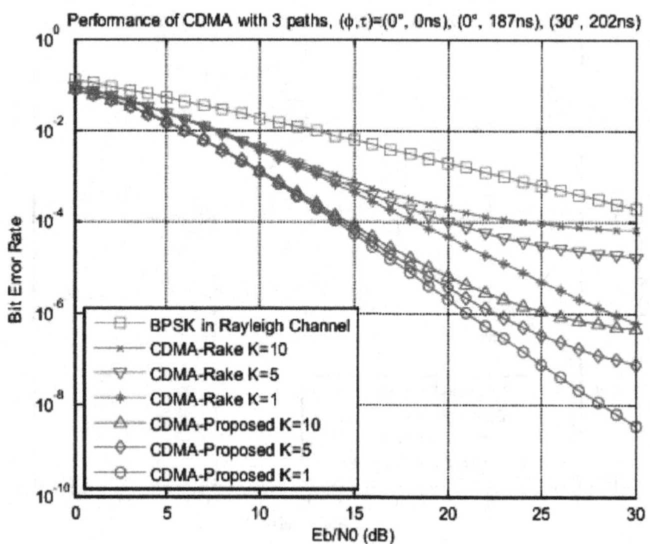

Figure 10. Data rate = 384 kps, $T_b = 2.6$ μs, $T_c = 40.6$ *ns, spread code length* $N = 64$*, number of multipaths* $L = 1$*, number of receive antennas* $N_t = 12$*, carrier phase error* $e = 0^o$*, signal arrive angle and delay* $(\varphi, \tau) = (0^o, 0ns), (0^o, 30ns), (0^o, 45ns)$*, antenna array* $g = 1$

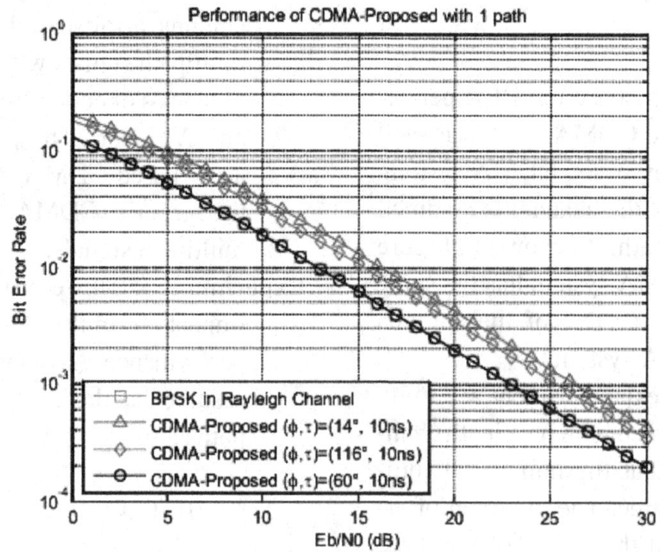

Experiment 4: The fourth experiment compares the BER performance of different antenna gains. In this experiment, we assume that the code length is $N = 64$. The arrive angles are assumed to be 14°, 116°, and 60°, respectively. The corresponding antenna gains of the three paths are assumed to be 0.679, 0.757 and 1, respectively. From Figure 10, we can see that when DOA is 14°, the BER performance is the worst. The reason is that the antenna array gain is only 0.679, which is the lowest antenna array gain among the three paths.

CONCLUSION

This paper aims to evaluate the performance of a multi-antenna CDMA system with a joint angular and time diversity gain. We proposed an antenna array based receiver structure for DS/CDMA wireless systems. We analyzed the performance of such a wireless system under multipath interference and multiuser interference. The uncoded BER closed form expression was derived as a function of SNR. The results demonstrate that with the help of joint angular and time diversity gain, performance of a CDMA system can be significantly improved. This research has illustrated that a CDMA system with joint angular and time diversity signal reception can be a viable candidate for futuristic wireless communications.

One of the future works includes the possible extension of the current study from the angular and time domains to angular, time and frequency domain diversity schemes. The addition of frequency domain diversity can be naturally fit into a popular OFDM modulation, and thus it may find a new way to optimize the design of a unified transceiver framework.

ACKNOWLEDGMENT

The authors would like to thank the financial support from National Science Council, Taiwan, under the research grand NSC99-2221-E-006-016-MY3. The authors would also like to thank the Editor-in-Chief of *International Journal of Wireless Networks and Broadband Technologies (IJWNBT)*, Professor Naveen Chilamkurti, to invite us to submit this paper to its Inaugural issue.

REFERENCES

Alamouti, S. M. (1998). A simple diversity technique for wireless communications. *IEEE Journal on Selected Areas in Communications*, *16*, 1451–1458. doi:10.1109/49.730453.

Alexiou, A., & Haardt, M. (2004). Smart antenna technologies for future wireless systems: Trends and challenges. *IEEE Communications Magazine*, *42*(9), 90–97. doi:10.1109/MCOM.2004.1336725.

Chen, H.-H., Chu, S.-W., & Guizani, M. (2008). On next generation CDMA technologies: The REAL approach for perfect orthogonal code generation. *IEEE Transactions on Vehicular Technology*, *57*(5), 2822–2833. doi:10.1109/TVT.2007.914055.

Jalloul, L. M. A., & Holtzman, J. M. (1994). Performance analysis of DS/CDMA with noncoherent M-ary orthogonal modulation in multipath fading channels. *IEEE Journal on Selected Areas in Communications*, *12*(5), 862–870. doi:10.1109/49.298060.

Kim, K. J., Kwon, S. Y., Hong, E. K., & Whang, K. C. (2000). Effect of tap spacing on the performance of direct-sequence spread-spectrum RAKE receiver. *IEEE Transactions on Communications*, *48*(6), 1029–1036. doi:10.1109/26.848565.

Naguib, A. F., & Paulraj, A. (1996). Performance of wireless CDMA with M-ary orthogonal modulation and cell site antenna arrays. *IEEE Journal on Selected Areas in Communications*, *14*(9), 1770–1783. doi:10.1109/49.545700.

Naguib, A. F., Paulraj, A., & Kailath, T. (1994). Capacity improvement with base-station antenna arrays in cellular CDMA. *IEEE Transactions on Vehicular Technology*, *43*(3), 691–698. doi:10.1109/25.312780.

Schilling, D. L., Pickholtz, R. L., & Milstein, L. B. (1990). Spread spectrum goes commercial. *IEEE Spectrum*, *27*(8), 40–45. doi:10.1109/6.58433.

Simanapalli, S. (1994). Adaptive array methods for mobile communications. In *Proceedings of the 44th International Conference on Vehicular Technology,* Stockholm, Sweden (Vol. 3, pp. 1503-1506).

Spagnolini, U. (2004). A simplified model to evaluate the probability of error in DS-CDMA systems with adaptive antenna arrays. *IEEE Transactions on Wireless Communications*, *3*(2), 578–587. doi:10.1109/TWC.2003.819020.

Swales, S. C., Beach, M. A., Edwards, D. J., & McGeehn, J. P. (1990). The performance enhancement of multibeam adaptive base station antennas for cellular land mobile radio systems. *IEEE Transactions on Vehicular Technology*, *39*(1), 56–67. doi:10.1109/25.54956.

Tarokh, V., Seshadri, N., & Calderbank, A. R. (1998). Space-time codes for high data rate wireless communicatoin: Performance analysis and code construction. *IEEE Transactions on Information Theory*, *44*, 744–765. doi:10.1109/18.661517.

Turyn, R. (1963). Ambiguity functions of complementary sequences. *IEEE Transactions on Information Theory*, *9*(1), 46–47. doi:10.1109/TIT.1963.1057807.

Viterbi, A. M., & Viterbi, A. J. (1993). Erlang capacity of a power controlled CDMA system. *IEEE Journal on Selected Areas in Communications*, *11*(6), 892–900. doi:10.1109/49.232298.

Yoo, T., & Goldsmith, A. (2006). On the optimality of multi-antenna broadcast scheduling using zero-forcing beamforming. *IEEE Journal on Selected Areas in Communications*, *24*(3), 528–541. doi:10.1109/JSAC.2005.862421.

This work was previously published in the International Journal of Wireless Networks and Broadband Technologies, Volume 1, Issue 1, edited by Naveen Chilamkurti, pp. 1-14, copyright 2011 by IGI Publishing (an imprint of IGI Global).

Chapter 2
Improving WLAN Performance by Modifying an IEEE 802.11 Protocol

Nurul I. Sarkar
Auckland University of Technology, New Zealand

ABSTRACT

One of the limitations of the IEEE 802.11 distributed coordination function (DCF) protocol is its low bandwidth utilization under medium-to-high traffic loads resulting in low throughput and high packet delay. To overcome performance problems, traditional IEEE 802.11 DCF ("DCF") protocol is modified to the buffer unit multiple access (BUMA) protocol. The BUMA protocol achieves a better system performance by introducing a temporary buffer unit at the medium access control (MAC) layer to accumulate multiple packets and combine them into a single packet (with a header and a trailer) before transmission. This paper provides an in-depth performance evaluation (by simulation) of BUMA for multiuser ad hoc and infrastructure networks. Results obtained show that the BUMA is more efficient than that of DCF. The BUMA protocol is simple and its algorithm (software) can be upgraded to 802.11 networks requiring no hardware changes. The BUMA protocol is described and simulation results are presented to verify the performance.

INTRODUCTION

IEEE 802.11 based wireless local area networks (WLANs) are gaining widespread popularity due to their simplicity in operation, robustness, low cost, well-defined standard (e.g. 802.11b/g) and user mobility offered by the technology.

DOI: 10.4018/978-1-4666-3902-7.ch002

A good wireless MAC protocol should provide an efficient mechanism for sharing a limited wireless channel bandwidth, together with simplicity of operation, high bandwidth utilization and network fairness (i.e. equality in channel access). Ideally low mean packet delay, high throughput, low packet drop ratio, and a good fairness under high-traffic loads are desired, but in reality the 802.11-based WLANs do not provide all the quality of service (QoS) provisions simultaneously.

Therefore, various MAC protocols have been developed to suit different applications where various tradeoff factors have been considered (Aad, Ni, Barakat, & Turletti, 2005; Chieochan, Hossain, & Diamond, 2010; Li & Chen, 2010; Luo, Rosenberg, & Girard, 2010). A study of the performance of 802.11 under high-traffic loads is required to assist efficient MAC protocol design for WLANs to achieve a better QoS in terms of high throughput and low packet delay in such systems.

The 802.11 standard defines two types of MAC schemes: Distributed Coordination Function (DCF) and Point Coordination Function (PCF). DCF is defined as a mandatory MAC protocol while PCF is optional. In this paper we focus on the DCF mode in 802.11 which has been widely deployed because of its simplicity. In the DCF, wireless stations (STAs) communicate with each other using a contention-based channel access method known as Carrier Sense Multiple Access with Collision Avoidance (CSMA/CA).

The performance of the DCF has been analyzed in numerous papers (Banchs, Serrano, & Azcorra, 2006; Bianchi, 2000; David, Ken, & Doug, 2007; Kuo, 2007). Clearly, the DCF has several limitations. Primarily, it does not perform well under medium-to-high traffic loads. If the number of active users increases, the mean packet delay, throughput, fairness and packet drop ratio of the 802.11 degrade significantly.

The service differentiation and fairness issues of 802.11 have been discussed in (Banchs & Vollero, 2006; Ferre, Doufexi, Nix, & Bull, 2004; Lin & Wu, 2007; Sarkar, 2006). The 802.11e task group is proceeding to build the QoS enhancements of the 802.11 (IEEE Standards Association, 2005). Although various innovative MAC protocols have been developed and reported in the computer networking literature, very few protocols satisfy simultaneously all the QoS provisions while retaining simplicity of implementation in real WLANs. This paper proposes an enhancement of DCF called buffer unit multiple access (BUMA) protocol to overcome the limitations of DCF mentioned above. The BUMA protocol is developed through minor modifications of the existing DCF.

The proposed BUMA protocol provides higher throughput, lower packet delay, lower packet dropping, and greater fairness under medium-to high loads than that of DCF. The better system performance is achieved by introducing a temporary buffer unit at the MAC layer for each active connection on the network, accumulating multiple packets (for example, three packets) and combining them into one large packet with a single header for transmissions. The BUMA can be used to improve the performance of 802.11 networks, including the 802.11b/g, and would be a good candidate for providing real-time multimedia services.

The remainder of this paper is organized as follows. We first provide an overview of DCF protocols. We then summarize past research on enhancements of the original 802.11 networks. The proposed BUMA protocol is then described. The simulation environment and parameter settings are presented. The performance of the BUMA is then compared with that of the DCF. Finally, the system implication is discussed and a brief conclusion concludes the paper.

THE BASIC DCF PROTOCOL

The basic operation of the DCF is illustrated in Figure 1. In the basic access mode, a station monitors the channel until an idle period equal to a DCF inter-frame space (DIFS) is detected before transmission of a frame. If the channel is found to be busy, the station defers (and continues listening to the channel) until the channel becomes idle for at least a DIFS. Then the station begins its backoff time to avoid collisions. After a successful backoff time the station transmits a packet.

The collision avoidance technique adopted in the DCF is based on a binary exponential backoff (BEB) method, which is implemented in each

Figure 1. Basic access method of DCF protocol

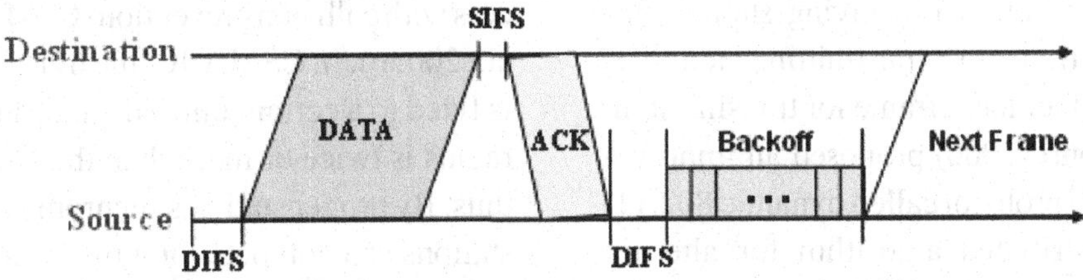

station by means of a parameter known as the *backoff counter*. The backoff time is used to initialize the backoff counter. This counter is decremented only when the medium is idle and is frozen when activity is sensed. The backoff counter is periodically decremented by one time slot each time the medium sensed is idle for a period longer than a DIFS. A station transmits a packet when its backoff counter is zero. The random backoff time can be computed using (1) (IEEE Standards Association, 1999):

Backoff time (BT) = Random () X aSlotTime (1)

Where, Random () is an integer randomly chosen from a uniform distribution over the interval [0, CW-1], CW is the contention window size.

At the first transmission attempt CW = CWmin, and it is doubled at each retransmission up to CWmax. In the IEEE 802.11 DSSS (direct sequence spread spectrum) specification, CWmin = 31 and CWmax = 1023. More details about backoff algorithm can be found in (IEEE Standards Association, 1999a; Yun, Ke-Ping, Wei-Liang, & Qian-Bin, 2006).

SHORTCOMINGS OF DCF PROTOCOL

Although DCF has been standardized and is gaining widespread popularity as a channel access protocol for WLANs, the protocol has several potential limitations. One of the limitations of the DCF protocol is the low bandwidth utilization under the medium-to-high loads, and consequently, it achieves low throughput, high packet delay and high packet drop ratios. Another deficiency of the DCF protocol is poor fairness in sharing a channel bandwidth among the active stations on the network. High transmission overhead (headers, inter-frame spaces, backoff time and acknowledgments) is also a fundamental problem in the DCF protocols (Xiao, 2004). The proposed BUMA protocol overcomes these shortcomings of DCF.

PRIVIOUS WORK ON THE ENHANCEMENT OF DCF

This section summarizes previous work on improvements to the original DCF protocol. Many network researchers have proposed methods to improve the protocol's performance. In this paper, for brevity, we refer to only a selected set of literature that is indicative of the range of approaches used to improve throughput, packet delay, and fairness performance.

Bharghavan et al. (1994) proposed a MAC protocol called multiple access with collision avoidance for wireless (MACAW) in alleviating the fairness problem of DCF. This fairness improvement is mainly due to the selection of a better backoff algorithm, called multiplicative increase and linear decrease (MILD), and some additional control packets in the system.

Xiao (2004) proposed two mechanisms, namely, concatenation and piggybacking in order to reduce the overhead of the DCF protocols. The idea is to concatenate multiple frames in a station's queue before transmission into the medium. Under the piggyback scheme, a receiving station piggybacks a data frame to a transmitting station as long as the receiver has a frame for transmission.

Cali and Conti (2000) proposed an improvement to the DCF protocol called dynamic 802.11, which is a distributed algorithm for altering backoff window size. By observing the status of the channel, a station obtains an estimate of the network traffic and uses this estimate to tune the backoff window sizes.

Bianchi, Borgonovo, Fratta, Musumeci, and Zorzi (1997) developed a MAC protocol for WLANs, C-PRMA. An algorithm for adaptive contention window size is proposed for DCF (Bianchi, 2000; Bianchi, Fratta, & Oliveri, 1996). The key concept is to dynamically select the optimum backoff window size based on an estimate of the number of contending stations. This optimization is performed through the measurement of channel activity by each station. They showed that by using this adaptive contention window size, instead of BEB, the DCF could be significantly improved, especially at high loads and large user numbers.

Bruno and Conti (2002) proposed an enhancement of the DCF called *p-persistent* 802.11. The improvement is mainly due to the selection of a better algorithm for selecting the backoff interval. Instead of the BEB used in the original DCF, the backoff interval of the *p-persistent* 802.11 is sampled from a geometric distribution with parameter *p*.

Natkaniec and Pach (2002) developed the MAC protocol called Priority Unavoidable Multiple Access (PUMA) to improve the performance of DCF. The key concept is to introduce a priority scheme for time bounded services. Both a special jam signal and an additional timer are required to support isochronous transmission. Network performance is improved using a new backoff algorithm called double increment double decrement (DIDD) (Natkaniec & Pach, 2000).

You et al. (2003a, 2003b) proposed two MAC methods for improving DCF, namely carrier sense multiple access with ID countdown (CSMA/IC) (You et al., 2003a) and carrier sense multiple access with collision prevention (CSMA/CP) (You et al., 2003b). In CSMA/IC, the transmission radius is fixed to a certain unified value and the sensing radius is twice or more than the transmission radius. By proper station synchronization, only two stations can compete for a medium at a time and the station with a packet that has a larger unique ID has priority. By exchanging synchronizing packets, CSMA/IC can be 100% collision free even in random access environments. In CSMA/CP, the wireless channel is partitioned into a control channel and several data channels. The key concept in achieving 100% collision free transmission is to prevent collisions in the control channel.

Kwon, Fang, and Latchman (2003) developed an efficient MAC algorithm called the fast collision resolution (FCR) to resolve the collisions by increasing the size of the contention window for both the colliding and deferring stations. All active stations redistribute their backoff timers to avoid possible future collisions.

Jagadeesan, Manoj, and Murthy (2003) proposed a MAC protocol suitable for wireless ad hoc networks known as Interleaved Carrier Sense Multiple Access (ICSMA). This protocol uses two data channels (channel 1 and 2) of equal bandwidth, and the handshaking process is distributed between the two channels. A station is allowed to transmit using either channel 1 or 2. For example, a source station may send an RTS packet to a destination station over channel 1, and the destination station responds with CTS on channel 2. ICSMA uses the same binary exponential backoff mechanism as DCF, including waiting times such as DIFS, and PIFS. The ICSMA performs better than the DCF protocol with exposed stations with respect to throughput, packet delay, and fairness.

Ozugur, Naghshineh, Kermani, and Copeland (1999) proposed a p_{ij}-persistent CSMA-based backoff algorithm for load balancing among the wireless links, improving the fairness performance of DCF. Each station calculates a link access probability p_{ij} using either a connection-based (i.e. the information on the number of connections it has with its neighboring stations) or a time-based method (based on the average contention period).

Jiang and Liew (2008) investigated methods of improving throughput and fairness of the DCF by reducing both the exposed and hidden station problems. They showed that the DCF network is not scalable due to exposed and hidden station problems, more access points (APs) do not yield higher total throughput. By removing these problems, it is possible to achieve a scalable throughput.

Wang and Garcia-Luna-Aceves (2003) proposed a hybrid channel access method to alleviate the fairness problem of DCF without sacrificing much throughput and simplicity. The protocol is based on both sender-initiated and receiver-initiated handshakes.

The previous approaches on the enhancement of DCF protocols reviewed in this section are grouped into four main categories shown in Table 1.

THE PROPOSED BUMA PROTOCOL

The proposed BUMA protocol described in this section differs from the earlier work. BUMA is implemented through minor modifications of DCF (IEEE Standards Association, 1999b). The design of BUMA was motivated by the key idea that a typical WLAN can increase throughput by sending a payload using fewer but longer packets (Choudhury & Gibson, 2007; Ergen & Varaiya, 2008; Xiao, 2004). With fewer packets used to deliver the same payload, proportionally less time is spent in the backoff state. Many network researchers have highlighted this aspect of network performance improvement (Chatzimisios, Vitsas, & Boucouvalas, 2002; Ergen & Varaiya, 2008; Ganguly et al., 2006; Wang, Liew, & Li, 2005).

In BUMA, for each active connection a temporary buffer unit is created at the MAC layer where multiple packets are accumulated and combined into a single packet (with a header and a trailer) before transmission. Assuming a realistic wireless Ethernet packet length of 1,500 bytes, the optimum length of the buffer unit was empirically determined to be that of three 1500-byte packets plus header and trailer. The optimization of buffer unit length is discussed later in this section.

The number of buffer units is determined by the number of active connections between the source and destination stations. Each link has its own buffer unit, and each buffer unit stores one or more packets where each packet appears as a MAC Protocol Data Unit (MPDU) in the MAC layer with the same destination address. Thus, the content of a buffer unit is a large packet that appears as a MAC Segment Data Unit (MSDU) in the MAC layer with a single header and a trailer. Now the question arises about the maximum length of the combined packet (i.e. length of an MSDU).

For both wired and wireless Ethernet LANs, the maximum length of a MAC frame is 2,346 bytes, which is a fragmentation threshold. The mean packet length is about 1,500 bytes with payload

Table 1. Categories of MAC approaches reviewed

Approach	Example of MAC proposals
Packet concatenation	802.11 DCF improvement (Xiao, 2004).
Optimization of contention window size and/or backoff algorithms.	MACAW (Bharghavan, 1994), PUMA (Natkaniec & Pach, 2002), FCR (Kwon et al., 2003), p-persistent 802.11 (Bruno et al., 2002), dynamic 802.11 (Cali & Conti, 2000), C-PRMA (Bianchi et al., 1997), Pij-persistent backoff (Ozugur et al., 1999).
Using a control channel for handshaking and/or collision prevention.	CSMA/CP (You et al., 2003b), CSMA/IC (You et al., 2003a), ICSMA (Jagadeesan et al., 2003), Hybrid channel access (Y. Wang & Garcia-Luna-Aceves, 2003).
Solving hidden and exposed station problems.	802.11 fairness improvement (Jiang & Liew, 2008).

length ranges from 46 to 1,460 bytes. In the optimized BUMA ($BUMA_{opt}$), the maximum length of a buffer unit is 4,534 bytes, accommodating three 1,500-byte packets plus a 34-byte envelope (MAC header and cyclic redundancy check (CRC)). In such cases, the MSDU would be fragmented into two frames before transmission since its length is greater than the fragmentation threshold.

When a station fills the buffer unit, it first schedules the packet and then puts the next set of packets in the empty buffer unit from the same link. Under medium-to-high traffic loads, each station will always have packets for transmission and the buffer unit will be filled up with packets quickly within a time interval. When traffic is low, BUMA (Figures 2 and 3) will perform as well as DCF by reducing the buffer unit length to one packet. DCF is effectively a special case of BUMA where the buffer unit length is one packet. Therefore, in the proposed method, the mean packet delay is bounded since a packet will not remain in the buffer permanently while waiting for the second and subsequent packets to arrive.

The basic operation and the frame structure of the BUMA protocol are illustrated in Figures 2 and 3, respectively. Buffer unit contains multiple MPDUs (Figure 3). The actual number of MPDUs in a buffer unit will depend on packet length supported by upper protocol layers. For instance, for the transmission of a 500-byte IP datagram, a maximum of nine MPDUs would be stored in a buffer unit of 4,500 bytes.

The proposed buffer unit mechanism has several benefits. Firstly, it transmits a greater payload (by scheduling a larger packet) and consequently achieves better throughput than DCF. Secondly, by adopting the buffer unit mechanism one can achieve higher bandwidth utilization and better fairness than in DCF because it wastes less potential transmission time in the backoff and channel contention processes. Referring to the example of the 500-byte IP datagram, instead of nine contention periods, only one contention period is needed to transmit nine IP datagrams. BUMA, therefore, dramatically reduces the average packet contention delay, especially for shorter packet lengths, while maintaining better throughput by transmitting a combined packet. Finally, the packet transmission overhead will be reduced

Figure 2. Basic operation of the BUMA protocol

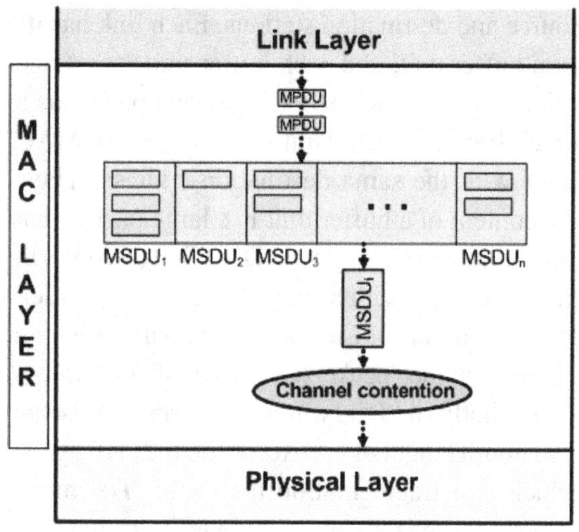

Figure 3. Frame structure of the BUMA protocol

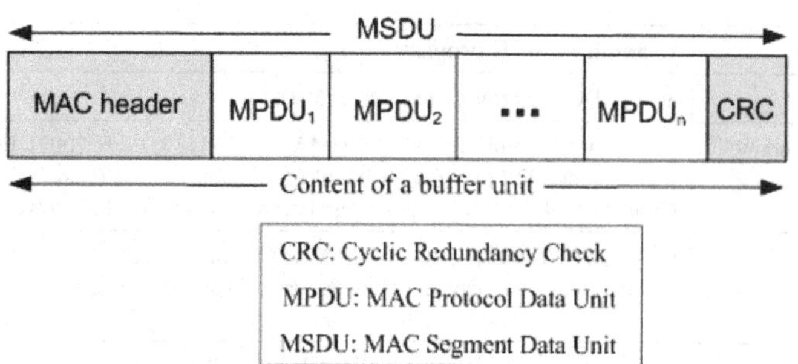

significantly. Without the buffer unit mechanism, each packet transmission requires a separate set of overheads, including headers, inter-frame spaces, backoff time, CRC and acknowledgements; in contrast, only one set of overheads would be used with the buffer unit mechanism. However, all these benefits come with a trade-off, a small processing delay at the stations. The transmission overheads and throughput analysis of BUMA are presented. The processing delay at the stations is also addressed in this section.

Although the BUMA protocol is based on the concepts of concatenation mechanism proposed in (Xiao, 2004), it differs in significant ways. The BUMA protocol is simple and does not require any additional control packets to deliver a combined packet (up to three packets), whereas (Xiao, 2004) requires three additional control frames (for example, a frame control type, concatenated frame count, and a total length field) to deliver a super-frame. In addition, we study the performance of BUMA protocol under high traffic loads, and have introduced two new metrics namely, packet drop ratio and fairness.

OPTIMIZATION OF BUFFER UNIT

Yin, Wang, and Agrawal (2005) investigated the optimal packet length that maximizes the DCF throughput under different channel conditions and traffic loads. A trade-off exists between a desire to reduce the overhead by adopting larger packet length, and the need to reduce packet error rates in the error-prone environments by using smaller packet length. For example, for an error-prone channel with a bit-error rate (BER) of 2×10^{-5}, the throughput reaches the maximal value of 0.6049 for a packet length of 900 bytes. However, the optimal packet length varies with the change of traffic load, and the optimal packet length under light loads is larger than the optimal packet length under high loads.

For an ideal channel condition, both throughput and packet delay increase with the packet length. Therefore, to achieve the best mean delay and throughput performance we have chosen a packet length of 1,500 bytes (a realistic figure close to the wired Ethernet). The mean throughput and packet delay of the BUMA protocol depends on the length of a buffer unit. Now the question arises as to the optimal buffer unit length required to achieve the best mean delay and throughput performance. We have determined the optimal length of the buffer unit empirically. We observe that the throughput performance of the BUMA protocol increases slightly with increasing buffer unit length and then saturates at buffer unit length of 10 or more packets. On the other hand, the lowest mean packet delay is achieved at a buffer unit length of three packets. By considering both the throughput and packet delays we have chosen three packets as the optimum buffer unit length.

STRENGTHS AND WEAKNESSES OF THE BUMA

The BUMA protocol provides better bandwidth utilization (Figure 4) than the DCF because it wastes less transmission capacity in the backoff state, and consequently, it achieves high throughput, low packet delay, low packet drop ratio, and good fairness especially under medium-to-high loads. In addition, the BUMA requires less overhead to send the same amount of payload than the DCF. This improvement is due to the BUMA protocol's scheduling strategy in which a single header and a trailer is required for scheduling multiple packets. Moreover, the BUMA protocol is simple and can easily be implemented in the DCF without changing any existing hardware.

Although the BUMA protocol provides a better fairness than the DCF, it does not provide 100% fairness in sharing a channel's bandwidth among the active stations on the network.

Figure 4. Throughput comparison of the 802.11b DCF, BUMAnonopt and BUMAopt protocols for a single user network

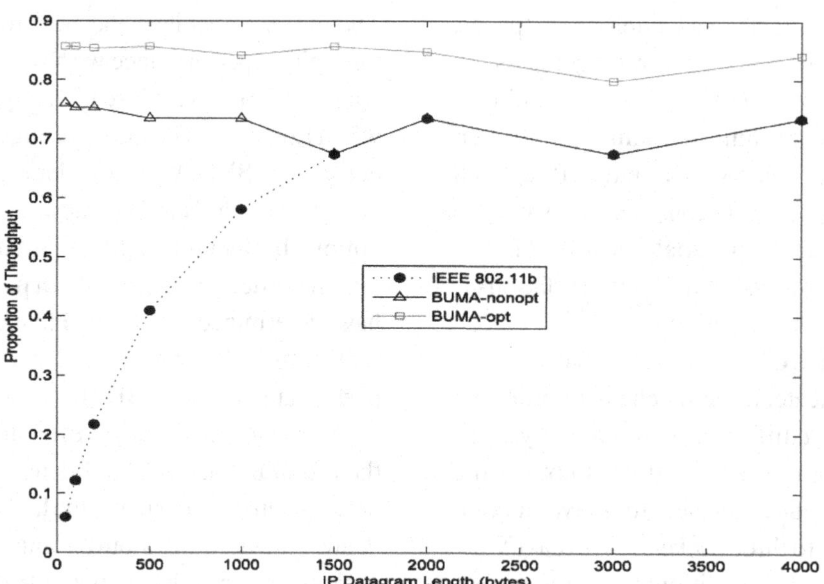

In the case of $BUMA_{nonopt}$, the maximum allowable MSDU is set to the wireless Ethernet fragmentation threshold (so that frames can be transmitted without fragmentation). As shown in Figure 4, $BUMA_{nonopt}$ achieves higher throughput than DCF for payload lengths $\leq$ 1,000 bytes i.e. the throughput gain is more significant for short payloads (see above Example). Sending a larger payload with the same set of overheads causes this improvement. However, for payloads > 1,000 bytes, $BUMA_{nonopt}$ does not improve on DCF since the length of the Ethernet fragmentation threshold limits the performance.

Now in the case of $BUMA_{opt}$, the optimum length of the buffer unit (4,534 bytes) becomes the maximum allowable MSDU. $BUMA_{opt}$ offers higher throughput than DCF irrespective of payload length. This is a notable result that clearly demonstrates the superiority of both $BUMA_{nonopt}$ and $BUMA_{opt}$ over DCF. Also, BUMA's throughput is almost independent of IP datagram length, unlike that of DCF.

By comparing $BUMA_{nonopt}$ and $BUMA_{opt}$, one can observe that $BUMA_{opt}$ offers 10 to 18% greater throughput than $BUMA_{nonopt}$ for payload lengths smaller than 4,000 bytes. This throughput improvement is as a result of $BUMA_{opt}$ transmits a slightly larger payload than $BUMA_{nonopt}$ with the same set of overheads.

SIMULATION ENVIRONMENT AND PARAMETERS

A simulation model was developed using ns-2 simulator [29] to study the performance of the proposed BUMA protocol. All stations communicate using identical half-duplex wireless radio based on the DCF, with data rate set at 11 Mbps. The RTS/CTS mechanism was turned off. Stations were stationary. The transmission and carrier-sensing ranges were set to 250 m and 550 m, respectively. The ad hoc on-demand distance vector (AODV) routing protocol and the two-ray ground propagation model were used. Streams of data packets generated at stations were modeled as independent Poisson processes with an aggregate mean packet generating rate λ packets/s. All data sources are user-datagram protocol (UDP) traffic streams with fixed packet length of 1500 bytes.

To simplify the simulation model, we consider a perfect radio propagation environment in which there is no transmission error due to interference and noise in the system, and no hidden and exposed station problems.

Simulation Parameters

Table 2 lists the parameter values used in the simulation of both the BUMA and 802.11 DCF protocols. Each simulation run lasted for 60 seconds simulated time in which the first 10 seconds was the transient period. The observations collected during the transient period are not included in the final simulation results.

MODEL VALIDATION

The ns-2 simulation results presented in this paper were verified in several ways. First, the simulation model was validated through radio propagation measurements from wireless laptops and APs for 802.11b (Sarkar & Sowerby, 2006; Siringoringo & Sarkar, 2009). A good match between simulation results and real measurements for N = 2 to 4 stations validates the simulation model. Second, ns-2 results were compared with the results obtained from OPNET Modeler (*OPNET Modeler*) and a good match between two sets of results further validated the simulation models.

Table 2. Simulation parameters

Parameter	Value
Data rate	11 Mbps
Basic Rate	2 Mbps
Wireless cards	802.11b
Slot duration	20 μs
SIFS	10 μs
DIFS	50 μs
Simulation time	60 seconds
Packet/Traffic type	UDP
Application	CBR
PHY modulation	DSSS
CWmin	31
CWmax	1023

PERFORMANCE RESULTS AND COMPARISON

We consider four important network performance metrics: (1) throughput; (2) packet delay; (3) fairness; and (4) packet drop ratio; for both individual stations and the overall network.

The throughput (measured in Mbps) is the mean rate of successful message delivery over a communication channel. The mean packet delay is defined as the average time (measured in seconds) from the moment the packet is generated until the packet is fully dispatched from that station. This packet delay includes queuing delay and medium access delay at the source station, and packet transmission time.

We define a metric for fairness measurement called mean deviation of throughput (MDT) as follows:

$$MDT = \frac{\sum \left|(T_i - \overline{T})\right|}{N} \tag{2}$$

where T_i is the throughput at station *i;* $\overline{T}$ is the network wide mean throughput; and N is the number of active stations.

MDT is the spread or variation of individual stations' throughput from the network wide mean throughput. For instance, a MAC protocol is said to be 100% fair if MDT is zero (i.e. $T_i = \overline{T}\ \forall\, i$). The value of MDT indicates the level of unfairness of a MAC protocol. Hence a MAC protocol with a smaller MDT is desirable. MDT was used to compare the fairness of BUMA and DCF.

We define another metric called packet drop ratio (P_{dr}) as follows.

$$P_{dr} = \frac{N_{pd}}{N_{tp}} \qquad (3)$$

where, N_{pd} is the total number of packets dropped at the network, which is the difference of the total transmitted packets and successfully received packets; and N_{tp} is the number of transmitted packets at the destination stations.

The P_{dr} is directly related to packet collision rates (i.e. high packet collisions at the destination stations result in higher packet drop ratio). This metric tells us about the capacity of a MAC protocol in delivering packets successfully to the destination stations. A MAC protocol with a low packet drop ratio is desirable.

We present the empirical results obtained from simulation runs for both the proposed BUMA and the DCF. We demonstrate the performance of the BUMA protocol by considering both ad hoc and infrastructure networks with UDP traffic operating under uniform loads. The simulation results report the steady-state behavior of the network and have been obtained with the relative error < 1%, at the 99% confidence level.

Throughput Performance

The network throughput versus offered load performance of BUMA and DCF with N = 40 stations for an ad hoc and infrastructure network is shown in Figure 5 (a) and (b), respectively. One can observe that BUMA provides higher throughput than DCF irrespective of network architecture, especially under medium-to-high traffic loads. For example, for an ad hoc network with N = 40 stations, BUMA throughput is about 45% higher than that of DCF at full loading (Figure 5a).

The network throughput versus the number of stations of BUMA and DCF for ad hoc and infrastructure networks is shown in Figure 6 (a) and (b), respectively. It is found that BUMA has higher throughput than DCF irrespective of network architecture for N= 10 to 100 stations. For example, for an ad hoc network with N = 20 stations, BUMA's throughput is about 45% higher than that of DCF at 80% load (Figure 6a). The main conclusion that can be drawn from Figures 5 and 6 is that BUMA's throughput (both for individual stations and network wide) is significantly better than that of DCF, especially under medium-to-high loads.

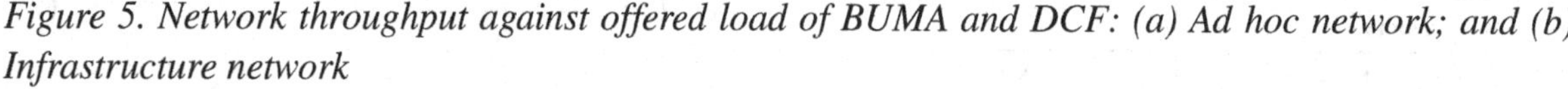

Figure 5. Network throughput against offered load of BUMA and DCF: (a) Ad hoc network; and (b) Infrastructure network

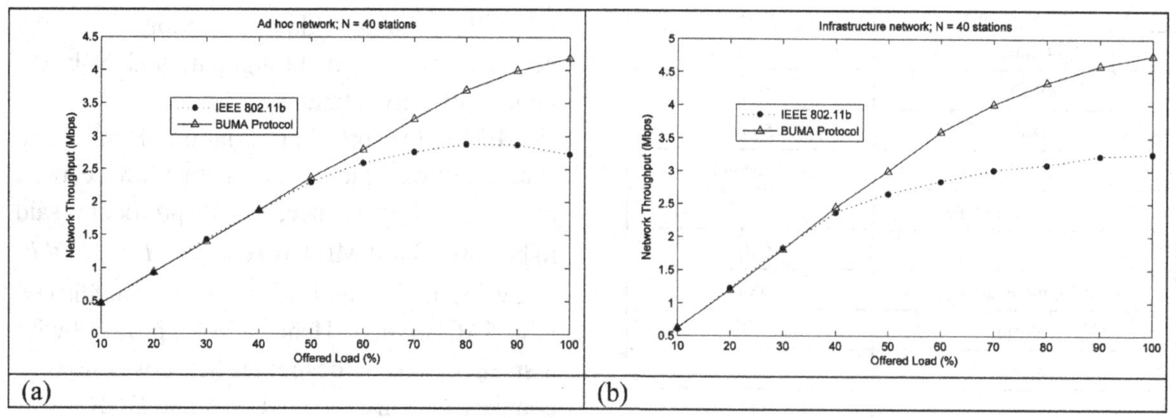

Figure 6. Network throughput versus number of active stations of BUMA and DCF: (a) Ad hoc network; (b) Infrastructure network

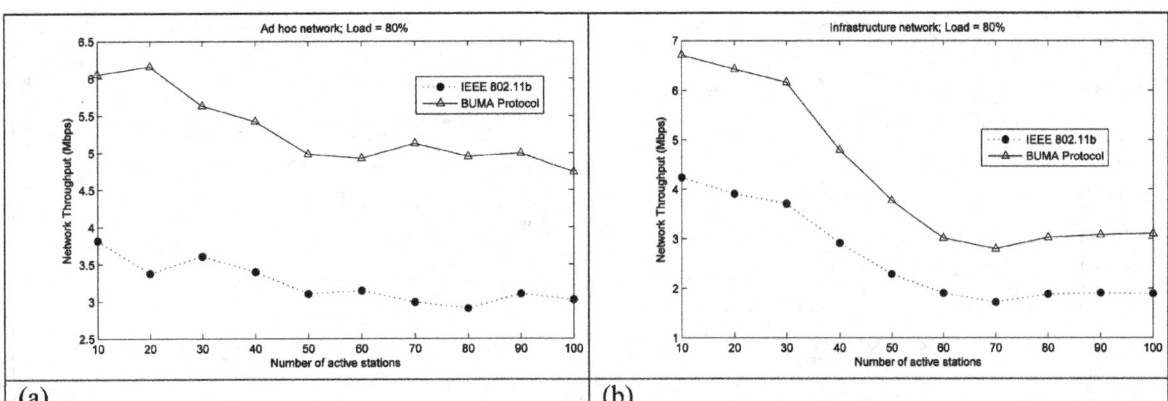

Packet Delay

The network mean packet delays of BUMA and DCF for N = 40 stations in ad hoc and infrastructure networks are shown in Figure 7 (a) and (b), respectively.

It is observed that BUMA achieves lower mean packet delay than DCF irrespective of network architecture, especially at load greater than 40%. For example, for an ad hoc network with N = 40 stations, BUMA's mean packet delay is about 96% lower than DCF's at 70% load (Figure7a).

The network mean packet delays versus the number of active stations of BUMA and DCF for ad hoc and infrastructure networks are shown in Figure 8 (a) and (b), respectively. It is observed that BUMA's mean packet delays are lower than DCF's for both ad hoc and infrastructure networks.

The main conclusion is that (Figures 7 and 8) stations using BUMA have a substantially lower mean packet delay than stations using DCF, especially under medium-to-high loads.

Fairness Performance

Figure 9 plots active links between source and destination stations (e.g. 0->1 indicates station 0 transmits a packet to station 1) versus MDT fairness

Figure 7. Mean packet delay versus offered load of BUMA and DCF: (a) Ad hoc network; and (b) Infrastructure network

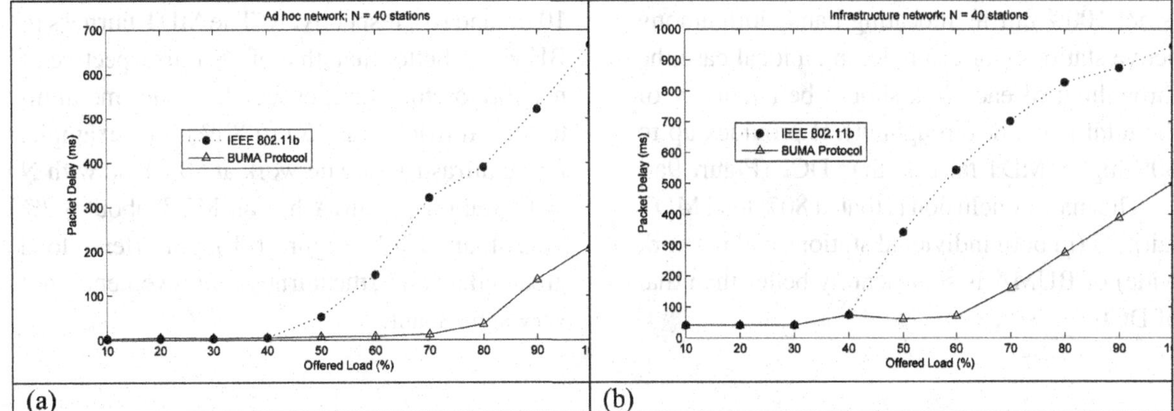

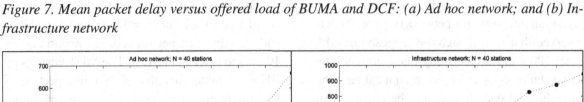

Figure 8. Mean packet delay versus number of active stations of BUMA and DCF: (a) Ad hoc network; and (b) Infrastructure network

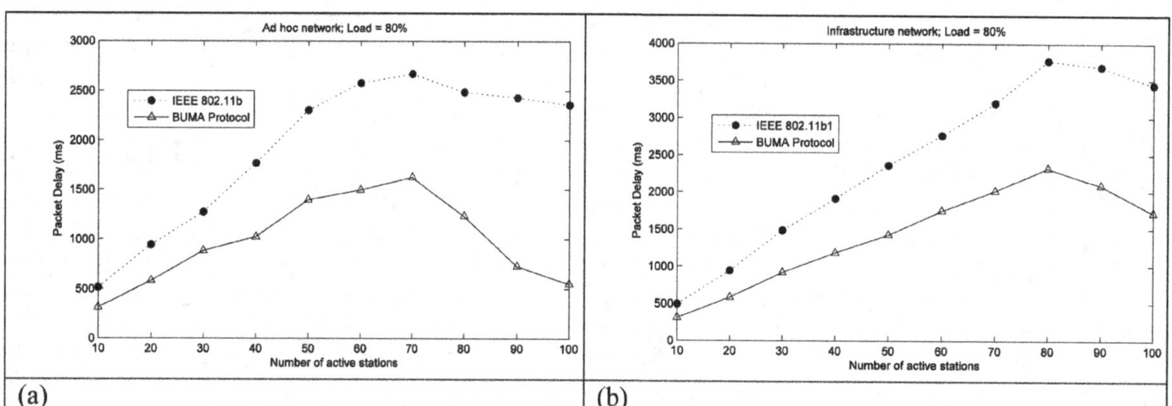

Figure 9. MDT fairness versus links of BUMA and DCF: (a) Ad hoc network; and (b) Infrastructure network

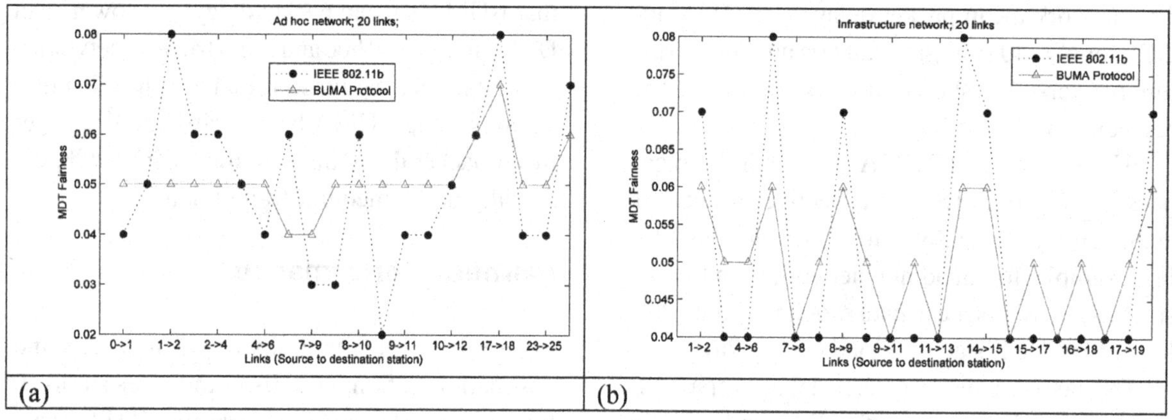

for N = 20 stations for an ad hoc network (Figure 9a) and an infrastructure network (Figure 9b). It is observed that even though the proposed BUMA is not 100% fair in allocating bandwidth among active stations (for example, in an ideal case the throughput of each link should be about 5% of the total network throughput), it provides up to 50% higher MDT fairness than DCF (Figure 9a).

The main conclusion is that at 80% load MDT fairness (in both individual stations and network wide) of BUMA is significantly better than that of DCF.

The effects of offered load on the fairness of BUMA and DCF for N = 10 stations for ad hoc and infrastructure networks are shown in Figure 10 (a) and (b), respectively. The MDT fairness of BUMA is better than that of DCF irrespective of network architecture, especially under medium-to-high traffic loads (50 to 80%). For example, for an infrastructure network at 50% load with N = 10 stations, BUMA has an MDT about 8.2% lower than DCF's (Figure 10b). For offered load greater than 80%, the fairness improvement is not very significant.

Figure 10. MDT fairness versus offered load of BUMA and DCF: (a) Ad hoc network; and (b) Infrastructure network

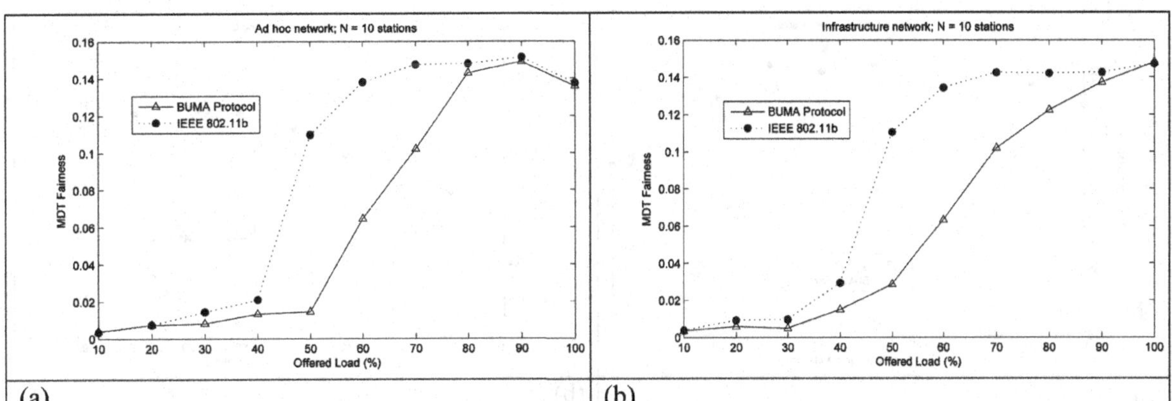

The effects of active stations on the MDT fairness of BUMA and DCF in ad hoc and infrastructure networks are demonstrated in Figure 11 (a) and (b), respectively. It is observed that BUMA has slightly better MDT fairness than that of DCF, especially for N > 80 stations at 80% load. However, this fairness improvement is not very significant.

The main conclusion (Figures 10 and 11) is that stations using BUMA achieve a slightly better MDT fairness than stations using DCF, especially under medium-to-high loads.

Packet Drop Ratio

The network-wide packet drop ratios of both the BUMA and 802.11b DCF protocols for N = 10 to 100 stations at 80% offered load for an ad hoc and infrastructure network are shown in Figure 12 (a) and (b), respectively.

We observe that fewer packets are dropped under the BUMA protocol than the DCF protocol. For example, the BUMA offers about 28.5% lower packet drop ratio than the DCF for N = 20 stations at 80% offered load (Figure 12a). This

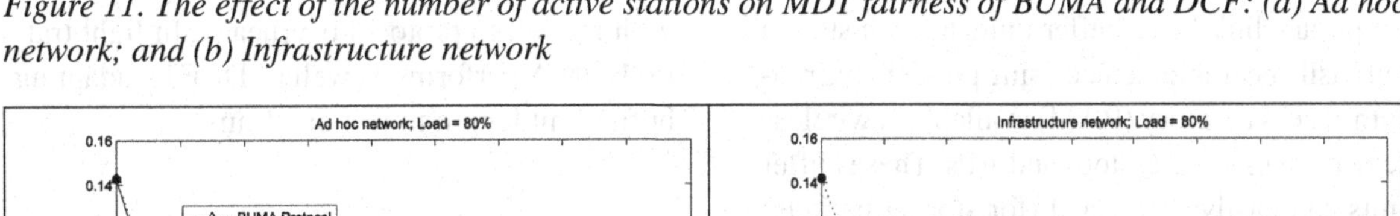

Figure 11. The effect of the number of active stations on MDT fairness of BUMA and DCF: (a) Ad hoc network; and (b) Infrastructure network

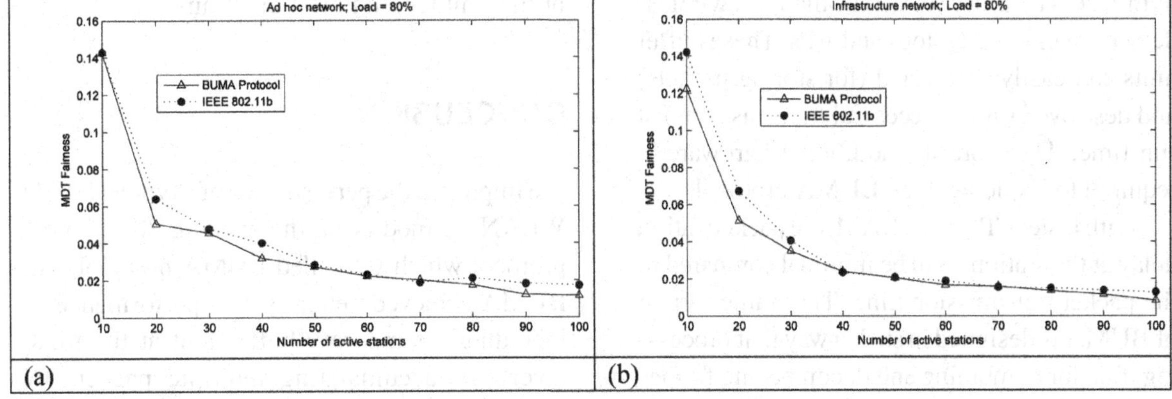

Figure 12. Packet drop ratio versus the number of active stations of BUMA and DCF: (a) Ad hoc network; and (b) Infrastructure network

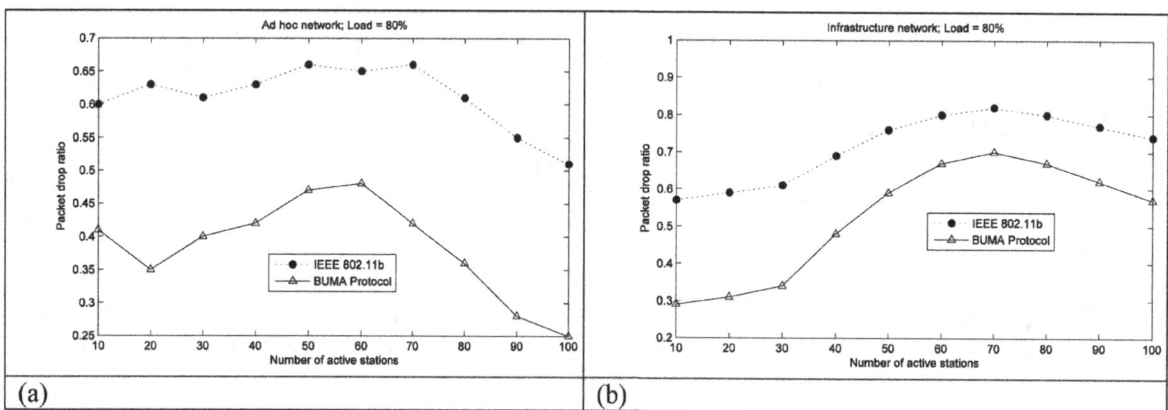

improvement in packet dropping is due to the BUMA protocol's channel access strategy in which relatively fewer contentions are faced by all active stations (i.e. fewer packet collisions) on the network, and consequently achieve high throughput performance than the DCF.

IMPLICATIONS FOR SYSTEM IMPLIMENTATION

The BUMA protocol is simple, easy to implement, and provides a low cost solution for improving WLAN performance. In the BUMA protocol, the temporary link layer buffer units at each station can easily be implemented using pre-existing random access memory (RAM) available in wireless devices, including laptops and APs. These buffer units can easily be created (for storing packets) and destroyed (after successful transmissions) at run time. Therefore, no additional hardware is required to implement the BUMA protocol.

With faster CPUs and RAM, the packetization delay at the stations will be minimal compared to the packet transmission time. The frame format of BUMA is designed in such a way that processing time for combining and decomposing frames is insignificant. The mechanisms for empty slot detection, slot synchronization, packet transmission, and packet reception can be implemented by firmware as is used in the 802.11 (IEEE Standards Association, 1997). When the destination station receives the BUMA frame, it decomposes the combined frame into normal frames, and acknowledges the last frame only. The destination station can easily identify boundaries of the combined frames using preambles and CRC.

In high traffic, BUMA performs better since the buffer unit can be filled up quickly with data from the upper protocol layers (e.g. IP datagram can be encapsulated up to the maximum length of the buffer unit) and hence carries larger payload with respect to protocol overheads. In light traffic BUMA performs as well as DCF by adapting buffer unit length to just one frame.

CONCLUSION

We improved the performance of a typical 802.11 WLAN by modifying the existing 802.11 DCF protocol which we called BUMA protocol. The BUMA achieved better system performance by including a temporary buffer unit at the MAC layer (for accumulating multiple packets for transmission). Simulation results have shown that the proposed BUMA achieved up to 45% higher throughput, 96% lower packet delay, 28.5% lower packet dropping, and 50% greater fairness than the DCF for both ad hoc and infrastructure networks under medium-to-high traffic loads.

The BUMA does not change the operation of PHY layer and thus it can easily be implemented in the 802.11 networks requiring no hardware changes and no additional costs. BUMA can be used to improve the performance of all variants of 802.11 WLANs, including the 802.11b/g. The models built using ns-2 simulator were validated using propagation measurements from wireless laptops and APs for an 802.11b WLAN. A good match between simulation results and measurements validated our simulation models.

The implementation aspect of BUMA has been discussed. RAM in wireless devices can be used to implement link layer buffer units at the stations without incurring any additional hardware costs. Fast modern processors and RAM render the station's packetization delay insignificant. A future paper will report on a rate adaptive QoS aware MAC protocol for multimedia WLANs.

REFERENCES

Aad, I., Ni, Q., Barakat, C., & Turletti, T. (2005). Enhancing IEEE 802.11 MAC in congested environments. *Computer Communications*, *28*(14), 1605–1617. doi:10.1016/j.comcom.2005.02.010.

Banchs, A., Serrano, P., & Azcorra, A. (2006). End-to-end delay analysis and admission control in 802.11 DCF WLANs. *Communications of the ACM*, *29*(7), 842–854.

Banchs, A., & Vollero, L. (2006). Throughput analysis and optimal configuration of 802.11e EDCA. *Computer Networks*, *50*(11), 1749–1768. doi:10.1016/j.comnet.2005.07.008.

Bharghavan, V. (1994). MACAW: A media access protocol for wireless LANs. In *Proceedings of the Annual Conference on Communications, Architectures, Protocols, and Applications* (pp. 212-225). New York, NY: ACM Press.

Bianchi, G. (2000). Performance analysis of the IEEE 802.11 distributed coordination function. *IEEE Journal on Selected Areas in Communications*, *18*(3), 535–547. doi:10.1109/49.840210.

Bianchi, G., Borgonovo, F., Fratta, L., Musumeci, L., & Zorzi, M. (1997). C-PRMA: A centralized packet reservation multiple access for local wireless communications. *IEEE Transactions on Vehicular Technology*, *46*(2), 422–436. doi:10.1109/25.580781.

Bianchi, G., Fratta, L., & Oliveri, M. (1996, October 15-18). Performance evaluation and enhancement of the CSMA/CA MAC protocol for 802.11 wireless LANs. In *Proceedings of the 7th IEEE International Symposium on Personal, Indoor and Mobile Radio Communications,* Taipei, Taiwan (pp. 392-396). Washington, DC: IEEE Computer Society.

Bruno, R., Conti, M., & Gregori, E. (2002). Optimization of efficiency and energy consumption in p-persistent CSMA-based wireless LANs. *IEEE Transactions on Mobile Computing*, *1*(1), 10–31. doi:10.1109/TMC.2002.1011056.

Cali, F., & Conti, M. (2000). Dynamic tuning of the IEEE 802.11 protocol to achieve a theoretical throughput limit. *IEEE/ACM Transactions on Networking*, *8*(6), 785–799. doi:10.1109/90.893874.

Chatzimisios, P., Vitsas, V., & Boucouvalas, A. C. (2002). Throughput and delay analysis of IEEE 802.11 protocol. In *Proceedings of the 5th IEEE International Workshop on Networked Appliances University,* Liverpool, UK (pp. 168-174). Washington, DC: IEEE Computer Society.

Chieochan, S., Hossain, E., & Diamond, J. (2010). Channel assignment schemes for infrastructure-based 802.11 WLANs: A survey. *IEEE Communications Surveys & Tutorials*, *12*(1), 124–136. doi:10.1109/SURV.2010.020110.00047.

Choudhury, S., & Gibson, J. D. (2007). Payload length and rate adaptation for multimedia communications in wireless LANs. *IEEE Journal on Selected Areas in Communications*, *25*(4), 796–807. doi:10.1109/JSAC.2007.070515.

David, M., Ken, D., & Doug, L. (2007). Modeling the 802.11 distributed coordination function in nonsaturated heterogeneous conditions. *IEEE/ACM Transactions on Networking*, *15*(1), 159–172. doi:10.1109/TNET.2006.890136.

Ergen, M., & Varaiya, P. (2008). Formulation of distributed coordination function of IEEE 802.11 for asynchronous networks: Mixed data rate and packet size. *IEEE Transactions on Vehicular Technology*, *57*(1), 436–447. doi:10.1109/TVT.2007.901887.

Ferre, P., Doufexi, A., Nix, A., & Bull, D. (2004). Throughput analysis of IEEE 802.11 and IEEE 802.11e MAC. In [). Washington, DC: IEEE Computer Society.]. *Proceedings of the IEEE Wireless Communications and Networking Conference*, *2*, 783–788.

Ganguly, S., Navda, V., Kim, K., Kashyap, A., Niculescu, D., & Izmailov, R. et al. (2006). Performance optimizations for deploying VoIP services in mesh networks. *IEEE Journal on Selected Areas in Communications*, *24*(11), 2147–2158. doi:10.1109/JSAC.2006.881594.

IEEE Standards Association. (1997). *IEEE 802.11 WG Standard for wireless LAN: Medium access control (MAC) and physical layer (PHY) specifications, IEEE 802.11 Standard*. Retrieved from http://standards.ieee.org/getieee802/download/802.11-2007.pdf

IEEE Standards Association. (1999a). *IEEE 802.11 Standard for Wireless LAN: Medium Access Control (MAC) and Physical Layer (PHY) Specification*. Retrieved from http://standards.ieee.org/about/get/802/802.11.html

IEEE Standards Association. (1999b). *IEEE 802.11b WG, Part II: Wireless LAN medium access control (MAC) and physical layer (PHY) specifications: High-speed physical layer extension in the 2.4 GHz band, IEEE 802.11b Standard*. Retrieved from http://standards.ieee.org/about/get/802/802.11.html

IEEE Standards Association. (2005). *IEEE P802.11e Amendment to IEEE Std 802.11, Part II: Wireless LAN Medium Access Control (MAC) and Physical Layer (PHY) Specifications: MAC Quality of Server (QoS) Enhancements*. Retrieved from http://standards.ieee.org/about/get/802/802.11.html

Jagadeesan, S., Manoj, B. S., & Murthy, C. S. R. (2003). Interleaved carrier sense multiple access: An efficient MAC protocol for ad hoc wireless networks. In *Proceedings of the IEEE International Conference on Communications*, Anchorage, AK (Vol. 2, pp. 1124-1128). Washington, DC: IEEE Computer Society.

Jiang, L. B., & Liew, S. C. (2008). Improving throughput and fairness by reducing exposed and hidden nodes in 802.11 networks. *IEEE Transactions on Mobile Computing*, *7*(1), 34–49. doi:10.1109/TMC.2007.1070.

Kuo, W.-K. (2007). Energy efficiency modeling for IEEE 802.11 DCF system without retry limits. *Computer Communications*, *30*(4), 856–862. doi:10.1016/j.comcom.2006.10.005.

Kwon, Y., Fang, Y., & Latchman, H. (2003). A novel MAC protocol with fast collision resolution for wireless LANs. In *Proceedings of the 22nd IEEE Computer and Communications Annual Joint Conference of INFOCOM* (Vol. 2, pp. 853-862). Washington, DC: IEEE Computer Society.

Li, Y., & Chen, I. (2010). Design and performance analysis of mobility management schemes based on pointer forwarding for wireless mesh networks. *IEEE Transactions on Mobile Computing*, *10*(3), 349–361. doi:10.1109/TMC.2010.166.

Lin, W.-Y., & Wu, J.-S. (2007). Modified EDCF to improve the performance of IEEE 802.11e WLAN. *Computer Communications*, *30*(4), 841–848. doi:10.1016/j.comcom.2006.10.013.

Luo, J., Rosenberg, C., & Girard, A. (2010). Engineering wireless mesh networks: Joint scheduling, routing, power control, and rate adaptation. *IEEE/ACM Transactions on Networking*, *18*(5), 1387–1400. doi:10.1109/TNET.2010.2041788.

Natkaniec, M., & Pach, A. R. (2000, July 3-6). An analysis of backoff mechanism used in IEEE 802.11 networks. In *Proceedings of the 5th IEEE Symposium on Computers and Communications*, Antibes-Juan les Pins, France (pp. 444-449). Washington, DC: IEEE Computer Society.

Natkaniec, M., & Pach, A. R. (2002, October 27-30). PUMA - a new channel access protocol for wireless LANs. In *Proceedings of the 5th International Symposium on Wireless Personal Multimedia Communications* (Vol. 3, pp. 1351-1355).

Ozugur, T., Naghshineh, M., Kermani, P., & Copeland, J. A. (1999, December 5-9). Fair media access for wireless LANs. In *Proceedings of the IEEE Global Telecommunications Conference*, Rio de Janeireo, Brazil (Vol. 1, pp. 570-579). Los Alamitos, CA: IEEE Press.

Sarkar, N. I. (2006, January 2-4). Fairness studies of IEEE 802.11b DCF under heavy traffic conditions. In *Proceedings of the First IEEE International Conference on Next-Generation Wireless Systems* (pp. 11-16). Washington, DC: IEEE Computer Society.

Sarkar, N. I., & Sowerby, K. W. (2006, November 27-30). Wi-Fi performance measurements in the crowded office environment: a case study. In *Proceedings of the 10th IEEE International Conference on Communication Technology*, Guilin, China (pp. 37-40). Washington, DC: IEEE Computer Society.

Siringoringo, W., & Sarkar, N. I. (2009). Teaching and learning Wi-Fi networking fundamentals using limited resources. In Gutierrez, J. (Ed.), *Selected readings on telecommunications and networking* (pp. 22–40). Hershey, PA: IGI Global.

Wang, W., Liew, S. C., & Li, V. O. K. (2005). Solutions to performance problems in VoIP over a 802.11 wireless LAN. *IEEE Transactions on Vehicular Technology*, *54*(1), 366–384. doi:10.1109/TVT.2004.838890.

Wang, Y., & Garcia-Luna-Aceves, J. J. (2003, March 16-20). Throughput and fairness in a hybrid channel access scheme for ad hoc networks. In *Proceedings of the IEEE Wireless Communications and Networking Conference* (pp. 988-993). Washington, DC: IEEE Computer Society.

Xiao, Y. (2004, March 21-25). Concatenation and piggyback mechanisms for the IEEE 802.11 MAC. In *Proceedings of the IEEE Wireless Communications and Networking Conference* (pp. 1642-1647). Washington, DC: IEEE Computer Society.

Yin, J., Wang, X., & Agrawal, D. P. (2005). Modeling and optimization of wireless local area network. *Computer Communications*, *28*(10), 1204–1213. doi:10.1016/j.comcom.2004.07.027.

You, T., Yeh, C.-H., & Hassanein, H. (2003a). CSMA/IC: A new class of collision-free MAC protocols for ad hoc wireless networks. In *Proceedings of the Eighth IEEE International Symposium on Computers and Communication* (pp. 843-848). Washington, DC: IEEE Computer Society.

You, T., Yeh, C.-H., & Hassanein, H. (2003b, May 11-15). A new class of collision prevention MAC protocols for wireless ad hoc networks. In *Proceedings of the IEEE International Conference on Communications*, Anchorage, AK (pp. 1135-1140). Washington, DC: IEEE Computer Society.

Yun, L., Ke-Ping, L., Wei-Liang, Z., & Qian-Bin, C. (2006). A novel random backoff algorithm to enhance the performance of IEEE 802.11 DCF. *Wireless Personal Communications*, *36*(1), 29–44. doi:10.1007/s11277-006-6176-8.

This work was previously published in the International Journal of Wireless Networks and Broadband Technologies, Volume 1, Issue 1, edited by Naveen Chilamkurti, pp. 15-31, copyright 2011 by IGI Publishing (an imprint of IGI Global).

Chapter 3
Insights from Experimental Research on Distributed Channel Assignment in Wireless Testbeds

Felix Juraschek
Freie Universität Berlin, Germany

Mesut Günes
Freie Universität Berlin, Germany

Matthias Philipp
Freie Universität Berlin, Germany

Bastian Blywis
Freie Universität Berlin, Germany

ABSTRACT

This article presents the DES-Chan framework for experimental research on distributed channel assignment algorithms in wireless mesh testbeds. The implementation process of channel assignment algorithms is a difficult task for the researcher since common operating systems do not support channel assignment algorithms. DES-Chan provides a set of common services required by distributed channel assignment algorithms and eases the implementation effort. The results of experiments to measure the channel characteristics in terms of intra-path and inter-path interference according to the channel distance on the DES-Testbed are also presented. The DES-Testbed is a multi-radio WMN with more than 100 nodes located on the campus of the Freie Universität Berlin. These measurements are an important input to validate common assumptions of WMNs and derive more realistic, measurement-based interference models in contrast to simplified heuristics.

INTRODUCTION

Channel assignment for multi-transceiver wireless mesh networks (WMNs) attempts to increase the network performance by decreasing the interference of simultaneous transmissions. Multi-transceiver mesh routers allow the communication over several wireless network interfaces at the same time. However, this can result in high interference of the wireless interfaces leading to a low network performance. With channel assignment, the reduction of interference is achieved by exploiting the availability of fully or partially non-overlapping channels. Channel assignment can be applied to

DOI: 10.4018/978-1-4666-3902-7.ch003

all wireless networks based on technologies that provide non-overlapping channels. Currently, wide-spread technologies are IEEE 802.11a/b/g, IEEE 802.11n, and IEEE 802.16 (WiMAX). With the relatively low cost for IEEE 802.11 hardware, the number of deployments based on this technology is increasing and channel assignment algorithms are gaining in importance.

Although channel assignment is still a young research area, many different approaches have already been developed (Si, 2009). These approaches can be distinguished into centralized and distributed algorithms. Centralized algorithms rely on a central entity, usually called channel assignment server (CAS), which calculates the network-wide channel assignment and sends the result to the mesh routers. In distributed approaches, each mesh router calculates its channel assignment based on local information. Distributed approaches can react faster to topology changes due to node failures or mobility and usually introduce less protocol overhead since communication with the CAS is not necessary. As a result, distributed approaches are more suitable once the network is operational and running. Another classification considers the frequency of channel switches on a network node. In fast channel switching approaches, channel switches may occur frequently, in the extreme for every subsequent packet a different channel is chosen. The limiting factor for dynamic algorithms is the long channel switching time with commodity IEEE 802.11 hardware, which is in the order of milliseconds. Slow channel switching approaches in contrast, switch the interfaces to a particular channel for a longer period, usually in the order of minutes or hours. Hybrid approaches combine both methods.

An important input for channel assignment algorithms are the particular channel characteristics of the used network technology. For instance, IEEE 802.11b/g offers in theory three non-overlapping channels, e.g. {1, 6, 11}, and all available channels in IEEE 802.11a use non-overlapping frequency spectrums. This means, that concurrent transmissions on these channels should not interfere with each other. In practice, experiments and measurement on different experimental platforms have shown, that the non-interfering characteristics do not hold for many reasons (Fuxjager, 2007; Subramaniam, 2008; Draves, 2004), an important one is the insufficient distance of less than 1 m between the antennas on a single network node. Since mesh routers are usually quiet compact, it is almost impossible to design a multi-radio mesh router with sufficient antenna distance. This is also the case with DES-Nodes of the DES-Testbed, on which the three WiFi antennas are mounted with a distance of about 30 cm. Therefore, we also expect to experience side-effects on the theoretical non-interfering channels, which are subject to experimentation in this article.

Next to the channel characteristics of the DES-Testbed, the focus of this article is on the experimentally driven research of distributed, slow channel switching algorithms on wireless testbeds. This process yields several challenges and pitfalls for the researcher. Since common operating systems are not designed to support channel assignment algorithms out of the box. Thus, the researcher has to deal with operating system specifics, drivers for the wireless interfaces, and the capabilities and limitations of the particular hardware. If more than one particular algorithm should be studied, the same problems and services have to be addressed multiple times. Among them are interface management, message exchange for node-to-node communication over the wireless medium, and the provision of data structures for network and conflict graphs.

A research framework for channel assignment algorithms can be beneficial for the implementation process in many ways. The framework introduces an abstraction for low-level and operating system specific tasks, for instance for the configuration of the wireless interfaces. In contrast to the implementation of one specific channel assignment algorithm, a framework should be as universal as possible in order to allow the implementation of a wide range of different algorithms. This is achieved by providing a set of basic services, which are common to multiple channel assignment algorithms.

The contribution of this article is two-fold. On the one hand, we present DES-Chan, a framework for distributed channel assignment in multi-transceiver WMNs that provides a wide range of common services distributed channel assignment algorithms. On the other hand, we present the results of experiments in order to measure the channel characteristics of the DES-Testbed. The gained results will serve as an important input for channel assignment algorithms. Additionally, the results allow to better estimate the increase in performance of channel assignment algorithms since the interference estimation of two channels is more close to reality than simplified models.

DES-TESTBED

The DES-Testbed is a multi-radio WMN located on the campus of the Freie Universität Berlin. Currently it consists of more than 100 indoor and outdoor nodes as shown in Figure 1. The hybrid DES-Nodes consist of a *mesh router* and a *sensor node* in the same enclosure, thus forming an overlapping WMN and WSN. The DES-Nodes are deployed in an irregular topology across several buildings on the campus, a snapshot of the network topology is depicted in Figure 2 with the DES-Vis 3D-visualization tool. Besides the DES-Testbed, several in-parallel IEEE 802.11 networks exist to provide network access to students and staff members on our campus. These networks are not under our control and thus contribute to the external interference. We treat this as a condition that is also likely to be expected in a real world scenario. For a description of the architecture of the DES-Testbed in full detail we refer to our technical reports (Günes, 2008a, 2008b).

Each DES-Node in the DES-Testbed (Figure 2) is equipped with three IEEE 802.11 WNICs. One of the interfaces is a Ralink RT2501 USB stick and the other two are Mini-PCI cards with an Atheros AR5413 chipset. The cards use the *rt73usb* and *ath5k* drivers, which are part of the Linux kernel. For the experiments presented in this chapter the Linux kernel 2.6.34 was used.

Figure 1. Indoor and outdoor DES-Nodes of the DES-Testbed. The left picture shows the DES-Node version 2. The multi-radio mesh router consists of an Alix2d2 board with three IEEE 802.11a/b/g Ralink- and Atheros-based radios. An additional sensor node is connected to the DES-Node via USB. The outdoor node comprises the same components as the indoor node, but uses the Alix3d2 board to fit into the certified enclosure.

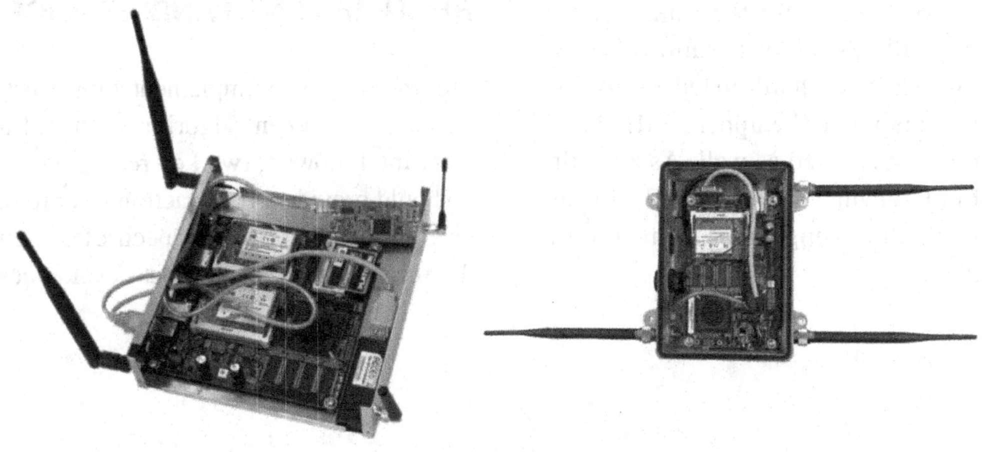

Figure 2. Snapshot of the DES-Testbed topology after a random channel assignment. The DES-Nodes are distributed over three buildings on the campus of the Freie Universität Berlin. Outdoor DES-Nodes are deployed to improve the connectivity between the adjacent buildings. Different colors are used for the different channels of the displayed links. The displayed channel assignment has been calculated with a simple random channel assignment algorithm based on DES-Chan.

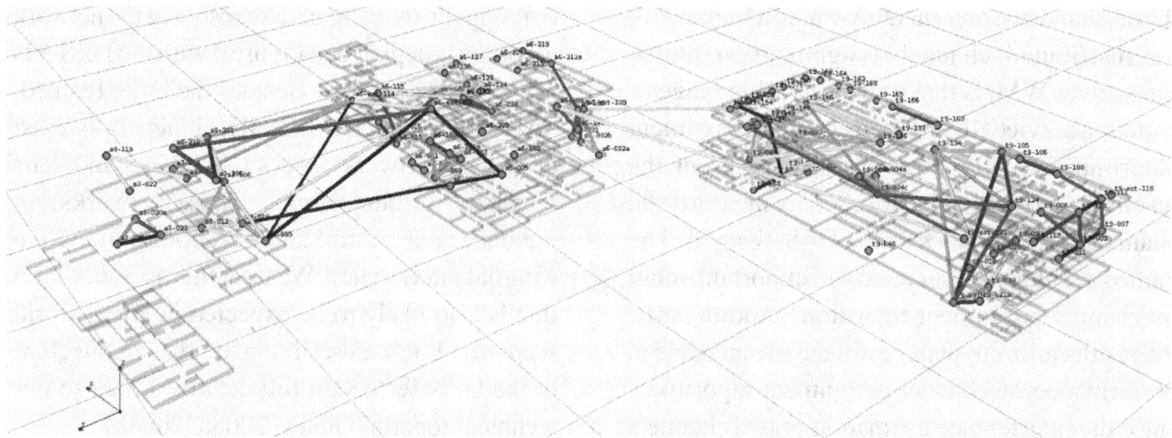

While the Ralink WNICs are IEEE 802.11b/g devices using the 2.4GHz band, the Atheros WNIC additionally support the IEEE 802.11a standard on 5GHz.

Although the 5GHz band theoretically offers 19 non-overlapping channels, only four of these can be used per default in the DES-Testbed. The Atheros cards only support IEEE 802.11a and not the IEEE 802.11h extension which adapts the standard to the European regulatory requirements. Since we are interested in the channel characteristics regardless of a specific regulatory domain, we configured a static regulatory domain database for the Linux kernel and removed all restrictions. Unfortunately, the *ath5k* driver has a hard-coded limitation for the ad-hoc mode in the upper 5 GHz band which had to be removed as well. As a result, all available 19 channels of IEEE 802.11a can be used for the following experiments on the DES-Testbed.

DES-CHAN

The DES-Chan framework has been developed for experiment-driven research on distributed channel assignment in real network environments. DES-Chan has been implemented as a Python framework and it is available at the website of the DES-Testbed http://www.des-testbed.net. This section presents the requirements for DES-Chan, its components, and the integration into the existing DES-Testbed management system.

REQUIREMENTS AND SERVICES

A framework for the implementation of distributed channel assignment algorithms should at least meet the following two key requirements. First, it should provide an abstraction layer to the low-level and operating system specific tasks, enabling the researcher to spend most development time on the algorithm logic. Second, common key services which are used by multiple algorithms have to be provided.

Naturally, all algorithms need a service for interface handling in order to change the wireless network interface card (WNIC) settings, for example to carry out channel switches. Additionally, the local topology has to be assessed with a neighborhood discovery service. Information of the local topology is needed as input for channel assignment decisions in distributed algorithms. A message exchange service for node to node communication between the instances of an algorithm is useful to exchange topology information messages and to implement a three-way handshake mechanism. A topology monitoring service periodically assesses the local topology and its state. Neighbors, links, and their respective quality can be monitored which enables the algorithms to adapt to topology changes by refining the channel assignment.

Interference models are used to estimate the local interference. They range from simple heuristics such as the m-hop neighborhood, where $2 \leq m \leq 3$, to measurement-based approaches for particular network topologies (Padhye, 2005; Reis, 2006). Even though the simple heuristics are not very realistic (Padhye, 2005), they are easy to implement and therefore widely used. Another common model is the conflict graph with algorithms trying to minimize the number of edges in order to minimize the network-wide interference (Katzela, 1996). Therefore, a framework should provide appropriate data structures and operations for modeling conflict graphs.

Architecture and Components

The DES-Chan framework comprises two main components as depicted in Figure 3. DES-Chan-Core is a Python library that provides common functions and data structures for channel assign-

Figure 3. The DES-Chan framework for channel assignment comprises the DES-Chan core and the Neighborhood Discovery service. The DES-Chan Core comprises the wrapper functions to configure the wireless network interfaces and the message exchange with neighboring nodes. The component also comprises common data structures for channel assignment and multiple interference metrics and models. The Neighborhood Discovery service allows to retrieve the local network information periodically and measures the quality of all discovered links using the ETX metric.

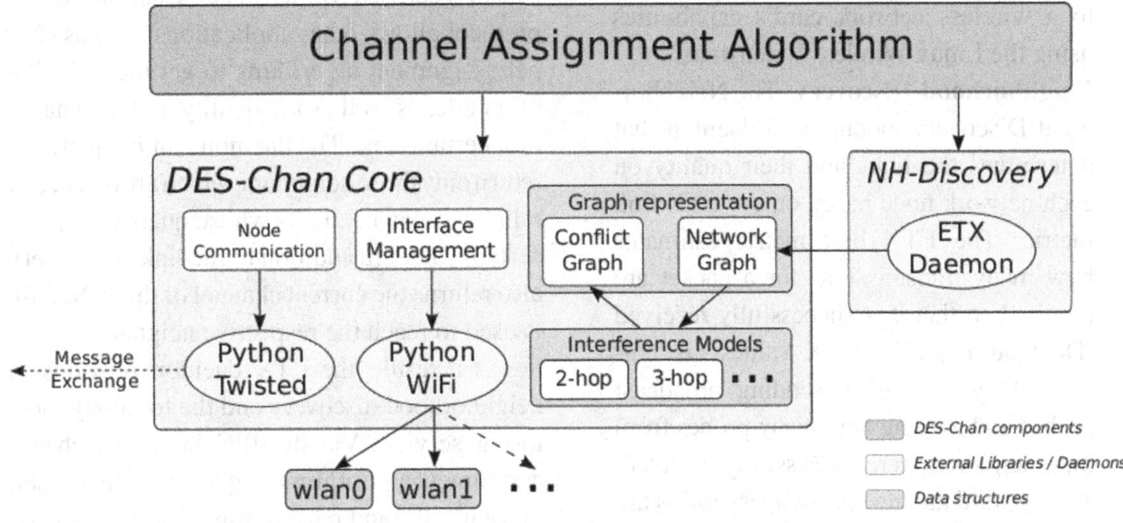

ment algorithms. It comprises the following services: interface management, node communication, graph representation, and interference models. The Neighborhood-Discovery module provides a basic service for each node to get information about all neighboring nodes.

As future algorithms will require additional services, DES-Chan has been designed to be extensible by additional modules. Therefore, a modular architecture has been chosen. As next, the components are described in detail.

- **Interface Management:** The interface management module acts as interface to the operating system and hides testbed-specific characteristics from the channel assignment algorithms. It provides various functions for configuring network interfaces and getting information about their state. Thus it is possible to set up and shut down interfaces, check whether an interface has been set up, and get information of unused interfaces. The module furthermore allows to tune an interface to a specific channel and thereby implements a crucial requirement for all channel assignment algorithms. The settings of the wireless interfaces are changed with Python WiFi (Joost, 2010), a library that provides read and write access to a wireless network card's capabilities using the Linux Wireless Extensions.
- **Neighborhood-Discovery:** The Neighborhood-Discovery module is a daemon that determines the links and their quality on each network node based on the ETX link metric. The ETX link metric estimates how many transmissions for a packet are required so that it is successfully received (De Couto, 2003). ETX values are calculated by each node sending broadcast probes and logging how many probes from their neighbors were successfully received. The forward and reverse delivery ratios are then used for the calculation of ETX, because for unicast communication in IEEE 802.11, an ACK frame has to be successfully received at the sender. The ETX value for a link is then calculated as

$$ETX = \frac{1}{df * dr}$$

where df is the forward delivery ratio and dr the reverse delivery ratio.

The ETX implementation etxd for the DES-Chan framework has been realized as a Linux daemon. By default, a probe interval of one second and a window size of 10 seconds is used. The values can be configured via command line arguments. The link quality values are averaged over a moving window, which spans 10 probe intervals. The daemon sends UDP probe packets on the broadcast addresses of the specified network interfaces and at the same time listens for incoming probes. In order to always provide up-to-date information, etxd dynamically adapts its configuration if network interfaces have been reconfigured, shut down, or brought up.

To offer the neighborhood and link quality information to other programs, etxd provides an inter process communication (IPC) interface that can be accessed via sockets. A simple, textual protocol allows other applications such as channel assignment algorithms to get the neighbors of a node, as well as the quality and the channel of a certain link. The daemon can be queried to return only those neighbors, which are reached via reliable links, i.e., links whose quality exceeds a certain value. In addition to the link quality, etxd also returns the current channel of the WNIC that is used to reach the respective neighbor.

As a result, the ETX daemon provides the neighborhood discovery and the topology monitoring service. Via the IPC interface, channel assignment algorithms can query the current state of their links and react to topology changes.

- **Node Communication:** Different channel assignment algorithms use different communication protocols and message formats. Thus, a flexible networking library is needed, that allows implementing various protocols with few efforts. The Python Twisted (Twisted, 2010) library serves as the foundation for the node to node communication in DES-Chan. It provides an asynchronous networking engine, targeted to ease the implementation of network programs.

The library provides an asynchronous networking engine and hides technical details like creating sockets and establishing connections from the developer. The core of Twisted is a global reactor object that can be instructed to monitor sockets and to connect to servers. The reactor implements an event loop that waits for events, such as an incoming client connection or a certain response from a server, and executes the associated callback functions. In contrast to threaded programming there is no concurrency, because only one function is active at a time. The reactor has control of the single program thread and if an event occurs, it hands the control to the associated callback. When the callback is finished, control is handed back to the Twisted event loop, which continues waiting for the next event.

Using this library, the researcher can quickly develop the required protocol implementation for exchanging messages among the network nodes, for instance in order to propagate changes in channel assignment or to carry out three-way handshakes prior to the actual channel switch.

- **Graph Representation:** DES-Chan provides several data structures and functions for graph representation. A data structure for the network graph contains a vertex for every network node and an edge for every link between two network nodes. The edges can be labeled with the number of the channel that is used for the communication. The NetworkGraph class also allows to store a list of channels for each edge, thereby supporting multiple links between node pairs. It is limited to undirected graphs, because it has been designed for IEEE 802.11 networks, which only have bi-directional links.

The NetworkGraph class offers intuitive methods for accessing vertices and edges, and provides several utility functions. It supports printing a human-readable adjacency matrix, which can be used for debugging purposes and for saving the graph in a file. Graph objects can also be stored in the DOT format, which is a simple textual graph description language that can be processed to generate an image of the graph. A function provides the possibility to merge two NetworkGraph objects together by creating the union of both vertex sets and edge sets. This can be used for example to merge network graphs that were obtained from different neighbors.

Additionally, a corresponding ConflictGraph data structure is provided. It allows to apply an interference model to a network graph, in order to calculate the interference between all link pairs in the corresponding network. The class maintains the relation between edges in the network graph and vertices in the conflict graph, and provides corresponding transformation methods. It allows manipulating the channel of the corresponding edge in the network graph and automatically updates the interference information in the conflict graph. This functionality is needed by algorithms that successively apply several channel assignments to find a configuration that minimizes interference. The ConflictGraph can be easily extended to implement other concepts, such as multi-radio conflict graph (Ramachandran, 2006).

- **Interference Models:** The DES-Chan framework supports the implementation of various interference models. Currently, only the two-hop interference model is implemented for reference, but the framework can be easily extended. The current implementation of the two-hop heuristic supports a binary notion of interference, i.e., two links either interfere or do not interfere, and a more accurate notion of the interference cost taking the spectral distance of two channels into account.

The following equation shows the interference cost function I for two frequencies f_1 and f_2. The parameter α denotes the minimum frequency difference of orthogonal channels.

$$I(f_1, f_2, \alpha) = \begin{cases} 0, & if\left|f_1 - f_2\right| \geq \alpha \\ 1 - \dfrac{\left(\left|f_1 - f_2\right|\right)}{\alpha}, & otherwise \end{cases}$$

For IEEE 802.11b/g three channels of the 2.4GHz frequency band are theoretically orthogonal. This can be modeled by setting α to 30MHz, which results in the set of orthogonal channels $\{1, 7, 13\}$.

Experiments are presented in order to validate the assumption of the orthogonality.

INTEGRATION INTO THE DES-TESTBED

The DES-Testbed Management system has been extended to support the DES-Chan framework (Günes, 2009). The 3D-visualization tool DES-Vis was modified so that links are colored differently depending on the utilized channel. This way, a graphical representation of the channel assignment can be displayed. A screenshot of an experiment with a random channel assignment shows the capabilities of the visualization tool in Figure 2.

Channel Characteristics of the DES-Testbed

This section documents several experiments that have been carried out on the DES-Testbed in order to validate common assumptions of WMN with results on a real multi-radio mesh network. The experimentally determined channel characteristics are an important input for channel assignment algorithms. For instance, co-channel interference can be measured and thus the existence of possible non-interfering channels can be validated. Additionally, the results of the experiments can be used to specify upper bounds for the expected performance increase by channel assignment algorithms.

Co-Channel Interference Measurements

One of the proposed measurement-based interference estimation schemes is the link interference ratio (LIR) as introduced in (Padhye, 2005). For two links $l_{u,v}$ and $l_{x,y}$ the LIR is defined as

$$LIR_{l_{u,v}, l_{x,y}} = \frac{T_{l_{u,v}}^{l_{u,v}, l_{x,y}} + T_{l_{x,y}}^{l_{u,v}, l_{x,y}}}{T_{l_{u,v}} + T_{l_{x,y}}}$$

where $T_{l_{u,v}}$ is the unicast throughput for link $l_{u,v}$ when only this link is active and $T_{l_{u,v}}^{l_{u,v}, l_{x,y}}$ is the unicast throughput for the link when the link $l_{x,y}$ is active simultaneously. The LIR expresses the interference of two links by relating the aggregate throughput of both links when they are active individually to the aggregate throughput when they are active simultaneously. A LIR value of 1 indicates that the two links do not interfere at all whereas a LIR value of 0.5 means that the aggregate throughput is halved when both links are active at the same time.

The LIR is suitable to investigate the impact of the channel distance on two simultaneous transmissions. Therefore, in a first experiment we measure the LIR of two links being adjacent to the same node for a varying distance to investigate the effect of intra-path interference. In a second experiment, two links with different sender and receiver pairs which are in each others interference range were chosen. This experiment will give insights on inter-path interference.

For both experiments we use two Atheros Mini-PCI cards with the *ath5k* driver. The auto-data-rate algorithm is used and RTS/CTS disabled. The channels of the links are sequentially set to all possible combinations on each frequency band. We perform the experiment for all channel combinations of the 2.4 GHz and frequency band, the 5 GHz frequency band, and finally using both bands simultaneously. We repeated the experiment so that we have at least 40 measurements for each channel distance.

- **Adjacent Traffic Flows:** First, we need to select a subset of the links between the mesh routers of the DES-Testbed. For this, we used the ETX daemon of the DES-Chan framework to identify high quality links in our testbed with an ETX value of 1. We then selected 5 node pairs which are connected by such high quality links. For each node pair, we selected one node as sender and the other as receiver, as depicted in Figure 4.

In order to measure the LIR we generate two UDP unicast flows from one of the routers (sender) to the other (receiver). Each flow is generated with iperf using 54 MBit/s for 30 seconds. After the flows, we start both flows another time simultaneously. We measure the individual and aggregate throughput and compute the LIR (Figure 5). We chose this scenario because it is common in multi-hop WMN where a node on a path forwards traffic to a destination. For simplicity, we reduced the set up to only two nodes, in which the sender and receiver utilize two radios each. Also, the advantage of multi-radio nodes lies in the capabilities to utilize more than one radio at the time and thus increase the throughput.

The results for the channel combinations on the 2.4 GHz band are depicted in Figure 5. Unfortunately, they show that in the DES-Testbed none of the channels of the 2.4 GHz band are non-interfering. The median of the LIR values is about 0.6 regardless of the used channel combination, which means that the aggregate throughput is almost halved when the two links are active simultaneously. Concluding from the results, a channel assignment with the highest possible spectral distance would only lead to a minor increase of the throughput. We credit these results to the near-antenna effect of the DES-Nodes. To avoid the near-antenna effect, the experimentally specified minimum distance between two antennas is about 1 m (Robinson, 2005). Since DES-Nodes are more compact, the three WiFi antennas are mounted with a distance of about 30 cm.

The results for a subset of all possible spectral channel distances of the 5 GHz band are depicted in Figure 6. It can be observed that the median of the LIR value increases with channel distances of up to 180 MHz . This rise of the LIR is much slower than expected, but the results show that the

Figure 4. Experiment setup for measuring the effect of spectral channel distance to the LIR with adjacent traffic flows. 5 node pairs of the DES-Testbed are selected to measure the LIR of their corresponding links.

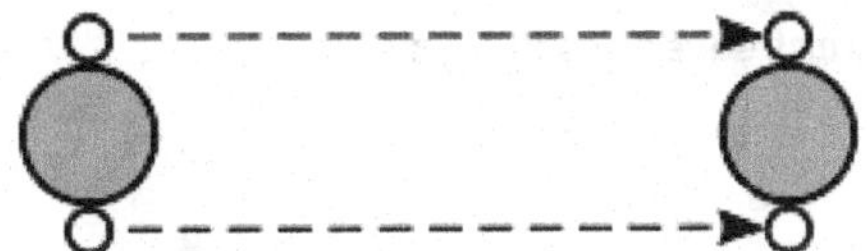

Mesh router
Wireless interface
UDP flow

Figure 5. Results of the LIR of adjacent flows for channel combinations on the 2.4 GHz band. The LIR of two links in respect to their spectral distance is shown. The median for all channel combinations is about 0.6.

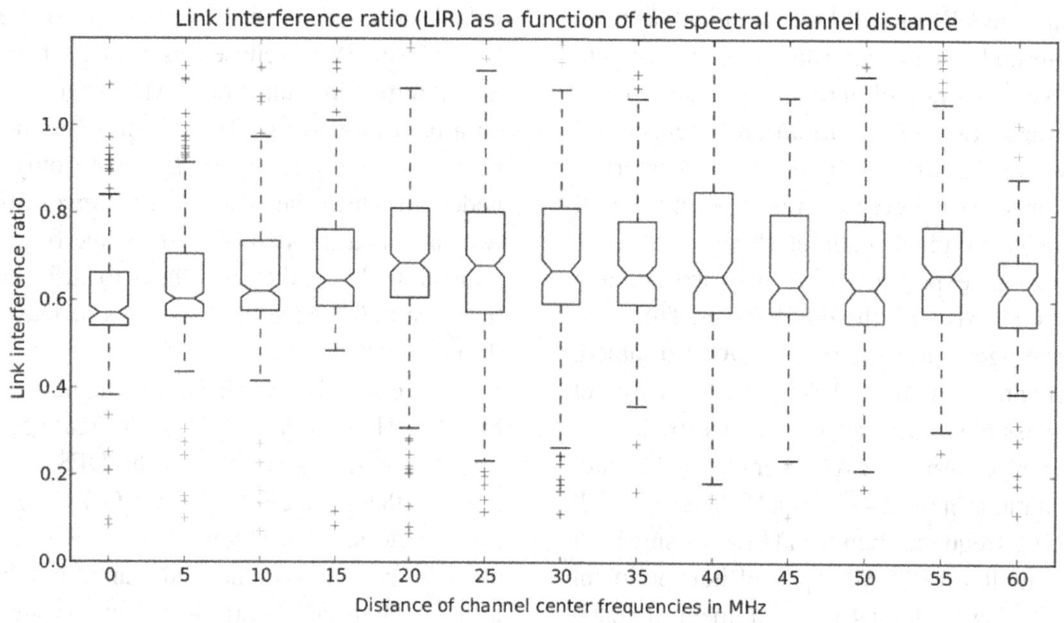

Figure 6. Results of the LIR of adjacent flows for channel combinations on the 5 GHz band. The median of the LIR value increases with channel distances of up to 180 MHz and then decreases again.

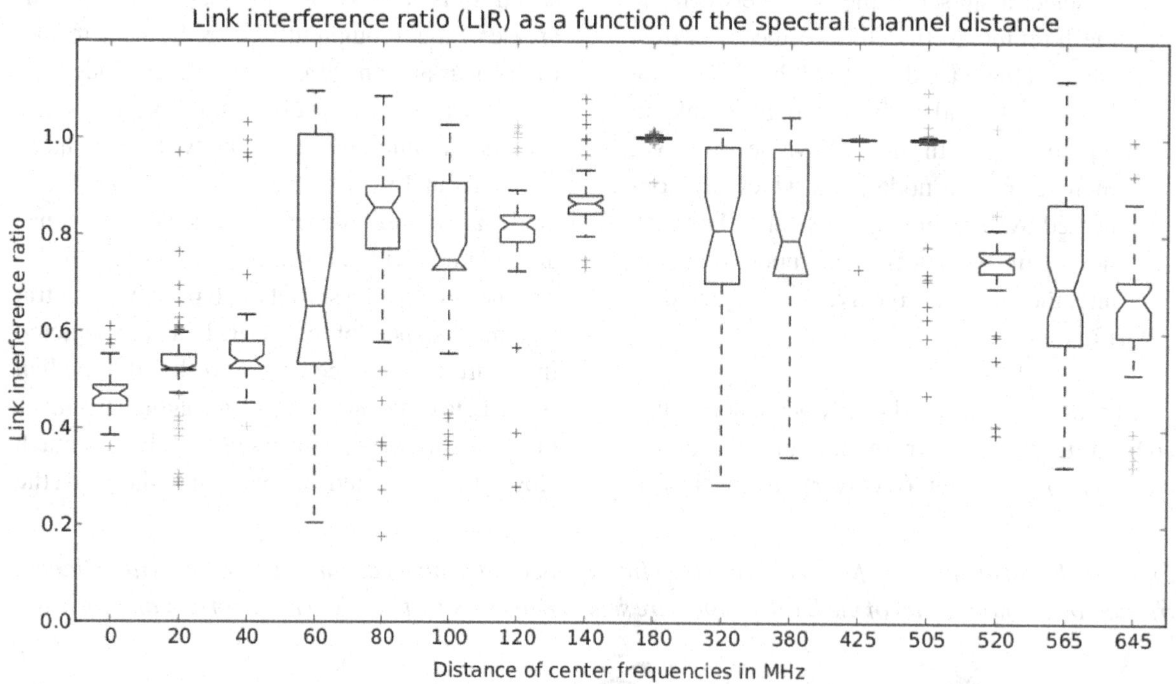

median is about 0.8 for a channel distance of at least 80 MHz. For a channel distance of 320 MHz and more the LIR decreases again, which we did not expect. In a first investigation, this seems to be related to the link quality. For the UDP flows, we monitored the received signal strength indicator (RSSI) values for each correctly received frame. We observed that the lower the RSSI values are, the lower the LIR values are for an increasing channel distance as shown in Figure 7 and Figure 8.

To display the results, we included the measured LIR values of two different node pairs u, v and j, k in Figure 7 and Figure 8. For the first node pair, the measured LIR values increase with a rising channel distance, which is as expected. We measured an average of about −65 dbm for all received frames. For the second node pair, the measured LIR values behave unexpectedly and start to drop already at a channel distance of 100 MHz below 0.5. We measured an average of about −89 dbm for all received frames, which is close to the threshold of the WNIC being able to receive a frame correctly. As a conclusion of this observation, we suspect that the huge difference in the RSSI values does hint at a very different link quality of these two link pairs. Unfortunately, the broadcast-based ETX-daemon does not seem to be appropriate to estimate link quality for unicast transmissions. We will further investigate this observation in future work.

In the last set of experiments for adjacent flows, we selected only channel combinations from both available 2.4 GHz and 5 GHz frequency bands. One sender/receiver pair of WNIC is tuned to channel c1 ∈ {1, 13} whereas the other is tuned to channel c2 ∈ {36, 64, 100, 140, 149, 165}. The results,

Figure 7. Results of the LIR of adjacent flows on links with a high RSSI value for channel combinations on the 5 GHz band. The average RSSI value for received frames was about -65 dBm. The LIR is close to 1 with a spectral distance of at least 60 MHz, which is as expected.

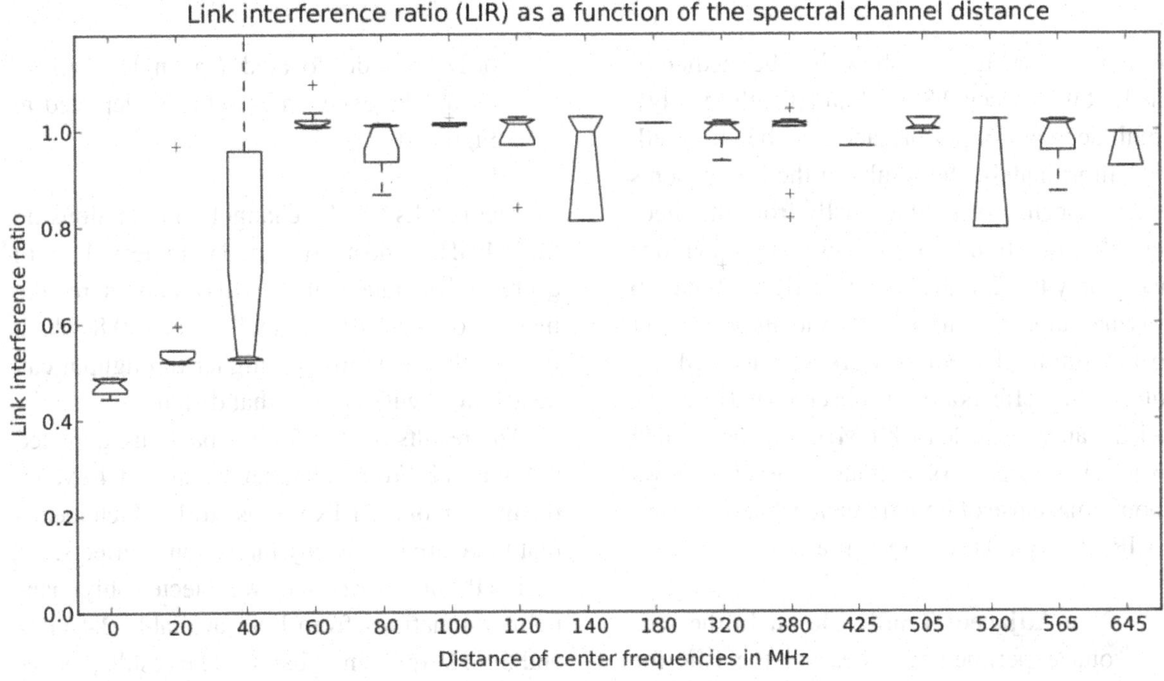

Figure 8. Results of the LIR of adjacent flows on links with a low RSSI value for channel combinations on the 5 GHz band. The average RSSI value for received frames was about -89 dBm. The LIR is less than 0.5 for channel combinations with a spectral distance of more than 100 MHz. We assume the low RSSI values to cause this effect, which we will further investigate in future work.

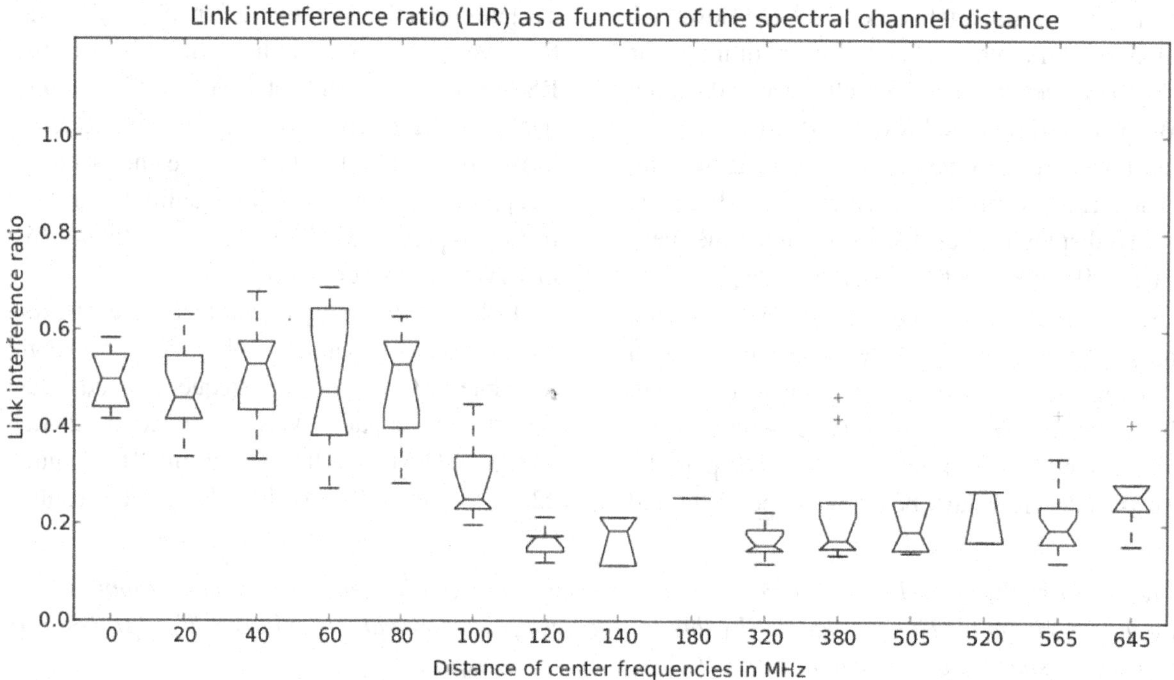

as depicted in Figure 9, show that the median of the LIR is between 0.8 and 1 and therefore only a small decrease of performance can be observed.

Unfortunately, the results of the experiments with adjacent flows differ vastly from the theoretical assumptions. For all channel combinations using only the 2.4 GHz band a LIR of about 0.6 was measured, which is only a minor improvement to the single channel network scenario. Minor interference effects are only observed with a channel distance of at least 80 MHz on the 5 GHz band. Therefore, two simultaneously active flows should make use of both frequency bands, where a LIR of about 0.8 was measured.

- **Non-Adjacent Traffic Flows:** In the second experiment we measure the LIR for two non-adjacent flows. For this, two pairs of DES-Nodes located in a single room are used. The experiment setup is depicted in Figure 10.

The results for the channel combinations on the 2.4 GHz band are depicted in Figure 11. With a channel distance of 30 MHz and more, the median of the LIR is usually above 0.8 which implies that a significant higher throughput can be achieved with at least that distance.

The results for the 5 GHz band are depicted in Figure 12. From a channel distance of 40 MHz, the median of the LIR is close to 1, which means that there are hardly any interference effects.

For the last experiment, we selected only channel combinations from both available 2.4 GHz and 5 GHz frequency bands. The results, as depicted in Figure 13, show that the median of the

Figure 9. Results of the LIR of adjacent flows for channel combinations on the 2.4 GHz and 5 GHz band. The median of the LIR for all channel combinations is about 0.8. This means that the links only exert minor interference effects on each other.

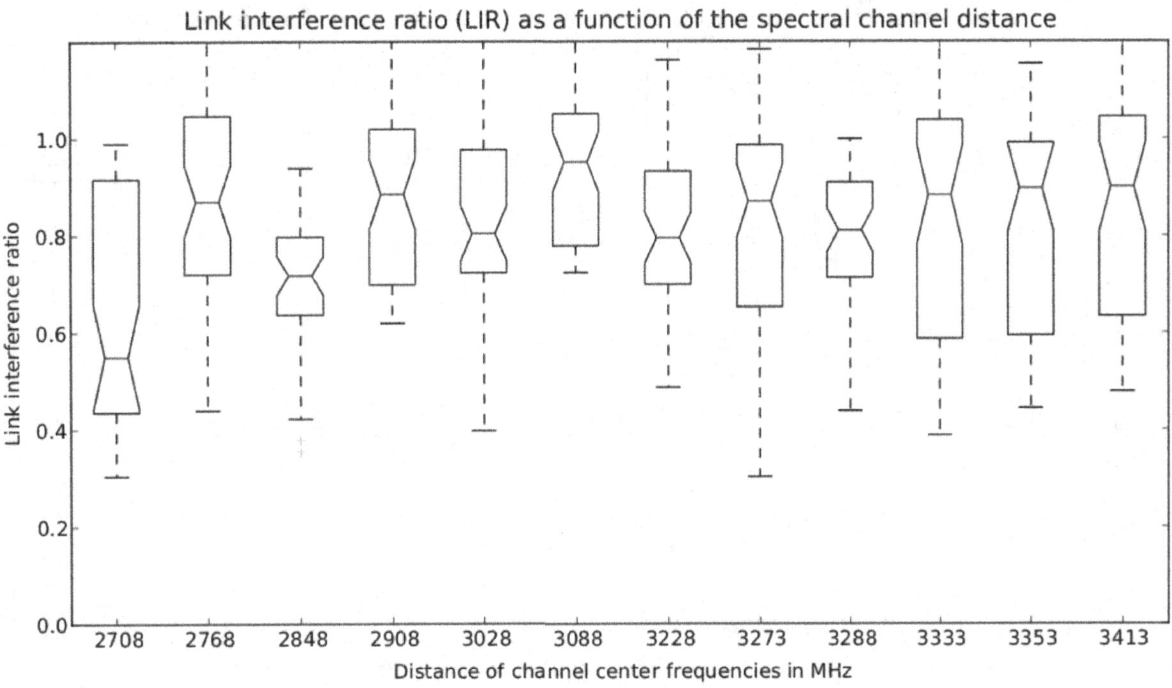

Figure 10. Experiment setup for measuring the effect of spectral channel distance to the LIR with non-adjacent traffic flows. Two node pairs are selected which are located in the same room. The LIR of the two links between the node pairs is measured.

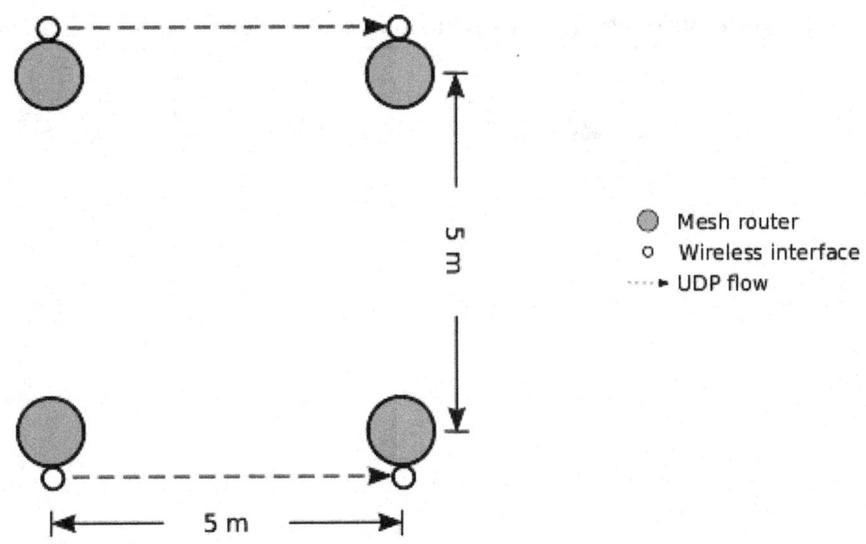

Figure 11. Results of the LIR of non-adjacent flows for channel combinations on the 2.4 GHz band. With a channel distance of 30 MHz, the median of the LIR is usually above 0.8 which implicates that a significant higher throughput can be achieved.

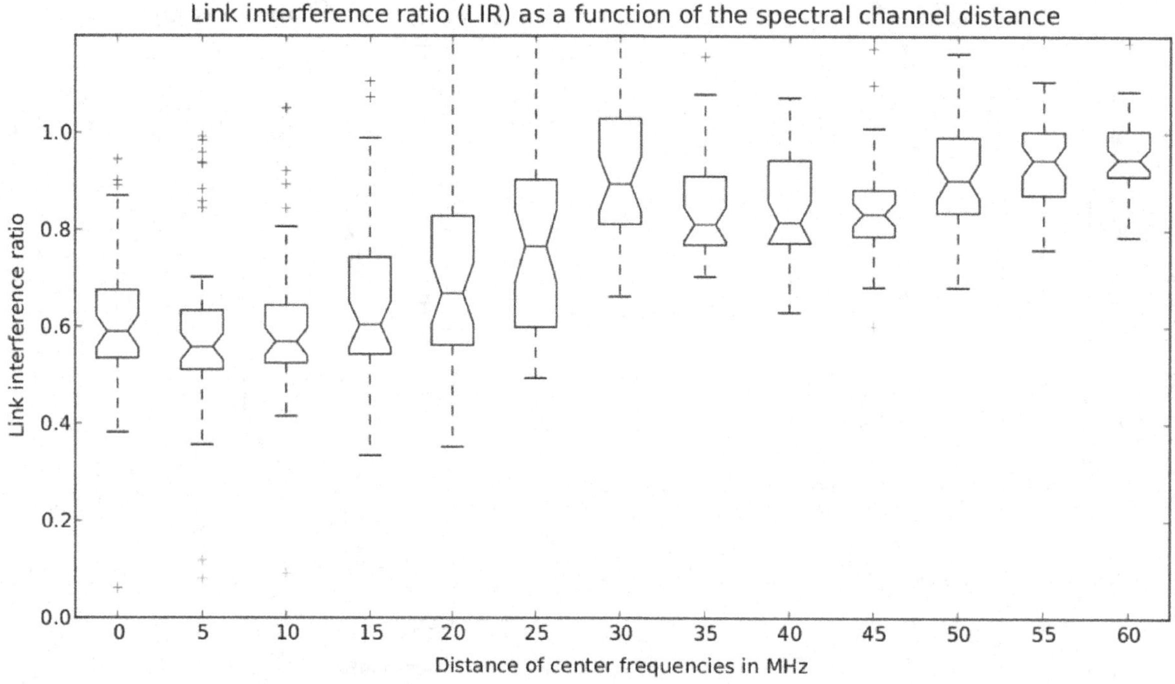

Figure 12. Results of the LIR of non-adjacent flows for channel combinations on the 5 GHz band. From a channel distance of 60 MHz, the median of the LIR is close to 1, which means that there are hardly any interference effects.

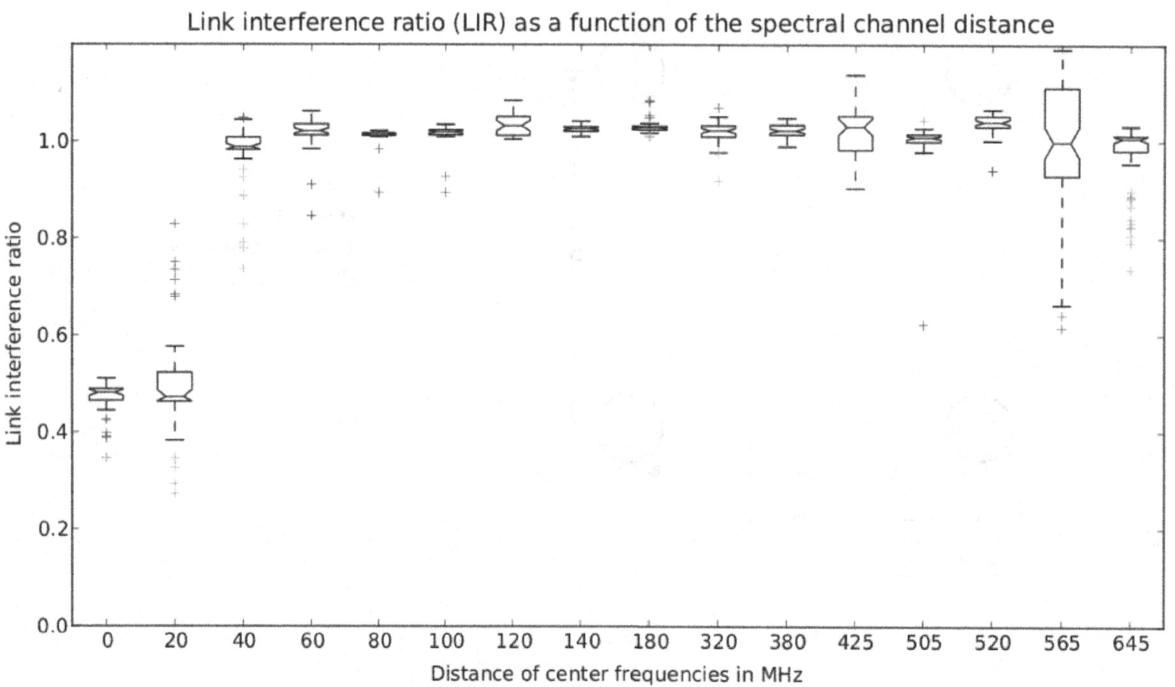

Figure 13. Results of the LIR of non-adjacent flows for channel combinations using the 2.4 GHz and 5 GHz band. The median of the LIR is close to 1 for all channel combinations and therefore no interference effect is observed.

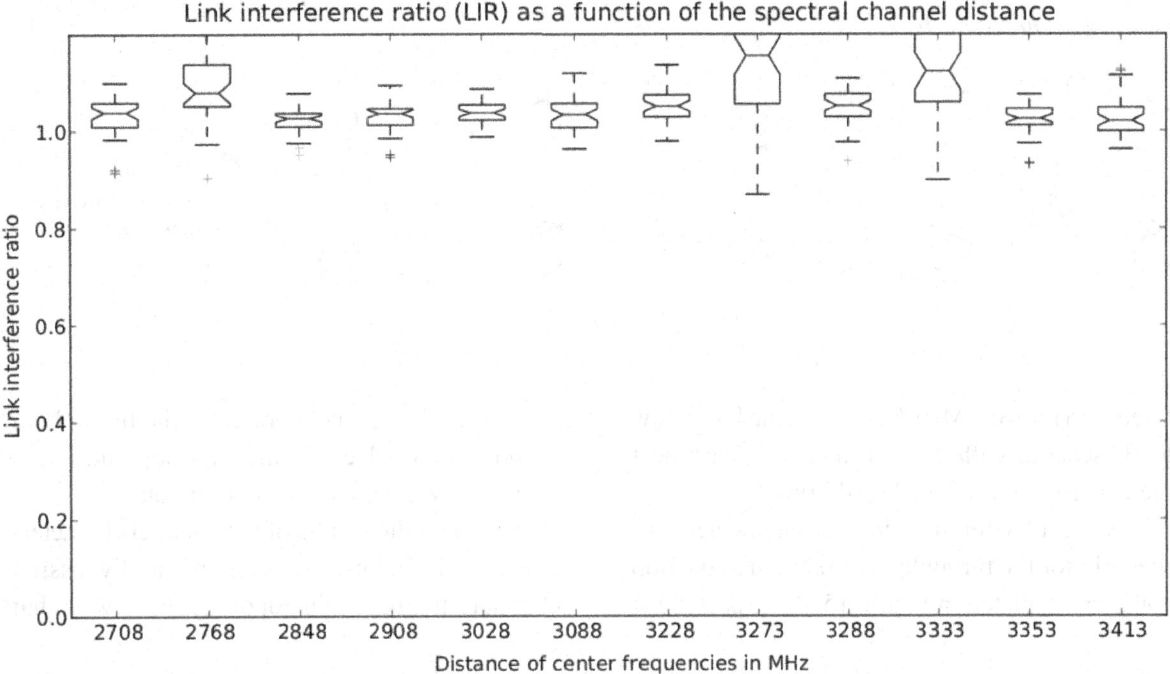

LIR is close to 1 for all channel combinations and therefore no interference effects are observed.

Although none of the channel combinations in the 2.4 GHz allow completely non-interfering transmissions, a minimum channel distance of 30 MHz should be used for simultaneously active flows in order to achieve the highest possible throughput. This results in three possible channels {1, 7, 13} for an efficient channel assignment. On the 5 GHz band a minimum spectral channel distance of 40 MHz is sufficient to experience only negligible interference effects. Using both bands simultaneously, hardly any interference effects could be measured with the median of the LIR being close to 1.

The results for non-adjacent flows show fewer impact of interference as the corresponding experiments with adjacent flows. We assume the main causes for these results being the bigger antenna distance for the experiments with non-adjacent flows (5 m to 0.3 m). To validate the assumption, we will perform experiments with adjacent flows and longer antenna cables therefore increasing the inter-antenna distance in future work.

Multi-Hop Path Interference

In this experiment, we validate the gained knowledge about the channel characteristics on the DES-Testbed with a manual channel assignment. For this we create a chain topology of five mesh routers, on which we can start traffic flows over up to four hops. First, we apply the same channel to all links in the chain topology, thus creating a single channel network scenario as depicted in Figure 14 (a). We then start an UDP flow with iperf from the first node of the chain to the second node. Afterwards we start an UDP flow from the first node to the third and so on, until the last node in the chain. We configured the WNIC with the

Figure 14. Experiment setup to measure throughput in a single- and multi-channel network. A subset of the mesh routers is selected to create a chain topology. In (a), we apply the same channel to all wireless links, thus creating a single channel network scenario. Based on the results of the previous experiments, we manually apply channels to the links which promise an increase of throughput compared to the single network scenario in (b).

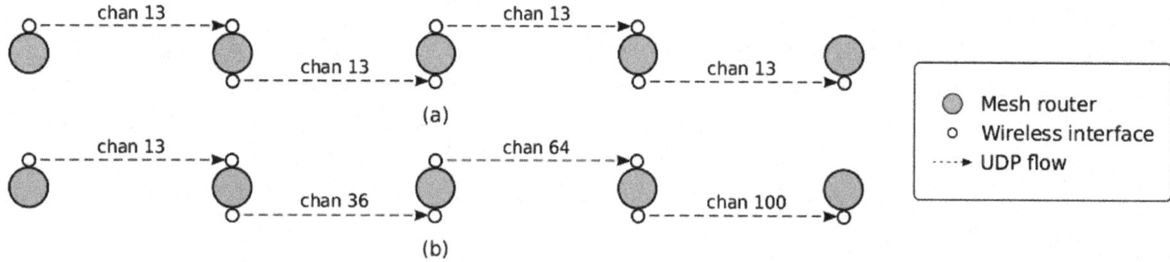

fixed data rate of 6 MBit/s and send the UDP flow for 30 seconds with the same data rate. We repeat the experiment for each hop 40 times.

The results for the single channel network scenario for the throughput in relation to the hop count are depicted in Figure 15. As expected for the single channel network case, the throughput is more than halved on the first hop and keeps dropping with an increasing hop-count.

Based on the results of the channel characteristics experiments, we then manually assign channels to the chain topology in a way that

Figure 15. Results for the single-channel path experiment. As expected for the single channel network case, the throughput is more than halved on the first hop and keeps dropping with an increasing hop-count.

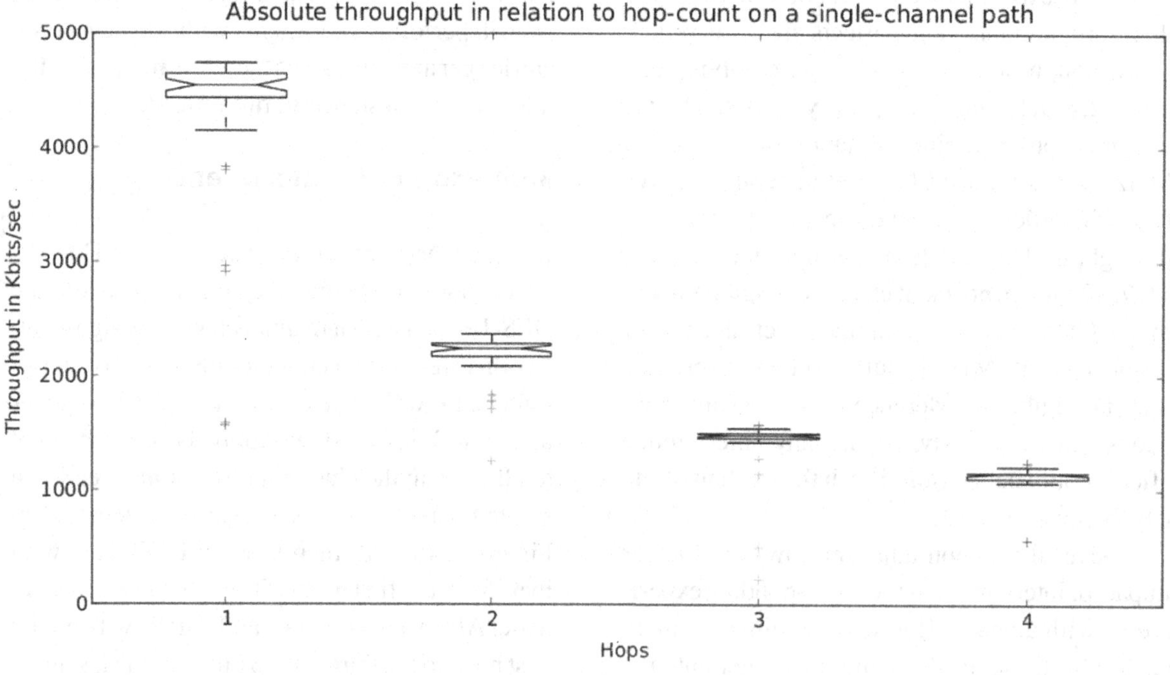

promises the biggest decrease of interference effects. As observed from the experiments on adjacent flows, both frequency bands should be used for the respective WNIC. We expect the throughput to be significantly higher compared to the single channel network. We apply the channels {13, 36, 64, 100} to the links as depicted in Figure 14 (b) with which we expect to exhibit only minor interference.

The results for the multi-channel network scenario are depicted in Figure 16. As expected, the manual channel assignment leads to a higher throughput if the hop-count is bigger than 1. It only drops slightly with the increasing hop-count which implicates that the interference effects have been reduced significantly with the chosen channel assignment. These results show that the experimentally determined channel characteristics also hold in multi-hop scenarios and underline the potential performance gain that can be achieved by proper channel assignment.

Related Work

Many approaches for distributed channel assignment have already been proposed (Si, 2009). In the greedy channel assignment algorithm by Ko (2007), one wireless interface of each node is switched to a common channel in order to ensure the network connectivity. For additional interfaces, a greedy algorithm selects the least interfering channel in the interference set using an interference cost function which takes the degree of overlap of two channels into account. The distributed greedy algorithm (DGA) assigns channels to links instead of interfaces and is

Figure 16. Results for the multi-channel path experiment. As expected, the manual channel assignment lead to a higher throughput if the hop-count is bigger than 1. It only drops slightly with the increasing hop-count which implicates that the interference effects have been reduced significantly with the chosen channel assignment.

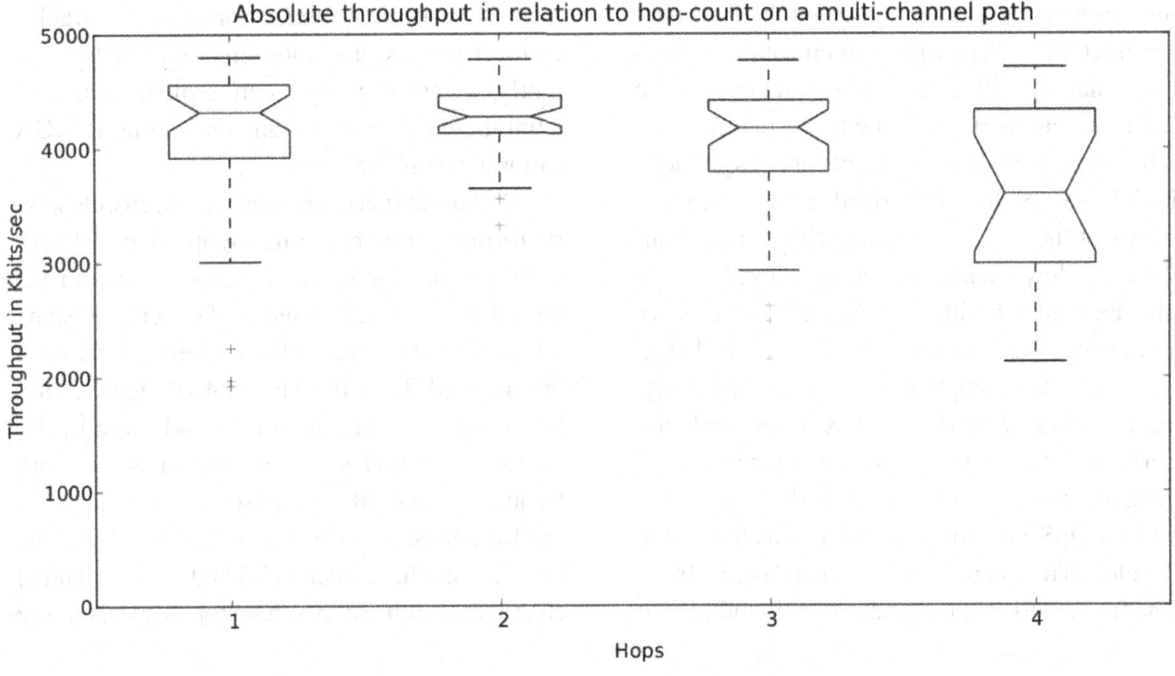

therefore topology preserving, meaning that all links are sustained during the channel assignment procedure (Subramanian, 2008). The link-based channel assignment approach by Sridhar et al. is similar to DGA with the difference that it additionally takes the expected traffic-load on the nodes into account (2009). Interference is modeled with a fixed interference range for all nodes and a weighted conflict graph is used to estimate the network-wide interference. The Skeleton Assisted Partition Free (SAFE) algorithm uses minimal spanning trees (MSTs) to preserve the network connectivity (Shin, 2006).

A fast channel switching algorithm is proposed by Kyasanur (2006a, 2006b). The set of network interfaces on each node are divided into fixed interfaces, which stay on a fixed channel, and switchable interfaces. If a node wants to communicate with a neighbor, it tunes one of the switchable interfaces to a channel of a fixed interface of the receiving node. The crucial part of this approach is the way how channels are assigned to the fixed interfaces. The NET-X framework was created to implement the algorithm in a wireless testbed environment based on the 2.4 Linux kernel (UIUC, 2010). The implementation of this particular algorithm requires several changes to the Linux network stack and to the driver of the wireless network interfaces. The services provided by NET-X are directly derived from the requirements of the implemented algorithm (Kyasanur, 2006b). Slow channel switching algorithms can also be realized with NET-X, although it is not in the focus and to the best of our knowledge, none has been implemented so far. Recently, experiments with the NET-X framework are performed to evaluate algorithms for assigning channels to the fixed interfaces (Raman, 2009) and for QoS-provisioning, extensions have been developed for queue management (Castro, 2009). The framework is not intended to as foundation to develop different distributed channel assignment algorithms, the focus is on algorithms to assign channels to the fixed interfaces. Data structures for network and conflict graphs are not provided. Also a topology monitoring functionality has not been in the scope of the framework.

Discussion and Outlook

With DES-Chan a wide range of different algorithms can be implemented, validated, and compared in a real network environment. DES-Chan does not require any changes of the Linux kernel or the wireless network interface drivers. It is therefore easy to integrate into existing wireless mesh network testbeds. The contribution of DES-Chan is two-fold. Firstly, DES-Chan provides an abstraction layer to the low-level and operating system specific tasks. This abstraction layer enables the researcher to spend most development time on the algorithm logic. Secondly, DES-Chan provides basic services and data structures often required for typical tasks in channel assignment algorithms. For instance, appropriate data structures have been provided for network graphs, conflict graphs, and interference models. Currently, we are working on an implementation of a link-based channel assignment similar to DGA (Subramanian, 2008).

The basic experiments and measurements were performed in order to gain insights on the network topology and the channel characteristics in the DES-Testbed. In a first series of experiments, the effects of the co-channel interference have been investigated. The LIR of two links is significantly lower for adjacent than for non-adjacent traffic flows. Nevertheless, using channels on both frequency bands also promises a higher throughput for adjacent traffic flows. The results of the experiments have been validated with a manual channel assignment in a chain topology spanning

4 hops. The throughput is significantly higher using the manual channel assignment compared to the single channel network scenario.

A comparison of the experiment results to the common assumptions of channel assignment algorithms yields some interesting deviations. First, the assumption of orthogonal channels as theoretically offered by IEEE 802.11 and considered in many channel assignment algorithms does not hold in practice. In contrast, if the experimental results are transferred to the channel assignment algorithms, the actual number of available channels is significantly reduced which may affect the performance of the algorithms. Second, the experiment results for adjacent and non-adjacent flows show different characteristics. Therefore, channel assignment algorithms should distinguish between adjacent and non-adjacent flows to optimally assign the available channels.

Future work will be focused on research of more realistic, measurement-based interference models. With these models, it is likely to close the gap between the graph-based and throughput-based results obtained from the experiments on the DES-Testbed. Additionally, we will implement further algorithms based on DES-Chan.

REFERENCES

Castro, M. C., Dely, P., Kassler, A. J., & Vaidya, N. H. (2009). Qos-aware channel scheduling for multi-radio/multi-channel wireless mesh networks. In *Proceedings of the 4th ACM International Workshop on Experimental Evaluation and Characterization* (pp. 11-18). New York, NY: ACM Press.

De Couto, D. S. J., Aguayo, D., Bicket, J., & Morris, R. (2003). A high-throughput path metric for multi-hop wireless routing. In *Proceedings of the 9th Annual International Conference on Mobile Computing and Networking* (pp. 134-146).

Draves, R., Padhye, J., & Zill, B. (2004). Routing in multi-radio, multi-hop wireless mesh networks. In *Proceedings of the 10th Annual International Conference on Mobile Computing and Networking* (pp. 114-128).

Fuxjager, P., Valerio, D., & Ricciato, F. (2007). The myth of non-overlapping channels: interference measurements in IEEE 802.11. In *Proceedings of the Fourth IEEE Annual Conference on Wireless on Demand Network Systems and Services* (pp. 1-8).

Günes, M., Blywis, B., & Juraschek, F. (2008). *Concept and design of the hybrid distributed embedded systems testbed* (Tech. Rep. No. TR-B-08-10). Berlin, Germany: Freie Universität Berlin.

Günes, M., Blywis, B., Juraschek, F., & Schmidt, P. (2008). *Practical issues of implementing a hybrid multi-NIC wireless mesh-network* (Tech. Rep. No. TR-B-08-11). Berlin, Germany: Freie Universität Berlin.

Günes, M., Blywis, B., Juraschek, F., & Watteroth, O. (2009). Experimentation made easy. In J. Zheng, S. Mao, S. Midkiff, & H. Zhu (Eds.), Ad Hoc Networks (LNCS 28, pp. 493-505).

Joost, R., & Robinson, S. (2010). *Python WiFi*. Retrieved from http://pythonwifi.wikispot.org/

Katzela, I., & Naghshineh, M. (1996). Channel assignment schemes for cellular mobile telecommunication systems. *IEEE Personal Communications*, *3*, 10–31. doi:10.1109/98.511762.

Ko, B.-J., Misra, V., Padhye, J., & Rubenstein, D. (2007). Distributed channel assignment in multi-radio 802.11 mesh networks. In *Proceedings of the IEEE Wireless Communications and Networking Conference* (pp. 3978-3983).

Kyasanur, P., So, J., Chereddi, C., & Vaidya, N. H. (2006). Multichannel mesh networks: Challenges and protocols. *Wireless Communications*, *13*(2), 30–36. doi:10.1109/MWC.2006.1632478.

Kyasanur, P., & Vaidya, N. H. (2006). Routing and link-layer protocols for multi-channel multi-interface ad hoc wireless networks. *Mobile Computing and Communications*, *10*(1), 31–43. doi:10.1145/1119759.1119762.

Padhye, J., Agarwal, S., Padmanabhan, V. N., Qiu, L., Rao, A., & Zill, B. (2005). Estimation of link interference in static multi-hop wireless networks. In *Proceedings of the 5th ACM SIGCOMM Conference on Internet Measurement* (pp. 28-28). New York, NY: ACM Press.

Python Twisted. (2010). *The python twisted documentation*. Retrieved from http://twistedmatrix.com/projects/core/documentation/howto/book.pdf

Ramachandran, K. N., Belding, E. M., Almeroth, K. C., & Buddhikot, M. M. (2006). Interference-aware channel assignment in multi-radio wireless mesh networks. In *Proceedings of the 25th IEEE International Conference on Computer Communications* (pp. 1-12).

Raman, V., & Vaidya, N. H. (2009). *Adjacent channel interference reduction in multichannel wireless networks using intelligent channel allocation* (Tech. Rep. No. 06-27074). Urbana, IL: University of Illinois at Urbana-Champaign.

Reis, C., Mahajan, R., Rodrig, M., Wetherall, D., & Zahorjan, J. (2006). Measurement-based models of delivery and interference in static wireless networks. *Communications of the ACM*, *36*(4), 51–62. doi:10.1145/1151659.1159921.

Robinson, J., Papagiannaki, K., Diota, C., Guo, X., & Krishnamurthy, L. (2005). *Experimenting with a multi-radio mesh networking testbed*. Paper presented at the First Workshop on Wireless Network Measurements.

Shin, M., Lee, S., & ah Kim, Y. (2006). Distributed channel assignment for multi-radio wireless networks. In *Proceedings of the IEEE International Conference on Mobile Adhoc and Sensor Systems* (pp. 417-426).

Si, W., Selvakennedy, S., & Zomaya, A. Y. (2009). An overview of channel assignment methods for multi-radio multi-channel wireless mesh networks. *Journal of Parallel and Distributed Computing*, *70*(5).

Sridhar, S., Guo, J., & Jha, S. (2009). Channel assignment in multi-radio wireless mesh networks: A graph-theoretic approach. In *Proceedings of the First International Communication Systems and Networks and Workshops* (pp. 1-10).

Subramanian, A. P., Gupta, H., Das, S. R., & Cao, J. (2008). Minimum interference channel assignment in multi-radio wireless mesh networks. *IEEE Transactions on Mobile Computing*, *7*(12), 1459–1473. doi:10.1109/TMC.2008.70.

Wireless Networking Group at UIUC. (2010). *Net-X channel assignment framework*. Retrieved from http://www.crhc.illinois.edu/wireless/netx.html

This work was previously published in the International Journal of Wireless Networks and Broadband Technologies, Volume 1, Issue 1, edited by Naveen Chilamkurti, pp. 32-49, copyright 2011 by IGI Publishing (an imprint of IGI Global).

Chapter 4
HTTP Traffic Model for Web2.0 and Future WebX.0

Vladimir Deart
Moscow Technical University of Communications and Informatics, Russia

Alexander Pilugin
Moscow Technical University of Communications and Informatics, Russia

ABSTRACT

This article presents a method of estimation for HTTP traffic quality service parameters mean delay and lost packets percentage. This method, based on statistic measurements, includes simulation and analytical modeling. Statistical HTTP traffic models presented earlier take into account typical features of WEB2.0 Internet traffic, which were used for the simulation model. Developed universal simulation models make it possible to research service quality parameters under setting network conditions over a wide range considering Internet development. The presented analytical method based on batch packet arrival model allows an accuracy estimation of mean HTTP-packets delay in Core Router by simple calculations. Objective results of HTTP traffic quality service parameters can be used in QoS standard development for WEB traffic and model QoE standard development.

INTRODUCTION

The modern IAPs provide a complex service including data transfer, VoIP, IPTV etc. There are standards, which normalize the main transfer packet parameters for a real-time traffic. Up to these days there are no such standards for a data traffic generated during Web browsing (access to web resources).

DOI: 10.4018/978-1-4666-3902-7.ch004

Web resource organization concept WEB2.0 was spread in modern networks. It helps users to get information access more quickly and use information put on a website by other users. According mentioned concept it can be built portals with dynamic pages or pages with aggregated newsletters from different sites. Thus using certain WEB2.0 concept changes not only user's relations to web access but HTTP-traffic structure in whole.

On one hand, network development leads to HTTP traffic change: transfer traffic volume increasing and data loading acceleration at the expense of parallel sessions. On the other hand, users, who have the access to new interactive services, want a provider to standardize the quality of their placing. The method presented solves both marked problems. Firstly, with the help of suggested method it is possible to simulate different HTTP traffic including typical characteristics of WEB2.0 traffic. Secondary, simulation and analytical modeling allow estimating HTTP packets service objective parameters such as packet delay in the queuing system and the percent of loss. These parameters estimation may be useful for the standard development, which regulate the HTTP traffic QoS.

The second part of the article devotes to an explanation how the HTTP-traffic statistical model was obtained and comparison of obtained results with other works. The third part of the article presents the simulation modeling results. The analytical modeling results are presented in the fourth part. The article conclusion contains resume and references.

RELATED WORKS

The analytical and simulation models in the article are based on statistic results described in Deart (2009). The measuring complex was presented and based on the open program components and database, which helped for traffic measurements in the provider access network. Deart (2009) showed how these results were processed and how it helped to get the statistical HTTP traffic model.

The developed model is possessed of some advantages in comparison with the model in the work of Shuai (2008). So far as interval approximation are used for a distribution of HTTP response size, HTTP responses are modeled more accuracy. Therefore, it could be shown the lower and tail range of values completely.

The obtained model takes into account the network HTTP traffic WEB2.0 features better by introduction of dependent generation mechanism of TCP sessions, which simulates a user work, and influence on TCP sessions opening of the user browser.

This statistical model is more accurate if it is compared with Mah (1997), Choi (1999), and Padhye (1998) as far as it describes the modern Internet web traffic and ties also amount of HTTP traffic with intensity of TCP sessions. The main parameters of HTTP traffic statistical model in Deart (2009) are opening TCP session intensity, quantity of GET requests in one TCP session, intervals between GET requests, GET request size in bytes and HTTP response size in bytes. Selected parameters are defined unambiguously a HTTP traffic generator according to a session principle and allow to use it in simulations. The modeling network topology is presented in Figure 1.

Simulation Model

HTTP-traffic simulation model was built with use of NS2 simulator and PackMIME module (Fall, 2010; Wiegle, 2007). The local network is presented by multiple clients simulating data exchange from Web servers. All users present

Figure 1.Modeling network topology

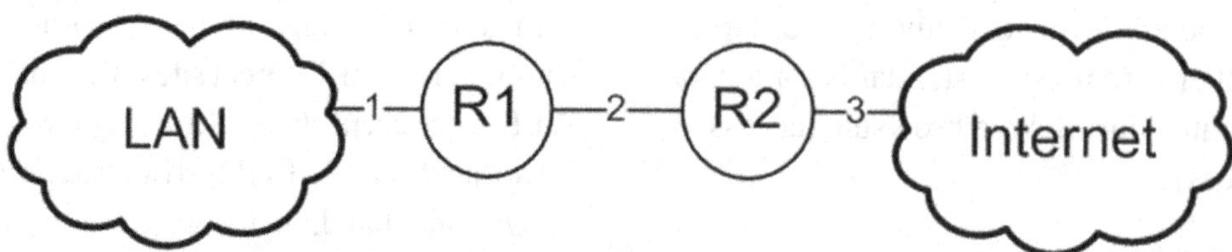

in access network have equal throughput before router R1 and single service Web traffic priority. Every client simulates the user's work with Web browser. The Internet is simulated as Web-servers placed in the network. Every Web-server replies by HTTP response for the coming request independently which client had sent it.

Quantity of clients and servers in each network were independent and not defined beforehand. Client and server multitude in simulation model are made by PackMIME component assistance and NS2 which use them in TCP session organization. Thus there are no relation between HTTP objects and pages in simulation model: all data downloaded by user are bind only with defined TCP session.

Quantitatively the client multitude in the LAN network is defined with TCP session intensity. Every session is made for appointment data volume and contains one or more GET requests. All appointment for transmission into session data by TCP protocol should be delivered from the server to the client. In case of data packet loss the retransmission occurs.

The analysis of Web traffic transmission throughput shows that the service quality are maximal influenced by the channel 2 capacity between Core (R2) and border (R1) router.

Therefore the channel 1 throughput in simulation model which contact Web clients and border router was determined equal 1 Gbps (channel 1), that is much more than general channel 2 throughput. The throughput of channel 2 in simulation model changes from 100 Mbps to 30 Mbps so that it can be possible to obtain different utilization. The channel 3 bandwidth was also 1 Gbps.

HTTP traffic has an asymmetric character: to the Web servers direction users send the official packets only signalized about TCP session installation and about successful data packet reception. Users transfer effective load to Web server direction consists of GET requests only with average sizes 650-700 byte (Deart, 2009). Then for every GET request servers send HTTP response to user with size 8-10 KB (Deart, 2009). It means that traffic from WEB servers to users is 10 times heavier than in opposite direction.

The simulation is done for downstream HTTP traffic coming from servers to users. The size of buffers in Routers considered as 100 packets and FIFO rule for queues was used. If overflow of buffer occurs, IP packets will be dropped.

The bandwidth of channel 2 is shared between real-time and HTTP traffic. Real-time applications have higher priority than HTTP traffic that is why we can consider that with increase of real-time traffic available bandwidth for HTTP traffic decrease.

The channel 2 is a bottleneck of simulating network and an output queue of Core Router 2 will define QoS in the whole Network.

Comparison of Simulation Results with M/M/1/100

Simulation results and results obtained from Little's formulae for the mean delay in the Router queue are shown on Figure 2. We can see that mean delay from simulation is much higher than calculated value. The same we can see on Figure 3 for packet loss ratio. It means we could not use M/M/1/100 model for estimation of QoS parameters of HTTP traffic transfer.

The detailed analysis of reasons for such big differences in calculated and simulated results shows that the most influence factor is a batch arrival of IP packets in HTTP responses.

Traffic Increase Influence on QoS

Let us assume that increasing number of simultaneous TCP sessions is directly shows the increasing of users in Access Network. Then a 25% increase of number of TCP sessions will mean 25% increase in number of users.

Simulation results for increased HTTP Traffic are shown on Figures 4 and 5. The ratio of mean packet delay to throughput of link 2 is presented

Figure 2. Mean packet delay

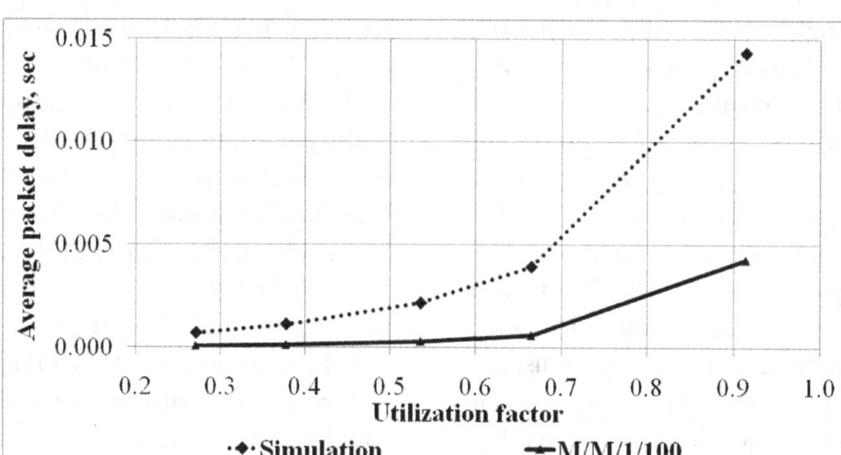

Figure 3. Loss packet ratio

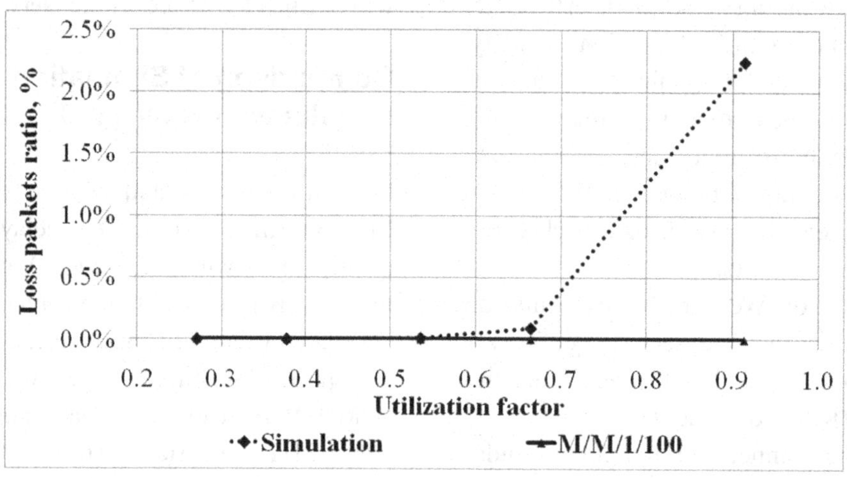

on Figure4 for three values of traffic intensity: regular traffic, traffic with 25% increase, and traffic with 50% increase. The regular traffic is based purely on measured values.

As it can be clearly seen on Figure 4 a 25% increase in traffic can produce three times increasing in mean packet delays. A 50% increase in traffic also produce increase of mean packet delay but due to the growth of packet loss the difference is not such big.

Increasing of traffic also causes packet loss ratio rising as it is shown on Figure 5. A 25% traffic increase results in increasing of packet loss ratio by 10 times. A 50% traffic increase causes increase of packet loss ratio by 15-20 times.

ANALYTICAL MODEL FOR QOS PARAMETERS CALCULATIONS

The analytical model has a great advantage that it can be used for a wide range of parameters without simulation of each parameter combination. We have chosen a batch arrival models for HTTP traffic description: $M^{[x]}/M/1$ and $M^{[x]}/D/1$ (Figure 6). The results of statistical measurements were used for the estimation of batch arrival process parameters. Both simulation results and measurement results (Deart, 2009) were investigated with a goal to get characteristics of a batch arrival process: intensity of batch arrival and batch size distribution.

Figure 4. Mean packet delays for increased traffic

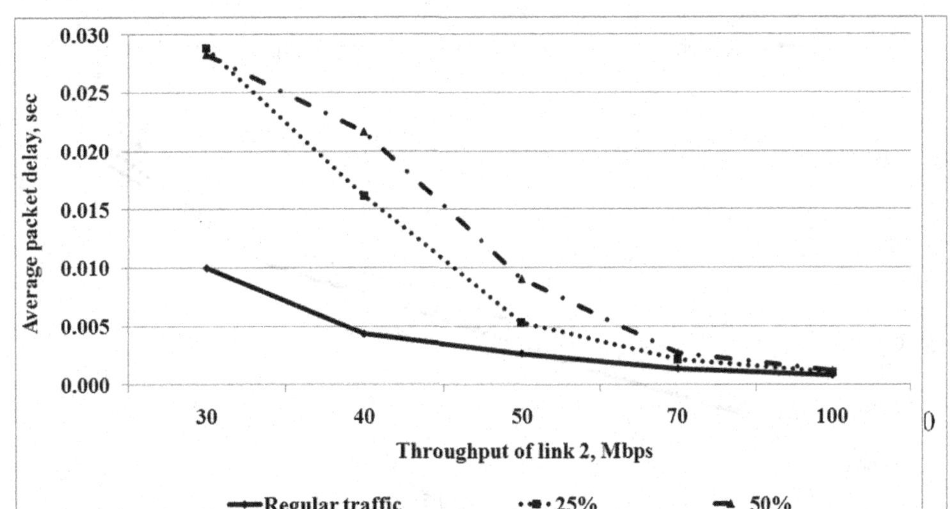

Figure 5. Packet loss ratio for increased traffic

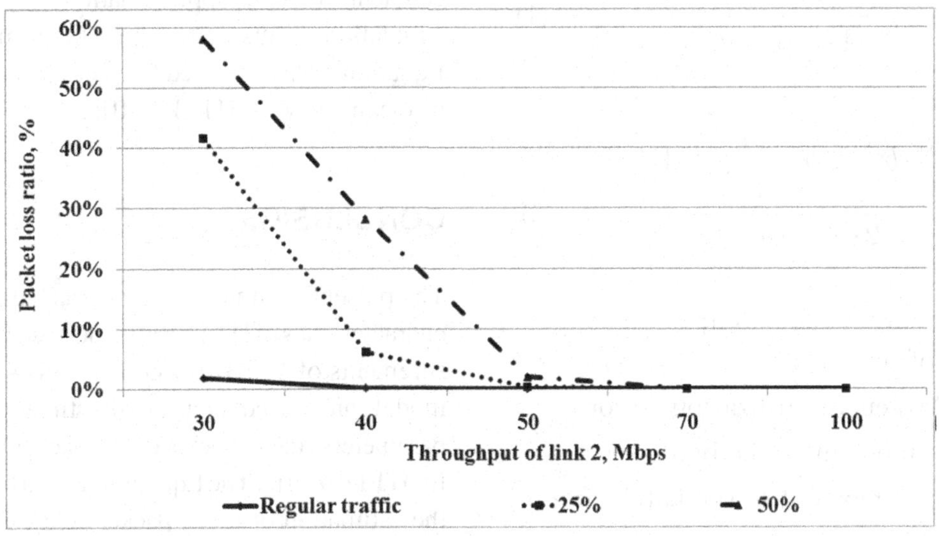

Results of this investigation show that a batch arrival process can be considered as a Poisson process as a variation is about 1 (0.7-1.3) and an autocorrelation function is streaming to 0. Mean batch arrival intensity is 700 batches per second. The batch size distribution can be approximate via negative binomial distribution with the mean number of four packets in one batch.

The measured packet size distribution contributed mostly to value of control packets (around 60 bytes) and value of 1460 bytes for data packets. For the simulation, a fixed size of a packet was used with the size 1460 bytes.

The Equations (1) and (2) for $M^{[x]}/M/1$ and for $M^{[x]}/D/1$ were derived from Cooper (1981) and produced as a simplification of $M^{[x]}/G/1$.

Figure 6. Comparison of average HTTP-packet delay for different models

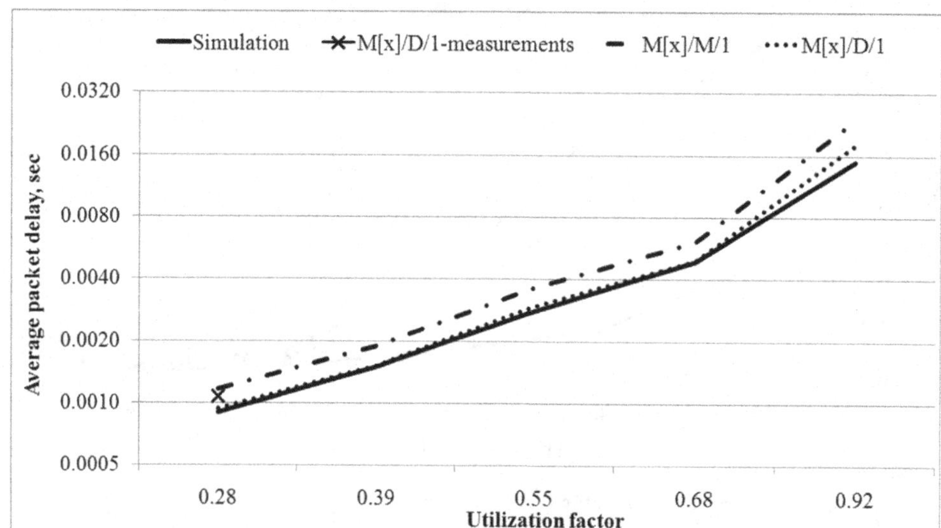

$$\omega = \frac{l^{(2)} + l^{(1)}}{2 \cdot \mu \cdot l^{(1)} \cdot (1 - \hat{\rho})} - \frac{1}{\mu} \qquad (1)$$

$$\omega = \frac{\lambda \cdot l^{(1)} \cdot b^{(2)} + b^{(1)} \cdot \left(\frac{l^{(2)}}{l^{(1)}} - 1 \right)}{2 \cdot (1 - \hat{\rho})} \qquad (2)$$

» =rate of arrival of batch;
μ = rate of service;
Á= $l^{(1)}$· Á=general utilization factor;
$l^{(1)}$ = first moment of the batch size;
$l^{(2)}$ = second moment of the batch size;
$b^{(1)}$ = first moment of the service time;
$b^{(2)}$ = second moment of the service time.

As it can be clearly seen from the Figure 6 a model $M^{[x]}/M/1$ gives upper level estimation of simulation results that exceeds them not more than 40%. Using Equation (2) for $M^{[x]}/D/1$ we can achieve better approximation not exceeding simulation results more than 10%. It means that Equation (2) can be used for accurate calculation of mean delay for HTTP traffic.

CONCLUSION

The presented simulation model is based on real characteristics of HTTP traffic derived from measurements of WEB2.0 Internet. This simulation model can be used for accurate estimation of QoS parameters (mean packet delay, packet loss ratio) for HTTP traffic. The Equation (2) can be used for the estimation of mean packet delay. These two models can be used for the construction of QoS standards for HTTP traffic. Taking in account that mostly trends of changes in HTTP traffic with forwarding from WEB2.0 to WEB3.0 and WEBX.0 will be the same we can assume that presented simulation and analytical models can be useful in changeable Internet World for a long time.

REFERENCES

Choi, H.-K., & Limb, J. O. (1999). A behavioral model of web traffic. In *Proceedings of the Seventh International Conference on Network Protocols* (pp. 327-334). Washington, DC: IEEE Computer Society.

Cooper, R. (1981). *Introduction to queueing theory*. Amsterdam, The Netherlands: North Holland Publishing.

Deart, V., Mankov, V., & Pilugin, A. (2009). HTTP traffic measurements on access networks: Analysis of results and simulation. In S. Balandin, D. Moltchanov, & Y. Koucheryavy (Eds.), *Proceedings of the 9th International Conference on Next Generation Wired/Wireless Advanced Networking and Second Conference on Smart Spaces* (LNCS 5764, pp. 180-190).

Fall, K., & Varadham, K. (2001). *The ns Manual: The VINT project documents*. Retrieved from http://www.cs.cornell.edu/people/egs/615/ns2-manual.pdf

Mah, B. (1997). An empirical model of HTTP network traffic. In *Proceedings of the Sixteenth Annual Joint Conference of the IEEE Computer and Communications Societies* (p. 592). Washington, DC: IEEE Computer Society.

Padhye, J., Firoiu, V., Towsley, D., & Kurose, J. (1998). Modeling TCP throughput: A simple model and its empirical validation. *Communications of the ACM*, *28*(4), 303–314. doi:10.1145/285243.285291.

Shuai, L., Xie, G., & Yang, J. (2008). Characterization of HTTP behavior on access networks in WEB2.0. In *Proceedings of the IEEE International Thermoelectric Society* (pp. 1-6).

Wiegle, M. (2007). *PackMime-HTTP: Web traffic generation in NS2*. Retrieved from http://www.isi.edu/nsnam/ns/doc/node552.html.

This work was previously published in the International Journal of Wireless Networks and Broadband Technologies, Volume 1, Issue 1, edited by Naveen Chilamkurti, pp. 1-14, copyright 2011 by IGI Publishing (an imprint of IGI Global).

Chapter 5
DMT Optimal Cooperative MAC Protocols in Wireless Mesh Networks with Minimized Signaling Overhead

Benoît Escrig
Universite de Toulouse, France

ABSTRACT

In this paper, a cooperative protocol is proposed for wireless mesh networks. Two features are implemented: on-demand cooperation and selection of the best relay. First, cooperation is activated by a destination terminal when it fails in decoding the message from a source terminal. Second, a selection of the best relay is performed when cooperation is needed. The robustness of wireless links are increased while the resource consumption is minimized. The selection of the best relay is performed by a splitting algorithm, ensuring a fast selection process, the duration of which is now fully characterized. Only terminals that improve the direct link participate in the relay selection and inefficient cooperation is avoided. The proposed protocol is demonstrated to achieve an optimal diversity-multiplexing trade-off. This study focuses on Nakagami- m *wireless channel models to encompass a variety of fading models in the context of wireless mesh networks.*

INTRODUCTION

One of the major properties of wireless mesh networks (WMNs) consists in the possibility of breaking long distances into a series of shorter hops. Apart from increasing the signal quality of the links, the mesh architecture allows the cooperative forwarding of data packets through intermediate terminals in the network. Cooperative communications provide an interesting contribution in this context. More precisely, they enable data transmission between two terminals through an alternate path when the direct wireless link is experiencing a deep fade. Cooperative communications can be envisioned at several network layers. However, implementing the forwarding scheme at the lowest layers renders the protocol more reactive to network conditions and mini-

DOI: 10.4018/978-1-4666-3902-7.ch005

mizes the transmission delay since each layer adds its own processing time and hence includes its own latency. Cooperative protocols are mainly implemented in two layers: cooperative transmissions are managed at the physical (PHY) layer whereas the set up of the cooperation is done at the medium access control (MAC) layer. At the PHY layer, cooperative communications increase the wireless link reliability. In a cooperative scenario, a source terminal S sends data to a destination terminal D through a direct path. One or several relay terminals help the transmission by receiving the source message and forwarding it to D through a relaying path (Figure 1). Hence the direct path is rendered more robust (Laneman & Wornell, 2000, 2003; Sendorais, Erkip, & Aazhang, 2003; Hunter & Nosratinia, 2006). However, this comes at the price of bandwidth consumption so that the system operates at diminished capacity[1]. One common way to compare cooperative transmission techniques is to compute the diversity-multiplexing tradeoff (DMT) (Zheng & Tse, 2002). The DMT analysis of a transmission scheme yields the diversity gain $d(r)$ achievable for a spatial multiplexing gain r. The diversity gain helps in quantifying the robustness of the S-D link and the multiplexing gain gives an hint on the capacity of the link. Both indicators should be maximized in order to get an optimal DMT curve. When $(N-1)$ relay candidates are involved in a cooperative scenario, the optimal DMT curve $d(r)$ is achievable by protocols that implement both on-demand relaying and a selection of the best relay (Bletsas, Khisti, & Win, 2008; Escrig, 2010): $d(r) = N(1-r)$ for $0 \leq r \leq 1$. In an on-demand relaying scenario (Laneman, Tse, & Wornell, 2004; Gomez, Alonso-Zarate, Verikoukis, Perez-Neira, & Alonso, 2007), the relay terminal transmits only when D fails in decoding the data transmitted by S. Thus, the bandwidth consumption due to cooperative transmissions is minimized. Moreover, when cooperation is needed, only the best relay terminal retransmits the source message (Bletsas, Khisti, Reed, & Lippman, 2006). This optimizes the robustness of the wireless link between the source terminal and the destination terminal through the property of spatial diversity while minimizing the resource consumption compared to the case of multiple relays. Hence, an optimal tradeoff between link robustness and bandwidth consumption is reached. This optimal tradeoff has been dem-

Figure 1. Cooperation scenario with three relay candidates: $(N-1) = 3$

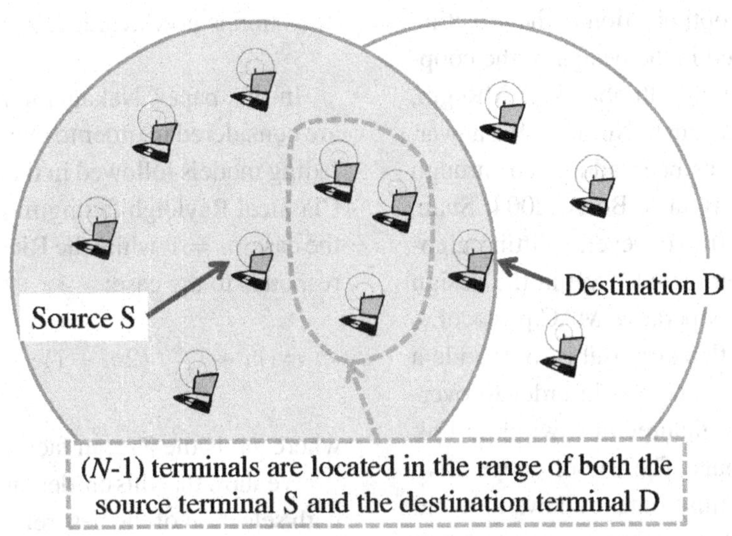
Destination D
Source S
(N-1) terminals are located in the range of both the source terminal S and the destination terminal D

onstrated by computing the DMT curve of the transmission scheme. Moreover, minimizing the number of relays also reduces the impact of cooperative communications on the rest of the network. Indeed, reducing the number of relays diminishes the contention area due to cooperative transmissions compared to the case of multiple relay transmissions. Note however that the DMT criterion fails in providing indications on the amount of bandwidth that is used at the MAC layer in order to implement the cooperative network. For instance in Laneman et al. (2003), the overhead induced by the allocation of space-time codes to relay terminals has not been taken into account. Practically, further optimization is required at the MAC layer in order to reduce the overhead due to the implementation of the cooperative network. In particular, the fast selection of appropriate relay terminals is a main issue the design of cooperative MAC protocols.

The selection process in Bletsas et al. (2008) and Escrig (2010) has some limitations. First, one relay is always chosen even if it cannot improve the direct path. Second, the duration of the selection step has not been optimized yet. Actually the amount of time devoted to this task cannot be predicted because the channel is accessed randomly and collisions occur between available relays (Bletsas et al., 2006; Gomez et al., 2007). More generally, the optimization of the selection has not been included in the design of the cooperative protocols (Azgin, Altunbasak, & AlRegib, 2005; Liu, Tao, Narayanan, Korakis, & Panwar, 2007a). This issue has been addressed through splitting algorithms (Qin & Berry, 2004; Shah, Mehta, & Yim, 2010). However, splitting algorithms have not been included yet in the design of a DMT optimal cooperative MAC protocol.

The rationale of this proposal is to provide a fast and opportunistic method in order to overcome the temporary failures of a wireless link between two terminals. The use of cooperative communications in this context allows to avoid the need to re-establish a whole route when one link is temporarily dropped. We start with the cooperative MAC protocols developed in (Bletsas et al., 2008; Escrig, 2010). They all implement on-demand cooperation and a selection of the best relay. It has been demonstrated that an optimal DMT is achieved when these two features are implemented. These protocols are improved with the following additional features:

- **Splitting Algorithm for Fast Relay Selection:** A splitting algorithm can find the best relay terminal, on average, within at most 2.507 slots even for an infinite number of relay candidates (Qin & Berry, 2004; Shah et al., 2010). Collision between relay candidates are not avoided but the contention time is fully characterized.
- **Pre-Selection of the Relay Terminals:** The efficiency of the selection process is guaranteed. Only terminals that are able to improve the direct transmission are preselected. The pre-selection is implemented by adding a condition on the relaying scheme. More precisely, the relay terminal should retransmit only when the relayed path has a mutual information greater than a given threshold (Chou, Yang, & Wang, 2007). Otherwise, the source terminal retransmits its message. Inefficient cooperation is now avoided.

In this paper, Nakagami-m fading channels are considered in order to encompass a variety of fading models followed in the context of WMNs. Classical Rayleigh fading model corresponds to the case $m = 1$ while the Rice fading model corresponds to the case:

$$m = (m+1)^2 \,/\, (2m+1) > 1$$

where m is the Ricean factor.

We show that this on-demand relaying protocol with selection of the best relay terminal provides an optimal performance in terms of DMT. This cooperative protocol has been designed in the context of IEEE 802.11-based mesh networks. Though restricted to this standard in this paper, we believe that our proposal can also be applied to other wireless systems such as wireless sensor networks, broadband wireless networks, and broadcast wireless systems. In particular, we show the optimality of the DMT when the relaying scheme is based on either a fixed amplify-and-forward (AF) method or a selective decode-and-forward (DF) method.

ON-DEMAND RELAYING WITH SELECTION OF THE BEST RELAY TERMINAL

System Model

We consider a slow Nakagami- m fading channel model. A half duplex constraint is imposed across each relay terminal, i.e. it cannot transmit and listen simultaneously. Transmissions are multiplexed in time, they use the same frequency band. Each channel gain h_{ij} is accurately measured by the receiver j, but not known to the transmitter i. Moreover, the channel gain h_{ij} is identical to the channel gain h_{ji}. This assumption is relevant since both channels are using the same frequency band. Statistically, the gain $|h_{ij}|$ is distributed according to a Nakagami- m distribution. In particular, the random variable $|h_{ij}|^2$ is gamma distributed with scale parameter θ_{ij} ($\theta_{ij} > 0$) and shape parameter m_{ij} ($m_{ij} > 0$). So, the probability density function $f_{|h_{ij}|^2}(x; m_{ij}, \theta_{ij})$ of the random variable $|h_{ij}|^2$ is

$$f_{|h_{ij}|^2}(x; m_{ij}, \theta_{ij}) = \frac{x^{m_{ij}-1}}{\theta_{ij}^{m_{ij}}\Gamma(m_{ij})}\exp(-\frac{x}{\theta_{ij}})$$

where $\Gamma(y)$ denotes the complete gamma function

$$\Gamma(y) = \int_0^{\infty} t^{y-1}\exp(-t)dt$$

The cumulative distribution function (CDF) of $|h_{ij}|^2$, denoted $F_{|h_{ij}|^2}(x; m_{ij}, \theta_{ij})$, is the regularized gamma function

$$F_{|h_{ij}|^2}(x; m_{ij}, \theta_{ij}) = \frac{\gamma(m_{ij}, \frac{x}{m_{ij}})}{\Gamma(m_{ij})}$$

where $\gamma(m_{ij}, \frac{x}{m_{ij}})$ is the lower incomplete gamma function such that

$$\gamma(s, x) = \int_0^{x} t^{s-1}\exp(-t)dt$$

Let P be the power transmitted by each terminal and σ_w^2 be the variance of the additive white gaussian noise (AWGN) in the wireless channel. We define $SNR = P / \sigma_w^2$ to be the effective signal-to-noise ratio.

We also restrict our study to a single source-destination pair. This pair may belong to any route in the network. Amongst terminals within the range of both the source terminal and the destination terminal, we focus on $(N-1)$ specific terminals. These terminals are available for implementing a cooperative transmission and they are not allocated to any other transmission. However, these $(N-1)$ terminals are likely to cause collision if they try to transmit data all at once. All other terminals are assumed to remain silent because they do not implement a cooperation functionality, or their cooperation functionality has been switched off. Hence, no extra interference occurs from neighboring terminals. This also contributes to reduce the impact of cooperative communications on the rest of the network. Clearly, the best spatial diversity gain is no more achievable since part of possible relays has been excluded from competition. However, this has been done with the purpose of reducing the contention level in the network. Further studies should address this tradeoff between spatial diversity and interference level. In any case, if a terminal should interfere with the cooperative transmission, the proposed protocol is implementing classical error recovery mechanisms.

Protocol Description

Cooperation Mode Activation

The cooperation mode is activated at terminal R_i, $1 \le i \le (N-1)$, upon reception of a data frame from any source terminal S. This triggers the relay selection process at the relay candidates. The data frame is stored when R_i is implementing the cooperation functionality and R_i is not already involved in any other transmission. When terminal D succeeds in decoding the data frame, it sends an acknowledgment frame (ACK). Otherwise, terminal D discards the data frame and sends a call for cooperation (CFC) (Gomez et al., 2007). This saves processing time without sacrificing the optimality of the DMT. Note that the data frame from S contains an additional control field on the source address. Hence, when the checksum on the entire frame is wrong and the checksum on the source address is good, the destination terminal is able to send a CFC with the source address. When the CFC frame is lost, the protocol implements classical error recovery mechanisms[2]. When a terminal R_i stores the source message, it waits for either an ACK frame or a CFC frame. If any of these two frames is not received within a given time-slot, the source message is discarded at terminal R_i. Hence, only terminals that have received both the data frame and the CFC frame trigger the relay selection process[3]. Moreover, only terminals that improve the direct path are allowed to compete for best relay terminal. To decide whether a terminal R_i is pre-selected or not, a suitability metric u_i is used: the mutual information of the cooperative link between the source and the destination. More precisely, the suitability metric of R_i is defined in (5) (resp. in (12)) when a fixed AF (resp. a selective DF) transmission scheme is used. This metric is also used in the splitting algorithm in order to evaluate the relay candidates. So the best terminal is the one that can achieve the best link capacity. A relay candidate is pre-selected if the capacity of the cooperative transmission through this relay is above a given threshold. This threshold is the target data rate R. Note that the computation of the mutual information in (5) and in (12) requires the knowledge of the channel gains h_{SR_i} and h_{R_iD}. These gains are estimated at terminal R_i using the signals corresponding to the data frame and the CFC frame respectively. Note also that the criterion can be simplified when using a selective DF transmission scheme. Indeed, choosing the best $I_{DF}^{(i)}$ in (12) is equivalent to choosing the best $|h_{R_iD}|^2$ coefficient.

Splitting Algorithm

Consider a time-slotted system with $(N-1)$ relay candidates. Each terminal R_i has a suitability metric u_i, defined as the mutual information of the cooperative transmission from S to D, through terminal R_i. The goal is to select the terminal with the highest metric. The metrics are continuous and i.i.d. with complementary CDF (CCDF) denoted by $F_c(u) = \Pr[u_i > u]$. Therefore, the $F_c(.)$ is monotonically decreasing and invertible. The algorithm is specified using three variables $H_L(k)$, $H_H(k)$, and $H_{\min}(k)$ (Qin & Berry, 2004). $H_L(k)$ and $H_H(k)$ are the lower and upper metric thresholds, respectively, such that a terminal R_i transmits at time slot k if and only if its metric u_i satisfies $H_L(k) < u_i < H_H(k)$. $H_{\min}(k)$ tracks the largest value of the metric known up to slot k above which the best metric surely lies.

- **Initialization:** In the first slot ($k = 1$), the parameters are initialized as follows: $H_L(1) = F_c^{-1}(1/N_r)$, $H_H(1) = \infty$, and $H_{\min}(1) = 0$. The parameter N_r denotes the number of possible relays and should be set to $(N-1)$. So, each terminal should know the value of N_r. When this value is not known at each terminal, it can be overestimated by the number of terminals in the range of D. Terminal D generally knows this number through upper layer protocols.
- **Transmission Rule:** At the beginning of each slot, each terminal locally decides to transmit if and only if its metric lies between $H_L(k)$ and $H_H(k)$.
- **Feedback Generation:** At the end of each slot, the destination terminal broadcasts to all terminals a two-bit feedback: (i) 0 if the slot was idle (when no terminal transmitted), (ii) 1 if the outcome was a success (when exactly one terminal transmitted), or (iii) e if the outcome was a collision (when at least two terminals transmitted).
- **Response to Feedback:** Let

$$split(a,b) = F_c^{-1}(F_c(a)/2 + F_c(b)/2)$$

be the split function. Then, depending on the feedback, the following possibilities occur:

If the feedback (of the kth slot) is an idle (0) and no collision has occurred so far, then set:

$$H_H(k+1) = H_L(k)$$

$$H_L(k+1) = F_c^{-1}((k+1)/N_r)$$

and $H_{\min}(k+1) = 0$ (Figure 2).

If the feedback is a collision (e), then set

$$H_L(k+1) = split(H_L(k), H_H(k))$$

$$H_H(k+1) = H_H(k)$$

and

$$H_{\min}(k+1) = H_L(k)$$

If the feedback is an idle (0) and a collision has occurred in the past, then set

$$H_H(k+1) = H_L(k)$$

$$H_L(k+1) = split(H_{\min}(k), H_L(k))$$

$$H_{\min}(k+1) = H_{\min}(k)$$

A selective DF transmission scheme is considered in order to illustrate the operation of the splitting algorithm in Figure 2, Figure 3, and Figure 4. The suitability metric u_i for terminal R_i is defined in (12). It can be noted that the channel coefficient $|h_{R_iD}|^2$ between terminal R_i

Figure 2. Threshold adjustments of the splitting algorithm at terminal R_i when the feedback is 0 (idle) and no collision has occurred so far

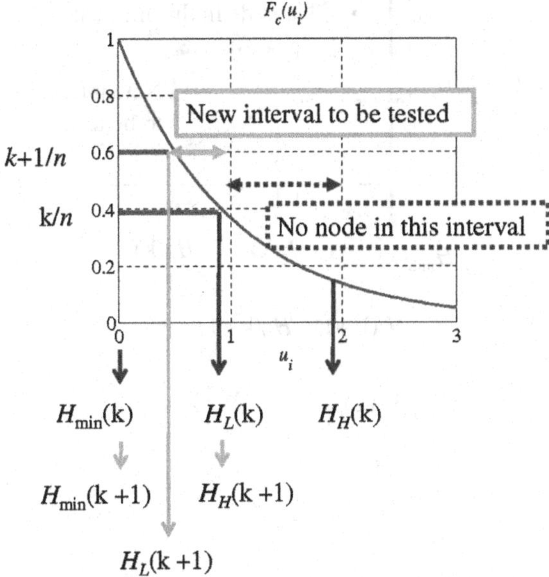

Figure 3. Threshold adjustments of the splitting algorithm at terminal R_i when the feedback is e (collision)

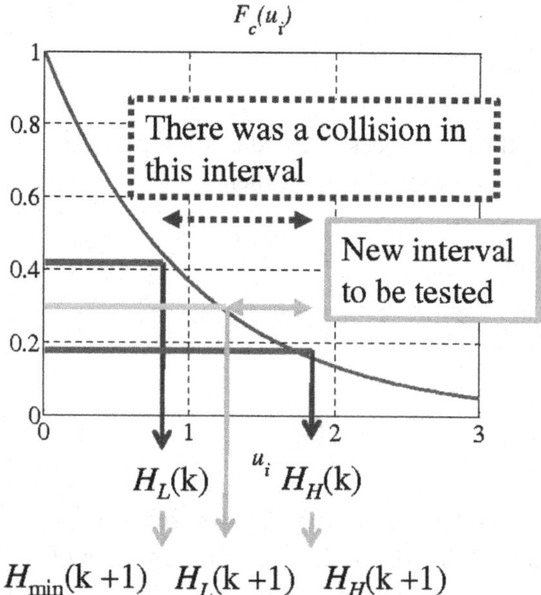

Figure 4. Threshold adjustments of the splitting algorithm at terminal R_i when the feedback is 0 (idle) and a collision has occurred in the past

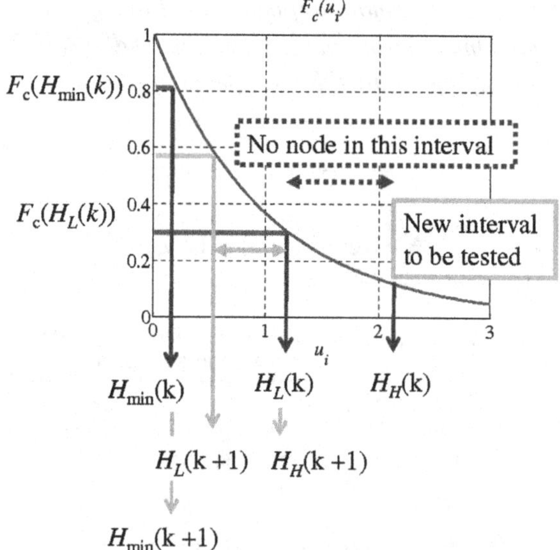

and the destination terminal D can be used directly as the suitability metric. The channel gain $|h_{R_iD}|$ is a random variable with a Nakagami-m distribution with parameter $m = 1$ (Rayleigh distributed variable) such that the variable $|h_{R_iD}|^2$ is exponentially distributed. We assume that the random variable $|h_{R_iD}|^2$ has a variance unity.

- **Termination:** The algorithm terminates when the outcome is a success (1).
- **Data Transmission:** When the destination terminal sends its last feedback, the best relay terminal sends a copy of the data frame using either a fixed AF forwarding scheme or a selective DF forwarding scheme. The destination receives the signal from the best relay terminal. When D succeeds in decoding the data frame, D sends an ACK frame (Figure 5). Otherwise, D remains silent and the timeout at the source terminal triggers a re-transmission. In section II.B of (Escrig, 2010) additional design constraints are given in order to implement this cooperative protocol in an IEEE 802.11-based network.

DMT Analysis of the On-Demand Cooperative Protocol

In this section, the DMT curve of the proposed protocol is studied. The DMT analysis focusses on the transmission part of the protocol, i.e. on the PHY layer. Indeed, DMT analyses do not take into account signaling overhead such as the one due to relay selection. Similarly, the relay selection has not been taken into account in the DMT analysis of Laneman et al. (2003) and the overhead required to distribute space-time codes to relay terminals has not been taken into account in the

Figure 5. Frame exchange sequence in the protocol using the basic IEEE 802.11 access method (S is the source terminal, D is the destination terminal, B is the best relay terminal, and R_i is a relay candidate).

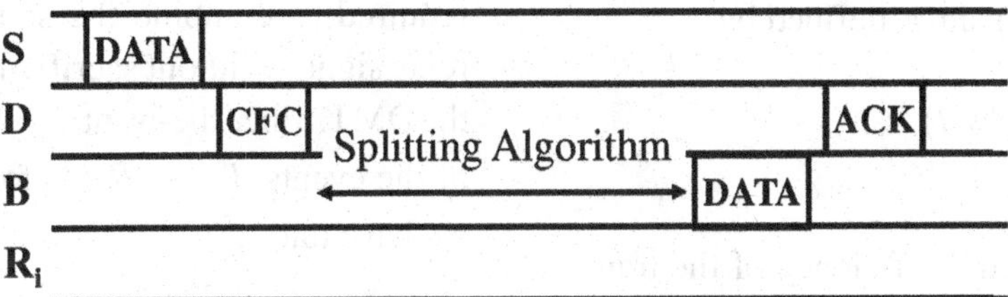

DMT analysis in Laneman et al. (2004). Further studies should provide a means to include MAC overhead in the capacity computing and then in DMT analyses. A first step toward this objective has been proposed in Li et al. (2009).

DMT Analysis for a Fixed AF Transmission Scheme

The channel models are characterized using the system model described in the previous section, and a time-division notation; frequency-division counterparts to this model are straightforward. A base-band-equivalent, discrete-time channel model is used for the continuous-time channel. Three discrete time received signals are defined in the following. Here, $y_{ij}(n)$ denotes the signal received by terminal j and transmitted by terminal i. During a first time-slot, D and the best relay terminal B are receiving signals from S

$$y_{SD}(n) = h_{SD}x(n) + w_{SD}(n) \tag{1}$$

$$y_{SB}(n) = h_{SB}x(n) + w_{SB}(n) \tag{2}$$

for $n = 1, 2, ..., T_M / 2$, where T_M denotes the duration of time-slots reserved for each message. When terminal D succeeds in decoding the data frame from S, no signal is transmitted by the best relay terminal B. Otherwise, B transmits a new signal using a fixed AF scheme, and D is receiving

$$y_{BD}(n) = h_{BD}[\beta y_{SB}(n)] + w_{BD}(n)$$

for

$$n = T_M / 2 + 1, ..., T_M$$

The noise $w_{ij}(n)$ between transmitting terminal i and receiving terminal j are all assumed to be i.i.d. circularly symmetric complex Gaussian with zero mean and variance σ_w^2. Symbols transmitted by the source terminal S are denoted $x(n)$. For simplicity, the same power constraint at both the source and the relay is imposed: $E[| x(n) |^2] \leq P$ and $E[| \beta y_{SB}(n) |^2] \leq P$.Afixed AF cooperation scheme is implemented. So the normalization factor β must satisfy

$$\beta^2 = P / (| h_{SB} |^2 P + \sigma_w^2)$$

It is assumed that the source and the relay each transmit orthogonally on half of the time-slots. We also consider that a perfect synchronization is provided at the block, carrier, and symbol level. The diversity order $d_{AF}(r)$ of the protocol using an fixed AF transmission scheme is defined by

$$d_{AF}(r) = \lim_{SNR \to \infty} - \frac{\log[p_{AF}^{out}(SNR, r)]}{\log(SNR)}$$

The probability $p_{AF}^{out}(SNR, r)$ is the outage probability for a signal to noise ratio SNR and a spatial multiplexing gain r defined by

$$r = \lim_{SNR \to \infty} R / \log_2(SNR)$$

where R is the spectral efficiency of the transmission (in b/s/Hz). For high SNR values, we use

$$R = r \times \log_2 SNR \quad (3)$$

Assuming that $(N-1)$ terminals are available, the protocol is in outage if all the relay terminals fail in improving the direct transmission. So the outage probability $p_{AF}^{out}(SNR, r)$ is

$$p_{AF}^{out}(SNR, r) = \Pr[I_{SD} \le R] \times \Pr[\bigcup_{i=1}^{N-1}(I_{AF}^{(i)} \le \frac{R}{2}) \mid I_{SD} \le R]$$

where I_{SD} is the mutual information of the direct transmission

$$I_{SD} = \log_2(1 + SNR \mid h_{SD} \mid^2) \quad (4)$$

and $I_{AF}^{(i)}$ is the mutual information of the relayed transmission using a fixed AF cooperation scheme at terminal R_i and implementing frame dropping at the destination terminal

$$I_{AF}^{(i)} = \frac{1}{2}\log_2[1 + f(SNR \mid h_{SR_i} \mid^2, SNR \mid h_{R_iD} \mid^2)] \quad (5)$$

where

$$f(x, y) = (xy) / (x + y + 1)$$

There is no $SNR \mid h_{SD} \mid^2$ term in $I_{AF}^{(i)}$ because the source message is now dropped at the destination terminal D when D fails in decoding the message from S. This saves the processing time required to combine the source signal and the relay signal, without sacrificing the optimality of the DMT. Since the event $I_{SD} \le R$ is independent of the events $I_{AF}^{(i)} \le R / 2$ for $1 \le i \le (N-1)$, we have that

$$p_{AF}^{out}(SNR, r) = \Pr[I_{SD} \le R] \times \Pr[\bigcup_{i=1}^{N-1}(I_{AF}^{(i)} \le \frac{R}{2})]$$

With (4), we have that

$$\Pr[I_{SD} \le R] = \Pr[\mid h_{SD} \mid^2 \le SNR^{r-1}]$$

for high SNR values. The random variables h_{SR_i} and h_{R_iD} being mutually independent, for high SNR, we have that

$$p_{AF}^{out}(SNR, r) \le \Pr[\mid h_{SD} \mid^2 \le SNR^{r-1}] \times \prod_{i=1}^{N-1} \Pr[f(SNR \mid h_{SR_i} \mid^2, SNR \mid h_{R_iD} \mid^2) \le SNR^{r}] \quad (6)$$

From Lemma 2 in (Bletsas et al., 2008), we have that

$$\lim_{SNR \to \infty} \frac{\log\{\Pr[\mid h_{SD} \mid^2 \le SNR^{r-1}]\}}{\log(SNR)} = m_{SD}(r-1) \quad (7)$$

when $\mid h_{SD} \mid$ is a Nakagami-m random variable with shape parameter m_{SD}.

The result of Lemma 4 in (Bletsas et al., 2006) is adapted for the other terms of (6)

$$\Pr[f(\rho a, \rho b) \le \rho^{r}] \le \Pr[min(a, b) \le \rho^{r-1} + \sqrt{\rho^{r-2}(\rho^{r} + 1)}]$$

Thus, we have that

$$\Pr[f(SNR \mid h_{SR_i} \mid^2, SNR \mid h_{R_iD} \mid^2) \leq SNR^r]$$
$$\leq \Pr[min(\mid h_{SR_i} \mid^2, \mid h_{R_iD} \mid^2) \leq SNR^{r-1}$$
$$+\sqrt{SNR^{r-2}(SNR^r + 1)}] \tag{8}$$

From (8) and Lemma 4 in (Bletsas et al., 2008) and the fact that

$$\sqrt{SNR^{r-2}(SNR^r + 1)} \to SNR^{r-1}$$

as $SNR \to +\infty$, we have that

$$\lim_{SNR\to\infty} \frac{\log\{\Pr[f(SNR \mid h_{SR_i} \mid^2, SNR \mid h_{R_iD} \mid^2) \leq SNR^r]\}}{\log(SNR)} \leq m_i(r-1) \tag{9}$$

where $\mid h_{SR_i} \mid$ and $\mid h_{R_iD} \mid$ are Nakagami-m random variables with shape parameter m_{SR_i} and m_{R_iD} respectively and

$$m_i = \min\{m_{SR_i}, m_{R_iD}\} \tag{10}$$

Using (7) and (9) in (6), we have that

$$\lim_{SNR\to\infty} -\frac{\log[p_{AF}^{out}(SNR, r)]}{\log(SNR)}$$
$$\leq (m_{SD} + \sum_{i=1}^{N-1} m_i)(1-r)$$

Hence, the diversity curve $d_{AF}(r)$, i.e. the DMT performance of the protocol using a fixed AF transmission scheme, is

$$d_{AF}(r) \geq (m_{SD} + \sum_{i=1}^{N-1} m_i)(1-r) \tag{11}$$

Equation (11) gives the lower bound of the DMT performance (Figure 6). Note that for the special case of Rayleigh fading, i.e $m_{SD} = m_i = 1$ for $1 \leq i \leq (N-1)$, we have that $d_{AF}(r) = N(1-r)$. So, when there are $(N-1)$ potential relay terminals, the proposed protocol achieves the optimal DMT curve reaching the two extreme points $d_{AF}(0) = N$ and $d_{AF}(1) = 0$. Note also that the only information provided by the DMT curve is that the data rate of the overall transmission scales

Figure 6. DMT curves of three protocols: the proposed protocol, the direct transmission, and the on-demand cooperation with one relay terminal. For the special case of Rayleigh fading, $m_{SD} + \sum_{i=1}^{N-1} m_i = N$.

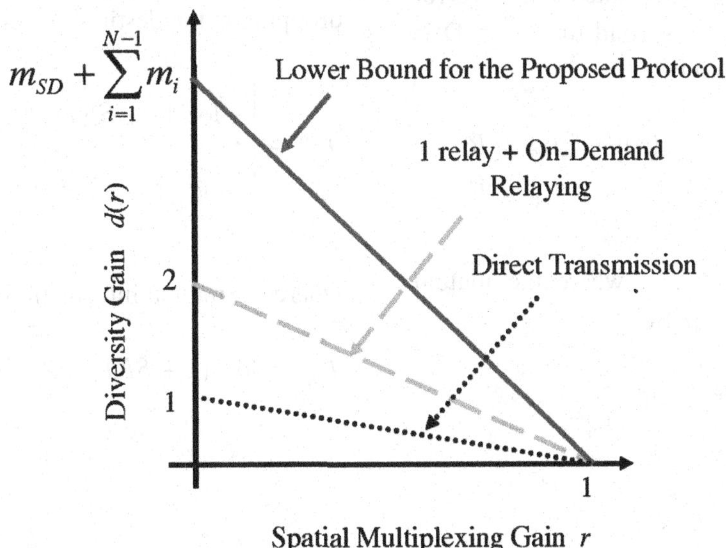

like a direct transmission, even in presence of a cooperative relaying. In particular, the overheard induced by the additional signaling frames (CFC, splitting algorithm) does not appear in (11) because the DMT analysis is just providing a rough estimate of the achieved multiplexing gain r, not a precise value. In particular, the spatial multiplexing gain scales like 1. This results is consistent with the one obtained with other on-demand cooperation techniques (Laneman et al., 2004).

DMT Analysis for a Selective DF Transmission Scheme

The same base-band-equivalent, discrete-time channel model is used. The first two received signals are defined in (1) and (2). When terminal D succeeds in decoding the data frame from S, no signal is transmitted by the best relay terminal B. Otherwise, the best relay terminal sends a new signal using a selective DF scheme, i.e. if and only if it has been able to decode the source message. The event that a relay R_i has successfully decoded the data transmitted by S with a spectral efficiency R is equivalent to the event that the mutual information of the channel between S and the relay R_i, I_{SR_i}, lies above the spectral efficiency R (Laneman et al., 2003; Bletsas et al., 2006). In that case, it can be considered that the estimation of signal $x(n)$, denoted $\hat{x}(n)$, is error free. Hence, during the second time slot, D is receiving a signal from B

$$y_{BD}(n) = \begin{cases} h_{BD}x(n) + w_{BD}(n), & \text{if } I_{SB} > R \\ 0, & \text{if } I_{SB} \leq R \end{cases}$$

for $n = T_M / 2 + 1, \ldots, T_M$,. where the mutual information I_{SB} is given by

$$I_{SB} = \log_2(1 + SNR \mid h_{SB} \mid^2)$$

The noise $w_{ij}(n)$ and the symbols $x(n)$ have been defined in the previous subsection. The same power constraint is also imposed at both the source and the relay: $E[\mid x(n) \mid^2] \leq P$. The source and the relay are assumed to transmit orthogonally on half of the time-slots. A perfect synchronization is assumed at the block, carrier, and symbol level. The diversity gain $d_{DF}(r)$ of the protocol is defined by

$$d_{DF}(r) = \lim_{SNR \to \infty} -\frac{\log[p_{DF}^{out}(SNR, r)]}{\log(SNR)}$$

The probability $p_{DF}^{out}(SNR, r)$ is the outage probability for a signal to noise ratio SNR and a spatial multiplexing gain r. For high SNR values, we use (3). When $(N-1)$ terminals are available, the protocol is in outage if all the $(N-1)$ candidates fail in improving the direct transmission

$$p_{DF}^{out}(SNR, r) = \Pr[I_{SD} \leq R] \times \Pr[\bigcup_{i=1}^{N-1} (I_{DF}^{(i)} \leq \frac{R}{2}) \mid I_{SD} \leq R]$$

where $I_{DF}^{(i)}$ is the mutual information of the relayed transmission using a selective DF cooperation scheme at terminal R_i and implementing frame dropping at the destination terminal

$$I_{DF}^{(i)} = \begin{cases} \frac{1}{2}\log_2(1 + SNR \mid h_{SD} \mid^2), & \text{if } I_{SR_i} \leq R \\ \frac{1}{2}\log_2(1 + SNR \mid h_{R_iD} \mid^2), & \text{if } I_{SR_i} > R \end{cases} \tag{12}$$

where the mutual information I_{SR_i} is defined by

$$I_{SR_i} = \log_2(1 + SNR \mid h_{SR_i} \mid^2)$$

and the mutual information I_{SD} is defined in (4). The probability $p_{DF}^{out}(SNR,r)$ can be expressed as the sum of $2^{(N-1)}$ terms

$$p_{DF}^{out}(SNR,r) = \sum_{j=1}^{2^{(N-1)}} P_j = \sum_{j=1}^{2^{(N-1)}} P_j^E \prod_{i=1}^{N-1} \Pr[\varepsilon_j^{(i)}] \quad (13)$$

where

$$P_j^E = \Pr[I_{SD} \le R] \times \Pr\{\bigcup_{i=1}^{N-1} [I_{DF}^{(i)} \le \frac{R}{2} \mid (\varepsilon_j^{(i)}, I_{SD} \le R)]\}$$

The event $\varepsilon_j^{(i)}$ equals the event $I_{SR_i} \le R$ or $I_{SR_i} > R$ according to the value of index j, $1 \le j \le 2^{(N-1)}$. The probability P_j in (13) is constituted with N components. The first component P_j^E is the probability denoted in (12) where each value of $I_{DF}^{(i)}$ is conditioned to the value of I_{SR_i}. The $(N-1)$ last terms in the product exhibit the probabilities that the I_{SR_i} are above or beyond the threshold R, for $1 \le i \le (N-1)$. We assume that there are N_j passive relay terminals such that $\varepsilon_j^k = [I_{SR_k} \le R]$ in P_j. We define the set K_j such that

$$K_j = \{k / \varepsilon_j^{(k)} = [I_{SR_k} \le R], 1 \le k \le (N-1)\}$$

with cardinality $|K_j| = N_j$. Thus, there are $(N-1) - N_j$ active relay terminals such that $\varepsilon_j^{(l)} = [I_{SR_l} > R]$ in P_j. We define the set L_j such that

$$L_j = \{l / \varepsilon_j^{(l)} = [I_{SR_l} > R], 1 \le l \le (N-1)\}$$

with cardinality $|L_j| = (N-1) - N_j$. Note also that $0 \le N_j \le (N-1)$. So, we have that

$$P_j^E = \Pr[I_{SD} \le R] \times \Pr\{\bigcup_{k \in K_j} [I_{DF}^{(k)} \le \frac{R}{2} \mid (\varepsilon_j^{(k)}, I_{SD} \le R)], \bigcup_{l \in L_j} [I_{DF}^{(l)} \le \frac{R}{2} \mid (\varepsilon_j^{(l)}, I_{SD} \le R)]\} \quad (14)$$

For the N_j passive relay terminals, we have that

$$I_{DF}^{(k)} \le \frac{R}{2} \mid \varepsilon_j^{(k)} \Leftrightarrow I_{DF}^{(k)} \le \frac{R}{2} \mid I_{SR_k} \le R \Leftrightarrow I_{SD} \le R$$

So, we have that:

$$\Pr[(I_{DF}^{(k)} \le \frac{R}{2} \mid \varepsilon_j^{(k)}) \mid I_{SD} \le R] = 1$$

Moreover, we have that:

$$\Pr[\varepsilon_j^{(k)}] = \Pr[I_{SR_k} \le R] = \Pr[\log_2(1 + SNR \mid h_{SR_k} \mid^2) \le R]$$

For the $(N-1) - N_j$ active relay terminals, we have that

$$I_{DF}^{(l)} \le \frac{R}{2} \mid \varepsilon_j^{(l)} \Leftrightarrow I_{DF}^{(l)} \le \frac{R}{2} \mid I_{SR_l} > R \Leftrightarrow \log_2(1 + SNR \mid h_{R_l D} \mid^2) \le \frac{R}{2}$$

and

$$\Pr[\varepsilon_j^{(l)}] = \Pr[I_{SR_l} > R]$$

For high SNR values, we have a simpler expression for (14)

$$P_j^E = \Pr[\log_2(1+SNR\,|\,h_{SD}\,|^2) \leq R] \\ \times \Pr\{\bigcup_{l\in L_j}[\log_2(1+SNR\,|\,h_{R_lD}\,|^2) \leq R]\}$$

The $|\,h_{R_lD}\,|^2$ random variables being mutually independent, we have that

$$P_j^E = \Pr[|\,h_{SD}\,|^2 \leq SNR^{r-1}] \\ \times \prod_{l\in L_j} \Pr[|\,h_{R_lD}\,|^2 \leq SNR^{r-1}] \quad (15)$$

using (3). The random variables $|\,h_{SR_k}\,|^2$ and $|\,h_{R_lD}\,|^2$ being mutually independent, we have that

$$\prod_{i=1}^{N-1}\Pr[\varepsilon_j^{(i)}] \leq \prod_{k\in K_j}\Pr[|\,h_{SR_k}\,|^2 \leq SNR^{r-1}] \quad (16)$$

So, using (16), we have that

$$P_j \leq \Pr[|\,h_{SD}\,|^2 \leq SNR^{r-1}] \\ \times \prod_{l\in L} Pr[|\,h_{R_lD}\,|^2 \leq SNR^{r-1}] \\ \times \prod_{k\in K} Pr[|\,h_{SR_k}\,|^2 \leq SNR^{r-1}] \quad (17)$$

The random variables $|\,h_{SD}\,|^2$, $|\,h_{R_lD}\,|^2$ for $l \in L_j$, and $|\,h_{SR_k}\,|^2$ for $k \in K_j$ are all gamma distributed variables. Let $|\,h\,|^2$ be one of these random variables, with shape parameter m. From Lemma 2 in (Bletsas et al., 2008), we have that

$$\lim_{SNR\to\infty}\frac{\log\{\Pr[|\,h\,|^2 \leq SNR^{r-1}]\}}{\log(SNR)} = m(r-1) \quad (18)$$

So, using (18) in (17), we have that

$$\lim_{SNR\to\infty}\frac{\log[P_j]}{\log(SNR)} \\ = (m_{SD} + \sum_{k\in K_j} m_{SR_k} + \sum_{l\in L_j} m_{R_lD})(r-1)$$

for every j, $1 \leq j \leq 2^{(N-1)}$, where m_{ij} is the shape parameter of the random variable $|\,h_{ij}\,|^2$. So, we have that

$$\lim_{SNR\to\infty}\frac{\log[P_j]}{\log(SNR)} \\ \geq (m_{SD} + \sum_{i=1}^{N-1} m_i)(r-1)$$

where m_i has been defined in (10). So, we have that

$$\lim_{SNR\to\infty} -\frac{\log[p_{DF}^{out}(SNR,r)]}{\log(SNR)} \\ \leq (m_{SD} + \sum_{i=1}^{v-1} m_i)(1-r)$$

Hence, the diversity curve $d_{DF}(r)$ of the protocol is lower bounded by the following expression

$$d_{DF}(r) \geq (m_{SD} + \sum_{i=1}^{N-1} m_i)(1-r) \quad (19)$$

Equation (19) gives the lower bound on the DMT performance of the protocol using the selective DF transmission scheme. Note again that for the special case of Rayleigh fading, i.e $m_{SD} = m_i = 1$ for $1 \leq i \leq (N-1)$, we have that $d_{DF}(r) = N(1-r)$.

SIMULATION RESULTS

The simulation results focus on the splitting algorithm. Simulation results showing the diversity order of the transmission scheme can be found in (Escrig, 2010) for the case of Rayleigh fading channels. These results are independent of the selection process. Figures 7 and 8 plot the number of slots required to select the best relay as a function of N the number of possible relays. Figure 8 plots the results for N going from 2 to 30 and Figure 7 plots the results for N going from 30 to 100. The results have been obtained through extensive MATLAB simulations using a Monte-Carlo approach. Ten thousands simulations have been run for each value of N. The selection of the best relay using a splitting algorithm is performed within 2.46 slots on average as long as the number of possible relays is greater than 30. Otherwise, the mean value of slots is lower. The standard deviation is 1.70 for N greater than 30. These results are consistent with the ones obtained in (Shah, Mehta, & Yim, 2009).

The slot duration includes two transmissions, one by the relay candidates and the other by the destination, and necessary gaps, as required, between these two transmissions. In IEEE 802.11-based networks, the duration of a slot may exceed 100 μsec (*Part 11: Wireless LAN medium access control (MAC) and physical layer (PHY) specifications, Tech. Rep. IEEE Std 802.11-2007*, 2007). So the duration of the selection should not exceed 250 μsec on average. Note however that the duration of the selection process has a standard deviation. Typically, the standard deviation is on the order of 2 slots. Comparatively, the other studies consider that some information exchange is needed to process the selection but the amount of bandwidth dedicated to this task is not computed (Beres & Adve, 2008). In (Bletsas et al., 2006), the protocol requires a contention period during which relay candidates may contend for access to the channel and two time slots to notify the selection of the best relay: one by the best relay and the other by the destination. The notification by the destination terminal is used to

Figure 7. Duration of the selection process (expressed in number of time-slots) as a function of the number of relay terminals N (case of N greater than 30): average number (top) and standard deviation (bottom). The dashed line on each figure shows the mean value.

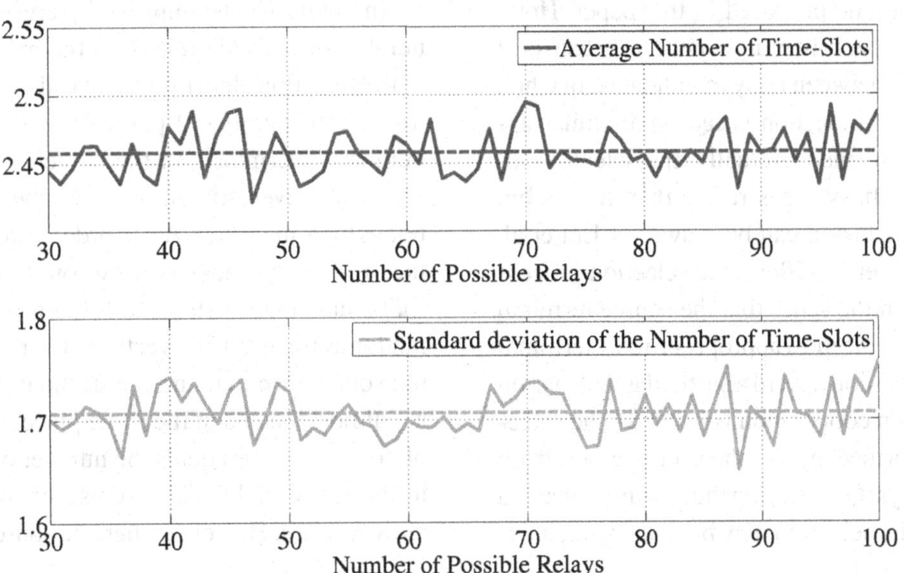

Figure 8. Duration of the selection process (expressed in number of time-slots) as a function of the number of relay terminals N (case of N lower than 30): average number (top) and standard deviation (bottom). The dashed line on each figure shows the mean value.

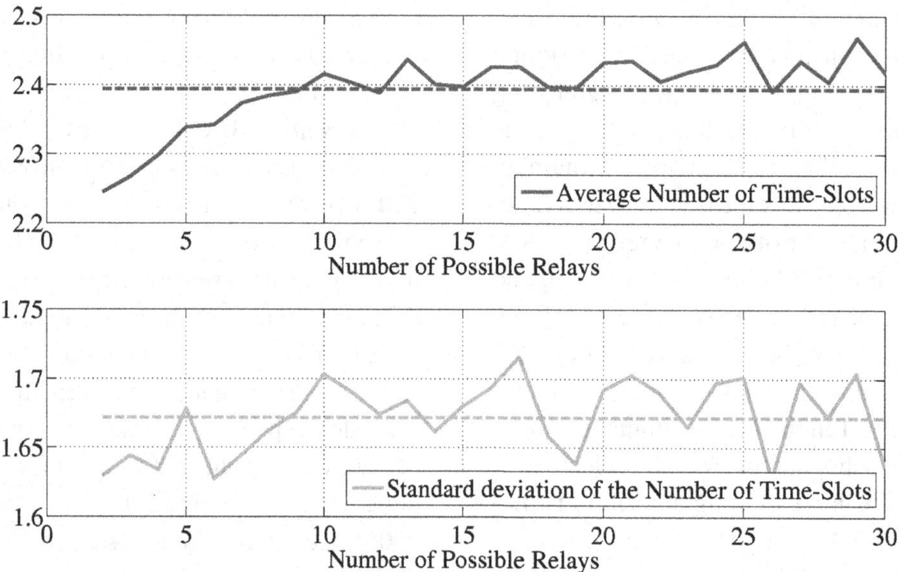

addressed the issues of hidden relays. Each relay candidate triggers a timer, the expiration of which triggers the transmission of a flag. The average duration of a timer can be made as small as 200 μsec. When the duration of the notification, 100 μsec in IEEE 802.11-based systems, is added to the average duration of the timers, the result is similar to the one proposed in this paper. However, the approach may not succeed because of the collisions between relays candidates. In Chou et al. (2007), a selection based on a similar approach is performed. Actually, the selection algorithm uses busy tones rather than timers but the same conclusion can be drawn. In Liu et al. (2007) and Li et al. (2009), the selection process is proactive in the sense that the source terminal already knows the most appropriate relay terminal for its transmission toward a particular destination terminal when cooperation is needed. The selection is performed by overhearing frames from possible relays. Quantifying the resources needed to perform the selection may be done by measuring the amount of time required to collect the channel state information about the possible relays. This corresponds to an awake time, the duration of which should be minimized. On the other hand, these protocols also need to transmit one or two signaling frames, so the overhead is comparable to the one induced by the proposed approach.

In Figure 10, the impact of overestimating the number of possible relays on the mean duration of the selection algorithm is investigated. Assuming that the number of possible relays equals the number of terminals in the range of terminal D amounts to overestimating the number of possible relays by some factor K. In order to compute this factor, two extremes cases are considered (Figure 9). Terminals are assumed to be uniformly (though randomly) distributed over the network area. When the source terminal and the destination terminal are almost located at the same place, the number of possible relays equals the number of terminals in the range of D. The overestimation factor is one ($K = 1$). However when the source terminal

Figure 9. Impact of terminal location on the performance of the splitting algorithm: source and destination terminals at the same place (left side), and source and destination terminals at the limit of their range (right side)

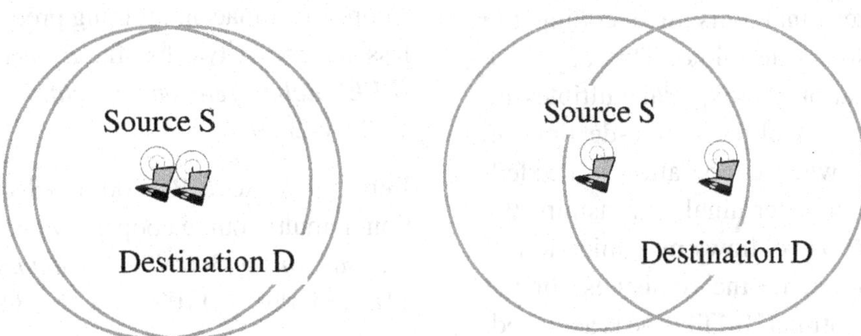

and the destination terminal are located at the limit of their range, the ratio between the coverage area of D and the area of possible relays is greater than one:

$$K = \pi / (2\pi / 3 - \sqrt{3} / 2)$$

This extreme case is the worst case, resulting in an overestimation factor K of 2.56. According to Figure 10, overestimating the number of relays increases both the mean duration of the selection process (from 2.46 to 3.21) and its standard deviation (from 1.70 to 1.81). Further studies should address this issue in order to design efficient implementations of the splitting algorithm.

CONCLUSION

The purpose of the study is the design of a DMT optimal access protocol in the context of IEEE 802.11 mesh networks. The designed protocol

Figure 10. Duration of the selection process (expressed in number of time-slots) as a function of the number of relay terminals N: average number (top) and standard deviation (bottom). Solid lines correspond to $K = 1$ and dotted lines correspond to $K = 2\pi / (\pi - 2)$.

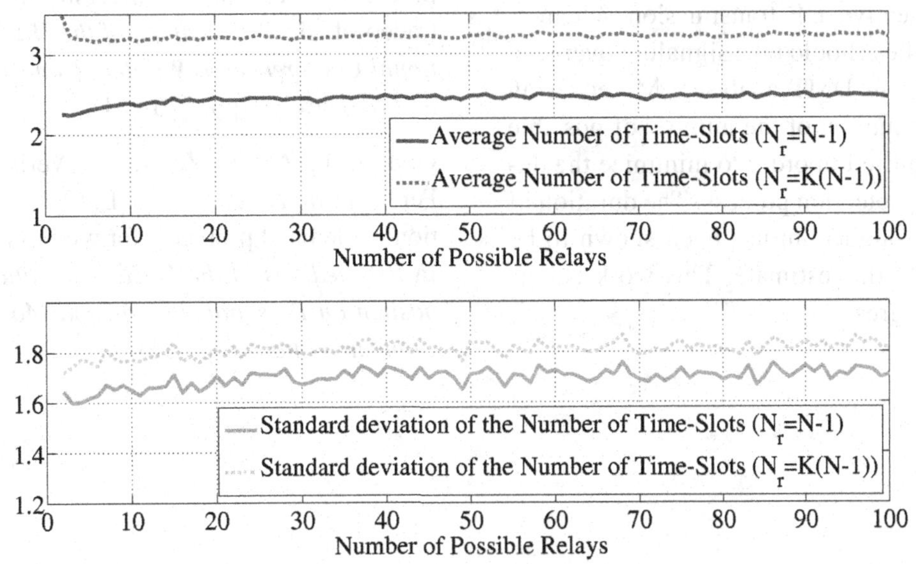

has two basic features: on-demand cooperation and selection of the best relay terminal. Cooperation is activated on-demand, i.e. only when a destination terminal fails in decoding the message of a source terminal. This approach allows maximization of the spatial multiplexing gain, i.e the capacity of the source-destination link. Moreover, when cooperation is needed, only the best relay terminal retransmits the source message. This allows maximization of the diversity order, i.e the robustness of the link. Hence, an optimal DMT curve is achieved. Three features are added in order to guarantee both a fast and an efficient relay selection. Using a splitting algorithm, the time required to select a best relay terminal is fully characterized. Moreover, only terminals that can improve the direct transmission are pre-selected. So inefficient cooperation is now avoided. Finally, the destination terminal discards the source message when it fails to decode it. This saves processing time without sacrificing the optimality of the DMT. When $(N-1)$ terminals are situated in the range of both a source terminal S and a destination terminal D, a diversity gain of N is provided while a spatial multiplexing gain of one is achieved when a Rayleigh fading channel is considered. Thus, the protocol implements a DMT optimal transmission scheme. The approach has been applied to both fixed AF and selective DF transmission schemes. Further studies should take signaling overhead into account in DMT analyses. Moreover an accurate estimation of the number of possible relays is required in order to minimize the duration of the selection process. The duration of the splitting algorithm has been shown to be dependent of this estimate. This work is currently in progress.

REFERENCES

Azgin, A., Altunbasak, Y., & AlRegib, G. (2005). Cooperative mac and routing protocols for wireless ad hoc networks. In. *Proceedings of the IEEE Global Telecommunications Conference*, *5*, 2854–2859.

Beres, E., & Adve, R. (2008). Selection cooperation in multi-source cooperative networks. *IEEE Transactions on Wireless Communications*, *7*(1), 118–127. doi:10.1109/TWC.2008.060184.

Bletsas, A., Khisti, A., Reed, D. P., & Lippman, A. (2006). A simple cooperative diversity method based on network path selection. *IEEE Journal on Selected Areas in Communications*, *24*(3), 659–672. doi:10.1109/JSAC.2005.862417.

Bletsas, A., Khisti, A., & Win, M. Z. (2008). Opportunistic cooperative diversity with feedback and cheap radios. *IEEE Transactions on Wireless Communications*, *7*(5), 1823–1827. doi:10.1109/TWC.2008.070193.

Chou, C.-T., Yang, J., & Wang, D. (2007). Cooperative mac protocol with automatic relay selection in distributed wireless networks. In *Proceedings of the IEEE Pervasive Computing and Communications Workshops* (pp. 526-531).

Escrig, B. (2010). On-demand cooperation mac protocols with optimal diversity-multiplexing tradeo-off. In *Proceedings of the IEEE International Conference on Wireless Communications and Networking* (pp. 576-581).

Gomez, J., Alonso-Zarate, J., Verikoukis, C., Perez-Neira, A., & Alonso, L. (2007). Cooperation on demand protocols for wireless networks. In *Proceedings of the IEEE International Symposium on Personal, Indoor and Mobile Radio Communications* (pp. 1-5).

Hunter, T. E., & Nosratinia, A. (2006). Diversity through coded cooperation. *IEEE Transactions on Wireless Communications*, *5*(2), 283–289. doi:10.1109/TWC.2006.1611050.

IEEE. Computer Society. (2007). *Part 11: Wireless lan medium access control (mac) and physical layer (phy) specifications* (Tech. Rep. No. 802.11-2007). Washington, DC: Author.

Laneman, J. N., Tse, D. N. C., & Wornell, G. W. (2004). Cooperative diversity in wireless networks: Efficient protocols and outage behavior. *IEEE Transactions on Information Theory*, *50*(12), 3062–3080. doi:10.1109/TIT.2004.838089.

Laneman, J. N., & Wornell, G. W. (2000). Energy-efficient antenna sharing and relaying for wireless networks. In. *Proceedings of the IEEE International Conference on Wireless Communications and Networking*, *1*, 7–12.

Laneman, J. N., & Wornell, G. W. (2003). Distributed space-time-coded protocols for exploiting cooperative diversity in wireless networks. *IEEE Transactions on Information Theory*, *49*(10), 2415–2425. doi:10.1109/TIT.2003.817829.

Li, Y., Cao, B., Wang, C., You, X., Daneshmand, A., Zhuang, H., et al. (2009). Dynamical cooperative mac based on optimal selection of multiple helpers. In *Proceedings of the IEEE Global Telecommunications Conference* (pp. 3044-3049).

Liu, P., Tao, Z., Narayanan, S., Korakis, T., & Panwar, S. S. (2007). Coopmac: A cooperative mac for wireless lans. *IEEE Journal on Selected Areas in Communications*, *25*(2), 340–354. doi:10.1109/JSAC.2007.070210.

Sendonaris, A., Erkip, E., & Aazhang, B. (2003). User cooperation diversitypart I: System description. *IEEE Transactions on Communications*, *51*(11), 1927–1938. doi:10.1109/TCOMM.2003.818096.

Shah, V., Mehta, N. B., & Yim, R. (2009). Relay selection and data transmission throughput tradeoff in cooperative systems. In *Proceedings of the IEEE Global Telecommunications Conference* (pp. 230-235).

Shah, V., Mehta, N. B., & Yim, R. (2010). Splitting algorithms for fast relay selection: Generalizations, analysis, and a unifed view. *IEEE Transactions on Wireless Communications*, *9*(4), 1525–1535. doi:10.1109/TWC.2010.04.091364.

Xiangping, R. Q. B. (2004). Opportunistic splitting algorithms for wireless networks. In *Proceedings of the Twenty-Third Annual Joint Conference of the IEEE Computer and Communications Societies* (Vol. 3, pp. 1662).

Zheng, L., & Tse, D. N. C. (2002). Diversity and multiplexing: A fundamental tradeoff in multiple antenna channels. *IEEE Transactions on Information Theory*, *49*(5), 1073–1096. doi:10.1109/TIT.2003.810646.

This work was previously published in the International Journal of Wireless Networks and Broadband Technologies, Volume 1, Issue 1, edited by Naveen Chilamkurti, pp. 56-72, copyright 2011 by IGI Publishing (an imprint of IGI Global).

Chapter 6
Performance Evaluation of a Three Node Client Relay System

Sergey Andreev
Tampere University of Technology, Finland

Olga Galinina
Tampere University of Technology, Finland

Alexey Vinel
Tampere University of Technology, Finland

ABSTRACT

In this paper, the authors examine a client relay system comprising three wireless nodes. Closed-form expressions for mean packet delay, as well as for throughput, energy expenditure, and energy efficiency of the source nodes are also obtained. The precision of the established parameters is verified by means of simulation.

INTRODUCTION

Wireless communication networks are becoming more widespread, as novel telecommunication standards emerge (Marks, Nikolich, & Snyder, in press; Nakamura, in press). The future of wireless communication, however, greatly depends on how successfully the disproportion between the required Quality of Service (QoS) and the limited system spectral resource is overcome. Meanwhile, the urge to increase system *spectral efficiency* gradually gives way to the task of the energy efficiency improvement. This is particularly true for small-scale handheld wireless devices due to the growing gap between available and required battery capacity (Lahiri, Raghunathan, Dey, & Panigrahi, 2002).

The problem of effective resource utilization is of primary importance for wireless systems, where a large population of users' shares limited spectral resource (Andreev, Koucheryavy, Himayat, Gonchukov, & Turlikov, 2010). Currently, layered system architecture dominates in network design, where each layer is treated independently following the concept of layer abstraction. Among these, Physical (PHY) layer is responsible for the transmission of raw data bits, whereas Media Access Control (MAC) layer arbitrates access of users to the shared wireless channel.

DOI: 10.4018/978-1-4666-3902-7.ch006

However, traditional layered architecture appears to be far less flexible and often implies inefficient resource utilization (Andreev, Galinina, & Vinel, 2010). To mitigate this discrepancy, a novel integral and adaptive approach is required. As a consequence, cross-layer techniques receive increasing attention from the research community (Andreev, Koucheryavy, Himayat, Gonchukov, & Turlikov, 2010) with primary focus being set on the joint consideration of MAC and PHY layers. New channel-aware solutions are introduced to achieve cross-layer benefits by taking advantage of wireless channel state information (CSI). They typically exploit extended MAC-PHY interaction and result in higher QoS, arrival flow, and channel state adaptability (Song, 2005; Miao, 2008; Kim, 2009).

As wireless users are becoming increasingly mobile, the focus of the latest research efforts shifts from throughput optimization (Song & Li, 2006) towards energy efficiency improvement at all layers of a wireless system (Anisimov, Andreev, Galinina, & Turlikov, 2010) from its architecture (Benini, Bogliolo, & de Micheli, 2000) to the adopted communication protocols (Schurgers, 2002). Recently, cooperative cross-layer approaches gain increasing international acclaim (Pyattaev, Andreev, Vinel & Sokolov, 2010; Pyattaev, Andreev, Koucheryavy, & Moltchanov, 2010). They exploit variability in CSI of wireless users and, as such, allow for additional performance gains thus constituting a promising research direction.

RESEARCH BACKGROUND

While more and more users are sharing the limited wireless resource and cellular networks are gradually shifting towards more aggressive frequency reuse scenarios (Marks, Nikolich, & Snyder, in press; Nakamura, in press), wireless interference becomes one of the major limiting factors that impair network performance growth. Wireless data transmission of a user, being unavoidably broadcast, necessarily impacts the transmission process of other users and consequently degrades the overall system energy efficiency. However, users may gain in their energy efficiencies by acting cooperatively (Cui, Goldsmith, & Bahai, 2004; Jayaweera, 2004). Such a spatial domain resource management is becoming increasingly important to improve the performance of the cell-edge users with a poor communication link (Andreev, Galinina, & Vinel, 2010).

On the other hand, cooperation typically implies extra energy expenditure as more data is transmitted over the air. Moreover, cooperative transmission may negatively impact packet delay, as data packets are sometimes relayed over a longer path. However, increasing delay could sometimes be compensated by reducing transmission data rate; and this is contrastingly known to increase user energy efficiency (Andreev, Koucheryavy, Himayat, Gonchukov, & Turlikov, 2010). As such, it is important to evaluate all the basic trade-offs behind wireless cooperation and indicate scenarios where it actually improves the performance of a cellular network.

Currently, studying the collaboration between neighboring users of a wireless system is highly significant. As energy expenditure to guarantee reliable data transmission exponentially grows with distance (Stuber, 2001), it is desirable to relay data over shorter intermediate hops (Rabaey, Ammer, da Silva Jr., & Patel, 2000). Consequently, client relay is believed to become a promising concept that would boost the performance of contemporary wireless cellular networks.

Enabling client relay, it is crucial to avoid scenarios when the use of this technology insufficiently increases the performance of the originating user (Haenggi & Puccinelli, 2005). As the result, the task of effective relay selection is often reduced to analyzing the trade-off between the source node benefits and the relay node losses. In this paper, we evaluate the performance of the simplest but nonetheless practical client relay network. We estimate mean packet delay for all the data sources within the considered system, as well as establish their throughput, energy expenditure, and energy efficiency.

SYSTEM MODEL

We consider a wireless cellular network enhanced with client relay capability borrowing the basic methodology from our previous paper (Andreev, Galinina, & Turlikov, 2010) and extending it. In what follows, we concentrate on the simplest network topology (Figure 1) comprising two source nodes and one sink node. We term user A the originator. The originator generates own data packets with the mean arrival rate λ_A. We also term user R the relay. The relay generates own data packets with the mean arrival rate λ_R. Additionally, the relay is capable of eavesdropping on the transmissions from the originator and may temporarily store the packets from A for the subsequent retransmission. The node B is termed the base station and receives data packets from both the originator and the relay. Below we detail the system model and present the set of main assumptions.

Assumption 1: System time is *slotted*. All the communicated data packets have equal size and the transmission of each one of them takes exactly one slot.

Assumption 2: The numbers of new data packets arriving to either the originator or the relay during consecutive slots are independent and identically distributed (i.i.d) random variables with the means λ_A and λ_R respectively. For the sake of analytical tractability we assume *Poisson* arrival flow of new packets in the rest of the text. The base station has no outgoing traffic.

Assumption 3: Both the originator and the relay have unbounded queues to store own data packets. Additionally, the relay has an extra memory location to keep a *single* data packet from the originator for the subsequent cooperative retransmission. Below we demonstrate that single memory cell is sufficient for the proposed client relay system operation.

Assumption 4: The communication system is centralized and is controlled by the base station. The fair *stochastic* round-robin scheduler operates at the base station and alternates source nodes accessing the wireless channel with equal probability (see Figure 2 as an example behavior of the system detailed in Figure 1, with no additional packet arrivals). In particular, if both the originator and the relay have pending data packets the subsequent slot is given to either of them with probability 0.5. The non-transmitting node stays idle during this slot. If either of the source nodes has no

Figure 1. Considered three node client relay system

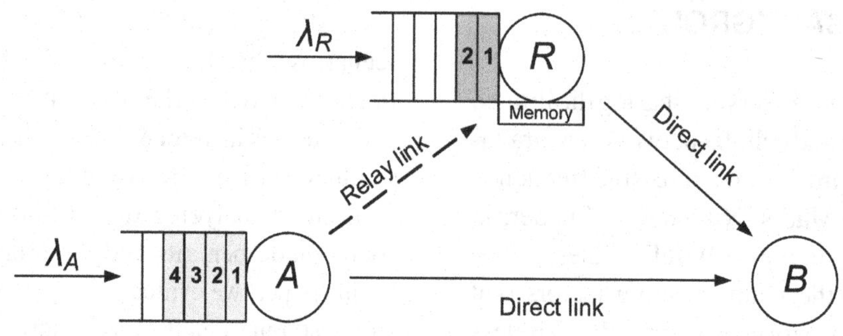

Figure 2. Example client relay system operation

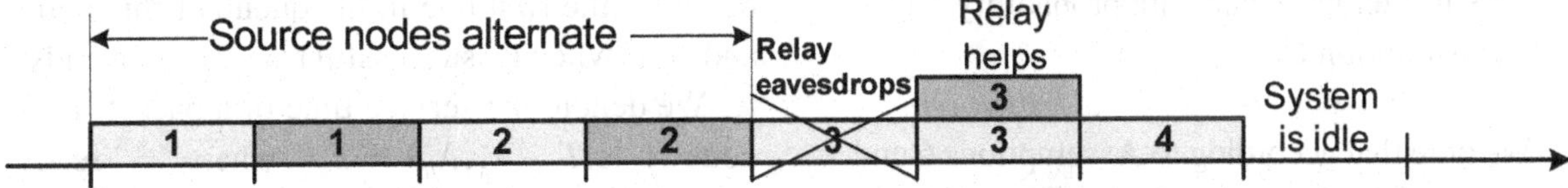

pending data packets the other one is given the subsequent slot with probability 1. This ensures efficient system time utilization. If neither source has pending data packets the entire system stays idle. We also assume that scheduling information about which node transmits in the subsequent slot is immediately available to both source nodes over a separate channel and consumes no system resources.

Assumption 5: The communication channel is error-prone and is based on the multi-packet reception channel model (Rong & Ephremides, 2009). The transmitted data packet is received by the destination successfully with the constant probability dependent only on the link type (direct or relay) and on which nodes are transmitting simultaneously.

We define the following non-zero success probabilities:

$$p_{AB} = \Pr\{packet\ from\ A\ is\ received\ at\ B \mid only\ A\ transmits\}$$

$$p_{RB} = \Pr\{packet\ from\ R\ is\ received\ at\ B \mid only\ R\ transmits\}$$

$$p_{AR} = \Pr\{packet\ from\ A\ is\ received\ at\ R \mid only\ A\ transmits\}$$

$$p_{CB} = \Pr\{packet\ from\ A\ is\ received\ at\ B \mid A\ and\ R\ cooperate\}$$

It is expected that $p_{AR} > p_{AB}$, as well as $p_{CB} > p_{AB}$.

We also assume that feedback information about the success/failure of each reception attempt by the base station is immediately available to both source nodes over a separate channel and consumes no system resources. If the packet is not received successfully, it is retransmitted by its source. The maximum number of allowable retransmission attempts is unlimited. The relay is incapable of simultaneous transmission and reception.

Assumption 6: At the first packet transmission attempt by the originator, the relay attempts eavesdropping on it with probability 1. According to Assumption 5, this eavesdropping attempt is successful with probability p_{AR}. If the packet is received successfully by the relay it is stored in the single memory location by replacing its previous contents. Also according to Assumption 5, this packet is at the same time successfully received at the base station with probability p_{AB}. If unsuccessful, the originator retransmits the same packet in the next available slot.

Assumption 7: At any retransmission attempt by the originator, the relay performs one of the following. If the packet which is being transmitted was already stored in its memory location the relay sends it simultaneously with the originator with probability 1 (Figure 2). As such, the relay tries to improve the performance of the originator. According to Assumption 5, this packet is successfully received at the base station with probability p_{CB}. Alternatively, if eavesdropping on the transmitted packet failed previously the relay attempts again with probability 1 (see Assumption 6).

We note that according to Assumptions 6 and 7 a single memory location for the eavesdropped data packets at the relay suffices for the considered client relay system operation. Moreover, the originator is unaware of the cooperative help from the relay and the relay sends no explicit acknowledgements to the originator by contrast to the approach from (Rong & Ephremides, 2009). This enables tailoring the proposed client relay model to the contemporary cellular standards (Marks, Nikolich, & Snyder, in press; Nakamura, in press). The relay improves the throughput of the originator by sacrificing its own energy efficiency. Extra energy is spent by the relay on the eavesdropping, as well as on the simultaneous packet transmissions with the originator.

Generally, the relay may sometimes decide not to eavesdrop on the transmissions from the originator or not to transmit a packet simultaneously subject to a particular client relay policy. In this paper, we restrict our further explorations to the baseline case when the relay is forced to eavesdrop on all the transmissions from the originator and to transmit a packet simultaneously whenever it is kept in the memory (see Assumptions 6 and 7). We leave any opportunistic cooperation for the future work.

Main Notations

The proposed analytical approach to the performance evaluation of the considered three node client relay system is based on the notion of the packet service time. The packet service time is the time interval from the moment when the tagged packet is ready for service to the moment its service ends (Jaiswal, 1968). More specifically, in the considered system the service time of the tagged packet from a node starts when this packet becomes the first one in the queue of this node and ends when its successful transmission ends.

We denote the service time of a packet from node A as $T_{AR}(\lambda_A, \lambda_R) \triangleq T_{AR}$, where '$\triangleq$' reads as "equal by definition". Additionally, we introduce the *mean* service time of a packet from node A as

$$\tau_{AR}(\lambda_A, \lambda_R) \triangleq \tau_{AR} = E[T_{AR}]$$

Further, we denote by $\tau_{AR}(\lambda_A, 0) \triangleq \tau_{A0}$ the mean service time of a packet from node A conditioning on the fact that $\lambda_R = 0$.

Symmetrically, we denote the service time of a packet from node R as $T_{RA}(\lambda_R, \lambda_A) \triangleq T_{RA}$ and the corresponding mean service time as

$$\tau_{RA}(\lambda_R, \lambda_A) \triangleq \tau_{RA} = E[T_{RA}]$$

Analogously, the conditional mean service time is introduced as $\tau_{RA}(\lambda_R, 0) \triangleq \tau_{R0}$ for $\lambda_A = 0$.

Note that as T_{R0} is distributed geometrically, for both system with cooperation (when $p_{AR} > 0$) and system without cooperation (when $p_{AR} = 0$) it holds the following:

$$\tau_{R0} = \frac{1}{p_{RB}} \qquad (1)$$

whereas only for the system without cooperation it holds:

$$\tau_{A0} = \frac{1}{p_{AB}} \qquad (2)$$

The derivation of τ_{A0} for the system with cooperation is a more complicated task and will be addressed below.

Denote the numbers of data packets in the queues of the nodes A and R at the beginning of a particular slot t by $Q_A^{(t)}$ and $Q_R^{(t)}$ respectively. As we observe the client relay system in stationary conditions, we omit the upper index t of variables $Q_A^{(t)}$ and $Q_R^{(t)}$.

Finally, we denote the queue load coefficient (Kleinrock, 1975) of node A as $\rho_{AR}(\lambda_A, \lambda_R) \triangleq \rho_{AR}$. By definition we have:

$$\rho_{AR} = \Pr\{Q_A \neq 0\} = \lambda_A \tau_{AR} \qquad (3)$$

In particular, queue load coefficient of node A conditioning on the fact that $\lambda_R = 0$ may be established as

$$\rho_{AR}(\lambda_A, 0) \triangleq \rho_{A0} = \lambda_A \tau_{A0}$$

Accounting for (2), for the system without cooperation ρ_{A0} further simplifies to $\rho_{A0} = \dfrac{\lambda_A}{p_{AB}}$.

Analogously, queue load coefficient of node R is denoted as $\rho_{RA}(\lambda_R, \lambda_A) \triangleq \rho_{RA}$. Also by definition we have:

$$\rho_{RA} = \Pr\{Q_R \neq 0\} = \lambda_R \tau_{RA} \qquad (4)$$

Symmetrically, queue load coefficient of node R conditioning on the fact that $\lambda_A = 0$ may be established as

$$\rho_{RA}(\lambda_R, 0) \triangleq \rho_{R0} = \lambda_R \tau_{R0}$$

Accounting for (1), for both systems with and without cooperation ρ_{R0} further simplifies to $\rho_{R0} = \dfrac{\lambda_R}{p_{RB}}$. The main notations we consistently use throughout this paper are summarized in Table 1.

Table 1. Main notations

Notation	Parameter Description
λ_A	Mean arrival rate of packets in node A
λ_R	Mean arrival rate of packets in node R
p_{AB}	Probability of successful reception from A at B when A transmits
p_{RB}	Probability of successful reception from R at B when R transmits
p_{AR}	Probability of successful reception from A at R when A transmits
p_{CB}	Probability of successful reception from A at B when A and R cooperate
τ_{AR}	Mean service time of a packet from node A
τ_{RA}	Mean service time of a packet from node R
ρ_{AR}	Queue load coefficient of node A
ρ_{RA}	Queue load coefficient of node R
q_A	Mean queue length of node A
q_R	Mean queue length of node R
δ_A	Mean packet delay of node A
δ_R	Mean packet delay of node R
η_A	Mean departure rate of packets from node A (throughput of A)
η_R	Mean departure rate of packets from node R (throughput of R)
ε_A	Mean energy expenditure of node A
ε_R	Mean energy expenditure of node R
φ_A	Mean energy efficiency of node A
φ_R	Mean energy efficiency of node R

General Statements

Consider the queue at node A. We remind that by definition $\rho_{AR} = \Pr\{Q_A \neq 0\}$ and set $\rho_{A0} > \rho_{R0}$ as an example. The following propositions may thus be formulated.

Proposition 1: For the queue load coefficient of node A it holds the following:

$$\rho_{AR} \leq \frac{\rho_{A0}}{1-\rho_{R0}} \tag{5}$$

Another important proposition may be formulated considering normalization condition of the respective system generating function or balance equations of the corresponding embedded Markov chain.

Proposition 2: For the queue load coefficients of nodes A and R it holds the following:

$$\rho_{AR} - \rho_{RA} = \rho_{A0} - \rho_{R0} \tag{6}$$

The proofs of Propositions 1 and 2 are not included into this text due to space constraints.

Proposition 3: For the queue load coefficient of node R it holds the following:

$$\rho_{RA} = \rho_{AR} - \rho_{A0} + \rho_{R0} \leq \frac{\rho_{A0}}{1-\rho_{R0}} - \rho_{A0} + \rho_{R0} \tag{7}$$

The proof of Proposition 3 follows immediately from (5) and (6).

The established upper bounds on ρ_{AR} and ρ_{RA} hold for both systems with and without cooperation. In what follows, we firstly study the system without cooperation and then extend the proposed analytical approach to the system with cooperation.

Non-Cooperative System Performance Evaluation

We study the behavior of node A within the framework of the queueing theory. As such, consider the queueing system associated with node A. Due to the fact that the queues of nodes A and R are mutually dependent, the notorious Pollazek-Khinchine formula (Kleinrock, 1975) may not be used to obtain the exact mean queue length of node A. We, however, apply this formula to establish the approximate value of the mean queue length of node A as:

$$\begin{aligned} q_A &\cong \lambda_A E[T_{AR}] + \frac{\lambda_A^2 E[T_{AR}^2]}{2(1-\lambda_A E[T_{AR}])} \\ &= \lambda_A \tau_{AR} + \frac{\lambda_A^2 E[T_{AR}^2]}{2(1-\lambda_A \tau_{AR})} \end{aligned} \tag{8}$$

where $\tau_{AR} = E[T_{AR}]$ is the mean service time of a packet from node A (the first moment of random service time T_{AR}) and $E[T_{AR}^2]$ is the respective second moment of the service time. Accounting for (3) and (8), we may write:

$$q_A \cong \rho_{AR} + \frac{\lambda_A^2 E[T_{AR}^2]}{2(1-\rho_{AR})} \tag{9}$$

We now demonstrate how to derive the unknown components of Equation (9). Consider the service time of the tagged packet from node A. We remind that the packet scheduler at the base station is stochastic, that is, it assigns the subsequent slot to node A with probability 0.5 if both source nodes are loaded. Therefore, in every slot for which $Q_R \neq 0$ and $Q_A \neq 0$ the packet from A is included into the system schedule with probability 0.5. We introduce the following auxiliary probability:

$$\gamma_A \triangleq \Pr\{Q_R \neq 0 \mid Q_A \neq 0\} = \frac{\Pr\{Q_R \neq 0, Q_A \neq 0\}}{\Pr\{Q_A \neq 0\}}$$

Clearly, the scheduler either assigns the subsequent slot to node R with probability $0.5\gamma_A$ or assigns it to node A with the complementary probability $1 - 0.5\gamma_A$.

Consider the probability of the event that $Q_R \neq 0$ and $Q_A \neq 0$ simultaneously. By the complete probability formula, we may write

$$\Pr\{Q_R \neq 0, Q_A \neq 0\} = \Pr\{Q_R \neq 0\} - \Pr\{Q_R \neq 0, Q_A = 0\}$$

On the other hand, by definition we have $\Pr\{Q_R \neq 0\} = \rho_{RA}$. Further, for the probability $\Pr\{Q_R \neq 0, Q_A = 0\}$ we obtain the following expression:

$$\Pr\{Q_R \neq 0, Q_A = 0\} = \Pr\{Q_A = 0\} - \Pr\{Q_R = 0, Q_A = 0\}$$

Using the definition of ρ_{AR}, we note that $\Pr\{Q_A = 0\} = 1 - \rho_{AR}$. Moreover, we also note that $\Pr\{Q_R = 0, Q_A = 0\} = 1 - \rho_{A0} - \rho_{R0}$.

Summarizing the above,

$$\Pr\{Q_R \neq 0, Q_A \neq 0\} = \rho_{AR} + \rho_{RA} - \rho_{A0} - \rho_{R0}$$

Additionally, from Proposition 2 it immediately follows that

$$\Pr\{Q_R \neq 0, Q_A \neq 0\} = 2 \cdot (\rho_{AR} - \rho_{A0})$$

Finally, we obtain:

$$0.5\gamma_A = 0.5 \cdot \frac{\Pr\{Q_R \neq 0, Q_A \neq 0\}}{\Pr\{Q_A \neq 0\}} = 1 - \frac{\rho_{A0}}{\rho_{AR}}$$

We may establish the following distribution for the service time of a packet from node A:

$$\Pr\{T_{AR} = n\} = p_{AB}(1 - 0.5\gamma_A)(1 - p_{AB}(1 - 0.5\gamma_A))^{n-1}$$

The above expression accounts for the fact that out of n slots spent to serve a packet from node A the last slot was assigned to node A and its transmission in this slot was successful. The previous $n-1$ slots were either not assigned to node A or its transmissions in these slots were unsuccessful.

Calculating the first and the second moment of the service time ($E[T_{AR}]$ and $E[T_{AR}^2]$), accounting for (9) and also using Little's formula in the form $q_A = \lambda_A \delta_A$, it is now easy to approximate the mean packet delay of node A as:

$$\delta_A \cong \frac{\rho_{AR}}{\lambda_A} + \frac{\lambda_A(2 - p_{AB}(1 - 0.5\gamma_A))}{2(1 - \rho_{AR})p_{AB}^2(1 - 0.5\gamma_A)^2}$$

The performance metrics of node R may be calculated analogously, due to the symmetric nature of the respective direct links. Accounting for

$$0.5\gamma_R = 1 - \frac{\rho_{R0}}{\rho_{RA}}$$

where

$$\gamma_R \triangleq \frac{\Pr\{Q_R \neq 0, Q_A \neq 0\}}{\Pr\{Q_R \neq 0\}}$$

the approximate mean packet delay of node R is given by:

$$\delta_R \cong \frac{\rho_{RA}}{\lambda_R} + \frac{\lambda_R(2 - p_{RB}(1 - 0.5\gamma_R))}{2(1 - \rho_{RA})p_{RB}^2(1 - 0.5\gamma_R)^2}.$$

The proposed analytical approach to the performance evaluation of the considered three node client relay system is also applicable for establishing the *exact* mean departure rate of packets from (throughput of) nodes A and R. In particular, the throughput of A is given by:

$$\eta_A = \begin{cases} \lambda_A, \ no \ saturation \\ \dfrac{1 - \lambda_R \tau_{R0}}{\tau_{A0}}, \ saturation \ for \ A \\ \dfrac{1}{2\tau_{A0}}, \ saturation \ for \ A, R \end{cases}$$

Similarly, the throughput of R may be derived by:

$$\eta_R = \begin{cases} \lambda_R, \ no \ saturation \\ \dfrac{1 - \lambda_A \tau_{A0}}{\tau_{R0}}, \ saturation \ for \ R \\ \dfrac{1}{2\tau_{R0}}, \ saturation \ for \ A, R \end{cases}$$

The above expressions may be further simplified accounting for Equations (1) and (2). Here, the saturation conditions are defined as follows:

- Saturation for A: $\left(\lambda_A \tau_{A0} + \lambda_R \tau_{R0} > 1\right)$ and at the same time $\left(\lambda_R \tau_{R0} < 0.5\right)$.
- Saturation for R: $\left(\lambda_A \tau_{A0} + \lambda_R \tau_{R0} > 1\right)$ and at the same time $\left(\lambda_A \tau_{A0} < 0.5\right)$.
- Saturation for A and R: $\left(\lambda_A \tau_{A0} > 0.5\right)$ and at the same time $\left(\lambda_R \tau_{R0} > 0.5\right)$.

Additionally, we may obtain the exact value of the mean energy expenditure of node A as:

$$\varepsilon_A = P_{TX} \eta_A \tau_{A0} + P_I (1 - \eta_A \tau_{A0})$$

together with the mean energy expenditure of node R as

$$\varepsilon_R = P_{TX} \eta_R \tau_{R0} + P_I \left(1 - \eta_R \tau_{R0}\right)$$

Here, P_{TX} is the average power that is spent by a node in the packet transmission state, whereas P_I is the average power that is spent by the same node in the idle state. As such, the mean energy efficiencies of nodes A and R readily follow and are given by expressions $\varphi_A = \frac{\eta_A}{\varepsilon_A}$ and $\varphi_R = \frac{\eta_R}{\varepsilon_R}$ respectively.

Cooperative System Performance Evaluation

In order to mathematically describe the system with cooperation, we firstly consider an important special case when the queue at node R is always empty. We establish the distribution of the number of slots required to serve a packet from node A. By using the obtained distribution, we then generalize the proposed approach for the case of non-empty queue at node R. All the respective performance metrics for the system with cooperation are marked by symbol '*' in the rest of the text.

Case 1: The queue at node R is always empty ($\lambda_R = 0$).

Analogously to the derivations in the previous section, we may express the sought distribution for the service time of a packet from node A as:

$$\Pr\{T^*_{A0} = n\} = X(1 - p_{CB})^{n-1}$$
$$-Y[(1 - p_{AB})(1 - p_{AR})]^{n-1}$$

where

$$X = \frac{p_{AR}(1 - p_{AB})p_{CB}}{1 - p_{CB} - (1 - p_{AB})(1 - p_{AR})}$$

and $Y = X - p_{AB}$.

Coming now to the mean service time, we have the following:

$$\tau^*_{A0} = \frac{p_{CB} + (1 - p_{AB})p_{AR}}{p_{CB}[p_{AB} + (1 - p_{AB})p_{AR}]} \quad (10)$$

Case 2: The queue at node R is *not* always empty ($\lambda_R > 0$).

Here we generalize the above analytical expressions for the most complex cooperative case with $\lambda_R > 0$. Omitting lengthy but straightforward derivations, we give the respective distribution for the service time of a packet from node A as:

$$\Pr\{T^*_{AR} = n\} = X(1 - 0.5\gamma^*_A)(1 - p_{CB}(1 - 0.5\gamma^*_A))^{n-1}$$
$$-Y(1 - 0.5\gamma^*_A)(1 - p_A(1 - 0.5\gamma^*_A))^{n-1}$$

where

$$0.5\gamma^*_A = 1 - \frac{\rho^*_{A0}}{\rho^*_{AR}}$$

and also

$$p_A = p_{AB} + p_{AR} - p_{AB} \cdot p_{AR}$$

for brevity. Here p_A is the probability to successfully receive a packet from A at either node R or at the base station.

Queue load coefficients of nodes A and R (ρ^*_{AR} and ρ^*_{RA}) may be calculated similarly to the respective parameters for the system without cooperation accounting for the fact that $\rho^*_{A0} \triangleq \lambda_A \tau^*_{A0}$, where the expression for τ^*_{A0} is given by (10).

Finally, calculating the second moment of the service time we derive the resulting expression for the approximate mean packet delay of node A as:

$$\delta^*_A \cong \frac{\rho^*_{AR}}{\lambda_A} + \frac{\lambda_A}{2(1 - \rho^*_{AR})(1 - 0.5\gamma^*_A)^2} \cdot \left[X \cdot \frac{2 - p_{CB}(1 - 0.5\gamma^*_A)}{p^3_{CB}} - Y \cdot \frac{2 - p_A(1 - 0.5\gamma^*_A)}{p^3_A}\right]$$

where X and Y were given above.

Accounting for (10), the resulting approximation for the mean packet delay δ^*_R of node R, as well as expressions for the throughput η^*_A and η^*_R of nodes A and R in the system with cooperation are similar to the respective metrics in the system without cooperation from the previous section. Analogously, the mean energy expenditure of node A in the considered case is given by:

$$\varepsilon^*_A = P_{TX}\eta^*_A\tau^*_{A0} + P_I(1 - \eta^*_A\tau^*_{A0})$$

whereas the mean energy expenditure of node R may be calculated as:

$$\varepsilon^*_R = P_{TX}\left(\eta^*_R\tau_{R0} + \eta^*_A \cdot \frac{1 - p_{AB}\tau^*_{A0}}{p_{CB} - p_{AB}}\right) + P_{RX}\left(1 - \eta^*_R\tau_{R0} - \eta^*_A \cdot \frac{1 - p_{AB}\tau^*_{A0}}{p_{CB} - p_{AB}}\right),$$

where P_{RX} is the average power that is spent by a node in the packet reception state. As before, the mean energy efficiencies of nodes A and R are given by expressions $\varphi_A^* = \frac{\eta_A^*}{\varepsilon_A^*}$ and $\varphi_R^* = \frac{\eta_R^*}{\varepsilon_R^*}$ respectively.

Numerical Results and Conclusions

In this section, we discuss some simulation results of the considered three node client relay system (Figures 3 and 4). Particularly, we focus on throughput, mean packet delay, energy expenditure, and energy efficiency of the source nodes. Partly following (Rong & Ephremides, 2009), the simulation parameters are set as: $p_{AB} = 0.3$, $p_{RB} = 0.7$, $p_{AR} = 0.4$, $p_{CB} = 0.5$, $P_{TX} = 1.0$, $P_{RX} = 0.9$, $P_I = 0.8$, $\lambda_R = 0.15$, whereas λ_A is varied across the system stability region. The plots compare the two scenarios: with cooperation (*cooperative*) and without cooperation (*non-cooperative*).

Figure 3. Dependency of throughput (left) and mean packet delay (right) of source nodes on mean packet arrival rate λ_A

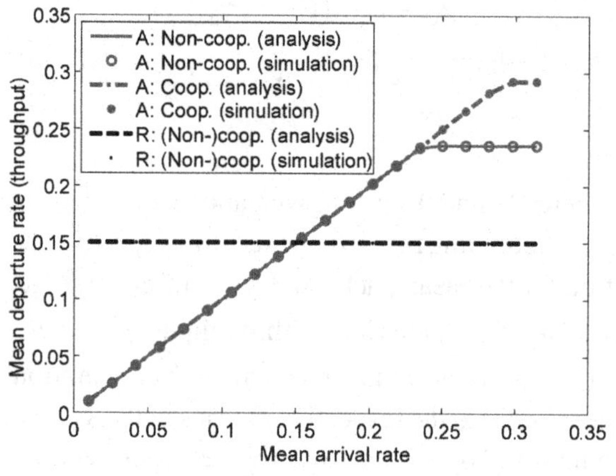

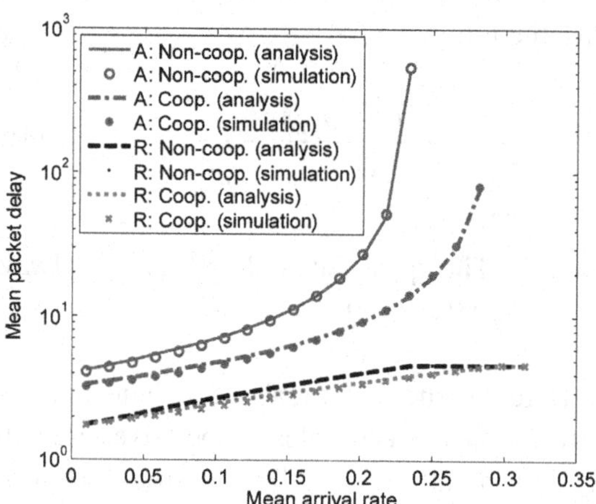

Figure 4. Dependency of energy expenditure (left) and energy efficiency (right) of source nodes on mean packet arrival rate λ_A

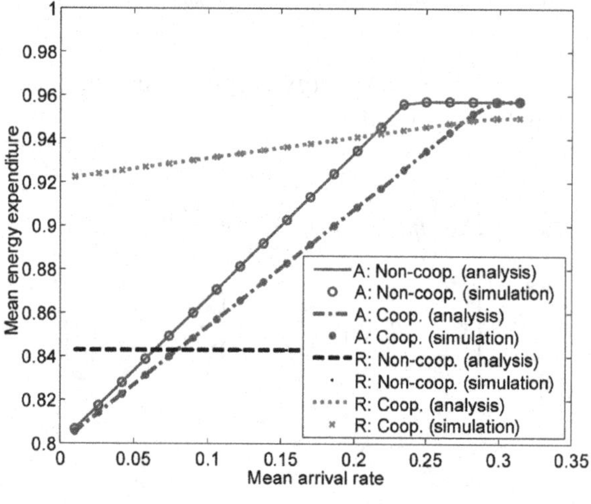

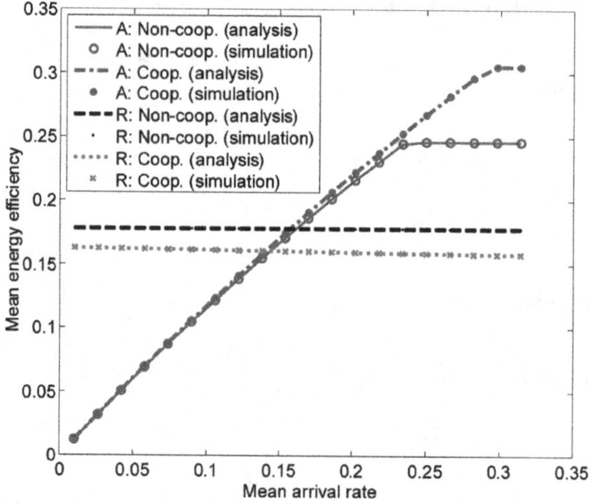

Evidently, the proposed analytical approach to the performance evaluation of the considered three node client relay system shows excellent agreement with simulation data. The obtained results allow for concluding upon the feasibility of the client relay technology. In particular, the originator throughput gain is up to 24% in saturation region. As such, client collaboration makes a promising technique to improve cell-edge user performance in the contemporary and future wireless cellular networks.

By contrast to known approaches where research is significantly simulation-based, this paper primarily introduces a formal mathematical model to assess the performance of a client relay network. The addressed client collaboration mechanisms may be implemented as part of networking equipment produced by Motorola, Intel, Nokia, etc. The proposed cooperation protocols could be tailored to next-generation telecommunication standards IEEE 802.16m (Marks, Nikolich, & Snyder, in press) and LTE-Advanced (Nakamura, in press).

ACKNOWLEDGMENT

This work is supported by Graduate School in Electronics, Telecommunications and Automation, Tampere Graduate School in Information Science and Engineering, Nokia Foundation, and HPY Research Foundation.

REFERENCES

Andreev, S., Galinina, O., & Turlikov, A. (2010). Basic client relay model for wireless cellular networks. In *Proceedings of the International Congress on Ultra Modern Telecommunications and Control Systems and Workshops* (pp. 909-915).

Andreev, S., Galinina, O., & Vinel, A. (2010). Cross-layer channel-aware approaches for modern wireless networks. In A. Vinel, B. Bellalta, C. Sacchi, A. Lyakhov, M. Telek, & M. Oliver (Eds.), *Proceedings of the Third International Workshop on Multiple Access Communications* (LNCS 6235, pp. 163-179).

Andreev, S., Koucheryavy, Y., Himayat, N., Gonchukov, P., & Turlikov, A. (2010, December). *Active-mode power optimization in OFDMA-based wireless networks.* Paper presented at the 6th IEEE Broadband Wireless Access Workshop, Miami, FL.

Anisimov, A., Andreev, S., Galinina, O., & Turlikov, A. (2010). Comparative analysis of sleep mode control algorithms for contemporary metropolitan area wireless networks. In S. Balandin, R. Dunaytsev, & Y. Koucheryavy (Eds.), *Proceedings of the 10th International Conference on Smart Spaces and Next Generation Wired/Wireless Networking* (LNCS 6294, pp. 184-195).

Benini, L., Bogliolo, A., & de Micheli, G. (2000). A survey of design techniques for system-level dynamic power management. *IEEE Transactions on Very Large Scale Integration*, *8*(3), 299–316. doi:10.1109/92.845896.

Cui, S., Goldsmith, A., & Bahai, A. (2004). Energy-efficiency of MIMO and cooperative MIMO techniques in sensor networks. *IEEE Journal on Selected Areas in Communications*, *22*, 1089–1098. doi:10.1109/JSAC.2004.830916.

Haenggi, M., & Puccinelli, D. (2005). Routing in ad hoc networks: A case for long hops. *IEEE Communications Magazine*, *43*, 112–119. doi:10.1109/MCOM.2005.1522131.

Jaiswal, N. (1968). *Priority queues*. New York, NY: Academic Press.

Jayaweera, S. K. (2004). An energy-efficient virtual MIMO architecture based on V-BLAST processing for distributed wireless sensor networks. In *Proceedings of the First Annual IEEE Communications Conference on Sensor and Ad Hoc Communications and Networks* (pp. 299-308).

Kim, H. (2009). *Exploring tradeoffs in wireless networks under flow-level traffic: Energy, capacity and QoS*. Unpublished doctoral dissertation, University of Texas, Austin.

Kleinrock, L. (1975). Queueing systems: *Vol. 1. Theory*. New York, NY: John Wiley & Sons.

Lahiri, K., Raghunathan, A., Dey, S., & Panigrahi, D. (2002). Battery-driven system design: A new frontier in low power design. In *Proceedings of the 15th IEEE International Conference on Design Automation and the 7th International Conference on Very Large Scale Integration Design* (pp. 261-267).

Marks, R. B., Nikolich, P., & Snyder, R. (in press). *IEEE Std 802.16m, Amendment to IEEE standard for local and metropolitan area networks – Part 16: Air interface for broadband wireless access systems – Advanced air interface*. Retrieved from http://ieee802.org/16/pubs/80216m.html

Miao, G. (2008). *Cross-layer optimization for spectral and energy efficiency*. Unpublished doctoral dissertation, Georgia Institute of Technology, School of Electrical and Computer Engineering, Atlanta.

Nakamura, T. (in press). *LTE release 10 & beyond (LTE-Advanced)*. Retrieved from http://www.3gpp.org/article/lte-advanced

Pyattaev, A., Andreev, S., Koucheryavy, Y., & Moltchanov, D. (2010, December). *Some modeling approaches for client relay networks*. Paper presented at the 15th IEEE International Workshop on Computer Aided Modeling Analysis and Design of Communication Links and Networks, Miami, FL.

Pyattaev, A., Andreev, S., Vinel, A., & Sokolov, B. (2010). *Client relay simulation model for centralized wireless networks*. Paper presented at the Federation of European Simulation Societies Congress, Prague, Czech Republic.

Rabaey, J., Ammer, J., da Silva, J., Jr., & Patel, D. (2000). PicoRadio: Ad-hoc wireless networking of ubiquitous low-energy sensor/monitor nodes. In *Proceedings of the IEEE Computer Society Annual Workshop on Very Large Scale Integration* (pp. 9-12). Washington, DC: IEEE Computer Society.

Rong, B., & Ephremides, A. (2009). On opportunistic cooperation for improving the stability region with multipacket reception. In *Proceedings of the 3rd Euro-NF Conference on Network Control and Optimization* (pp. 45-59).

Schurgers, C. (2002). *Energy-aware wireless communications*. Unpublished doctoral dissertation, University of California, Los Angeles.

Song, G. (2005). *Cross-layer optimization for spectral and energy efficiency*. Unpublished doctoral dissertation, Georgia Institute of Technology, School of Electrical and Computer Engineering, Atlanta.

Song, G., & Li, Y. (2006). Asymptotic throughput analysis for channel-aware scheduling. *IEEE Transactions on Communications, 54*(10), 1827–1834. doi:10.1109/TCOMM.2006.881254.

Stuber, G. (2001). *Principles of mobile communication*. Boston, MA: Kluwer Academic Publishers.

This work was previously published in the International Journal of Wireless Networks and Broadband Technologies, Volume 1, Issue 1, edited by Naveen Chilamkurti, pp. 73-84, copyright 2011 by IGI Publishing (an imprint of IGI Global).

Chapter 7
BER Fairness and PAPR Study of Interleaved OFDMA System

Sabbir Ahmed
Ritsumeikan University, Japan

Makoto Kawai
Ritsumeikan University, Japan

ABSTRACT

For an OFDMA system, the role of interleavers is analyzed to ensure fairness of BER performance among the active users and investigate their respective PAPR properties. In this paper, the authors consider a generic system and show that for a slowly changing multipath channel, individual user's BER performance can vary, implying that the propagation channel effect is unfairly distributed on the users. Applying different types of frequency interleaving mechanisms, the authors demonstrate that random interleaving can ensure BER fairness on an individual user basis but the associated system overhead for de-interleaving is very high. In this context, the authors introduce the application of cyclically shifted random interleaver and demonstrate its effectiveness in achieving BER fairness (dispersion in individual users BER reduced by 94% compared to no interleaving at 20dB SNR) with little system overhead. The authors also explore the comparative performance of different interleavers for scenarios with varying number of total subcarriers and subcarriers per user. Based on the scenario specific results, the authors conclude that for a heavily loaded system, i.e., relatively low number of subcarriers per user, cyclically shifted random interleavers can effectively ensure uniform performance among active users with reduced system complexity and manageable PAPR.

1. INTRODUCTION

Future generation wireless communication systems demand multiple access schemes with capabilities of high data transmission even in very harsh propagation environment. In this context, Orthogonal Frequency Division Multiple Access or OFDMA, a multiple access scheme based on the well known Orthogonal Frequency Division Multiplexing (OFDM) technique, has drawn significant research interests. OFDM is known for its robust performance against inter-symbol interference (ISI) which is a common phenomenon in high speed data transmission, especially

DOI: 10.4018/978-1-4666-3902-7.ch007

under multipath propagation environment (Prasad, 2004). In addition, since OFDM employs overlapped carriers, its spectrum efficiency is also very good. Considering these, OFDMA appears to be a strong candidate as a future multiple access method. In fact, it has already been standardized for the PHY layer of IEEE 802.16 wireless metropolitan area networks (Marks, Stanwood, & Chang, 2004) and the downlink of 3GPP-LTE (http://www.3gpp.org).

In OFDMA, the task of subcarrier allocation amongst the active users is a crucial resource allocation issue. Based on the concept of OFDM, in OFDMA, the total available bandwidth is at first divided into narrowband orthogonal subchannels each having its own carrier known as subcarriers. Then users are assigned subcarriers (for that matter subchannels also) depending on different criteria. Literature survey reveals that the resource allocation problem is being investigated from different perspectives, e.g., joint power control, rate control and scheduling, data rate, type of data traffic, throughput and delay performances, fairness in throughput achievement, computational complexity, condition of the propagation channel and so forth (Kulkarni, Adlakha, & Srivastava, 2005; Cao, Tureli, & Liu, 2004; Toktas, Biyikoglu, & Yilmaz, 2009).

High peak to average power ratio (PAPR) is another important issue for any OFDM based system (Jiang & Wu, 2008). High PAPR in OFDMA signal means the possibility of BER due to nonlinearity in power amplifier is also high. Due to its significant influence on the system performance, the problem of PAPR from the perspective of subcarrier allocation has also been addressed by some recent research works. In particular, Wang and Chen (2004) analyzed the relationship between asymptotic distribution of peak power of uplink OFDMA system and subcarrier allocation scheme and He, Xiao, and Li (2008) and Xiao, Peng, and Li (2007) proposed low complexity solutions for reducing high PAPR.

In this study, based on our previous works (Ahmed & Kawai, 2010, 2011) considering the downlink of a generic 64-IFFT OFDMA system, we investigate the impact of subcarrier allocation on the physical layer performance of the system on an individual user basis. We show that for a slowly changing multipath propagation environment, i.e., channel co-efficients remaining unchanged over many OFDMA symbols, if subcarriers are allocated on a contiguous basis there may be significant amount of dispersion in the individual users BER performances. We refer to this disparity or non-uniformity as lack of fairness in BER. If BER performance is not fair, it implies that the SNR values required by individual users to attain a common acceptable BER varies over a very wide range, which in turn means the power control scheme needs to be strictly adjusted on an individual user basis. We replace this strict power control requirement by a central frequency interleaving mechanism and show that it can achieve BER fairness amongst the users. We show that interleaving the subcarriers in a randomized fashion achieves better fairness compared to fixed equidistant interleaving. But the associated overhead in the corresponding deinterleaving process is quite high and as such may seem infeasible to some extent. In this context, for achieving BER fairness with little system overhead, we introduce the application of cyclically shifted random interleaver and compare its performance with fixed and random interleavers. Cyclically shifted random interleaver is a modified version of a pure random interleaver that needs less memory space and low-side information transmission. We show that when BER fairness, reduction in SNR requirement for acceptable BER and system overhead all are taken into account, cyclically shifted random interleaver outperforms the others.

In order to look into the scalability issue, we then explore the comparative performance of these interleavers where different pertinent system parameters, e.g., number of total subcarriers, number of active users and number of subcarriers per user assume different values. In this context, at first we investigate a larger system, i.e., higher number of total subcarriers with higher number of users such that the number of subcarriers per user remains constant. Then, we look into the issue where the system moves from the region of low to high number of total subcarriers per users.

Finally, for generic OFDMA, we examine the PAPR property of different interleavers and report that cyclically shifted interleaver has identical PAPR property to that of random interleavers and hence low complexity PAPR reduction scheme like He and Li (2008) and Peng and Li (2007) can be applied without increasing the system complexity too much.

For all the scenarios considered, we conclude by suggesting use of cyclically shifted random interleaver in achieving better BER fairness amongst users compared to other interleavers with low system complexity; and manageable PAPR especially when the system is heavily loaded.

The rest of the paper is arranged as follows. In Section 2, we show the system model and define the different performance parameters. Section 3 explains the concept of different types of interleaving along with their corresponding de-interleaving strategies. Section 4 describes the simulation model, scenarios and lists the important parameter values. In Section 5, performance evaluation is carried out by presenting simulation results along with their explanations. Finally in Section 6, a summary of the presented work along with some comments about future research plan is given.

2. SYSTEM DESCRIPTION

The transmission model of a synchronized downlink OFDMA system is depicted in Figure 1. Here, baseband modulated symbols from every user are at first serial to parallel converted. In Figure 1, $[d_1^K, d_2^K, \ldots, d_M^K]$ denotes the M number of parallel data symbols from the k-th user. In the next stage, the symbols from all the users are fed into

Figure 1. The transmission model for the downlink of OFDMA

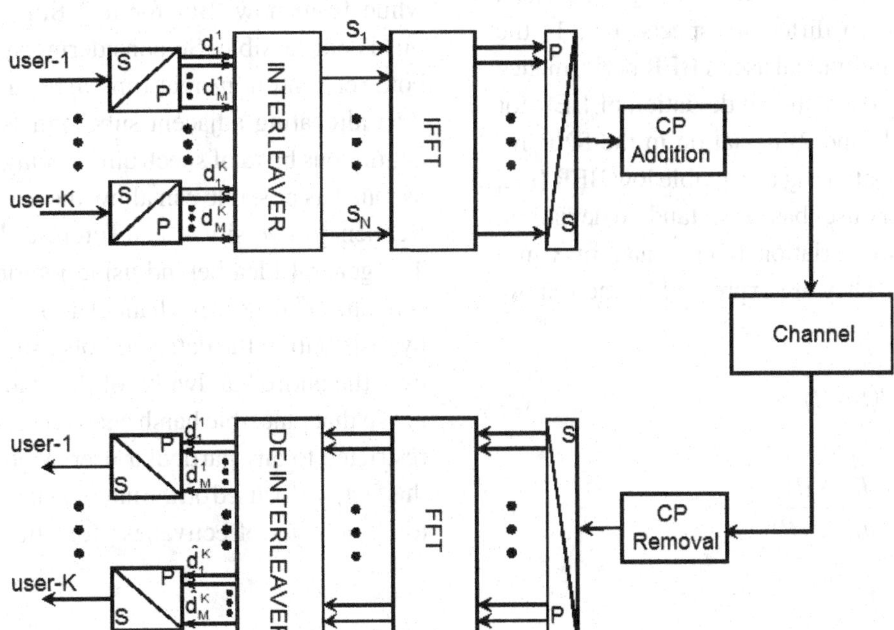

a frequency interleaver block. Subcarrier allocation amongst the users is carried out by this interleaver block. We denote the subcarriers by their indices given by $[S_1, S_2, ... S_N]$, where $N=KxM$. After interleaving, multicarrier modulation is performed through IFFT Then the time domain symbols are converted from parallel to serial. At this stage, the complex baseband OFDMA signal for the k-th user, where the user are allocated a continuous spectrum of subcarrier (referred to as contiguous allocation) can be expressed as given by Equation (1).

$$x^k[n] = \sum_{m=1}^{M} d_m^k e^{j\frac{2\pi}{N}n[(k-1)M+m]} \quad (1)$$

In the next stage, cyclic prefix (CP) is added and the signal is transmitted At the receiver, the reverse operations of the transmitter, i.e., CP removal, serial to parallel conversion, FFT, de-interleaving and finally parallel to serial conversion on individual user basis are done. In Figure 1, $[\hat{d}_1^K, \hat{d}_2^K, \ldots, \hat{d}_M^K]$ represents the estimated modulated symbols of the k-th user.

In order to numerically quantify the notion of fairness in BER performance amongst the users, we consider two different aspects, i.e., 1) the dispersion in individual user's BER performance by estimating the standard deviation of BER for all SNR levels and 2) deviation in the SNR requirement for attaining acceptable low BER (i.e., $<10^{-3}$) for every user based on standard deviation, co-efficient of variation (COV) and max-min ratio of SNR. They are expressed by the following relations

$$SNR_{dev} = std(SNR) \quad (2)$$

$$SNR_{cov} = \frac{std(SNR)}{mean(SNR)} \quad (3)$$

$$SNR_{\max-\min\ ratio} = \frac{\max(SNR)}{\min(SNR)} \quad (4)$$

Finally, PAPR is defined by,

$$PAPR_{dB} = 10\log_{10}\frac{\max|X[n]|^2}{E|X[n]|^2} \quad (5)$$

3. INTERLEAVING TECHNIQUES

The function of subcarrier spacing amongst users is performed by the frequency interleaver. The objective of interleaving stems from the fact that users in different locations may experience different fading characteristics depending on the geographical terrain and nature of obstacles, especially in large macro-cell type of scenarios. This actually implies that some subchannels may suffer more than the others due to the nature of spatial and frequency selective fading. In delay tolerant networks, this problem can be overcome by delaying data transmission of users experiencing bad channels expecting that channels that are bad at this moment probably will be better in a while from now. But for real time application this is not feasible. So considering systems with both real and non- real time applications, other than allocating adjacent subchannels forming a continuous band of spectrum, spacing them apart as much as possible can allow the exploitation of frequency diversity (Cao, Tureli, & Liu, 2004). The general idea behind using interleavers here is to attain fairness in channel allocation to users by distributing the data symbols from every user over the entire bandwidth of the available channel so that plausible harsh channel segment is not restricted to any particular user. Motivated from this fact, we focused on the interleaving mechanism to explore its effectiveness for attaining better performance on an individual user basis. We investigated contiguous allocation along with three types of interleaver as discussed in the following, i.e., 1) fixed interleaver, 2) random interleaver and 3) cyclically shifted interleaver.

A. Contiguous Allocation

Contiguous or sub-band allocation refers to allocating channels to a user where the subcarriers are all adjacent. Though very simple from the implementation perspective, this type of allocation does not exploit the benefits of frequency diversity. Equation 6 represents such allocation scheme.

$$\begin{bmatrix} AS^1 \\ AS^2 \\ \cdots \\ AS^K \end{bmatrix} = \begin{bmatrix} S_1 & S_2 & \cdots & S_M \\ S_{M+1} & S_{M+2} & \cdots & S_{2M} \\ \cdots & \cdot\cdot & \cdots & \cdot\cdot \\ S_{M(K-1)+1} & S_{M(K-1)+2} & \cdots & S_{MK} \end{bmatrix} \tag{6}$$

here, AS^K refers to subcarriers that are allocated to the K-th user.

B. Fixed Equidistant Interleaver

In fixed equidistant interleaving, users are assigned subcarriers in such a manner so that the allocated subcarriers are equally and maximally spaced. The main objective is to achieve maximum subcarrier separation between every data symbol on a per user basis. Referring to Figure 1, the subcarrier allocation can be visualized by the matrix representation given in Equation 6.

$$\begin{bmatrix} AS^1 \\ AS^2 \\ \cdots \\ AS^K \end{bmatrix} = \begin{bmatrix} S_1 & S_{1+K} & \cdots & S_{1+(M-1)K} \\ S_2 & S_{2+K} & \cdots & S_{2+(M-1)K} \\ \cdots & \cdot\cdot & \cdots & \cdot\cdot \\ S_K & S_{K+K} & \cdots & S_{K+(M-1)K} \end{bmatrix} \tag{7}$$

Here the data symbols, of the K-th user, i.e., $[d_1^K, d_2^K, \ldots, d_M^K]$ will be mapped on the subcarriers that are represented with the indices given by $[S_K, S_{2K}, \ldots, S_{MK}]$. In Figure 2(a), the operation of this interleaver is illustrated.

Figure 2. Strategies of interleaving: (a) fixed equidistant (b) random

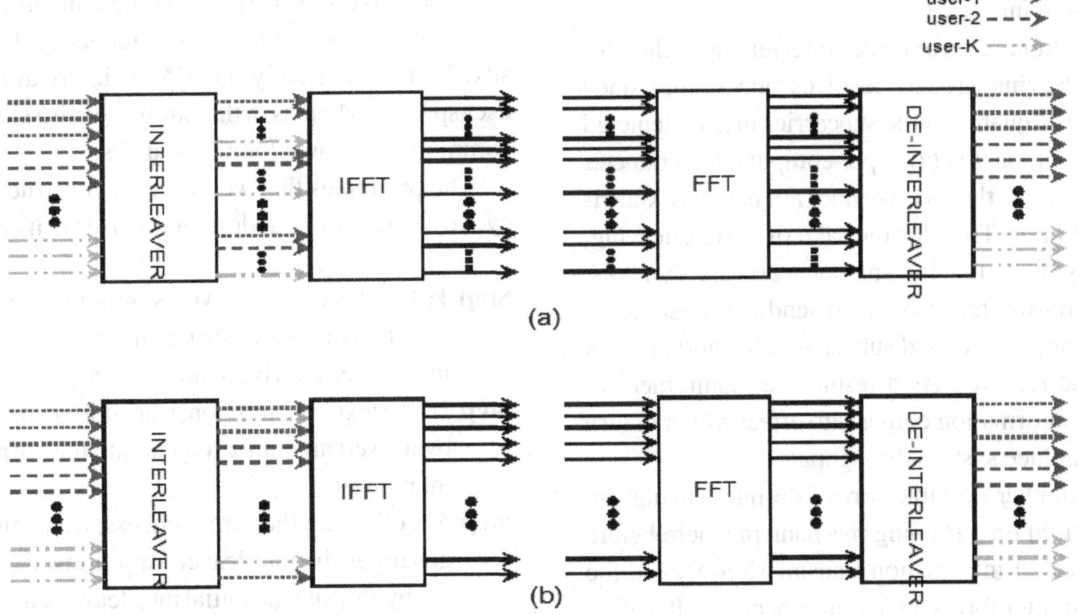

The de-interleaver component on the receiver side uses the matrix of Equation 6 for performing the opposite operation. Since this matrix is a known quantity that depends only on the value of K and M, there is no uncertainly on the receiver side as far as de-interleaving is concerned.

C. Random Interleaver

For random interleaving, the subcarriers of Equation 6 is selected on a random basis for every OFDMA symbol transmission as depicted in Equation 7 where, $r[1]$ to $r[MK]$ are unique integer random values between 1 and MK.

$$\begin{bmatrix} AS^1 \\ AS^2 \\ \cdots \\ AS^K \end{bmatrix} = \begin{bmatrix} S_{r[1]} & \cdots & \cdots & S_{r[M]} \\ S_{r[m+1]} & \cdots & \cdots & S_{r[2M]} \\ \cdots & \cdot\cdot & \cdots & \cdot\cdot \\ S_{r[M(K-1)+1]} & \cdots & \cdots & S_{[MK]} \end{bmatrix} \tag{8}$$

Rather than achieving maximum subcarrier separation, the objective here is to attain fair sharing of the channel by exploiting randomness in the allocation mechanism. Figure 2(b) depicts a possible subcarrier allocation scenario with random interleaver.

Compared to fixed interleaving, the de-interleaving process involves uncertainty since unlike Equation 6 the subcarrier matrix depicted in Equation 7 is not a pre-computable parameter and as such the receiver does not have any knowledge on it. Thus for the sake of de-interleaving, along with the information carrying symbols, the transmitter also has to send the most recent randomly generated subcarrier allocation matrix to the receiver. As a result, the requirement of side-information comes into effect which in turn can reduce system throughput.

Another possible way of de-interleaving can be based on a training mechanism where before the actual information transmission begins; the transmitter forms a random subcarrier allocation matrix, sends it to the receiver along with some pilot data and works on the feedback from the receiver in order to find a suitable subcarrier allocation matrix. This trial may run for a predefined number of times until BER comes down below a given threshold. Once the target BER is achieved, the current subcarrier allocation matrix can be used for a certain time without sending further side information. In this study, we assume that the receiver gets the information of the current subcarrier allocation matrix on every OFDMA symbol transmitted without any error irrespective of the channel state. This may be ensured by using a control channel with proper channel and error correction coding.

D. Cyclically Shifted Random Interleaver

The concept of cyclically shifted random interleaving was first associated with Interleaved Division Multiple Access system (IDMA) (Kusume & Bauch, 2008). But the objective and the method of applying this interleaver in this study is different from its implementation in IDMA. For example, in IDMA, the main objective of using interleaver is to perform user separation but here in our case we use it as a tool for achieving frequency diversity. More importantly, in IDMA, interleaver is user specific whereas in this study we use a single common interleaver for all the users.

The principles that we follow in constructing cyclic shifting of a random interleaver is listed:

Step 1: On the first OFDMA symbol transmission, transmitter constructs the initial random interleaver matrix, denoted by π

Step 2: On next transmission, transmitter shifts π by a fixed amount τ to generate the current interleaver π_1

Step 3: Following the same process, transmitter generates the interleaver for n transmission π_{n-1} by shifting the initial interleaver π by τ_{n-1}

Now for de-interleaving the initial random interleaver matrix is needed by the receiver which is sent only on the first transmission and the current shifting value τ_{n-1} on every subsequent transmission. We assume that they are received by the receiver without any error. Compared to random interleaver scheme, since the on and transmission of the random interleaver matrix is needed only once, it reduces computational overhead, memory and storage space requirement and amount of side information. As a result, the overall system complexity along with probability of errors in de-interleaving is also reduced. As an example for side information reduction issue, for a system with 4 users (i.e., K=4) and 4 symbols/user/OFDMA symbol (i.e., M=4), with 4-bit representation of subcarrier indices we would require 64-bits to represent the entire interleaving matrix. Thus 64-bits side information needs to be transmitted on every OFDMA symbol transmission when pure random interleaving is used. However, for cyclically shifted random interleaving, this 64-bits side information needs to be sent only on the first OFDMA symbol transmission. For second transmission, shift amount information τn-1 consisting of a maximum of log2MK, i.e., 4-bits needs to be transmitted. Similarly for every subsequent transmission at most 4-bits of side information is essential. As a result, if we consider a communication session consisting of 100 OFDMA symbols, random interleaver requires 100x64=6400-bits of side information whereas cyclically shifted random interleaver only requires 64+(100-1)x4=460-bits which implies around 92% storage reduction.

4. SIMULATION SCENARIOS

Based on the transmission model depicted in Figure 1 we carried out our simulation We considered a slowly changing 3-tap For it, we allowed the randomly generated channel coefficient values to change only after every 25% of the total number of OFDMA symbol transmission. By this model, we represent a propagation environment where the number of obstacles is very few and relative movement of receivers and obstacles is also negligible (Goldsmith, 2005). On the receiver side, we assumed that perfect channel state information is available and zero forcing equalizer is employed to further mitigate inter-symbol interference. For fairness study, we first considered a generic 64-IFFT OFDMA system with 16 users. Then we looked in to the issue when the system becomes large from the perspective of higher number of users but with the same data rate per user as before. For this we considered increasing the number of users along with the number of total subcarriers keeping the number of subcarriers per user same. Finally, we investigated the situation where the number of subcarriers per user changes. For this we considered increasing the total number of subcarriers keeping the number of users constant. The study scenarios mentioned above are briefly listed:

Scenario 1: Generic 64-IFFT system with a fixed number of total 16 users.

Scenario 2: The number of total subcarriers (64, 128 and 512) along with users (16, 32 and 128) changes such that number of subcarriers per user remain same.

Scenario 3: The number of total subcarriers changes (64, 128 and 512) while the number of users remain same (16), i.e., the number of subcarriers per user changes.

The simulation parameters are listed in Table 1.

5. PERFORMANCE EVALUATION

We organize our simulation results into 2 parts. In the first part, we show the results associated with fairness issue for the 3 different scenarios mentioned in the previous section. And in the second part we show the PAPR related observations.

Table 1. Simulation parameters

Modulation scheme	QPSK
Number of subcarriers	64, 128, 512
Number of users	16, 32, 128
Channel	3 Tap slowly changing
Length of CP	25% of OFDMA symbol length
Equalizer	Zero forcing

A. Fairness in Bit Error Rate

Scenario 1

The BER simulation result is shown in Figure 3 where the effects on individual user's BER are depicted for contiguous allocation, fixed interleaving, random interleaving, and cyclically shifted random interleaving respectively. Every curve on each of the four plots represents the BER performance of a single individual user. From this figure, in general, it is noticeable that for contiguous subcarrier allocation, the BER performance curves are very widely spread, i.e., individual user performance varies over a wide range. For example, for an acceptable BER value of 10-3, there is a difference of about 24dB in SNR values between the best and the worst user performances. This scenario can be explained by considering a situation where a single or a group of users is geographically located in such a manner that the allocated subchannels experience fading which is both adverse in magnitude and long in duration compared to all other users. Thus users are getting unfair channel condition. Now, taking a look on the plots depicting results of any of the three interleaving schemes, it is evident that the deviation of individual user performances from each other is much less here. And also, the worst individual BER performance is either equal or better than that of no interleaving. Though it comes at the cost of a degraded BER performance for

Figure 3. (Case:1) Bit Error Rate performance of different interleaving mechanisms

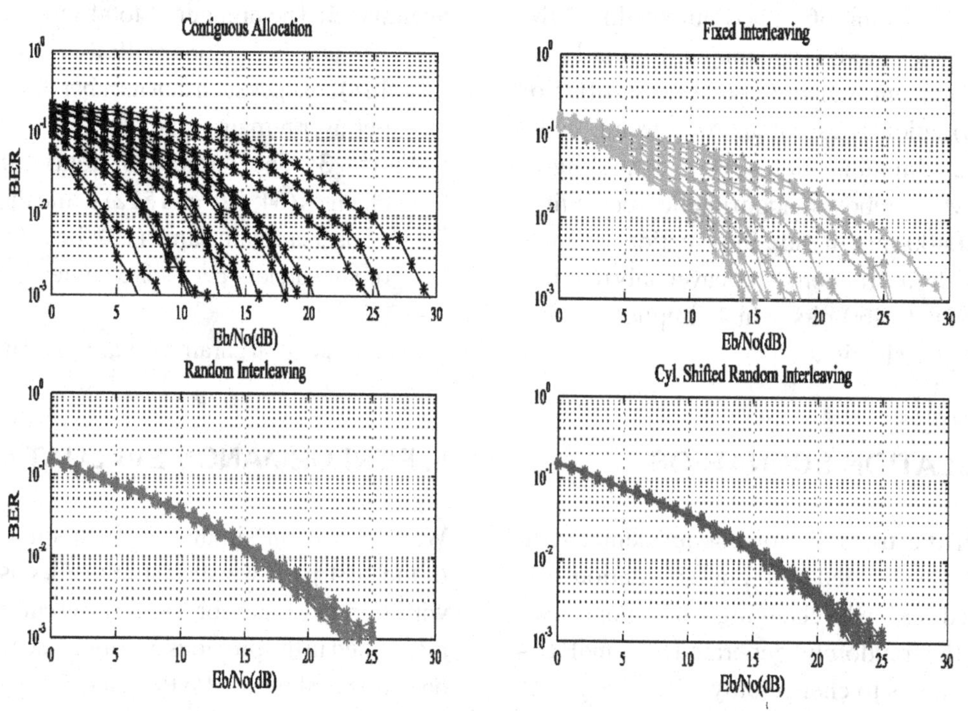

mostof the other users compared to the case of no interleaving. This suggests that for a multipath channel affecting a particular user's or a group of users' subchannels more adversely than the others, interleaving improves the BER fairness amongst the users as far as channel allocation is concerned. Again comparing BER performance for the three interleaving schemes, it is obvious that both random and cyclically shifted random interleavers are superior compared to fixed interleaving.

The data of Figure 3, Figure 4, and Figure 5 depict the fairness issue in BER performance from two different perspectives, i.e., dispersion in the BER of individual user and variation in SNR for achieving acceptable BER respectively. Figure 4 shows the standard deviation of individual user's BER at every level of SNR. It is obvious that the dispersion in individual performances is very high for no interleaving or fixed interleaving compared to random and cyclically shifted random interleaving. For example, at an SNR level of 20dB, the standard deviation values for no interleaving and cyclically shifted interleaving is around 0.0113 and 4.4422×10^{-4} respectively which means the later has been reduced by approximately 94%.

Now, it is more practical to look into the situation where individual user's BER comes down under an acceptable value, e.g., 10^{-3}. To achieve this BER, if the required SNR varies over a wide range from user to user, it implies that from the perspective of its condition, channel has not been allocated in a fair manner. Figure 5 depicts the relative performance of different interleaving schemes from this point of view. It shows the standard deviation, co-efficient of variance and the max-min ratio of SNR in achieving a BER ≤10-3. Just for clarification the max-min ratio is the ratio of SNR required by the worst performing user and the best performing user in achieving the target BER of ≤10-3. As can be seen from this figure, both no interleaving and fixed interleaving have considerably higher values for all three parameters compared to random or cyclically shifted random interleaving.

Figure 4. (Scenario:1) Individual user's Bit Error Rate performance deviation

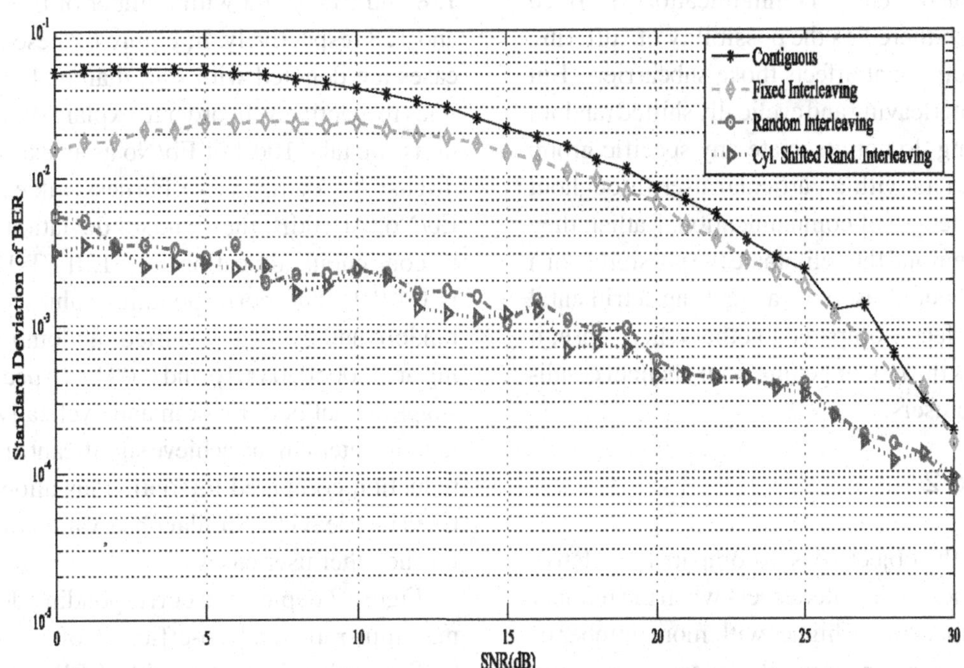

Figure 5. (Scenario:1) Signal to Noise Ratio variation

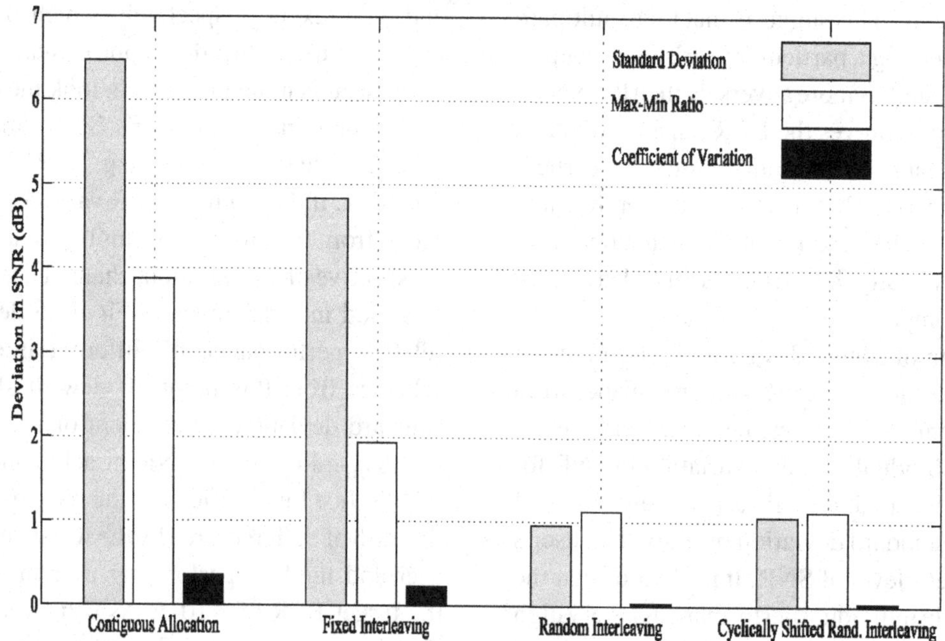

The superior performance of random or cyclically shifted random interleaving over fixed interleaving can be attributed to the fact that for the later, subcarriers allocated to a particular user throughout the entire communication are fixed and as such there lies the possibility of suffering from a fading that affects those subcarriers. But random interleaving and cyclically shifted random interleaving do not associate any specific group of subcarriers with a particular user throughout the entire length of communication. Rather, they change them and thus eliminate the possibility of a user or a group of users always getting detrimental channel effects compared to others. In a sense, it distributes the risk of getting poor channel condition on all users.

Scenario 2

In case 2, the objective is to compare the relative performance of the interleavers when the number of total subcarrier is higher with more number of users. This scenario is similar to case 1, the only difference being higher number of total users is being considered. For this, we consider systems with different IFFT sizes with different number of users such that the number of subcarriers per user remains same. We consider IFFT sizes of 64, 128 and 512 along with number of users 16, 32 and 128 respectively. In Figure 6, these different cases are denoted with star, diamond and circle lines respectively. In order to explain the observations, we take 10dB of Eb/No as an example and discuss the comparative fairness values for this Eb/No. At 10dB, the standard deviation of BER of contiguous allocation for 512IFFT-128Users is 0.2191. The corresponding values of fixed, random and cyclically shifted random interleaving are 0.0050, 0.0016 and 0.0017 respectively. It suggests that both random and cyclically shifted random interleaving achieve significant dispersion reduction compared to contiguous allocation or fixed interleaving. Similar observations are found for the other user cases.

Figure 7 depicts the corresponding deviation, max-min ratio and co-efficient of variation in SNR in achieving acceptable BER, i.e., in the region of ≤10-3 for 128IFFT-32Users and

Figure 6. (Scenario:2) Individual user's Bit Error Rate performance deviation

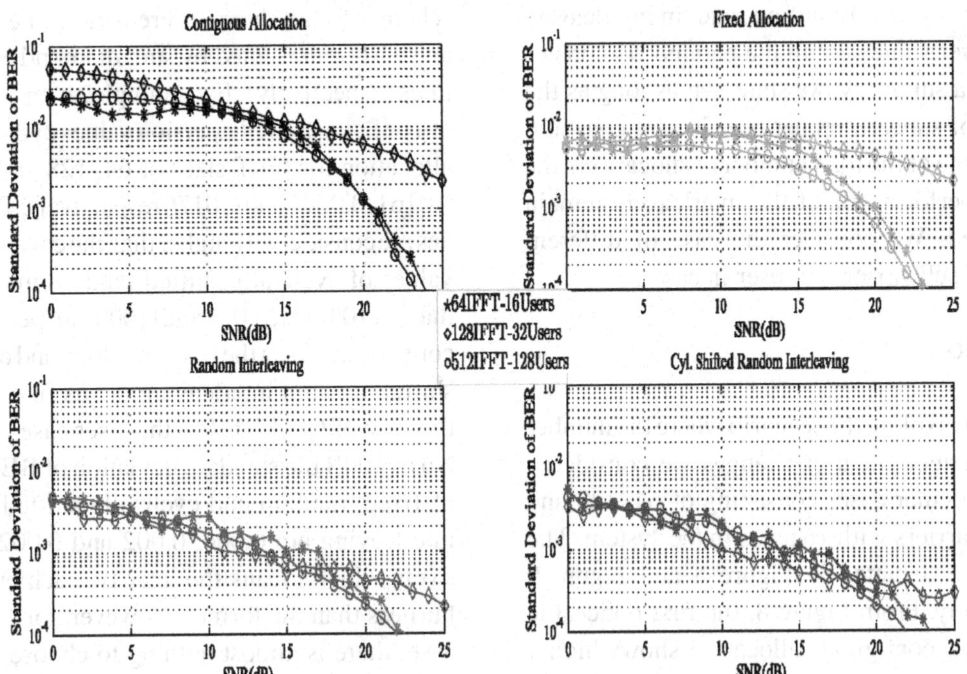

Figure 7. (Scenario:2) Signal to Noise Ratio variation

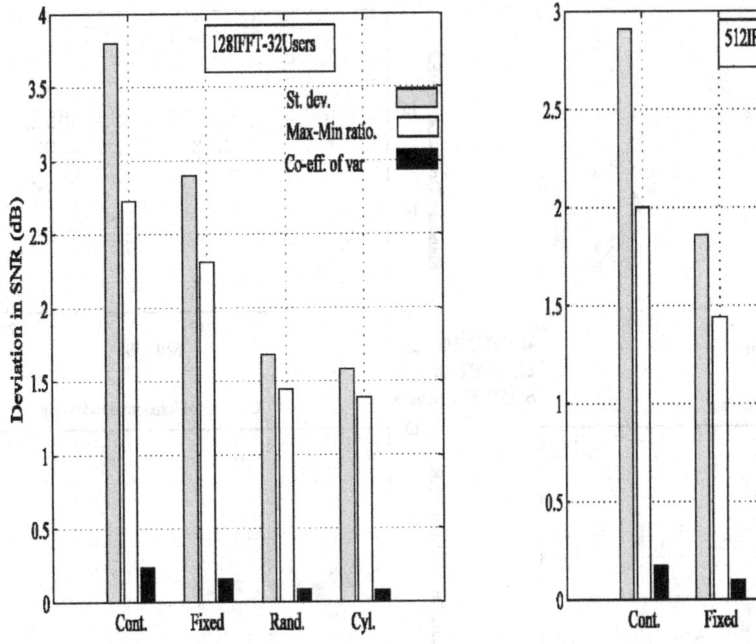

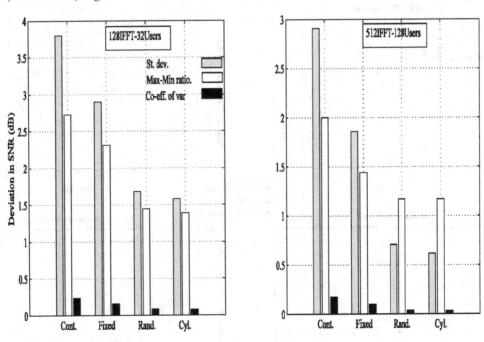

512IFFT-128Users cases. As seen from this figure random or cyclically shifted random interleavers show lower values for all the parameters considered. As a summary, we state that as long as the number of subcarriers per user is kept constant, irrespective of the total number of subcarriers the relative performance of the interleavers remain unchanged. We consider the case of different values of subcarriers per user in case 3.

Scenario 3

In order to explore the scenario where the number of subcarriers per user changes we considered three particular cases consisting of 64, 128 and 512 subcarriers with 16 users in the system. The results are depicted in Figure 8 and Figure 9 respectively. From Figure 8, the first noticeable fact is that contiguous allocation shows higher dispersion of BER compared to all interleaving schemes for all three different cases, i.e., 64, 128 and 512 (represented by star, diamond and circle lines respectively). For example, at an SNR level of 10dB, the BER dispersion values of contiguous allocation for IFFT sizes of 64, 128 and 512 are 0.0161, 0.0227 and 0.0176 respectively. These values are considerably higher than the corresponding values of cyclically shifted random interleaving that read 0.0022, 0.001 and 0.0004 respectively. To compare amongst the fixed, random, and cyclically shifted random interleaving, let us first focus on the case of 64-IFFT. For this user case, at an Eb/No of 10dB the standard deviation of BER values of fixed, random and cyclically shifted random interleaving are 0.006, 0.002 and 0.002 respectively which means the later two achieve better fairness than the former. However, for 128-IFFT case, there is almost nothing to choose between

Figure 8. (Scenario:3) Individual user's Bit Error Rate performance deviation

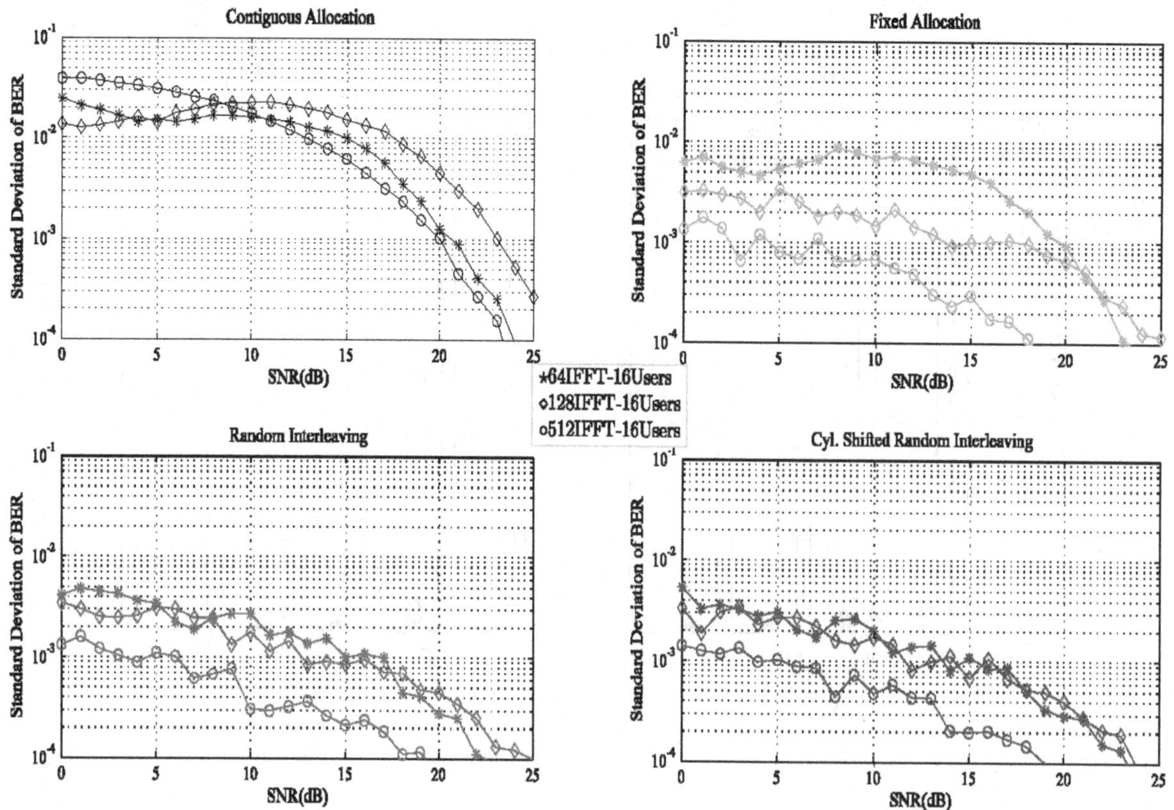

Figure 9. (Scenario:3) Signal to Noise Ratio variation

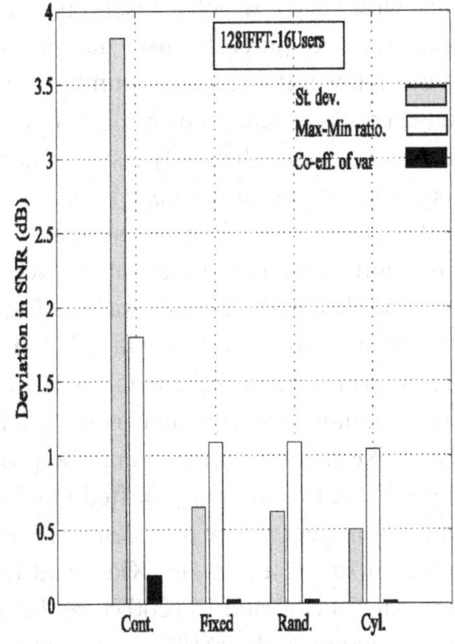

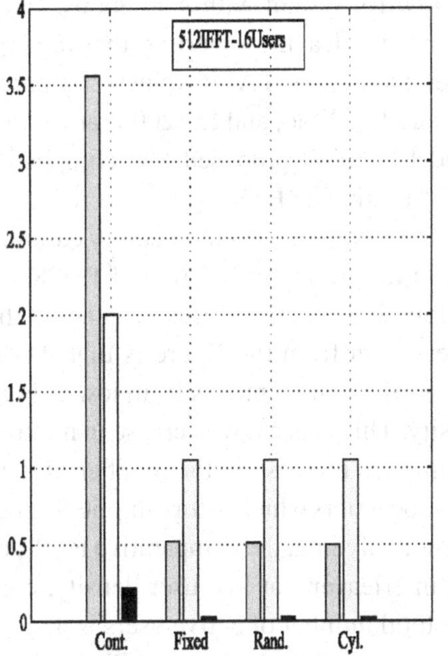

these three interleavers as the deviation values are very close, i.e., 0.0014, 0.0017 and 0.0017. And for 512-IFFT case, the same results are seen is almost nothing to choose between these three interleavers as the deviation values read 0.0006, 0.0003 and 0.0004 respectively. Thus it implies that as the number of subcarriers per user is low or in other terms number of users is very high, random and cyclically shifted random interleavers perform better than fixed interleaving. But when the number of subcarriers per user increases all interleavers tend to perform almost the same. One reason for this could be when the total number of subcarriers per users increases the possibility of some subcarriers getting bad channel effect is evenly distributed on all users.

Similar observation is found from Figure 9 where dispersion in SNR is shown for 128 and 512 IFFT cases. As we can see contiguous allocation shows higher values of standard deviation, co-efficient of variance and the max-min ratio of SNR compared to all the interleaving schemes. And all the interleavers have almost same type of performances.

The summary from the fairness study results is that, for all scenarios, any kind of interleaving ensures uniform BER performance compared to that of contiguous allocation. And amongst the three interleaving schemes, considerable difference can be observed when the system has relatively low number of subcarriers per user. That means if the total number of subcarriers is fixed, which is a practical situation, individual users may experience non-uniform performance as more and more users join. In this situation random and cyclically shifted random interleavers perform better in attaining BER fairness compared to fixed interleaving. However, when number of subcarriers per user is high all three interleaving schemes show almost similar performance.

B. Peak to Average Power Ratio

We also examined the comparative PAPR of different interleavers at different levels of user densities. Unlike our focus of the downlink, some previous related studies dealt with uplink single user case, e.g., Wang and Chen (2004) analyzed the asymptotic distribution of peak power of uplink OFDMA system with contiguous, fixed and random interleaving reporting that random interleaver has higher PAPR values compared to others and He, Xiao, and Li (2008) and Xiao, Peng, and Li (2007) proposed low complexity solutions for reducing PAPR.

Figure 10 shows the complementary cumulative distribution function (CCDF) of PAPR for selected four different user cases. In general, the first observation from the figure is that PAPR increases as the system moves from low to high user density. This is because increase in number of users implies increase in the number of data carrying subcarriers which in turn implies higher PAPR Now analyzing, the comparative PAPR of different interleavers, at low user density, e.g., when the total number of active users is 4, contiguous allocation has the least PAPR whereas random and cyclically shifted random interleavers have almost the same PAPR distribution which is higher than the others dealing with the uplink single user scenario. This is because in downlink with very few active users, the number of subcarriers required for user data transmission is much lower than the total number of available subcarriers- a situation similar to the uplink single user case when only a small chunk of the total number of subcarriers is used at a time. But as more users join the system, the PAPR distributions of all the four interleaver schemes become close and eventually same at fully loaded condition. This is due to the fact that the level of freedom in choosing subcarriers gets diminished since the number of free subcarriers reaches zero. Thus like random interleaving, the problem of high PAPR for cyclically shifted interleaving is only prominent at low user density. We feel the solution suggested in He, Xiao, and Li (2008) and Xiao, Peng, and Li (2007) can be effective in managing high PAPR of cyclically shifted random interleavers also, since their PAPR is similar. PAPR at varying user density is shown in Figure 10.

Figure 10. Peak to Average Power Ratio at varying user density

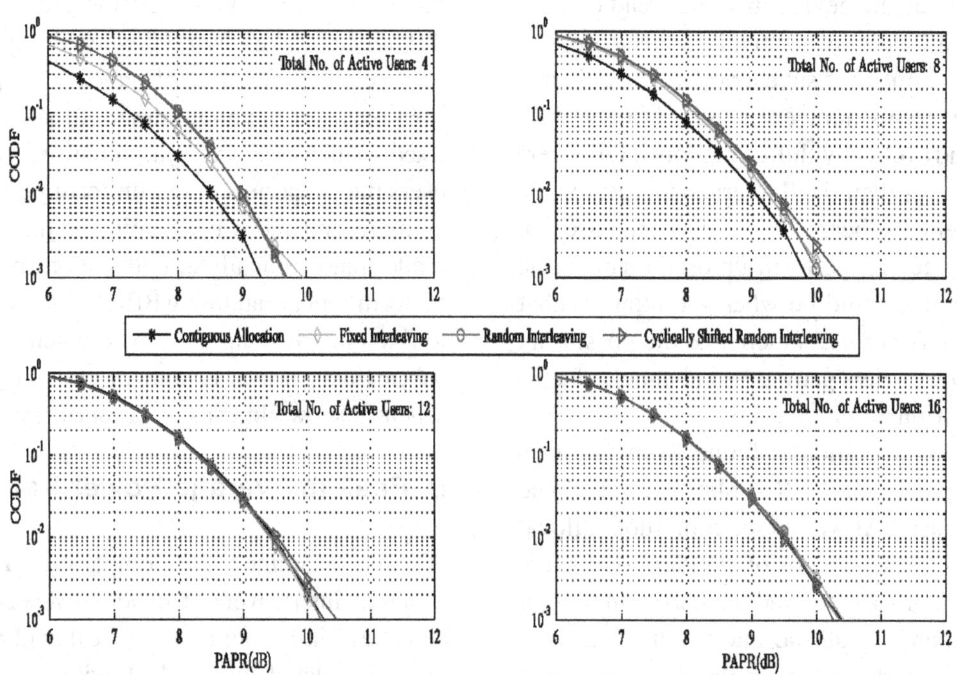

CONCLUSION

Considering an OFDMA system, the effect of subchannel interleaving amongst users on the individual users' BER performance was analyzed. For a slowly changing multipath channel, it was shown that individual users can be susceptible to unfair channel condition resulting in widely varying BER performances. By presenting the performance of different types of interleaving schemes in different type of scenarios, we demonstrated that interleaving in general plays a positive role in ensuring uniformity amongst individual user's BER performance. In addition we also showed that when the system is heavily loaded random and cyclically shifted random interleavers perform better than fixed interleaver. But when associated system overhead is also taken into account, we affirm that cyclically shifted interleaver is the best choice for attaining BER fairness amongst the active users.

In future, we are interested in exploring the possibilities of applying spreading code correlation characteristics for constructing robust interleaver permutation tables that can ensure improved overall performance. We would also like to undertake practical experiments in order to verify the simulation results.

ACKNOWLEDGMENT

This work was supported in part by Global COE Program for Education, Research and Development of Strategy on Disaster Mitigation of Cultural Heritage and Historic Cities, MEXT, Japan.

REFERENCES

Ahmed, S., & Kawai, M. (2010). Interleaver-based subcarrier allocation schemes for BER fairness in OFDMA. In *Proceedings of the 9th International Conference on Wireless Networks* (pp. 262-266).

Ahmed, S., & Kawai, M. (2011). Effects of interleavers on BER fairness and peak to average power ratio in OFDMA System. In *Proceedings of the 6th International Symposium on Wireless and Pervasive Computing* (pp. 1-6).

Cao, Z., Tureli, U., & Liu, P. (2004). Optimum subcarrier assignment for OFDMA uplink. In *Proceedings of the Thirty-Seventh Asilomar Conference on Signals, Systems and Computers* (Vol.1, pp. 708-712).

Goldsmith, A. (2005). Path loss and shadowing. In Goldsmith, A. (Ed.), *Wireless communications* (pp. 27–41). Cambridge, UK: Cambridge University Press.

He, X., Xiao, Y., & Li, S. (2008). PAPR reduction of random interleaved uplink OFDMA system. In *Proceedings of the 3rd International Conference on Communications and Networking in China* (pp. 478-481).

Jiang, T., & Wu, Y. (2008). Overview: Peak-to-average power ratio reduction techniques for OFDM signals. *IEEE Transactions on Broadcasting*, *54*(2), 257–268. doi:10.1109/TBC.2008.915770.

Kulkarni, G., Adlakha, S., & Srivastava, M. (2005). Subcarrier allocation and bit loading algorithms for OFDMA-based wireless networks. *IEEE Transactions on Mobile Computing*, *4*(6), 652–662. doi:10.1109/TMC.2005.90.

Kusume, K., & Bauch, G. (2008). Simple construction of multiple interleavers: Cyclically shifting a single interleaver. *IEEE Transactions on Communications*, *56*(9), 1394–1397. doi:10.1109/TCOMM.2008.060420.

Marks, B. R., Stanwood, K., & Chang, D. (2004). *ANSI/IEEE std. 802.16-2004: Local and metropolitan area networks part 16: Air interface for fixed broadband wireless access systems (Revision of IEEE Std. 802.16-2001)*. Washington, DC: IEEE Computer Society.

Prasad, R. (2004). Basics of OFDM and synchronization. In Prasad, R. (Ed.), *OFDM for wireless communications systems* (pp. 117–147). Boston, MA: Artech House.

Toktas, E., Biyikoglu, E.-U., & Yilmaz, A. O. (2009). Subcarrier allocation in OFDMA with time varying channel and packet arrivals. In *Proceedings of the European Wireless Conference* (pp. 178-183).

Wang, H., & Chen, B. (2004). Asymptotic distributions and peak power analysis for uplink OFDMA signals. In. *Proceedings of the IEEE International Conference on Acoustics, Speech, and Signal Processing*, *4*, 1085–1088.

Xiao, Y., Peng, Y., & Li, S. (2007). PAPR reduction for interleaved OFDMA with low complexity. In *Proceedings of the 6th International Conference on Information, Communications and Signal Processing* (pp. 1-4).

This work was previously published in the International Journal of Wireless Networks and Broadband Technologies, Volume 1, Issue 2, edited by Naveen Chilamkurti, pp. 1-15, copyright 2011 by IGI Publishing (an imprint of IGI Global).

Chapter 8
Lifetime Maximization in Wireless Sensor Networks

Vivek Katiyar
National Institute of Technology Hamirpur, India

Narottam Chand
National Institute of Technology Hamirpur, India

Surender Soni
National Institute of Technology Hamirpur, India

ABSTRACT

One of the fundamental requirements in wireless sensor networks (WSNs) is to prolong the lifetime of sensor nodes by minimizing the energy consumption. The information about the energy status of sensor nodes can be used to notify the base station about energy depletion in any part of the network. An energy map of WSN can be constructed with available remaining energy at sensor nodes. The energy map can increase the lifetime of sensor networks by adaptive clustering, energy centric routing, data aggregation, and so forth. In this paper, the authors describe use of energy map techniques for WSNs and summarize the applications in routing, aggregation, clustering, data dissemination, and so forth. The authors also present an energy map construction algorithm that is based on prediction.

1. INTRODUCTION

A lot of applications require acquisition of data from physical environment. This leads to development of a new kind of network of tiny sensor nodes. These sensor nodes have ability of sensing, transmitting and forwarding data wirelessly to another node of same or different type. These types of networks are known as wireless sensor networks (WSN). A lot of research can be seen in WSNs like development of energy efficient routing protocol, clustering, data aggregation, etc. Battle field surveillance (Bokareva et al., 2006) habitat monitoring (Hart & Martinez, 2006) medical (Yan, Xu, & Gidlund, 2009) smart home (Hussain, Schaffner, & Moseychuck, 2009) and sports (Espina, Falck, Muehlsteff, Yilin, Adan, & Aubert, 2008) are some of the major application areas of sensor networks.

DOI: 10.4018/978-1-4666-3902-7.ch008

In WSNs, the sensor nodes are very constrained in terms of battery power. Sensor nodes in WSNs have non-rechargeable batteries. At the same time, it is not easy to replace batteries because WSNs are deployed generally in inhospitable environments like forests, sea and battlefields. The only way to make a WSN alive for longer time is to make the efficient use of available battery power of sensor nodes. Power optimization must be taken into account at each layer of network model including physical and application layer. Since a large fraction of the energy of a sensor node is consumed in data transmission, so most of the energy efficient protocols are designed at network layer. A large number of protocols for energy efficient clustering (Heinzelman, Chandrakasan, & Balakrishnan, 2002; Liu & Lin, 2005), routing (Al-Karaki & Kamal, 2004; Shah & Rabaey, 2002; Al-Karaki, Ul-Mustafa, & Kamal, 2004), and data aggregation (Rajagopalan & Varshney, 2006; Kalpakis, Dasgupta, & Namjoshi, 2003) exist in the literature. Heinzelman et al. (2002) proposed an energy efficient clustering protocol that selects clusterheads based on probability. In Liu and Lin (2005) authors introduced a re-clustering strategy and a redirection scheme for cluster-based WSNs in order to address the power-conserving issues in such networks, while maintaining the merits of a clustering approach. A good survey on the routing algorithms for WSNs has been presented in Al-Karaki and Kamal (2004). Some examples of the energy efficient routing algorithms are Energy Aware Routing (EAR) (Shah & Rabaey, 2002) and Virtual Grid Architecture routing (VGA) (Al-Karaki, Ul-Mustafa, & Kamal, 2004). Rajagopalan et al. (2006) presented data aggregation techniques for WSNs. Kalpakis et al. (2003) have proposed maximum lifetime data gathering with aggregation (MLDA) algorithm to obtain data gathering schedule with maximum lifetime where sensors aggregate incoming data packets.

To optimize the energy consumption in sensor nodes, Zhao et al. (2002) have designed a residual energy scan for whole sensor network to monitor the energy consumption in every part of the sensor network. Mini et al. (2004, 2005) extended their work of energy scan construction and named it as Energy map. An energy map is a scan of available remaining energy at sensor nodes in WSNs. Energy maps can also be useful in increasing lifetime of sensor network by adaptive clustering, energy centric routings, data aggregation, etc. With the help of energy map we can determine if any part of the sensor network is about to fail in near future due to depleted energy. In Zhao, Govindan, and Estrin (2002) authors described the aggregation based approach for energy map construction. In this approach, a composite scan is created by combining all local scans by sensor nodes. Another approach for energy map construction used in Mini et al. (2004, 2005) is based on prediction. In prediction based approach, a sensor node can predict its energy consumption based on its past history. Based on that prediction, energy map can be constructed. In the mechanism proposed by Mini et al. (2004, 2005) every sensor node need not to send energy information to the monitoring node, it can just send its available energy and parameters of energy consumption model. In this way authors minimized the cost of energy map construction. Song et al. (2006) proposed an energy map construction approach based on non-linear manifold learning algorithm. In general, we should devise some efficient methods for energy map construction so that utility of the energy information compensate the amount of energy spent in this process.

Energy map is a very important phenomenon in the context of WSN because the energy information is usually very crucial to develop a good sensor network protocol stack such as clustering algorithms and routing protocols. In Song,

Guizania, and Sharif (2007) authors presented a clustering algorithm that takes clustering decision based on energy information in the sensor network. The algorithm also dynamically selects the clusterheads based on residual energy information. A routing algorithm named Energy Centric Routing (ECR) has been proposed by Al-Karaki et al. (2007). This algorithm takes routing decision based on residual energy in the network. The work proposed in Niculescu and Nath (2002) is an example of a routing protocol that could take advantage of the energy map. In this work author described the trajectory based forwarding protocol that is a new forwarding algorithm suitable for routing packets along a predefined curve. Many other applications of energy map can be found in data gathering (Goussevskaia et al., 2005) and data dissemination (El Rhazi & Pierre, 2007). Some other possible applications are reconfiguration algorithms, query processing and data fusion that could take advantage of the energy map. In fact, we cannot think of an application or an algorithm that is not benefited with the use of energy map.

This paper presents applications of energy map in various fields of WSNs like data aggregation, routing, clustering, data dissemination, etc. By using the energy map, a user may be able to determine if any part of the network is about to suffer in the near future due to depleted energy (Zhao, Govindan, & Estrin, 2002). We also propose an energy map construction algorithm for WSN. If a sensor node can predict its future energy consumption based on past history, the clusterhead will be able to prolong the lifetime of sensor node by distributing load. But the effectiveness of this concept depends on how accurate the prediction is.

Remainder of the paper is organized as follows. Section 2 describes the energy map. Approaches for energy map construction like prediction based, aggregation based, statistical based, etc. are discussed in Section 3. Section 4 describes the applications of energy map in various fields of WSN. Section 5 presents the energy map construction algorithm that is based on prediction. We conclude the paper in Section 6.

2. ENERGY MAP

Just like a weather map or air traffic radar images, a scan of wireless sensor network can describe the geographical distribution of network resources or activity of a sensor field. We can also draw an energy scan for a sensor network in which we group the sensor nodes according to their residual energy. This energy scan is also known as energy map. An energy map for a typical WSN is shown in Figure 1. In other words, an energy map is a scan of available remaining energy at sensor nodes in WSNs. Energy map can help in detecting the network failure due to depleted energy in next rounds. In this way, the energy map can be useful in deployment of additional nodes in the regions where the energy of the sensors is likely to deplete soon.

A node is assigned to collect the scans from all parts of the sensor network. This node is known as monitoring node. Selection of monitoring node depends upon the application. Generally speaking the selection of monitoring node is done from the region having comparatively more residual energy. Nodes near the monitoring node probably spend more energy because they are used more frequently to relay packets to the monitoring node. Energy map construction methods should be energy efficient so that utility of the energy information compensates the amount of energy spent in this process.

Figure 1. Energy map for a typical wireless sensor network

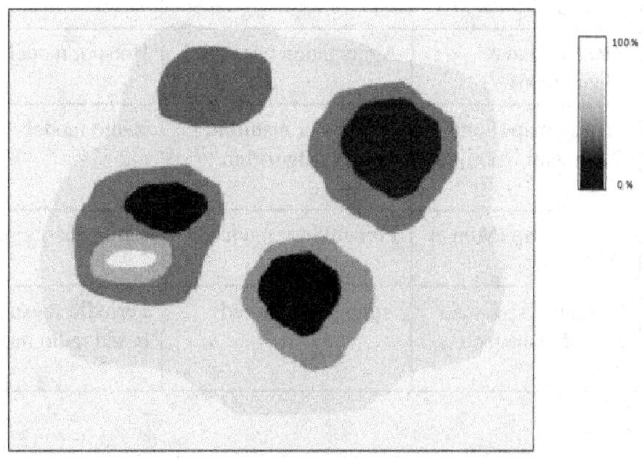

3. CONSTRUCTION OF ENERGY MAP

In this section we describe the popular and recent approaches for energy map construction such as aggregation based, prediction based, statistical based, contour map based, etc. Residual energy scan has been computed by many authors with different names as eScan (Zhao, Govindan, & Estrin, 2002), Energy map (Mini et al., 2004, 2005; Song & Guizani, 2006), Continuous Residual Energy Monitoring (CREM) (Han & Chan, 2005) and Iso-Map (Li & Liu, 2010).

In this section we present a detailed survey of approaches along with their strengths and shortcomings. The survey is also summarized in Table 1.

Residual Energy Scan

Zhao et al. (2002) first tried to design a monitoring scheme for sensor networks in the same manner as SCAN (Reddy, Estrin, & Govindan, 2000) that provides a multicast based continuous monitoring infrastructure. This scheme is based on residual energy scan for monitoring sensor network health. Any network failure due to energy depletion can be avoided. An aggregation based approach is used for energy map construction. First each node calculates its local scan with residual energy and its location. Sensor node reports to the monitoring node only when its energy level drops significantly since last time it reported its scan. All local scans are aggregated with help of aggregation tree. Figure 2 shows the aggregated scan from two local scans. Aggregated scan actually is a tuple consisting of nodes forming polygon and residual energy level. Authors have also designed some aggregation rules for in-network aggregation.

The pioneering work of Zhao et al. can be thought of as revolution in the field of energy efficient WSNs, however some problem exists. Sometimes there may be so much message exchange for energy scan construction that will not justify the energy saved because there was no

Table 1. Energy map construction

Algorithm	Approach for construction	Energy dissipation model	Energy data collection	Topology maintenance	Location awareness
eScan (Zhao et al., 2002)	Aggregation based	Uniform and Hotspot model	Local	Yes	Yes
Energy map (Mini et al., 2005)	Prediction based	State-based model	Global	No	No
Energy map (Mini et al., 2005)	Sampling based	State-based model	Global	No	No
Iso-Map (Li & Liu, 2010)	Contour map based	Contour map based	Global	Yes	No
Energy map (Mini et al., 2004)	Statistical based	State-based model	Global	No	No
CREM (Han & Chan, 2005)	Aggregation based	Hotspot model	Local	Yes	Yes
Energy map (Song & Guizani, 2006)	Non-linear manifold learning algorithm based	Radio model	Global	Yes	Yes
Energy map (Mini et al., 2004)	Probabilistic model	Finite energy model	Global	Yes	No
ECscale (Al-Karaki & Al-Mashaqbeh, 2007)	Aggregation based	Periodic sensing based radio model	Local	Yes	Yes

Figure 2. Representation and aggregation of energy scans (Zhao, Govindan, & Estrin, 2002)

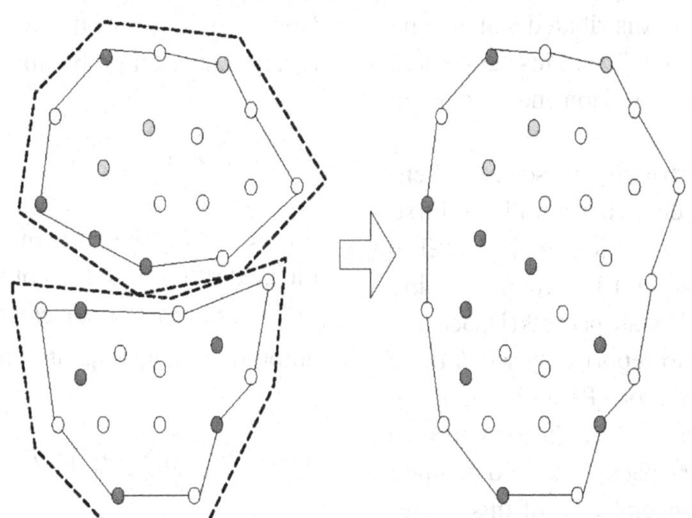

clear cut assumption of hierarchal structure. Moreover there is no topology control mechanism. Nodes near to the base station will consume more energy. To overcome these shortcomings and reduce the cost of collecting residual energy scan of sensor network, the authors proposed a monitoring tool called *digest* which is an aggregate of some network properties (Zhao, Govindan, & Estrin, 2003).

Another improvement in eScan approach has been proposed by Han and Chan (2005) that follows a hierarchical approach for continuously collecting residual energy scans to construct the final scan at base station. Continuous Residual Energy Monitoring (CREM) divides whole sensor network into clusters, represented by clusterheads. A backbone is constructed with the help of clusterhead nodes. In this approach a local energy scan is constructed per cluster basis. Then, with the help of this backbone (topology tree) all local scans are collected at base station and final energy scan is constructed. One of the major advantages of CREM over previous approaches is topology maintenance. Topology maintenance is very much important because clusterhead nodes have to transmit and receive continuously for energy scan construction and hence will deplete its energy resource at a much higher rate.

Aggregation base approaches generally suffer from heavy transmission traffic and sometimes a large computational overhead on each sensor node. Li and Liu (2010) proposed an energy efficient contour mapping scheme named Iso-Map. Iso-Map constructs a contour map that reduces the network traffic and computation overhead by selecting some nodes intelligently to report energy data. These nodes are called as isoline nodes. Isoline nodes are the sensor nodes residing on isolines around the contour region. Isoline sends data to sink and sink constructs the contour map which is delineated by isolines of different isolevels.

A scheme that abstracts the energy concentration in the sensor field into prescribed levels like a topographic map has been proposed by Al-Karaki et al. (2007). The scheme provides an aggregated view of the residual energy levels of different regions in the sensor field instead of detailed information of residual energy at individual sensor nodes. The scheme is named as ECscale. Algorithm is also benefited by Blocking Island (BI) paradigm (Frei, 2000) that divides the network graph into multiple sub-graphs representing a set of nodes belonging to a certain energy level. ECscale has three energy levels, namely high, medium, and low. When there is any change in energy state of a particular node, the energy level in ECscale changes accordingly. The levels of EC-scale are constructed in a distributed manner and using only localized algorithms. This aggregated view reduces the communication and processing cost significantly.

The process of constructing ECscale of a sensor field can be briefly described as follows. First, each node determine its residual energy level in a periodic fashion and with its group ID using special packets, called Update packets (Upackets). A sensor node needs to report only when there is a significant energy drop. Periodic reporting consumes much energy. Clusterheads can also aggregate two or more Upackets depending upon reported levels. The general goal of this aggregation or abstraction procedure is to reduce the communication cost of collecting Upackets on the expense of discarding some detailed information about the individual nodes. After this procedure, an aggregation tree is constructed with root at the Base Station (BS). This tree is refreshed periodically to adapt to network dynamics and sensor node failures.

Energy Map

Mini et al. (2004, 2005) proposed energy map construction with prediction based, sampling based and statistical based techniques. Here, we discuss these approaches one by one.

In Mini et al. (2005) an approach to form energy map by predicting the energy consumption of sensor node is proposed. It predicts the energy consumption based on past history of sensor nodes. According to the model, if any sensor node efficiently predicts its energy consumption in future, then a significant amount of message flow to collect the energy information will be reduced. At the same time, effectiveness of the approach depends upon the accuracy of the prediction model.

To predict the energy dissipation, probabilistic model based on Marcov chain is used. Sensor nodes with *M* operation modes are modeled as Marcov chain with *M* states. In this model, there is a fixed probability of sensor nodes transiting from one state to other in next time step. The n-step transition probability can be defined as:

$$p_{ij}^{(n)} = \sum_{k=1}^{M} p_{ik}^{(r)} p_{kj}^{(n-r)}$$

for 0< r < n. where i is the current state and j is the next state. With help of transition probability $p_{ij}^{(n)}$, each sensor node can find out the expected amount of energy spent in next T times.

$$E^T(i) = \sum_{s=1}^{M} \left(\sum_{t=1}^{T} p_{is}^{(t)} \right) \times E_s$$

Using this $E^T(i)$, each node can calculate the energy dissipation rate ΔE. Now, each sensor node sends remaining energy and ΔE to monitoring node. Monitoring node collects energy information from all nodes and constructs a final energy map of entire sensor network. Simulation results show that this approach saves a significant amount of energy in energy map construction in comparison with naive approach.

Another approach proposed by Mini et al. (2005) constructs energy map using sampling technique. The basis of sampling technique is that sometimes neighboring nodes in WSN spent their energy similarly. Every node needs not to send its energy information. Energy dissipation can be found from neighboring nodes using sampling technique. Marcov rule is considered as special case of sampling model. In Marcov model energy information is sent only when error between sensor node energy and corresponding value at monitoring node exceeds a threshold, while in sampling model energy information is sent with probability *p* whenever error tends to the threshold.

When any node sends energy information to monitoring node, it computes the energy consumption of the neighboring node by using interpolation and last consumption rate. The probability is defined as

$$p_{neighbor} = (1-d) + d \times \left(1 - \frac{k}{k+n_i}\right)$$

Where n_i is the interpolation used for this node. Now, monitoring node updates energy consumption rate of all neighboring nodes of a particular node sending energy information by following formula:

$$c_{estimate} = c_{neighbors} \times p_{neighbor} + c_{node} \times (1 - p_{neighbor})$$

A statistical model to forecast the available energy is defined in Mini et al. (2004). Energy map can be constructed with forecasted values. The energy drop of a sensor node can be represented as time series (Brockwell & Davis, 2002). In this work author uses the ARIMA model (Box & Jenkins, 1976) to predict the future values of time series. In ARIMA model, the used time series should be stationary. Next issue in this model is to identify AR terms. An autoregressive model is simply a linear regression of the current value against one or more prior values of the series. Moving Average (MA) should also be found out.

A time series T_t can be represented by ARIMA (*p, d, q*) where *p* is order of autoregressive model, *d* is number of differencing required to achieve stationarity and *q* is order of moving average, if we find stationary time series after differencing time series (T_t), *d* times.

In another work of Mini et al. (2004) an approach for energy map construction under finite energy budget is discussed. Finite energy budget means that each node will spend a fixed amount of energy for energy map construction. Authors assume number of packets as metric for energy budget. This approach of energy map construction is based on probability. The ideal situation of energy map construction under finite energy budget is when error is constant during all times. For this, the main challenge is to find out the best time when any node will send its energy information. At each second, sensor node should check whether it should send the energy information or not. This decision can be made according to probability *p*. In other words, *p* decides the frequency of sending energy information and hence the total amount of energy spent in energy map construction. Probability *p* can be found as follows:

$$p = \frac{P_{total} - P_{used}}{T_{total} - T_{used}}$$

Where T_{total} is estimated lifetime, T_{now} is the current time, P_{total} is the number of energy information packets each node can send, and P_{used} the number of packets the node has already used.

Song et al. (2006) proposed an energy map construction approach based on non-linear manifold learning algorithm such as ISOMAP (Tenenbaum, Silva, & Langford, 2000). Proposed approach finds the dynamic energy consumption pattern from the collected sensor network data. As manifold learning is being used for dimensionality reduction, a submanifold embedded in the much larger observation space can be found. And hence, a correlation among the data sets can also be found. In this work authors devised to mine sensor energy data from large scale sensor network and developed a framework called energy map for extracting the energy distribution of the whole sensor network.

For *N* sensors in sensor network, the sensor data is collected as X_i for any sensor *i*. The multidimensional data is expressed as $X_i = \{x_i(1), x_i(2),..., x_i(M)\}$ where M is the M_{th} sample of sensor *i*. Now dimensional domain Y_i contained in Euclidean space R^d would be found as $Y_i = \{y_i(1), y_i(2),..., y_i(M)\}$. When all classical techniques for manifold learning fail because of non-linear structure, ISOMAP (Tenenbaum, Silva, & Langford, 2000) algorithm can be used for developing a framework that inverts X_i to Y_i.

4. APPLICATIONS OF ENERGY MAP IN WSNS

Energy map is very important for WSNs because the energy information is usually required to develop good protocols for routing and clustering. Recent applications of energy map in routing decision can be seen in Al-Karaki and Al-Mashaqbeh (2007) and Niculescu and Nath (2002). An algorithm is proposed in Song, Guizani, and Sharif (2007) that makes clustering decision in WSNs with the help of energy map. Many other applications of energy map can be found in data gathering (Goussevskaia et al., 2005) and data dissemination (El Rhazi & Pierre, 2007). Some other possible applications are reconfiguration algorithms, query processing and data fusion that could take advantage of the energy map. In fact, we cannot think of an application or an algorithm that is not benefited with the use of energy map because battery power is the main bottleneck in the efficiency of algorithms in WSN.

Recent and important applications of energy map are described in this section. The application summary is also presented in Table 2.

Clustering

An adaptive clustering algorithm for energy efficiency based on energy information (energy map) is proposed in Song, Guizani, and Sharif (2007). This algorithm uses the data mining techniques to get the location and topology information from live sensor nodes. Algorithm takes clustering decision based on residual energy of nodes. It selects the clusterheads based on inferred location by mining sensor energy data and residual energy level of each node. Clusterhead role is rotated among sensor nodes for energy efficiency.

This algorithm uses the manifold learning algorithm to get the network topology information by applying data mining on sensor energy data. Manifold learning algorithm is generally used for

Table 2. Application summary of energy map

S. No.	Application Area	Algorithms	Remarks
1	Clustering	Adaptive Clustering Algorithm (Song, Guizani, & Sharif, 2007)	• Uses data mining techniques to get the location and topology information from remaining of sensor nodes
		Centralized Clustering Scheme (El Rhazi & Pierre, 2009)	• A tabu search heuristic is used to solve the clustering problem
2	Routing	Trajectory Based Forwarding (TBF) (Niculescu & Nath, 2002)	• Forwarding algorithm suitable for routing packets along a predefined curve • Hybrid technique combining source based routing and Cartesian forwarding
		Energy Centric Routing (Al-Karaki & Al-Mashaqbeh, 2007)	• Scheme is based on monitoring residual energy distributions at different parts of the network through a mechanism called Energy Centric scale (ECscale).
3	Data Aggregation	Data Collection Algorithm (El Rhazi & Pierre, 2007, 2009)	• Based on distributed clustering and uses Energy map to reduce energy consumption
		Energy Efficient Data Aggregation (Min, Yi, & Hong, 2008)	• Based on chain construction methodology
4	Data Dissemination	Trajectory and Energy-Based Data Dissemination (TEDD) (Goussevskaia et al., 2005)	• Combines the information provided by energy map with concepts presented in TBF to determine root in dynamic fashion

dimensionality reduction, i.e., finding the dissimilar data from the samples. The main advantage of this algorithm is that no location information is required for clustering and hence makes sensor network deployment easier and flexible.

El Rhazi and Pierre (2009) proposed a centralized clustering scheme based on energy map in wireless sensor networks. Authors use this clustering scheme for data collection in order to reduce energy consumption. A tabu search heuristic is used to solve the clustering problem modeled as hypergraph partitioning because it is assumed that the numbers of clusters and clusterheads are unknown before clusters are created. Tabu search is capable of determining feasible clusters. A feasible cluster is defined as a set of nodes that fulfill some cluster building constraints defined in El Rhazi and Pierre (2009). The nodes ensuring zone coverage are called active nodes. Some risks regarding tabu search method may also be introduced by accepting solutions that are not necessarily optimal i.e., cycle risk. This risk can be avoided by keeping track of the solutions that have been considered in the past i.e., forming a tabu list. The size of the tabu list has a direct impact on the quality of the solution. Hence, it is important to analyze its impact, in order to adjust its value accordingly. Tabu search methods can be improved with two mechanisms like diversification and intensification. Tabu search-based clustering algorithm is scalable and provides quality solutions in terms of clustering cost and execution time. One major drawback of this algorithm is that it needs a lot of time to solve optimization problem and also costs additional for communicating the node information.

Routing

Any routing protocol can save a significant amount of energy if it chooses nodes in route based on residual energy. The protocol proposed in Niculescu and Nath (2002) is an example of a routing protocol that could take advantage of the energy map. In his work author described the Trajectory Based Forwarding (TBF) protocol that is a new forwarding algorithm suitable for routing packets along a predefined curve. TBF is a hybrid technique combining source based routing (Johnson & Maltz, 1996) and Cartesian forwarding (Finn, 1987). In TBF the path is indicated by the source, but without actually specifying all the intermediate nodes like source based routing. The decisions taken at each node are greedy, but not based on distance to destination like Cartesian Forwarding. In TBF, the algorithm embeds the trajectory in each packet, and the intermediate nodes make the forwarding decisions based on their distances from the desired trajectory. The trajectory could be planned in order to pass through regions with more energy, if the protocol had the information about the energy map, thus preserving or avoiding regions of the network with small energy reserves.

An energy centric routing scheme that maximize the lifetime of WSNs is proposed in Al-Karaki and Al-Mashaqbeh (2007). Al-Karaki et al. (2007) investigated the problem of energy-centric data routing using a cluster-based solution. This scheme is based on monitoring residual energy distributions at different parts of the network through a mechanism called Energy Centric scale (ECscale). In this work authors combines the ECscale scheme with clustering scheme introduced by Al-Karaki et al. (2005) and BI paradigm proposed in Frei (2000) to perform energy-efficient routing in WSNs in a unified manner. Both the exact and heuristic solutions are proposed. The exact solution for energy centric routing is formulated as an Integer Linear Program (ILP). ILP selects some subgraphs that have higher remaining energy to form an optimal path hence minimize total routing cost. One drawback of ILP is that it can solve only small and medium sized networks, for large sensor networks it would be computationally infeasible.

Three heuristic based energy centric routing in WSNs are proposed based on ECscale. These approach route packets efficiently to the BS while avoiding regions with low energy reserves. The difference between these scalable heuristics lies in the way they search for a route over Virtual Grid Architecture (VGA). In VGA, a source can use many routes to reach the BS.

Data Aggregation

An application of energy map in data collection from sensor nodes is proposed in El Rhazi and Pierre (2007, 2009). The proposed data collection algorithm is based on distributed clustering and uses energy map to reduce energy consumption. The data collection algorithm using energy map is divided into following phases. In query distribution phase, the network is divided into set of clusters so that energy consumption is reduced and hence network lifetime is improved. Energy map is used in cluster building so that network coverage is maximum. Cluster formation is followed by negotiation process and then data collection step. The network model proposed for data collection algorithm is multi-hop communication model i.e., nodes cannot send measurements directly to the collector nodes.

During data collection clusterhead monitors the following activities: coordinating the data collection within its cluster, filtering redundant measurements, computing aggregate function and sending results to collector node. Author also compared the simulation results to Tiny Aggregation (TAG) (Madden, Franklin, Hellerstein, & Hong, 2002) a generic aggregation service of network of TinyOS motes. The aggregation approach based on energy map outperforms TAG.

An energy efficient data aggregation protocol for WSN by using energy map is proposed by Min et al. (2008). This aggregation method is based on chain construction methodology proposed in PEGASIS (Lindsey, Raghavendra, & Sivalingam, 2002). The chain is a series of sensor nodes. Sensor nodes transmit its sensing data to a neighbor along this chain. The problems in PEGASIS by using this chain that affect the energy efficiency adversely are delay, unexpected long transmission time and non directional transmission to the sink. A new chain construction algorithm is proposed to resolve these problems. The proposed algorithm forwards packets to the sink node depending upon the decision based on energy map. In the algorithm, an energy map is constructed and data is forwarded through the nodes having more residual energy.

Data Dissemination

Cooperation among neighboring nodes is done through efficient data communication among them. Monitoring node also communicates important information with a set of sensor nodes. This kind of data communication is often called data dissemination. An energy-efficient data dissemination algorithm for WSNs, called Trajectory and Energy-Based Data Dissemination (TEDD), is proposed in Goussevskaia et al. (2005). This algorithm combines the information provided by energy map with concepts presented in TBF (Niculescu & Nath, 2002) to determine root in dynamic fashion.

An algorithm is proposed for generating trajectories dynamically based on the energy map of the network that pass through regions where nodes residual energy is comparatively higher. TEDD is based on curve fitting problem that pass through the set of most suitable nodes selected by the monitoring node for disseminating the data packets sent by it. There are various criteria for best set of curves as the amount of energy available at the forwarding nodes and the amount of nodes that receive the packet disseminated by the monitoring node i.e., area at which dissemination is aimed. Hence the trajectory generation procedure is designed to be driven by the requirements of the application.

5. PREDICTION BASED ENERGY MAP CONSTRUCTION

Here, we propose a prediction based approach for energy map construction. This approach is based on prediction based cooperation between the sensor node and the sink node. Both sensor node and sink node will use the same prediction model and same energy data for prediction. Sensor nodes need not to send the energy information to the sink every time when there is an energy drop. Sink node can predict the energy drop in each sensor and an energy map can be constructed based on the predicted value of energy in each sensor node. Thus a significant amount of energy can be saved if sensor node and sink node efficiently predict the amount of energy sensor node will spend in future.

Assumptions

1. Sensor node can send energy information only once in a period.
2. Both the sensor node and sink node will use the same prediction model.
3. Sink node has sufficient computing power, energy and storage.
4. There is reliable data transmission between sensor node and sink node. The sink node broadcasts its maximum acceptable prediction error ε, to all senor nodes. Two data queues, prediction energy queue, $PEQ_{sensor,i}$ at each sensor and corresponding $PEQ_{\text{sink},i}$ at sink node for each sensor i is constructed. The length of the $PEQ_{sensor,i}$ and $PEQ_{\text{sink},i}$ are equal and specified by prediction algorithm. Initially $PEQ_{\text{sink},i} = PEQ_{\text{sensor},i}$, i.e., both $PEQ_{\text{sink},i}$ and $PEQ_{sensor,i}$ store the same predicted values of energy of sensor node i at sink node and sensor node respectively. Actual energy value (*AEV*) at any period t is compared with $PEQ_{sensor,i}[t]$, the tth item in sensor prediction queue. If the difference between the $PEQ_{sensor,i}[t]$ and AEV is greater than ε, the $PEQ_{sensor,i}[t]$ is replaced with AEV and the value of $PEQ_{sensor,i}[t]$ is sent to the sink. Sink node will update the tth item in its $PEQ_{\text{sink},i}$ queue. Now the prediction model uses the updated energy information for further predictions.

For this work we have assumed the sensor network to be static and homogeneous and replacement of batteries is unfeasible and impossible. To analyze the performance of proposed system, we have implemented a grey system theory based energy map construction scheme on ns-2. Compared to other prediction based approaches, grey system theory has been found light weight.

We simulated our approach for 100 sensor nodes having $1J$ of initial energy for 1000s. Figure 3 shows that by using grey system theory based prediction we have to send only three packets (167, 515, 789) when the prediction error is less than maximum acceptable error, ε, while naive approach will send eight packets (120, 173, 399, 502, 639, 781, 875, 985). This shows a significant reduction in number of packets sent for energy map construction, hence saving in energy.

Figure 3. Energy packet transmission for energy map construction

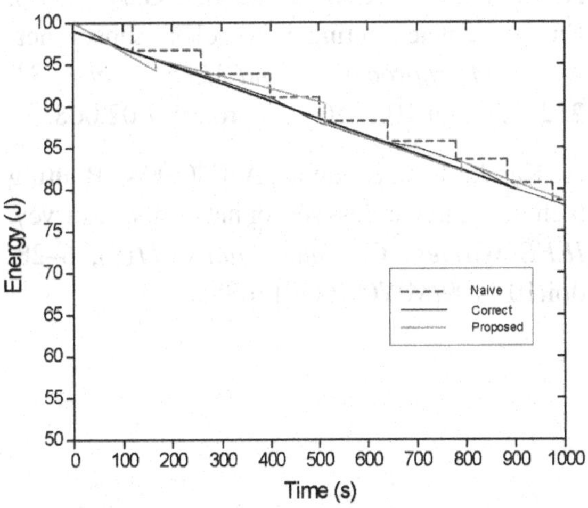

6. CONCLUSION

Since energy is the most critical resource in wireless sensor networks, it is very important to conserve it. The information about the energy status at every part of the network may be very useful in efficient utilization of energy resource. Energy map is a graphical representation of available energy i.e., residual energy in every part of the sensor network. In this paper we have discussed various approaches of energy map construction. Further, we describe the importance of energy map for WSNs as it can be useful in various research areas like energy efficient clustering, data aggregation, routing, data dissemination, etc. Clustering and routing decisions can be made depending upon information provided by energy map. Energy map has a lot of potential to maximize the benefits of WSNs in various fields of real life applications by minimizing the energy consumption and maximizing the throughput.

We have also proposed an energy map construction algorithm that is based on prediction. Simulation result proves the effectiveness of the proposed approach. It reduces up to 30% message exchange for energy map construction than naive approach.

REFERENCES

Al-Karaki, J. N., & Al-Mashaqbeh, G. A. (2007). Energy-centric routing in wireless sensor networks. *Microprocessors and Microsystems*, *31*, 252–262. doi:10.1016/j.micpro.2007.02.008.

Al-Karaki, J. N., & Kamal, A. E. (2004). Routing techniques in wireless sensor networks: A survey. *IEEE Wireless Communications*, *11*(6), 6–28. doi:10.1109/MWC.2004.1368893.

Al-Karaki, J. N., & Kamal, A. E. (2005). End-to-end support for statistical quality of service in heterogeneous mobile ad hoc networks. *Computer Communications*, *28*(18), 2119–2132. doi:10.1016/j.comcom.2004.07.038.

Al-Karaki, J. N., Ul-Mustafa, R., & Kamal, A. E. (2004, April). Data aggregation in wireless sensor networks - Exact and approximate algorithms. In *Proceedings of the IEEE Workshop on High Performance Switching and Routing*, Phoenix, AZ (pp. 241-245).

Bokareva, T., Hu, W., Kanhere, S., Ristic, B., Gordon, N., Bessell, T., et al. (2006). *Wireless sensor networks for battlefield surveillance.* Retrieved from http://www.cse.unsw.edu.au/~tbokareva/papers/lwc.html

Box, G. E. P., & Jenkins, G. M. (1976). *Time series analysis: Forecasting and control*. San Francisco, CA: Holden-Day.

Brockwell, P. J., & Davis, R. A. (2002). *Introduction to time series and forecasting* (2nd ed.). New York, NY: Springer.

El Rhazi, A., & Pierre, S. (2007). A data collection algorithm using energy maps in sensor networks. In *Proceedings of the Third IEEE International Conference on Wireless and Mobile Computing, Networking and Communications* (p. 64).

El Rhazi, A., & Pierre, S. (2009). A tabu search algorithm for cluster building in wireless sensor networks. *IEEE Transactions on Mobile Computing*, *8*(4), 433–444. doi:10.1109/TMC.2008.125.

Espina, J., Falck, T., Muehlsteff, J., Yilin, J., Adan, M. A., & Aubert, X. (2008). Wearable body sensor network towards continuous cuff-less blood pressure monitoring. In *Proceedings of the 5th International Summer School and Symposium on Medical Devices and Biosensors* (pp. 28-32).

Finn, G. (1987). *Routing and addressing problems in large metropolitan-scale internetworks* (Tech. Rep. No. ISI/RR-87-180). Los Angeles, CA: University of Southern California.

Frei, C. (2000). *Abstraction techniques for resource allocation in communication networks.* Unpublished doctoral dissertation, Swiss Federal Institute of Technology, Lausanne, Switzerland.

Goussevskaia, O., Machado, M. D. V., Mini, R. A. F., Lourerio, A. A. F., Mateus, G. R., & Nogueira, J. M. (2005). Data dissemination based on the energy map. *IEEE Communications Magazine, 43*(7), 134–143. doi:10.1109/MCOM.2005.1470845.

Han, S., & Chan, E. (2005). Continuous residual energy monitoring in wireless sensor networks. In J. Cao, L. T. Yang, M. Guo, & F. Lau (Eds.), *Proceedings of the Second International Symposium on Parallel and Distributed Processing and Applications* (LNCS 3358, pp. 169-177).

Hart, J. K., & Martinez, K. (2006). Environmental sensor networks: A revolution in the earth system science. *Earth-Science Reviews*, 177–191. doi:10.1016/j.earscirev.2006.05.001.

Heinzelman, W. B., Chandrakasan, A. P., & Balakrishnan, H. (2002). Application specific protocol architecture for wireless microsensor networks. *IEEE Transactions on Wireless Networking, 1*(4), 660–670. doi:10.1109/TWC.2002.804190.

Hussain, S., Schaffner, S., & Moseychuck, D. (2009). Applications of wireless sensor networks and RFID in a smart home environment. In *Proceedings of the Seventh Annual Communication Networks and Services Research Conference* (pp. 153-157).

Johnson, D. B., & Maltz, D. A. (1996). Dynamic source routing in ad hoc wireless networks. In Imielinski, T., & Korth, H. (Eds.), *Mobile computing* (p. 353). Boston, MA: Kluwer Academic. doi:10.1007/978-0-585-29603-6_5.

Kalpakis, K., Dasgupta, K., & Namjoshi, P. (2003). Efficient algorithms for maximum lifetime data gathering and aggregation in wireless sensor networks. *Computer Networks, 42*(6), 697–716. doi:10.1016/S1389-1286(03)00212-3.

Li, M., & Liu, Y. (2010). Iso-Map: Energy-efficient contour mapping in wireless sensor networks. *IEEE Transactions on Knowledge and Data Engineering, 22*(5), 699–710. doi:10.1109/TKDE.2009.157.

Lindsey, S., Raghavendra, C., & Sivalingam, K. M. (2002). Data gathering algorithms in sensor networks using energy metrics. *IEEE Transactions on Parallel and Distributed Systems, 13*(9), 924–935. doi:10.1109/TPDS.2002.1036066.

Liu, J.-S., & Lin, C.-H. R. (2005). Energy-efficiency clustering protocol in wireless sensor networks. *Ad Hoc Networks, 3*(3), 371–388. doi:10.1016/j.adhoc.2003.09.012.

Madden, S. R., Franklin, M. J., Hellerstein, J. M., & Hong, W. (2002, December). TAG: Tiny aggregation service for ad-hoc sensor networks. In *Proceedings of the Fifth Symposium on Operating Systems Design and implementation*, Boston, MA.

Min, H., Yi, S., & Hong, J. (2008). Energy-efficient data aggregation protocol for location-aware wireless sensor networks. In *Proceedings of the International Symposium on Parallel and Distributed Processing with Applications* (pp. 751-756).

Mini, R. A. F., Loureiro, A. A. F., & Nath, B. (2004, October). Energy map construction for wireless sensor network under a finite energy budget. In *Proceedings of the 7th ACM International Symposium on Modeling, Analysis, and Simulation of Wireless and Mobile Systems* (pp. 165-169).

Mini, R. A. F., Loureiro, A. A. F., & Nath, B. (2004). The distinctive design characteristic of a wireless sensor network: The energy map. *Computer Communications, 27*, 935–945. doi:10.1016/j.comcom.2004.01.004.

Mini, R. A. F., Loureiro, A. A. F., & Nath, B. (2005). Prediction-based energy map for wireless sensor networks. *Ad Hoc Networks Journal*, *3*, 235–253. doi:10.1016/j.adhoc.2004.07.008.

Niculescu, D., & Nath, B. (2002). *Trajectory-based forwarding and its applications* (Tech. Rep. No. DCS-TR-488). New Brunswick, NJ: Rutgers University.

Rajagopalan, R., & Varshney, P. K. (2006). Data-aggregation techniques in sensor networks: A survey. *IEEE Communications Surveys & Tutorials*, *8*(4), 48–63. doi:10.1109/COMST.2006.283821.

Reddy, A., Estrin, D., & Govindan, R. (2000). Large scale fault isolation. *IEEE Journal on Selected Areas in Communications*, *18*(5), 733–743. doi:10.1109/49.842989.

Shah, R. C., & Rabaey, J. (2002, March). Energy aware routing for low energy ad hoc sensor networks. In *Proceedings of the IEEE Wireless Communications and Networking Conference*, Orlando, FL (pp. 350-355).

Song, C., & Guizani, M. (2006). Energy map: Mining wireless sensor network data. In. *Proceedings of the IEEE International Conference on Communications*, *8*, 3525–3529.

Song, C., Guizani, M., & Sharif, H. (2007). Adaptive clustering in wireless sensor networks by mining sensor energy data. *Computer Communications*, *30*, 2968–2975. doi:10.1016/j.comcom.2007.05.027.

Tenenbaum, J. B., Silva, V. D., & Langford, J. C. (2000). A global geometric framework for nonlinear dimensionality reduction. *Science*, *290*, 2319–2323. doi:10.1126/science.290.5500.2319.

Yan, H., Xu, Y., & Gidlund, M. (2009, January). Experimental e-health applications in wireless sensor networks. In *Proceedings of the WRI International Conference on Communications and Mobile Computing* (pp. 563-567).

Zhao, Y. J., Govindan, R., & Estrin, D. (2002, March). Residual energy scans for monitoring wireless sensor networks. In *Proceedings of the IEEE Wireless Communications and Networking Conference* (pp. 356-362).

Zhao, Y. J., Govindan, R., & Estrin, D. (2003). *Computing aggregates for monitoring wireless sensor networks* (Tech. Rep. No. 02-773). Los Angeles, CA: University of Southern California.

This work was previously published in the International Journal of Wireless Networks and Broadband Technologies, Volume 1, Issue 2, edited by Naveen Chilamkurti, pp. 16-29, copyright 2011 by IGI Publishing (an imprint of IGI Global).

Chapter 9
Doubly Cognitive Architecture Based Cognitive Wireless Sensor Networks

Sumit Kumar
International Institute of Information Technology, Hyderabad, India

Deepti Singhal
International Institute of Information Technology, Hyderabad, India

Garimella Rama Murthy
International Institute of Information Technology, Hyderabad, India

ABSTRACT

Scarcity of spectrum is increasing not only in cellular communication but also in wireless sensor networks. Adding cognition to the existing wireless sensor network (WSN) infrastructure has helped. As sensor nodes in WSN are limited with constraints like power, efforts are required to increase the lifetime and other performance measures of the network. In this article, the authors propose Doubly Cognitive WSN, which works by progressively allocating the sensing resources only to the most promising areas of the spectrum and is based on pattern analysis and learning. As the load of sensing resource is reduced significantly, this approach saves the energy of the nodes and reduces the sensing time dramatically. The proposed method can be enhanced by periodic pattern analysis to review the strategy of sensing. Finally the ongoing research work and contribution on cognitive wireless sensor networks in Communication Research Centre (IIIT-H) is discussed.

1. INTRODUCTION AND MOTIVATION

Cognitive networking speaks for an intelligent communication system, consisting of both the wire-line and/or the wireless connections, that is aware of its transmission environment, both internal and external, and acts adaptively and autonomously to attain its intended goal(s). A unit of Cognitive networking system i.e., Cognitive radios dynamically selects spectrum, waveform design, time diversity, and spatial diversity options. It can even make changes at higher layers, for example, by modifying the medium access protocols or changing its routing behavior based

DOI: 10.4018/978-1-4666-3902-7.ch009

on the network topology (Fette, 2006; Mahmoud, 2007). This implies that all the network nodes and the end devices with cognitive capabilities are self-aware and context-aware all of the time. The interest in cognitive networking is mainly driven from the need to manage the increasing complexity and the efficient utilization of available resources to deliver applications and services as economically as possible. A wireless sensor network (WSN) is one of the areas where there is very high demand for cognitive networking. Wireless Sensor Network is a collection of sensor nodes working in a co-operative manner. Each node has certain processing capability, RF transceiver, power source apart from sensing and actuating units (Raghavendra, 2007; Sohraby, Minoli, & Znati, 2007). There are several constraints among which the constraint of resources (spectrum and power) is the most appealing one in a WSN. Although in WSN the nodes are constraint in resources mainly in terms of battery power but these days there is scarcity increasing in terms of spectrum availability also. Traditionally all the WSN works in ISM band (2.4 GHz), but in the same band we have many competing technologies working simultaneously like WLAN 802.11 a/b/g and ZigBee 802.15.4, Wi-Fi, Bluetooth. Hence in such an environment where all these competing technologies are working simultaneously, it becomes difficult to find free spectrum and transmit without an error. Also at the same time the licensed mobile communication bands are almost free for 85% of the time (Akyildiz, Lee, Vuran, & Mohanty, 2006; Cavalcanti, Das, Wang, & Challapali, 2008). Hence there are two motives: either to find a free channel in the unlicensed band and do wireless transmission or to find a free channel in the licensed band and do communication over that. Also it will be strictly required that whenever the licensed user comes back in the licensed band, the cognitive user backs-off from the channel and switch to another free channel without creating any difficulty to the primary user. Several other things are also important such as selecting the most suitable channel among the free channels and fair allocation of free channels among several competing nodes.

Adding cognition to the existing WSN infrastructure brings several benefits. A Cognitive Wireless Sensor Network (CWSN) enables current WSN to overcome the spectrum scarcity as well as node energy problem. A relatively better propagation characteristic is an added advantage. By adaptively changing the systems parameters like transmitted power, operating frequency, modulation, pulse shape, symbol rate, coding technique and constellation size a wide variety of data rates and QoS can be achieved which improves the power consumption and network life time in a WSN (Rondeau & Bostian, 2009).

There are two similar looking terms in the context of cognitive radio: cognitive radio WSN and cognitive WSN, which appear similar but are quite different from each other. A cognitive radio WSN is a WSN with each node having cognitive radio capability and nothing more than that. This means it has cognitive capabilities in the physical layers only. But this does not fulfill our purpose as the WSN demands cognitive radio capabilities with cognitive networking among them which can take benefit from this cognition. Hence the concept for cognitive WSN which involves cognition not only in the physical layer but a cross layered approach. Intuitively this cross layered approach is more beneficial for the WSN.

2. DOUBLY COGNITIVE APPROACH FOR CWSN

During continuous research in the area of cognitive radio and wireless sensor networks we came up with an idea which is based on pattern recognition as well as multi-objective optimization for cognitive WSN. The basic idea of doubly cognitive WSN is explained in Figure 1. Underlying notion of our idea is to progressively allocate the spectrum sensing resources to only

Figure 1. Doubly cognitive approach

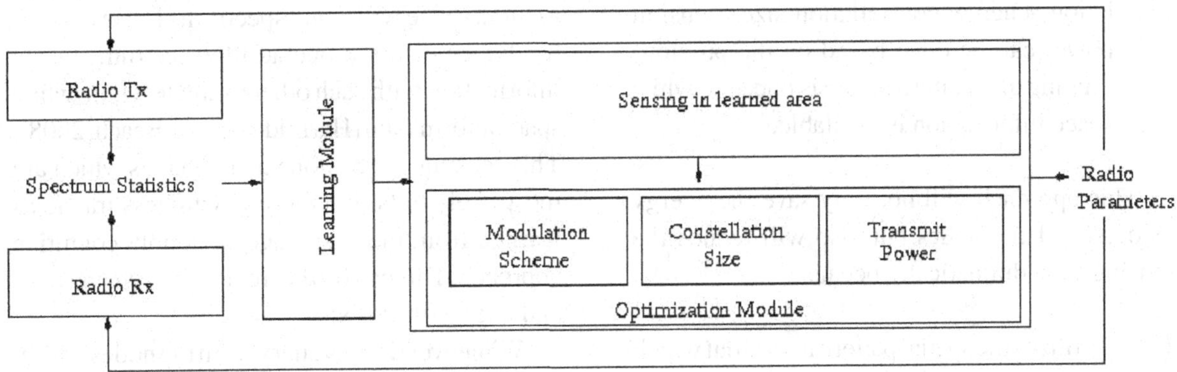

the most promising areas of the spectrum. This translates in a reduction of energy consumption in sensing resources and time needed to accurately identify spectrum holes, in contrast with the conventional approaches that allocate the sensing budget over the entire spectrum uniformly all the time. This conventional approach is very energy and time consuming in a cognitive WSN where there is very hard constraint for energy as well as sensing time. Doing sensing only in the most promising area can not only save the energy of the spectrum sensing nodes (cluster heads) but also save the spectrum sensing time. It can also reduce the time to react and transmit/broadcast the information to other nodes. This idea can work very well in the scenarios where there is a multichannel cognitive radio network (it generally happens in WSN working along with WLAN, Wi-Fi, and Bluetooth). Hence multiple spectrum bands are sensed to identify transmission opportunities.

Doubly cognitive idea is based on a cognitive module that will analyze data, recognize patterns, and use them for supervised training/learning. During the training phase it will continuously collect the information for white space availability, traffic density both spatially and temporally. Also the radio frequency environmental information i.e., fading, interference, SNR, location, visual clues on each channel for a sufficient long time in the area in which the CWSN is supposed to be deployed will be collected. Once it will have a long time repository of the above information then it can be used as a training pattern for supervised learning of the cognitive spectrum sensing machine which can be trained to

1. Sense for the spectrum only in the bands or channels where there is a pattern of less traffic spatially/temporally and/or there is a pattern of having white space spatially/temporally (Figure 2).

Figure 2. Spectrum hole information

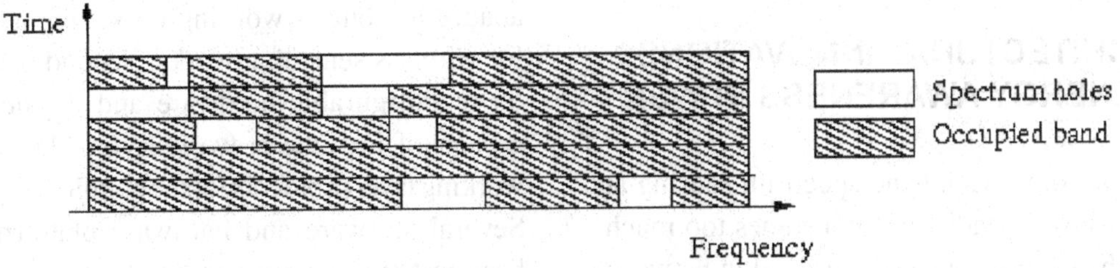

2. Optimization of the radio module i.e., modulation scheme, constellation size, transmit power can be done based on the previous training and pattern analysis once the white space information is available.

This approach will not only save the energy recourses of the nodes but also will reduce the sensing time dramatically because

1. It utilizes the spatial pattern to find out which channels to sense.
2. Temporal patterns to find out when to sense.

After this first stage of spectrum sensing the second obvious stage of multi-objective optimization is done for the radio frequency environmental parameters. This multi-objective optimization is based on the learning which was achieved from the earlier optimized settings of the radio transceiver done in several types of radio frequency environment situations.

Some of the effective/ suitable learning techniques for this purpose are Case-based decision theory (CBDT) (Gilboa & Schmeidler, 2001), Neural Networks (Le, Rondeau, Maldonado, Scaperoth, & Bostian, 2006; Fehske, Gaeddert, & Reed, 2005), Hidden Markov Models (HMM) (Kanal & Sastry, 1978), Fuzzy Logic (Baldo & Zorzi, 2007), and Genetic algorithms (Rondeau, Le, Rieser, & Bostian, 2004). This simple and intuitive work can be enhanced by doing the pattern analysis with respect to white space availability as well as environmental parameters after a sufficiently long time again and again to review the strategy of sensing and optimization.

3. ARCHITECTURAL INNOVATIONS FOR ENERGY AWARENESS IN WSN

Instead of doing wide band spectrum sensing at a single cluster head which consumes too much energy the task of sensing is divided in between several cluster-heads so that each of them senses a chunk of the wideband spectrum. These cluster-heads then share (co-operate) the spectrum sensing information with each other to create whole white space information (Harrold, Faris, & Beach, 2008). This sensing can be done on the bands which are most likely to be vacant or having less traffic as learned from the first stage of doubly cognitive approach. This can further reduce the sensing time and energy in WSN.

Whenever there is query from the nodes within a cluster, the cluster head looks for the latest updates of white space information and directs the nodes to do communication over those free channels. During the communication, the cluster heads can keep themselves busy in updating their white space information also. If in between the communication there is some change in the white space information over the channel which is engaged by the node doing communication then the cluster head asks that node to back-off form that channel and directs him to a new channel according to the white space information.

4. RESEARCH WORK GOING IN COMMUNICATION RESEARCH CENTRE, IIIT HYDERABAD

The major research focuses of CRC (IIIT-H) are:

1. Developing efficient algorithms of spectrum sensing for Cognitive WSN.
2. Implementing such algorithms on reconfigurable hardware.

Three teams are working in CRC, IIIT Hyderabad. First one is working on setting a test bed for wireless sensor networks. Second is working on reconfigurable hardware and physical layer aspects. Third one is working on Mac and networking layer protocols for the cognitive networks. Several software and hardware platforms have been tested and analyzed. Open source software platforms are really useful but because of lack of proper documentation they take much time in analysis and development.

Open source software platform for physical layer research include GNU Radio system, IRIS (Implementing Radio in Software), KUAR (The Kansas University Agile Radio).Other software platforms which are not open source but have proper documentations include Matlab, LabView. We are working on both GNU Radio system and Matlab. Similarly open source platforms for MAC layer research includes NS2 (Network Simulator 2), TOSSIM.

Among hardware platforms we have chosen reconfigurable hardware options which include USRP (Universal Software radio peripheral) by Ettus Research LLC and MSR Software Radio Academic kit by Microsoft Research. The community for GNU Radio system and USRP is well developed and active hence we are primarily working on them. The goal is to emulate a fully functional Cognitive Wireless Sensor Network on hardware.

ACKNOWLEDGMENT

This work is supported by Ministry of Communication and Information technology, Government of India under the project "Mobile and Static Cognitive Wireless Sensor Networks".

REFERENCES

Akyildiz, I. F., Lee, W.-Y., Vuran, M. C., & Mohanty, S. (2006). Next generation/dynamic spectrum access/cognitive radio wireless networks: A survey. *International Journal of Computer and Telecommunications Networking*, *50*(13), 2127–2159.

Baldo, N., & Zorzi, M. (2007, January). Fuzzy logic for cross-layer optimization in cognitive radio networks. In *Proceedings of the 4th IEEE Consumer Communications and Networking Conference* (pp. 1128-1133).

Cavalcanti, D., Das, S., Wang, J., & Challapali, K. (2008). Cognitive radio based wireless sensor networks. In *Proceedings of the 17th International Conference on Computer Communications and Networks* (pp. 1-6).

Fehske, A., Gaeddert, J., & Reed, J. H. (2005). A new approach to signal classification using spectral correlation and neural networks. In *Proceedings of the First IEEE International Symposium on New Frontiers in Dynamic Spectrum Access Networks* (pp. 144-150).

Fette, B. (2006). *Cognitive radio technology*. Oxford, UK: Newnes.

Gilboa, I., & Schmeidler, D. (2001). *A theory of case-based decisions*. Cambridge, UK: Cambridge University Press. doi:10.1017/CBO9780511493539.

Harrold, T. J., Faris, P. C., & Beach, M. A. (2008, September 18). Distributed spectrum detection algorithms for cognitive radio. In *Proceedings of the IET Seminar on Cognitive Radio and Software Defined Radios: Technologies and Techniques* (pp. 1-5).

Kanal, L. N., & Sastry, A. R. K. (1978). Models for channels with memory and their applications to error control. *Proceedings of the IEEE*, *66*(7), 724–744. doi:10.1109/PROC.1978.11013.

Le, B., Rondeau, T. W., Maldonado, D., Scaperoth, D., & Bostian, C. W. (2006). Signal recognition for cognitive radios. In *Proceedings of the Software Defined Radio Forum Technical Conference.*

Mahmoud, Q. (2007). *Cognitive networks: Towards self-aware networks*. New York, NY: Wiley-Interscience.

Raghavendra, C. S. (2007). *Wireless sensor networks*. New York, NY: Springer.

Rondeau, T. W., & Bostian, C. W. (2009). *Artificial intelligence in wireless communications*. Boston, MA: Artech House.

Rondeau, T. W., Le, B., Rieser, C. J., & Bostian, C. W. (2004). Cognitive radios with genetic algorithms: Intelligent control of software defined radios. In *Proceedings of the Software Defined Radio Forum Technical Conference* (pp. 3-8).

Sohraby, K., Minoli, D., & Znati, T. F. (2007). *Wireless sensor networks: Technology, protocols, and applications*. New York, NY: Wiley-Interscience. doi:10.1002/047011276X.

This work was previously published in the International Journal of Wireless Networks and Broadband Technologies, Volume 1, Issue 2, edited by Naveen Chilamkurti, pp. 30-35, copyright 2011 by IGI Publishing (an imprint of IGI Global).

Chapter 10
Synchronous Relaying in Vehicular Ad-Hoc Networks

D. Moltchanov
Tampere University of Technology, Finland

J. Jakubiak
Tampere University of Technology, Finland

A. Vinel
Tampere University of Technology, Finland

Y. Koucheryavy
Tampere University of Technology, Finland

ABSTRACT

In this paper, the authors propose a simple concept for emergency information dissemination in vehicular ad-hoc networks. Instead of competing for the shared wireless medium when transmitting the emergency information, the authors' proposed method requires nodes to cooperate by synchronizing their transmissions. The proposed scheme is backward compatible with IEEE 802.11p carrier sense multiple access with collision avoidance. The authors also briefly address some of the implementation issues of the proposed scheme.

1. BACKGROUND AND MOTIVATION

In July 2010 the IEEE 802.11p standard (IEEE Computer Society, 2010) has been ratified, which specifies physical (PHY) and medium access control (MAC) extensions to provide wireless access in vehicular environments. The 802.11p MAC layer adopts the enhanced distributed channel access (EDCA) based on carrier sense multiple access with collision avoidance (CSMA/CA). According to the EDCA protocol the medium is slotted and stations compete for channel resources using different interframe spacing (IFS) intervals and contention window (CW) sizes. If the channel is free during the whole duration of the IFS interval a node is allowed to transmit. If the medium becomes occupied during the sensing, a node randomly chooses the backoff counter and defers its transmission attempt. When the channel is free the counter is decremented each time slot. The sizes of both the IFS interval and the CW depend on the type of the message to be transmitted, i.e., the higher the priority of a message is the shorter these values are. Assigning priorities in this way the MAC protocol ensures that the critical messages (i.e., those, notifying about safety-of-life hazards, accident in a close proxim-

DOI: 10.4018/978-1-4666-3902-7.ch010

ity of a node, etc.) have relative priority over the infotainment-related ones. This results in lower delivery delays and higher delivery probabilities of critical information. Nodes having packets of the same priority compete for transmission using the same IFS and CW.

In this paper we are concerned with the dissemination speed and coverage of emergency messages, i.e., those, having the utmost importance. Having safety in mind, for a protocol delivering such kind of information we can stipulate two requirements. First of all, emergency information needs to be spread out to all vehicles in a certain neighborhood, h, of a car detecting the hazard. This geographic area may include infrastructure nodes that may disseminate notifications further using wide area wireless networks such as cellular systems. Secondly, emergency messages need to be disseminated as quickly as possible. CSMA/CA protocol which has been standardized in IEEE 802.11p is far from the optimal solution for these problems due to the inherent properties of random access.

Consider an arbitrary road environment crowded with vehicles equipped with the IEEE 802.11p on-board units (OBU) and concentrate on what happens just after a detection of a hazard. According to this scenario a number of vehicles compete for wireless media trying to disseminate critical information. As soon as one of them gets access then some nodes in its coverage area become notified and in their turn start to compete for the channel for further transmissions. Different probabilistic approaches to divide the numerous retransmissions of different nodes in time domain to prevent broadcast storm and to increase the efficiency of random accessing are proposed. Among them Emergency Message Dissemination in Vehicular environment (EMDV) (Torrent-Moreno, Mittag, Santi, & Hartenstein, 2009) is one of the well-known location-based forwarding. However, the efficiency of all these approaches depends on the performance of one-hop IEEE 802.11p broadcasting, which is briefly evaluated.

Assuming that the CW is constant the probability of choosing a slot in CSMA/CA protocol is approximated by $2/(W+1)$, where W is the size of the CW (see e.g., Bianchi, 2000). The probability that one or more nodes choose a certain slot i together with a given node and the packet transmission would result in a collision is given by $\tau(W,N)=1-(1-2/[W+1])^{N-1}$, where N is the overall number of nodes. Taking into account that a packet can be lost as a result of insufficient signal strength with probability p and noticing that it is independent of the collision probability we get the following equation for probability of incorrect packet reception in exactly one transmission attempt.

$$r(w, N, p) = 1 = 1 - \left(1 - \frac{2}{W+1}\right)^{N-1} (1-p) \quad (1)$$

Behavior of $r(W,N,p)$ as a function of the number of competing nodes, CW size, and probability p is illustrated in Figure 1. Analyzing it one may observe two unwanted dependencies existing in $r(W,N,p)$. First of all, for small CW size, W, and negligible loss probabilities caused by channel conditions (p=0), the probability of incorrect packet reception is close to 1. Since the window size is rather small for the distribution of emergency information the protocol is expected to perform poorly in the congested environments. Traffic accidents frequently lead to jams increasing the density of nodes in the local neighborhood. Since no acknowledgments are sent by the receivers of the broadcast messages their successful delivery is not guaranteed. Retransmissions may improve loss performance of the system; however, this improvement comes at the expense of higher delivery delays.

The second upsetting feature of IEEE 802.11p MAC protocol is that the presence of a number of nodes having the same emergency data to distribute does not positively affect the probability of packet delivery. Transmitting nodes do not

Figure 1. Probability of incorrect reception as a function of W, N, p

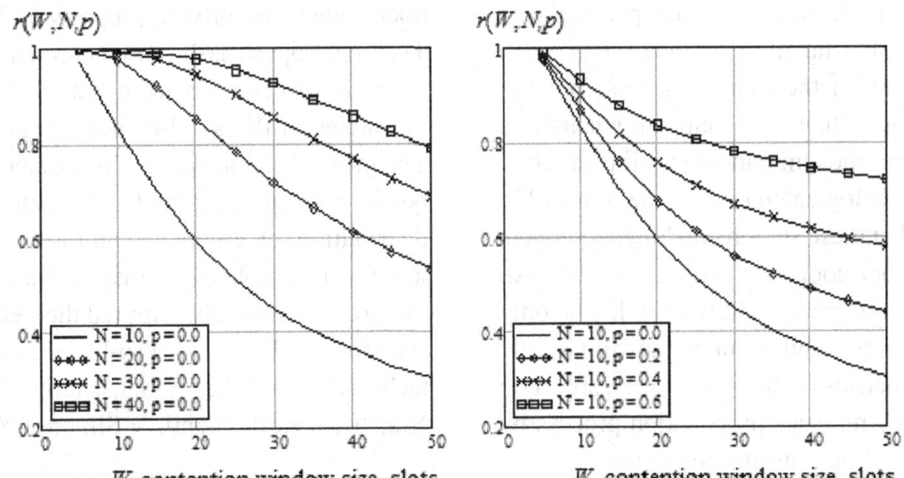

cooperate on delivering of a message but fecklessly compete with each other. Finally, the so-called synchronization effect between stations competing for the resources may degrade performance of different forms of forwarding. This happens when two or more nodes perform retransmissions immediately after receiving the packet. Since their IFSs are the same they all start transmitting simultaneously and their packets collide.

The problems discussed above are inherent for city centers. They are also common for highways, where the traffic usually slows down once an accident happens. Even when the traffic flow is not affected, cars are known to clip on the road leading to the worst possible scenario for 802.11p MAC protocol. Considering the goals of the protocol distributing emergency information this situation is unreasonable and should be avoided. The protocol should exploit denseness of networking nodes distributing the same information meaning that higher density should lead to better performance. In what follows, we propose the concept of synchronous relaying which effectively avoids the described situation by exploiting the density of nodes. Instead of competing nodes help each other to deliver messages. For our scheme more nodes are in the coverage area of each other the better the performance of the system. The most important fact about our approach is that for an external observer the system still operates according to the current version of IEEE 802.11p.

2. SYNCHRONOUS RELAYING

There are two ways to overcome highlighted deficiencies of 802.11p MAC protocol. The straightforward approach is to get rid of random MAC and use coordinated access mechanisms. However, the lack of network-wide coordination point in VANETs makes this task too complicated. Choosing local coordination points (e.g., infrastructure nodes or random cars) would lead to significant signaling overhead and may still not provide sufficient performance due to diverse dynamic nature of VANETs. Finally, this approach is not philosophically compatible with the recently standardized IEEE 802.11p MAC protocol.

Another approach is to modify the random access protocol such that special rules are applied when emergency information is disseminated. IEEE 802.11p working group adopted priority-based EDCA CSMA/CA mechanisms, which were originally developed for wireless local area networks within IEEE 802.11e group. However, as it has been shown above the performance of the protocol is far from optimal when many nodes compete for the channel access. What is more important is that it also happens when these nodes distribute the same emergency information. As a result, it is logical to modify the protocol in such a way that these stations no longer compete for resource but cooperate with each other. An appropriate cooperation strategy will not only affect collision probability but may also positively affect other metrics of the system, e.g., speed up the emergency message propagation process by increasing one-hop transmission range.

A. The Concept

Assume that a certain node has got new emergency data to distribute. It then tries to get access to the channel by competing with a certain number of other nodes having the same information. Once one of them gets access all the nodes in its coverage area become notified. To redistribute this information further they do not compete for the medium any longer. Instead, they all start transmitting the received message synchronously in a certain time slot avoiding collisions. Assuming perfect synchronization the resulting transmission heard on the channel by other node is also more reliable as all the nodes add power to the signal. Notice that according to our proposal relaying nodes should not modify the content of messages. However, this feature is not expected to be crucial as long as the reason for distribution of emergency information and location of the hazard are provided in the body of message by its originator.

Note that the proposed concept allows increasing performance of the protocol in terms of both speed of dissemination and probability that a given node in the coverage area of a network successfully receives an emergency message. The later property stems from the fact that the increase in the number of nodes in the local contention environment having emergency information increases the probability of successful packet reception by other nodes due to amplifying property of the scheme. The first property is due to synchronous operation of the protocol decreasing the number of competing nodes in the local contention environment. The proposed concept and its variants discussed below not only achieve better performance for distribution of emergency information but allow for its simple performance analysis. This is important given rather limited theoretical results available for IEEE 802.11p-based dissemination techniques (see e.g., Vinel, Dudin, Andreev, & Xia, 2010; Resta, Santi, & Simon, 2007).

B. Implementation Variants

According to the simplest scheme all the nodes hearing the ongoing transmission of emergency information are required to distribute it just after the current transmission is over. This approach ensures that the emergency data are distributed as quickly as possible and do not encounter any backoff delays or competition from the lower priority traffic. In addition, this scheme would eliminate the need for additional synchronization at the MAC layer as stations access the channel immediately after the ongoing transmission is over. Thus, no modifications to the current message format are needed. Notice that this scheme blocks any emergency messages notifying about other hazards that might have been detected later. This scheme also places strict requirements on the speed on information processing at the network nodes. Below we discuss two schemes that address these shortcomings and do not completely block lower priority traffic and other emergency messages.

Another approach is to use fixed time intervals expressed in a certain number of slots between redistribution of emergency information. Consider a set of nodes which have received the ongoing transmission of emergency information. They sense the media for a certain number of slots, i, where i is constant. If the media is free they begin synchronized relaying. In this scheme there is non-zero probability that transmission of lower priority or other emergency packets may occupy the channel first. In this case nodes waiting for relaying freeze their counters for transmission time of these packets. The amount of time slots before relaying can be chosen at the design phase. In this case, there is no need to modify the structure of IEEE 802.11p packets to carry additional information. However, if there is more than one flow of emergency messages to be distributed in the local contention environment this would result in permanent collisions. Alternatively, the originator of the emergency message may set *i* to a certain value depending on the environment around the hazard. For example, for high density areas lower value of *i* should be chosen. This would ensure greater rate of information distribution by lowering the probability of channel access for nodes having low priority traffic. Information about the density of nodes can be obtained using beaconing defined in IEEE 802.11p. This value should not be changed as message is relayed further. Observe that the message format of IEEE 802.11p packets needs to be modified to include the value of the counter associated with the packet. It is also important to note that a certain node may participate in emergency data delivery associated with different hazards. This is done by storing backoff counters associated with different emergency packets and performing the described procedure for each packet independently.

Finally, the originator of the emergency message may simply draw a certain value of backoff counter probabilistically, use it for accessing the channel and advertise it in the sent message. This scheme is not fundamentally different from the constant counter value discussed above. However, it better illustrates the backward compatibility of the proposed approach with CSMA/CA-based IEEE 802.11p mechanism. In addition, it allows deeply understanding the reason for performance gains obtained using the proposed scheme. Indeed, the contention environment with *N* nodes having different type of traffic degenerate to the contention environment with (*N*-*n*) nodes, where *n* is the number of nodes which currently participate in the distribution of emergency information, i.e., they received the message and their backoff counter for this message has not reached 0 yet. The rest of the nodes are those that have not received the emergency message yet and have low priority traffic or other emergency messages to transmit. Operation of the proposed schemes is illustrated in Figure 2, where T_1 and T_2 are message propagation times.

C. Synchronization Issues

Observing Figure 2 one may notice that synchronization between nodes at both MAC and physical layers is required. The synchronized operation at the MAC layer is achieved by counting the number of slots since the last received message. Indeed, only those nodes which have heard the message correctly are required to participate in the dissemination process. However, due to different propagation times between modes in VANETs ($T_1 \neq T_2$ in Figure 2) additional efforts ensuring synchronization of the beginning of slots at the physical layer are required. This information can be obtained using current ongoing transmissions

Figure 2. The proposed schemes, left: i=0, right: i~U(0,W)

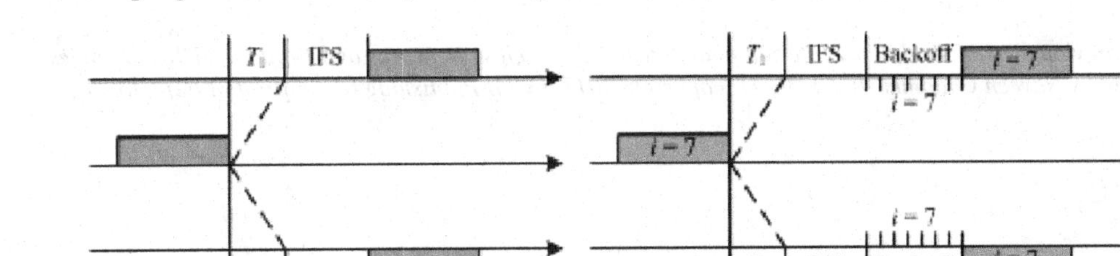

and beaconing performed by the nodes. The problem is partially alleviated by usage of OFDM in 802.11p that does not impose strict requirements on time synchronization.

This question has been recently addressed in context of cooperative relaying in cellular systems (see e.g., Lin, Sammour, Sfar, Charlton, Chitrapu, & Reznik, 2009; Ping, Zheng, Kai, Lu, & Wu, 2009). Also notice that only those nodes which directly hear each other need to be synchronized. Assuming perfect synchronization at the MAC and physical layers, analysis of our scheme is trivial and for this reason is not presented here.

3. CONCLUSION

The simple concept of emergency information dissemination presented here is backward compatible with IEEE 802.11p Carrier Sense Multiple Access with Collision Avoidance. In this scheme stations do not compete for the bandwidth when trying to relay the same emergency message, but instead cooperate with each other by synchronizing their transmissions, thus reducing the propagation delay. The proposed scheme can prove to be particularly useful in dense traffic conditions, e.g., in city centers or highways during rush hours, where each node has a significant number of stations within its antenna range.

REFERENCES

Bianchi, G. (2000). Performance analysis of the IEEE 802.11 distributed coordination function. *IEEE Journal on Selected Areas in Communications*, *18*(3), 535–547. doi:10.1109/49.840210.

IEEE. Computer Society. (2010). IEEE 802.11: Amendment 6: Wireless access in vehicular environments. Washington, DC: IEEE Computer Society.

Lin, Z., Sammour, M., Sfar, S., Charlton, G., Chitrapu, P., & Reznik, A. (2009, April). MAC v. PHY: How to relay in cellular networks. In *Proceedings of the IEEE Conference on Wireless Communications and Networking*, Budapest, Hungary (pp. 1-6).

Ping, G., Zheng, J., Kai, N., Lu, T., & Wu, W. (2009, April). A timing synchronization scheme for space-time cooperative relay OFDM system. In *Proceedings of the Conference on Wireless and Optical Communication Networks*, Cairo, Egypt (pp. 1-5).

Resta, G., Santi, P., & Simon, J. (2007, September). Message propagation in vehicular ad hoc networks. In *Proceedings of the ACM International Symposium on Mobile Ad Hoc Networking and Computing*, Montreal, QC, Canada (pp. 140-149).

Torrent-Moreno, M., Mittag, J., Santi, P., & Hartenstein, H. (2009). Vehicle-to-vehicle communication: Fair transmit power control for safety-critical information. *IEEE Transactions on Vehicular Technology*, *58*(7), 3684–3703. doi:10.1109/TVT.2009.2017545.

Vinel, A., Dudin, A., Andreev, S., & Xia, F. (2010, September). Performance modeling methodology of emergency dissemination algorithms for vehicular ad-hoc networks. In *Proceedings of the 7th International Symposium on Communication Systems, Networks and Digital Signal Processing*, Newcastle, UK (pp. 397-400).

This work was previously published in the International Journal of Wireless Networks and Broadband Technologies, Volume 1, Issue 2, edited by Naveen Chilamkurti, pp. 36-41, copyright 2011 by IGI Publishing (an imprint of IGI Global).

Chapter 11
Mobile WiMAX Bandwidth Reservation Thresholds:
A Heuristic Approach

Sondes Khemiri
University of Paris 6, France

Khaled Boussetta
University of Paris 13, France

Nadjib Achir
University of Paris 13, France

Guy Pujolle
University of Paris 6, France

ABSTRACT

This paper addresses the issue of wireless bandwidth partitioning of a Mobile WiMAX cell. The authors consider a Complete Partitioning strategy, where the wireless bandwidth capacity of a cell is divided into trunks. Each partition is strictly reserved to a particular type of connection. Four IEEE 802.16e 2005 service classes are distinguished: UGS, rtPS, nrtPS, and ErtPS. The authors consider mobility and differentiate new call request from handoffs. In addition, the authors take into consideration the Adaptive Modulation and Coding (AMC) scheme, through the partition of the cell into different areas associated to a particular modulation and coding scheme. The purpose of the paper is to determine, using an analytical model and a heuristic approach, the nearly optimal sizes of the partition sizes dedicated to each type of connection, which is characterized by its service class, type of request and modulation, and coding scheme.

INTRODUCTION

This work focuses on IEEE 802.16e-2005 WiMAX network with Wireless MAN OFDMA physical layer. This technology, also referred as a mobile WiMAX, aims to offer a quadruple-play service to multiple mobile subscribers (MSS), which can have access anytime and anywhere to various application types, like file downloading, video streaming, emails and VoIP with and without silence suppression. In order to guarantee the quality of service required by these applications, the standard defines five service classes. Namely: UGS for VoIP without silence suppression, rtPS for video streaming, nrtPS for file downloading, ErtPS for voice with silence suppression and

DOI: 10.4018/978-1-4666-3902-7.ch011

finally BE for web and mailing applications. For notation simplicity, we will refer to UGS, rtPS, nrtPS, ErtPS and BE as a class 1, 2, 3, 4 and 5 respectively. Let C={ 1,2,3,4,5}. In our work we will only study class 1,2,3 and 4 as BE service does not requires any QoS guarantees.

WiMAX network is like a cellular network, as we see in Figure 1 there is a base station (BS) that communicates with many subscribers station (SS) in down-link (from the base station) and up-link (to the base station).

IEEE 802.16 supports very high bit rates in both up-link and down-link in a large coverage.

This is why this technology is suitable for stringent QoS requirements of multimedia applications and it can handle such services as VoIP, VoD, or IP connectivity.

In order to guarantee these QoS constraints, the WiMAX provider must use an efficient call admission control (CAC) mechanism and bandwidth allocation. This work is interested on dimensioning a WiMAX network by giving a method of how to configure parameter settings of such CAC for an optimal operating system which satisfies QoS constraints.

In this paper, we focus on Complete Partitioning as a wireless bandwidth allocation strategy for a mobile WiMAX network. The purpose of the paper is to determine, using an analytical model and a heuristic approach, the nearly-optimal sizes of the partition sizes which are dedicated to each type of connection in a aim to fulfill two objectives: (1) statistically guarantee that the connection blocking probabilities remains under a given threshold, and (2) maximize the average gain of the wireless link. As BE service does not requires any QoS guarantees, we distinguish in our model four IEEE 802.16e 2005 service classes: UGS, rtPS, nrtPS and ErtPS. We also differentiate new call request from hand off ones and we integrate in the system model the Adaptive Modulation and Coding (AMC) scheme.

The rest of this paper is organized as follows: the state of the art is presented in the next section. We then describe the most important concepts defined by IEEE 802.16e-2005 standard in physical and MAC layer that we used in our proposal. The system modeling and the problem formalization is also detailed. Our incremental Partition size algorithm is then presented. Finally, the evaluation of this proposal is discussed. Finally, the last section summarizes the paper.

STATE OF THE ART

In order to understand the allocation bandwidth strategy that we used for each service class in our proposal, this section gives an overview of the mostly used policies that a provider uses to allocate bandwidth between service classes.

Figure 1. Mobile WiMAX architecture

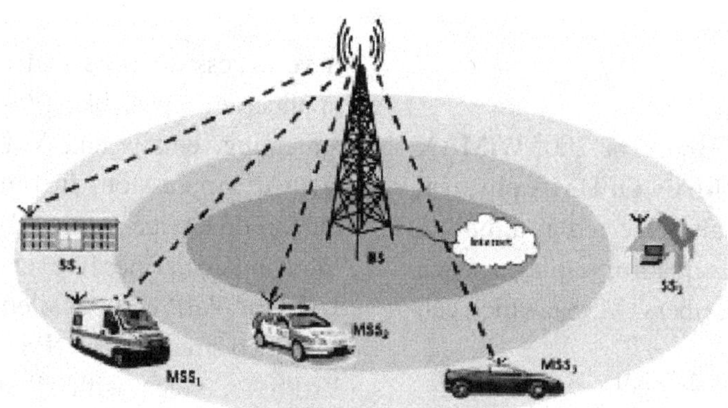

Bandwidth Sharing Strategies: Background

To maintain a quality of service required by the constraining and restricting services, we use different strategies of bandwidth allocation and admission control.

Many allocation bandwidth policies have been developed in order to give for different classes a certain amount of resource. In this work we focus on heuristic or non-optimal strategies namely Complete Sharing (CS), Upper Limit (UL), complete partitioning (CP), Guaranteed Minimum (GM) and Trunk Reservation (TR) policies:

If we suppose that we have two service classes 1 and 2, there are different ways to share the system capacity between these two classes. We will explain the main idea of each strategy described in Figure 2.

Complete Sharing (CS)

The bandwidth is shared amongst the different classes of services. So classes are in competition. In other words, we consider an offered capacity system equal to C and 2 types of service class (class 1 and 2). If class 1 uses I units then the remaining bandwidth C-I may be used by class 1 or 2.

Upper Limit (UL)

This policy is similar to CS one but it imposes threshold for each class in order to eliminate the case that one class can dominate the use of the resource. This threshold (t1 and t2) consists of the upper limit of the possible number of calls that one of each class may use.

Complete Partitioning (CP)

This policy allocates a set of resources for every service class, these resources can only be used by that class.

Guaranteed Minimum (GM)

As we see in the scheme the resource is divided into different parts. The policy gives each class their own part of bandwidth. If this partition is used up, classes can use the remained resource which is shared by all other classes.

Trunk Reservation (TR)

There is no divide of resource. In fact, class $i \in \{1, 2\}$ as we see in Figure 2 may use resources in a system until the remaining unused resource is equal to a certain threshold $r_i \in \{1, 2\}$ bandwidth units.

Figure 2. Heuristic CAC policies

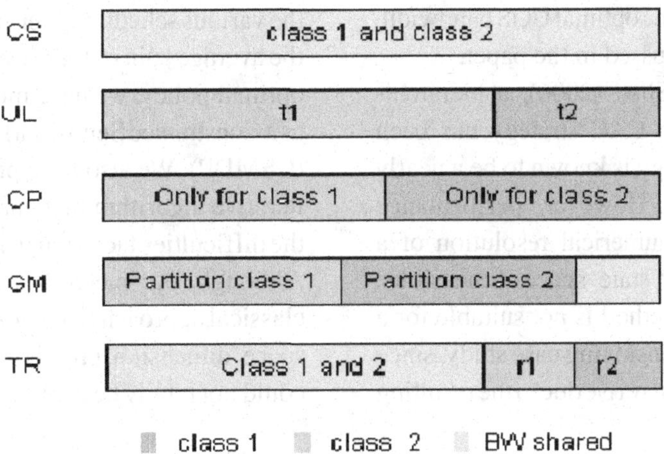

Several comparisons have been made between these policies and with optimal solution. But the most important challenge is to explain the method that thresholds imposed by GM, UL and CP strategies are computed or determined.

The following section will summarize some works which have been developed in order to propose an allocation bandwidth strategy in WiMAX networks taking into account the bandwidth sharing policies described in this part.

Related Works

To the best of our knowledge, most of the existing works have focused on the proposal of bandwidth allocation and CAC policies (Baynat, Nogueira, Maqbool, & Coupechoux, 2009; Altman, 2002; Wang & Li, 2005; Hou, Ho, & Shen, 2006; Dusit & Ekram, 2006; Khemiri, Boussetta, Achir, & Pujolle, 2007b) for the fixed WiMAX technology.

In Wang and Li (2005), the proposed bandwidth allocation strategy is based on the sharing with Upper Limits strategy (Choudhury, Leung, & Whitt, 1995). The latter one belongs to the coordinate convex CAC policies. Consequently, performances parameters could be computed analytically. Nevertheless, in this work, only the maximum number of bandwidth units which can be allocated to aggregated UGS traffics was considered. nrtPS and rtPS calls are in competition for the remaining radio resources, without any protection mechanism. In addition, the methodology to determine the value of the optimal UGS bandwidth threshold was not discussed in the paper.

In Hou, Ho, and Shen (2006), a hierarchical Trunk Reservation CAC strategy has been considered. This strategy is known to be a nearly optimal CAC policy. However, performance metrics requires the numerical resolution of a Markov chain of four state space dimensions. Such computational method is not suitable for a realistic WiMAX dimensioning case study, since numerical difficulties may rise due to the resulting huge state space dimension. In addition, computing the optimal trunk reservation thresholds for hierarchical Trunk Reservation policy is not easy, and it was not investigated in the paper.

In Dusit and Ekram (2006), the authors presented a joint bandwidth allocation and CAC policy for radio resource management in IEEE 802.16-based multi-service Broadband Wireless Access (BWA) networks. The aim of the proposal is to guarantee both the packet-level and the connection-level QoS requirements for different types of services while maximizing the system utility. In this work the total available bandwidth is shared among the different types of services using a Complete Partitioning approach. The authors developed an interesting mathematical model, which takes into account the specificities of the IEEE 802.16 technology. Different sophisticated aspects, like Adaptive Modulation and Coding (AMC) were considered in the model. From their mathematical modeling, they formulated the optimization problem that needs to be solved in order to obtain the optimal partition sizes, but unfortunately without providing a particular resolution methodology. Partition sizes values were considered in the performance evaluation section, but without further justifications.

In Khemiri, Boussetta, Achir, and Pujolle (2007b), we have investigated the computational of the optimal CAC policy for an IEEE 802.16 Wireless MAN, which satisfies the following two objectives: (1) statistically guarantee the QoS of the various scheduling services, and (2) maximize the average gain of the wireless link. To find such optimal policy, we have modeled our CAC agent as a Constrained Semi-Markov Decision Processes (CSMDP). We also have proposed an alternative iterative algorithm that can be used to overcome the difficulties faced when resolving the CSMDP. Although, our method is more robust than the classical approach (Linear Programming) to state space dimension problem, numerical problems could not totally be avoided for large state spaces, like those observed in realistic dimensioning case studies. Consequently, in Khemiri, Boussetta, Achir, and Pujolle (2007a), we considered the Guaranteed Minimum bandwidth sharing strategy. The latter belongs also to the coordinate convex CAC policies, which means that the average performances of the system could be obtained analytically. Our main contribution in Khemiri, Boussetta, Achir, and Pujolle (2007a) was the proposal of a heuristic method, based on Tabu search approach, to find the optimal bandwidth reservation thresholds. The present paper differs from Khemiri, Boussetta, Achir, and Pujolle (2007a) by the fact that we consider mobile WiMAX instead of fixed WiMAX technology. We also consider here Complete partitioning rather than Guaranteed Minimum bandwidth sharing strategy. Finally, we integrate here Adaptive Modulation and Coding (AMC) scheme into our problem formulation.

All these works were done assuming fixed WiMAX. As far as we know, little works on bandwidth allocation have concentrated on the mobile WiMAX version (IEEE 802.16e Standard). A notable work is Tung, Tsang, Lee, and Ko (2008), where an adjustable Quadra Threshold Bandwidth Reservation scheme (QTBR) based on trunk reservation (TR) policy and a QoS aware bandwidth allocation algorithm (QoSABAA) for mobile WiMAX networks was proposed, in this proposal, classes may use resources in a system until the remaining unused resource is equal to a certain threshold. The main objective of the work was to maximize the radio efficiency (utilization ratio). But, as in previously mentioned works, the authors have not proposed a methodology to set the bandwidth reservation thresholds.

Moreover, contrarily to our work, ErtPS class service was not integrated in the study. We focus also on the possible differentiation between service classes, while in most of the work, as in (Tarhini & Chahed, 2008, Rong, Qian, Lu, Hsiao-Hwa, & Mohsen, 2008; Shafaq & Ratan, 2007; Liping, Fuqiang, Yusheng, & Nararat, 2007; Elayoubi & Fourestié, 2008) where the CAC policy proposed is the Complete Sharing (CS).

In this article we will focus on this problem and we will propose a method to compute size of partition for complete partitioning strategy in a mobile WiMAX system in order to maximize a system gain.

MOBILE WIMAX OVERVIEW

As we said previously, our aim in this work is: to compute thresholds fixed by a given allocation bandwidth strategy in a mobile WiMAX system, in order to maximize a system gain. This section presents an overview of the most important concepts defined by IEEE 802.16e-2005 standard in physical and MAC layer that we used in our proposal.

WiMAX PHY Layer

In order to understand the system and the scenario studied in our work, we will give in this section details about PHY layer and we will focus specially on specified concepts that we have taken into account in our system namely, the specification of the PHY layer studied, the OFDMA multiplexing scheme and the permutation scheme for sub-channelization from which we deduce the bandwidth unit allocated to accepted calls in the system and the Adaptive Modulation and Coding scheme (AMC).

Generality

The IEEE 802.16 defines five PHY layers which can be used with a MAC layer to form a broadband wireless system.

As we see in Figure 3, these PHY layers provide a large flexibility in terms of bandwidth channel, duplexing scheme and channel condition. These layers are described as follows:

1. **WirelessMAN SC:** In this PHY layer single carriers are used to transmit information for frequencies beyond 11GHz in a Line of sight (LOS) condition.
2. **WirelessMAN SCa:** It is a single carrier too but it is used for frequencies between 2 GHz and 11GHz.
3. **WirelessMAN OFDM (Orthogonal Frequency Devision Multiplexing):** Based on a Fast Fourier Transform (FFT) with a size of 256 point. It is used for point multipoint link in a non-LOS condition for frequencies between 2 GHz and 11GHz.
5. **WirelessMAN OFDMA (OFDM Access):** Also refereed as mobile WiMAX, it is also based on a FFT with a size of 2048 point. It is used in a non LOS condition for frequencies between 2 GHz and 11GHz.
6. **Finally a WirelessMAN SOFDMA (SOFDM Access):** OFDMA PHY layer has been extended in IEEE 802.16e to SOFDMA (scalable OFDMA) where the size is variable and can take different values: (128, 512, 1024, and 2048).

As we see in previous paragraph many combinations of configuration parameters like band frequencies, channel bandwidth and duplexing techniques are possible. This is why WiMAX Forum tries to insure interoperability between terminals and base station by creating a WiMAX system profiles. It is a set of fixed configuration parameters to insure the functionality of all terminals and the interoperability between base stations. In our system we will use parameters given by mobile WiMAX system profile.

The WiMAX PHY layer has also the responsibility of slot allocation and framing over the radio channel. The minimum time-frequency resource that can be allocated by a WiMAX system to a given link is called a slot. Each slot consists of one sub-

Figure 3. Physical layer

PHY Layer	description	Function	Frequency	LOS/ NLOS	Duplexage
WirelessMAN SC	single carriers	PTP	10-66 GHz	LOS	TDD,FDD
WirelessMAN SCa	single carriers	PTP	2-11 GHz	NLOS	TDD,FDD
WirelessMAN OFDM*	.*Fixed WiMAX: 802.16d-2004* .FFT** with a size of 256 point	PMP	2-11 GHz	NLOS	TDD,FDD
WirelessMAN OFDMA***	.*Mobile WiMAX: 802.16e-2005* .FFT with a size of 2048 point	PMP	2-11 GHz	NLOS	TDD
WirelessMAN SOFDMA****	.OFDMA extension .Variable number of subcarrier (128, 512, 1024, 2048)	PMP	2-11 GHz	NLOS	TDD

*Orthogonal Frequency Devision Multiplexing
** Fast Fourier Transform
*** OFDM Access
**** Scalable OFDMA

channel over one, two, or three OFDM symbols, depending on the particular sub-channelization scheme used.

WiMAX defines several sub-channelization schemes. The sub-channelization could be adjacent, i.e., sub-carriers are grouped in the same frequency range in each sub-channel or distributed and sub-carriers are pseudo-randomly distributed across the frequency spectrum.

- **Full Usage Sub-Carriers (FUSC):** Each slot is 48 sub-carriers by one OFDM symbol.
- **Down-Link Partial Usage of Sub-Carrier (PUSC):** Each slot is 24 sub-carriers by two OFDM symbols.
- **Up-Link PUSC and TUSC Tile Usage of Sub-Carrier:** Each slot is 16 sub-carriers by three OFDM symbols.
- **Band Adaptive Modulation and Coding (BAMC):** Each slot is 8, 16, or 24 sub-carriers by 6, 3, or 2 OFDM symbols.

In this work we will focus on the last permutation scheme, i.e., only AMC scheme will be investigated in this article.

The Adaptive Modulation and Coding Scheme (AMC)

In order to adapt the transmission to the time varying channel conditions that depends on the radio link characteristics WiMAX presents the advantage of supporting the link adaptation called Adaptative Modulation and Coding scheme (AMC). It is an adaptive modification of the combination of modulation, channel coding types and coding rate also known as burst profile that takes place in the physical link depending on a new radio condition. Table 1 shows examples of burst profiles in mobile WiMAX (there are 52 in IEEE802.16e-2005 (IEEE Std 802.16e 2005).

Table 1. Burst profile examples: (CC) Convolutional Code, (RS) Reed-Solomon

Profile	Modulation	Coding scheme	Rate
0	BPSK	(CC)	$\frac{1}{2}$
1	QPSK	$(RS + CC/CC)$	$\frac{1}{2}$
2	QPSK	$(RS + CC/CC)$	$\frac{3}{4}$
3	16 QAM	$(RS + CC/CC)$	$\frac{1}{2}$
6	64 QAM	$(RS + CC/CC)$	$\frac{3}{4}$

In fact when a subscriber station tries to enter to the system, the WiMAX network undergoes various steps of signalization: the Down-link channel is scanned and synchronized, in this step after the synchronization the SS obtains information about PHY and MAC parameters corresponding to the DL and UL transmission from control messages that follow the preamble of the DL frame. Based on this information negotiations are established between the SS and the BS about basic capabilities like maximum transmission power, FFT size, type of modulation, and sub-carrier permutation support.

In this negotiation the BS takes into account the time varying channel conditions by computing the signal to noise ratio (SNR) and then decides which burst profile must be used for the SS. In fact, the down-link SNR is provided by the mobile to the base station, using the channel quality feedback indicator, with feedback on the down-link channel quality. For the up-link, the base station can estimate the channel quality, based on the received signal quality.

Indeed different modulation schemes will be employed in the same network in order to maximize throughput in a time-varying channel.

When the distance between the base station and the subscriber station increases the signal to the noise ratio decreases due to the path loss. This is why modulation must be used depending on the station position starting from the lower efficiency modulation (near the BS) to the higher efficiency modulation (far away the BS). In our work we will consider the transmitted signal will be composed by 3 burst profiles 1, 3 and 6.

WiMAX MAC Layer and QoS Overview

The primary task of the WiMAX MAC layer is to provide an interface between the higher transport layers and the physical layer.

The IEEE 802.16-2004 and IEEE 802.16e-2005 MAC design includes a convergence sublayer that can interface with a variety of higher-layer protocols, such as ATM,TDM Voice, Ethernet, IP, and any unknown future protocol.

Support for QoS is a fundamental part of the WiMAX MAC-layer design. QoS control is achieved by using a connection-oriented MAC architecture, where all down-link and up-link connections are controlled by the serving BS.

Before any data transmission happens, the BS and the MS establish a unidirectional logical link, called a connection, between the two MAC-layer peers. Each connection is identified by a connection identifier (CID), which serves as a temporary address for data transmissions over the particular link.

WiMAX also defines a concept of a service flow. A service flow is a unidirectional flow of packets with a particular set of QoS parameters and is identified by a service flow identifier (SFID).

The QoS parameters could include traffic priority, maximum sustained traffic rate, maximum burst rate, minimum tolerable rate, scheduling type, ARQ type, maximum delay, tolerated jitter, service data unit type and size, bandwidth request mechanism to be used, transmission PDU formation rules, and so on. Service flows may be provisioned through a network management system or created dynamically through defined signaling mechanisms in the standard. The base station is responsible for issuing the SFID and mapping it to unique CIDs.

Mobile WiMAX is emerging as one of the most promising 4G technology. It has been developed keeping in view the stringent QoS requirements of multimedia applications. Indeed, the IEEE 802.16e 2005 standard defines five QoS scheduling services that should be treated appropriately by the base station MAC scheduler for data transport over a connection:

1. Unsolicited Grant Service (UGS) is dedicated to real-time services that generate CBR or CBR-like flows. A typical application would be Voice over IP.
2. Real-Time Polling Service (rtPS) is designed to support real-time services that generate delay sensitive VBR flows, such as MPEG video or VoIP (with silence suppression).
3. Non-Real-Time Polling Service (nrtPS) is designed to support delay-tolerant data delivery with variable size packets, such as high bandwidth FTP.
4. Extended Real-Time Polling Service (ErtPS) is expected to provide VoIP services with Voice Activation Detection (VAD).
5. Best Effort (BE) service is proposed to be used for all applications which do not require any QoS guarantees.

In order to guarantee the QoS for these different service classes Call Admission Control (CAC) and resource reservation strategies are needed by the IEE 802.16e system.

As in cellular networks, the IEEE 802.16 Base Station MAC layer is in charge to regulate and control bandwidth allocation. Therefore, incorporating a Call Admission Control (CAC) agent becomes the primary method to allocate network resources in such a way that the QoS user constraints could be satisfied. Before any connection establishment, each SS informs the BS about its QoS requirements. The BS CAC agent has the responsibility to determine whether a connection request can be accepted or should be rejected. The rejection happens if its QoS requirements cannot be satisfied or if its acceptance may violate the QoS guarantee of ongoing calls.

Remember that this work answers to the following question: How to configure parameter settings of a CAC and bandwidth allocation mechanisms in order to obtain an optimal operating system which satisfies QoS constraints. We need to model our system to answer this question.

SYSTEM MODELING

System Description

We consider a network relying on IEEE 802.16e 2005 technology with WirelessMAN OFDMA physical layer. The system provides a quadruple-play service to multiple mobile subscribers (MSS), which can have access anytime and anywhere to various application types, like file downloading, video streaming, emails and VoIP. In order to guarantee the quality of service required by these applications, the service provider has to distinguish four service classes. Namely: UGS for VoIP, rtPS for video streaming, nrtPS for file downloading and ErtPS for voice without silence suppression. For notation simplicity, we will refer to UGS, rtPS, nrtPS and ErtPS as a class 1, 2, 3 and 4, respectively. Let T = {1, 2, 3, 4}.

In order to adapt the transmission rate according to the channel quality we consider that the system supports an Adaptive Modulation and Coding (AMC) scheme. Precisely, we assume that the cell is divided into Z areas where a same modulation and coding scheme is used for all calls ongoing in a specific area. Without loss of generality, we assume that Z={ 1,2,3} and that the areas border shape are concentric perfect circles. Figure 4 illustrates the system, with a typical example of modulation schemes associated to each area.

In this study, we consider that the bandwidth is a crucial resource that has to be shared and that the wireless link is the main bottleneck. The total offered bandwidth capacity of the cell S is supposed to be a fixed number of bandwidth units, which we note C. This capacity is divided into Z fixed partitions, noted C_j j ∈ Z. Each one is exclusively associated to a particular modulation and coding scheme.

We assume that the bandwidth allocation is Granted Per Connection (GPC). Thus, for each MSS an amount of bandwidth units is assigned to each established connection. Precisely, any class j ∈ T connection request associated to an MSS located in an area i ∈ Z will be assigned a given

Figure 4. System description

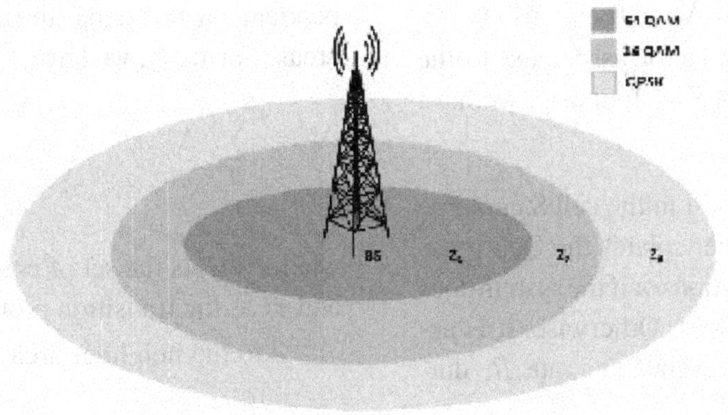

number of bandwidth units, noted s_{ij}. This assigned capacity has to be carefully chosen so that the connection could be established and maintained with satisfied QoS requirements in the down-link and up-link directions.

As rtPS, nrtPS and ErtPS connections generate variable bit rate traffic, and then the allocated bandwidth capacity could be variable over the time. Nevertheless, in order to guarantee the QoS, a minimum capacity is required to accept any corresponding request. For any class $j \in T$ request arriving in an area $i \in Z$ we note $\underline{s}_{ij}$ the associated minimum required bandwidth units. Similarly, we assume that a given maximum bandwidth capacity is sufficient to fulfill the Qos constraints of the supported classes. Therefore, any class $j \in T$ request arriving in an area $i \in Z$ will be assigned at maximum $\overline{s_{ij}}$ bandwidth units. So, formally, $\forall i \in Z$ and $\forall i \in T$ we have $s_{ij} \in \left[\underline{s_{ij}}, \overline{s_{ij}}\right]$ Here, it is worth to note that since UGS connections generate constant bit rate traffic then $s_{i1} = \underline{s_{i1}} = \overline{s_{i1}}$.

Traffic Model and System Parameters

Since an IEEE 802.16e 2005 cell's coverage can reach a diameter of several kilometers, we could reasonably assume an infinite population traffic model. Precisely, class $j \in T$ connection requests arrive to the cell according to a Poisson process with an aggregate rate λ_j^{nc}. Non uniform traffic patterns are obtained by setting the arrival rates in each area $i \in Z$ with: $\lambda_{ij}^{nc} = \alpha_{ij}\lambda_j^{nc}$ where α_{ij} is a spatial distribution factor associated to the class j and the area $i \in Z$. Obviously, we have $\sum_{i \in Z} \alpha_{ij} = 1,$.

Any class $j \in T$ request in the cell S could be rejected if its acceptance violates the QoS guarantee of ongoing connections or if the system does not have enough resources. Otherwise, it is accepted and generates a revenue at a rate R_j during the connection holding time, which is supposed to be exponentially distributed with mean $\left[\mu_j^h\right]^{-1}$, (The revenue rate corresponds for instance, to the amount of money per second earned by the system for carrying this call).

Mobility

As the considered technology supports mobility, we will distinguish between two types of connection requests: (1) new calls (nc) entering to the system, and (2) Handoff (ho) demands associated to ongoing connections of MSS moving from an area $i \in Z$ to another adjacent area $l \in Z$ or from a cell to a neighbor one. Precisely, we assume that the sojourn duration of an MSS in an area $i \in Z$ is exponentially distributed with mean $\left[\mu_i^s\right]^{-1}$. A connection of class $j \in T$ can leave an area $i \in Z$ if (1) the call is terminated or (2) if the MSS moves to an adjacent area (i.e., handoff). Hence, the duration time of a class $j \in T$ connection in an area $i \in Z$ is exponentially distributed with mean:

$$\left[\mu_{ij}\right]^{-1}, where, \mu_{ij} = \mu_i^s + \mu_j^h$$

Let λ_{ij}^{ho} and Λ_{ij}^{ho} be respectively, the arrival and the departure handoff rates associated to traffic class $j \in T$ in the area $i \in Z$. As illustrated in Figure 5, there is a dependency between these rates and among the different areas. Precisely, in one hand, the arrival rate in a given area is dependent on the departure rate from neighbors' areas. Formally, we have,

$$\lambda_{ij}^{ho} = \sum_{i \in V(i)} \Lambda_{lj}^{ho}.t_{li} \quad (1)$$

where, V(i) is the set of neighbor areas to $i \in Z$ and t_{li} is the transition probability from the area $i \in A$ to the neighbor area $l \in V(i)$. Transition

Figure 5. Possible hand off

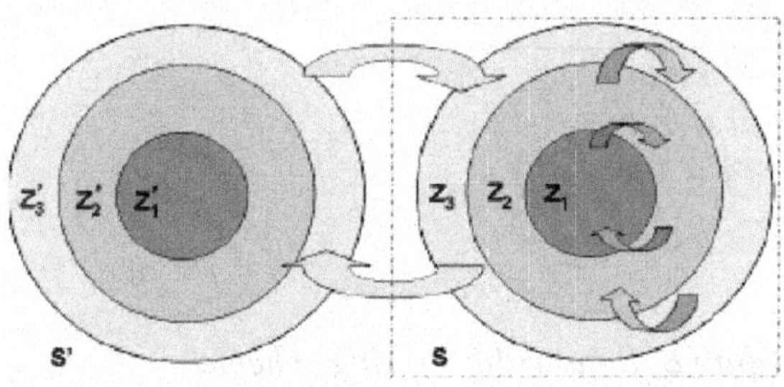

possibilities between areas are illustrated in Figure 6. One can see that Z_2 is the only neighbor to Z_1, while Z_3 has Z_2 and all adjacent cells Z_3 areas as neighbors.

On the other hand, the departure handoff rate for class $j \in T$ connections from an area $i \in Z$ can be computed as follow:

$$\Lambda_{ij}^{ho} = \frac{\mu_i^s}{\mu_{ij}}(\lambda_{ij}^{nc}\left(1 - B_{ij}^{nc}\right) + \lambda_{ij}^{ho}\left(1 - B_{ij}^{ho}\right)) \quad (2)$$

Where B_{ij}^{nc} and B_{ij}^{ho} are respectively, the nc and ho blocking probabilities of class $j \in T$ calls in the area $i \in Z$.

Figure 6. Possible transition between zones

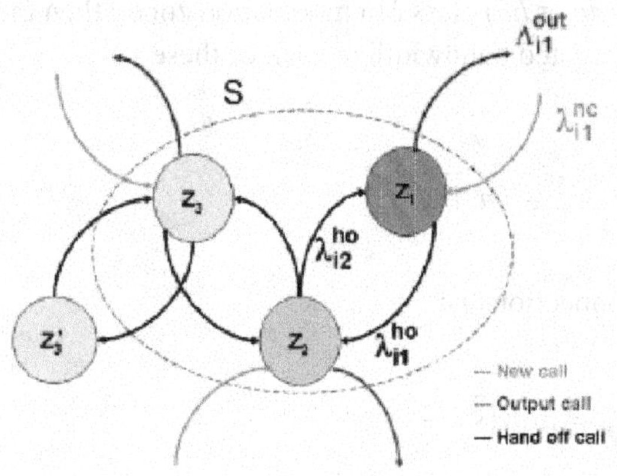

BANWIDTH SHARING SCHEME STRATEGY

Problem Formulation

Following this system modeling, we consider that the packet-level's QoS parameters (loss rates, delays, jitters etc.) of a class j connection are statistically guaranteed in the wireless link part of the WiMAX network as long as the allocated bandwidth s_{ij} remains in the $\left[\underline{s_{ij}}, \overline{s_{ij}}\right]$.

Where R is the average revenue of the cell. Formally,

$$R = \sum_{j \in T} R_j \sum_{i \in Z} \frac{\lambda_{ij}^{nc}}{\mu_{ij}}\left(1 - B_{ij}^{nc}\right) + \frac{\lambda_{ij}^{ho}}{\mu_{ij}}\left(1 - B_{ij}^{ho}\right) \quad (3)$$

$\forall lj \in T$, 2_j^{nc} and 2_j^{ho} are the given nc and ho tolerated blocking thresholds, respectively.

Complete Partitioning (CP)

In realistic case studies, the value of the thresholds β_j^{nc} and β_j^{ho} are usually set by the network operator to very low values. Typically, in the magnitude of 10^{-3} and 10^{-4}, respectively. Therefore, any CAC policy Π which can satisfy the constraints of the Figure 7 will generate an average revenue which will be very close to the one obtained by the optimal CAC policy Π^*. Hence, in this paper we do not seek to find Π^*, but rather a simple and computational CAC policy which satisfy the constraints of the Figure 7.

Among the CAC policies which can be considered, we choose to focus on the coordinate convex category. This choice is motivated by the fact that the steady states probabilities of the resulting system have closed form expression. This is very suitable since the blocking probabilities

Figure 7. Problem 1: Optimal CAC problem

Problem 1 Optimal CAC problem

$$\begin{aligned} \texttt{maximize} \quad & R \\ \texttt{subject to} \quad & B_{ij}^{nc} \leq \beta_j^{nc}, \forall i \in Z \forall j \in T \\ & B_{ij}^{ho} \leq \beta_j^{ho}, \forall j \in Z \forall i \in T \end{aligned}$$

can be derived analytically, avoiding by this way the difficulties raised by numerical computations.

There are several policies which belong to the coordinate convex category. In this paper, we choose the Complete Partitioning (PC) strategy, which now has the capacity to guarantee the QoS constraints of each traffic class. This is simply achieved by allocating a specific bandwidth partition to the exclusive use of each traffic class.

The application of CP for our case study is illustrated in Figure 6. The capacity C is divided into 3 partitions. Each partition C_i is assigned to the calls associated to the area Z_i. This capacity is also partitioned into 4 trunks of capacity C_{ij} ($i \in Z$ and $j \in T$, each one being reserved to the traffic class j. Finally, each partition C_{ij} is divided into two trunks, noted C_{ij}^{nc} and C_{ij}^{ho}, allocated respectively to nc and ho class j connections in area i.

Bandwidth Sharing in a Partition

When a new request of type *k* (new call or handover) class *j* arrives in area *i*, the system checks if there is enough available bandwidth in partition C_{ij}^{k} to accept this connection (Figure 8). Precisely, the system has to find a minimum of $\underline{s}$ij bandwidth units in the corresponding partition to accept the request. Otherwise, the latter is blocked. Since we assume complete fairness among calls from same types, the bandwidth allocated in a partition to all ongoing calls is the same. Hence, all ongoing class

Figure 8. Bandwidth sharing scheme

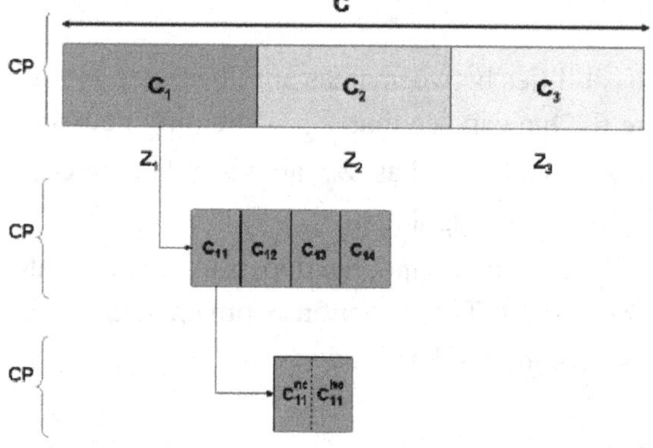

j calls in modulation zone *i* are allocated the same bandwidth quantity s_{ij}. Therefore, if *n* connections of type *k* (*nc* or *ho*) class *i* and modulation zone *j* are actives while a new request from the same category arrives, the latter is accepted only if

Otherwise, the request is blocked. Note that when a call leaves the system, the available bandwidth is shared in fair way among ongoing calls. Formally, if *n* is the number of left calls of type *k* (*nc* or *ho*) class *i* in modulation zone *j* then the allocated bandwidth to each of these

$$\max(\underline{s}_{ij}, \frac{C_{ij}^{k}}{(n+1)}) \geq s_{ij}$$

connections is:

$$s_{ij} = \min(\overline{s}_{ij}, \frac{C_{ij}^{k}}{n})$$

Optimal Partitions Size

Assuming the CP strategy, each partition C_{ij}^{k},

$$(i,j,k) \in (Z \times T) \times \{nc, ho\}$$

can be modeled as an Erlang B queue model. Therefore, the call blocking probabilities can be analytically expressed as:

$$B_{ij}^{k} = \frac{1}{G_{ij}^{k}} \frac{\left(\frac{\lambda_{ij}^{k}}{\mu_{ij}}\right)^{\left\lfloor \frac{C_{ij}^{k}}{s_{ij}} \right\rfloor}}{\left\lfloor \frac{C_{ij}^{k}}{s_{ij}} \right\rfloor !} \quad (4)$$

where the normalization constant is:

$$G_{ij}^{k} = \sum_{n=0}^{\left\lfloor \frac{C_{ij}^{k}}{s_{ij}} \right\rfloor} \frac{\left(\frac{\lambda_{ij}^{k}}{\mu_{ij}}\right)^{n}}{n!}$$

Equation 4 shows clearly that the call blocking probabilities are dependent on the partition sizes. Thus, problem 1 can be associated to a dual problem, where the aim is to find the optimal partitions reservation vector:

$$\vec{c}^{\,*} = \{C_{ij}^{k}, (i,j,k) \in (Z \times T) \times \{nc, ho\}\}$$

solution to the optimization problem seen in Figure 9.

Finding $\vec{c}^{\,*}$ is not a trivial task. Especially, when the dimension of the solutions space is large. The latter is dependent on C. In our case study, the dimension of the reservation vector is 24 and in realistic WiMAX systems the capacity C is quite large (e.g., 2000 bandwidth units). As indicated in the related works section, few works have focused on this problem or proposed a scalable solution to determine the partitions size. In the next section we detail our proposal to derive a nearly-optimal reservation vector.

Incremental Partitions Size Algorithm (IPSA)

Our Incremental Partitions Size Algorithm is executed following two steps. In the first step, the objective is to find a vector $\vec{c}^{\,0}$ which can satisfy the blocking probabilities constraints, but without necessarily ensuring that the vector capacity is equal to C. In other words, we seek to determine the smallest partition sizes required to satisfy the constraints of Figure 7.

Finding the smallest required partition sizes for nc calls is trivial using Equation 4. One can simply increment the capacity,

$$C_{ij}^{nc}, \forall (i,j) \in (Z \times T)$$

until that the resulting nc blocking probabilities, given by Equation 4 are below the thresholds β_{j}^{nc}. This algorithm can also be used for the case of ho calls. However, for the latter case, the difficulty is that the handover arrival rates,

Figure 9. Problem 2: Optimal partitions size problem

Problem 2 Optimal partitions size problem

$$\begin{array}{ll} \texttt{maximize} & R \\ \texttt{subject to} & B_{ij}^{nc} \le \beta_{j}^{nc}, \forall i \in Z \forall j \in T \\ & B_{ij}^{ho} \le \beta_{j}^{ho}, \forall i \in Z \forall j \in T \\ & \sum_{i \in Z} \sum_{j \in T} C_{ij}^{nc} + C_{ij}^{ho} \le C \end{array}$$

$$\lambda_{ij}^{ho}, \forall (i,j) \in (Z \times T)$$

are not initially known. Indeed, as indicated in Equation 1, Equation 2 and Equation 4, the handoff rates, the blocking probabilities and the partition sizes are related.

We propose to solve this dependency problem, using a fixed point algorithm. Precisely, after having obtained the smallest required partition sizes for nc calls, we derive an initial estimation of the departure rates using Equation 2 and assuming that,

$$\lambda_{ij}^{ho} = 0, \forall (i,j) \in (Z \times T)$$

Then, λ_{ij}^{ho} are updated using Equation 1. The resulting values are then injected in Equation 4, which serves to find the smallest required partition sizes for ho calls. These three-steps are repeated, considering the recent values of λ_{ij}^{ho} until that each handoff blocking probability,

$$B_{ij}^{ho}, \forall (i,j) \in (Z \times T)$$

converges toward a stable value.

If the total capacity of the smallest required partition sizes is above C, then we consider that Figure 9 is infeasible for C and with our algorithm. Else, if the total capacity of the partitions reservation vector obtained in this first step is equal to C, then it's considered as our solution to Figure 9. Otherwise, we proceed with the execution of the second step of our algorithm. Here, since the blocking constraints are satisfied, the objective is now to share the remaining capacity, noted C_r, among the partitions so that the average revenue of the system is maximized.

According to Equation 3 and given that our sharing policy is based on Complete Partitioning, maximizing the revenue R leads to maximize the quantities,

$$R_j \left(\lambda_{ij}^k / \mu_{ij}\right)\left(1 - B_{ij}^k\right), (i,j,k) \in (Z \times T) \times \{nc, ho\}$$

Thus, the basic idea to share judiciously the remained capacity C_r is to select the tuple (i,j,k) which minimizes the quantity,

$$R_j \left(\lambda_{ij}^k / \mu_{ij}\right) B_{ij}^k, (i,j,k) \in (Z \times T) \times \{nc, ho\}$$

The capacity C_{ij}^k associated to the elected tuple (i,j,k) is incremented by a capacity of $\underline{s}_{ij}$ bandwidth units (at the condition that $C_r \geq \underline{s}_{ij}$).

The new value of C_{ij}^k is then used to update the blocking probabilities expressed by Equation 4. The algorithm pursue its iterations by selecting again a tuple (i,j,k) which minimizes the quantity,

$$R_j \left(\lambda_{ij}^k / \mu_{ij}\right) B_{ij}^k, (i,j,k) \in (Z \times T) \times \{nc, ho\}$$

These incremental assignment iterations are repeated as long as $\exists i \in Z$ and $\exists j \in T$ such that $C_r \geq \underline{s}_{ij}$ (Figure 10).

RESULTS

Parameters and Scenarios

We evaluated our proposal assuming a single mobile WIMAX cell. The physical layer is based on WirelessMAN OFDMA with three burst profiles (1,3,5) (IEEE Std 802.16e 2005) corresponding to QPSK $(codingrate = 1/2)$, 16 QAM $(codingrate = 1/2)$ and 64 QAM $(codingrate = 2/3)$ modulation schemes. We considered a system capacity of 70 Mbps with a bandwidth unit set to 32kbps. That is, C=2200. We chose the parameters in Table 2.

Following the previous table we fixed the minimal and maximal assignable bandwidth units as seen in Table 3.

Figure 10. Partitioning bandwidth algorithm

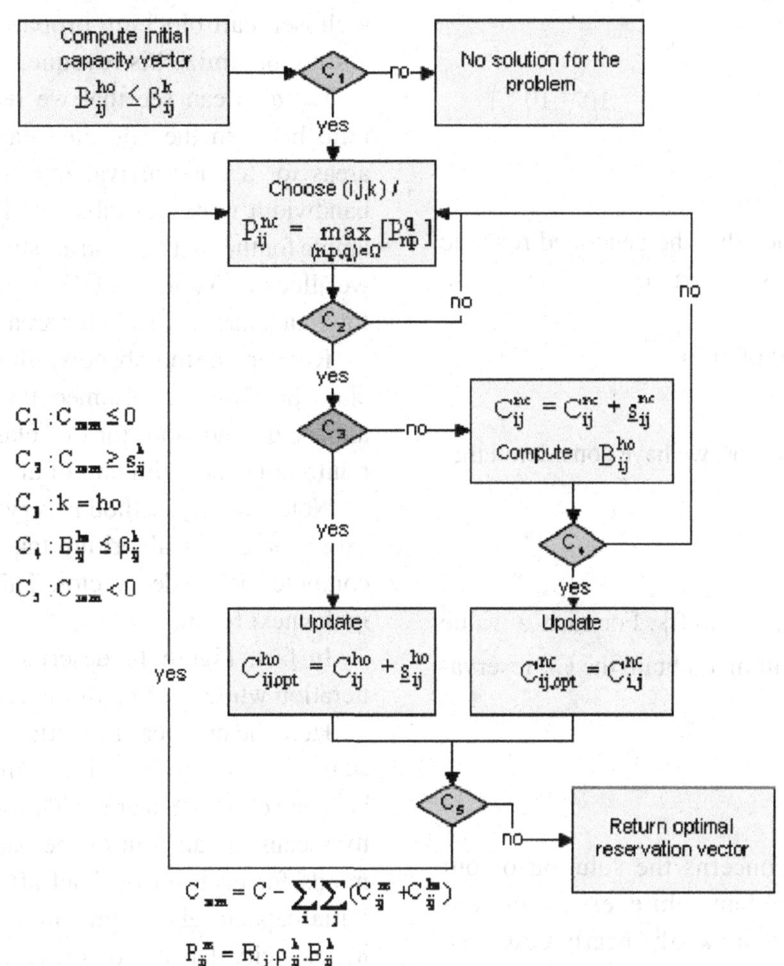

Table 2. System parameters

Parameters	Area $i = 1$	Area $i = 2$	Area $i = 3$
$\mu_{i1} = \mu_{i2} = \mu_{i3} = \mu_{i4}$	0.5	0.5	0.5
$(\underline{s}_{i1}, \underline{s}_{i2}, \underline{s}_{i3}, \underline{s}_{i4})$	$(1, 8, 4, 2)$	$(2, 16, 8, 4)$	$(3, 24, 12, 6)$

Table 3. QoS parameters

Classes	Area 1	Area 2	Area 3
UGS (1)	$[32, 128]$ kbps	$[64, 256]$ kbps	$[96, 384]$ kbps
rtPS (2)	$[256, 1024]$ kbps	$[512, 2048]$ kbps	$[768, 3072]$ kbps
nrtPS (3)	$[128, 512]$ kbps	$[256, 1024]$ kbps	$[384, 1536]$ kbps
ErtPS (4)	$[64, 256]$ kbps	$[128, 512]$ kbps	$[192, 768]$ kbps

The call blocking probability constraints were set to

$$\left(\beta_1^{nc},\beta_2^{nc},\beta_3^{nc},\beta_4^{nc}\right)=\left(10^{-4},10^{-3},10^{-3},10^{-3}\right)$$

and $\forall j\in T,\beta_j^{ho}=10^{-1}\beta_j^{nc}$.

We also assumed that the generated revenue rate for each classes is equal to

$$\left(R_j\right)_{j\in T}=\left(1,0.1,0.05,0.5\right)$$

In our test scenarios, we have considered that

$$\forall\left(i,j\right)\in Z\times T,\lambda_{ij}^{nc}=\lambda$$

where $\lambda\in\left\{1,1.1,1.2...,3.9\right\}$. For each λ value we used our algorithm to obtain the CP reservation vector $\vec{c}^{\,*}$.

Results

The first result concerns the solution of our optimization problem while executing our algorithm. Table 4 shows the nearly CAC reservation vector obtained while varying the new call arrival rate. Each line corresponds to the reservation vector that a provider need to use in order to share his system bandwidth to obtain a chosen call blocking probability constraints and to maximize his revenue.

We also can see that we retrieve the same ratio between the allocated capacity for the 3 areas for a done arrival rate and the minimal bandwidth units described in Table 3. For example for the first column and the three first lines we allocate 13 units for UGS in area 1 the double for area 2 and the triple for area 3.

RtPS are the most bandwidth allocated service class because we assumed that the minimum allocated bandwidth for this class is the highest minimum bandwidth unit of all classes.

Note that this method is very fast because the execution of our algorithm takes 15 seconds to compute the needed vector. This could be seen in the next result.

In fact, Figure 11 describes the number of iteration while varying the arrival rate.

Here the number of iteration is the number of steps needed to share the remaining capacity between classes and areas. The remaining capacity means the amount of the bandwidth that reminds for each traffic load after executing the initial capacity algorithm. This is why the remaining capacity depends on the arrival rate.

When traffic load (or arrival rate) increases, the remaining capacity computed in the initial capacity algorithm decreases and the second

Table 4. CAC vector reservation

		New call				Handoff			
λ_{ij}^{nc}	Area	UGS	rtPS	nrtPS	ErtPS	UGS	rtPS	nrtPS	ErtPS
1	1	13	96	48	26	14	80	40	20
1	2	26	192	96	52	28	208	104	56
1	3	39	288	144	78	42	288	144	78
2	1	13	96	48	26	10	72	36	20
2	2	26	192	96	52	30	224	104	60
2	3	39	288	132	78	42	288	144	84
3	1	14	96	44	26	10	80	32	20
3	2	28	192	88	52	32	224	104	60
3	3	42	288	132	78	42	288	144	84
3.9	1	14	96	44	26	10	72	36	20
3.9	2	28	176	88	52	32	224	112	60
3.9	3	39	264	132	78	45	312	156	84

Figure 11. Iteration

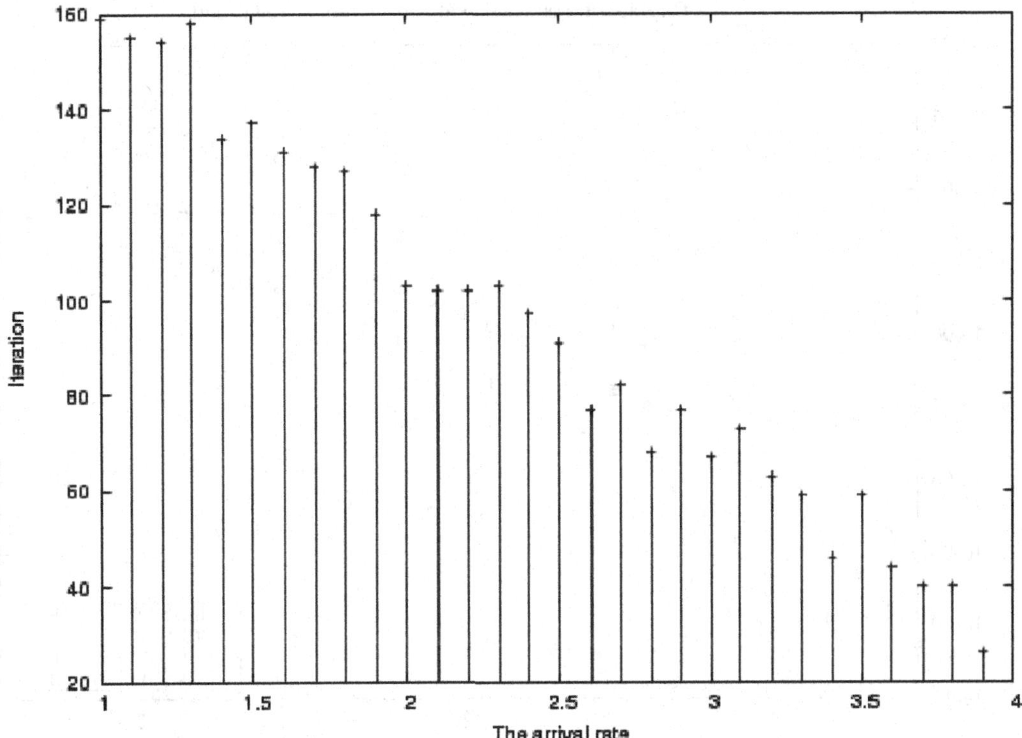

step of our algorithm, which share this capacity in order to maximize the revenue, requires fewer steps than with a low traffic.

Next, we evaluate the system performance. We first plot the call blocking probability while varying the arrival rate for the four (nc, ho) classes (1,2,3,4) in the area 1. These results are obtained for a system bandwidth shared using a computed nearly optimal reservation capacity vector obtained by our algorithm.

First of all, threshold constraints are fulfilled for all classes and all call type as illustrated in Figures 12 and 13, i.e., secondly the call blocking probability increases with arrival rate.

Note that the same results are obtained for new call and for all areas in Z but for a luck of place we choose to present only results for hand-off and new calls in modulation area 1. The fluctuation of the probability is very insignificant because we plot results on a log scale. The other significant fluctuation could be explained by the variation of the nearly optimal capacity allocation with the arrival rate as we sow in the first table result.

The second performance parameter studied concerns the bandwidth utilization ratio. Table 5 presents the bandwidth utilization ratio while varying the arrival rate.

We can easily see that the bandwidth utilization ratio is near the 50%, of course it is not the optimal use of the bandwidth nevertheless this CAC policy is known to be a tractable one because it is simple to model the system. Our purpose is to provide a method or a tool for a WiMAX service provider to compute the threshold for each area and each service classes that satisfies the call blocking probabilities constraints and maximizes his revenue.

Figure 12. UGS rtPS nrtPS ErtPS Call blocking probability in the area 1 for new call type

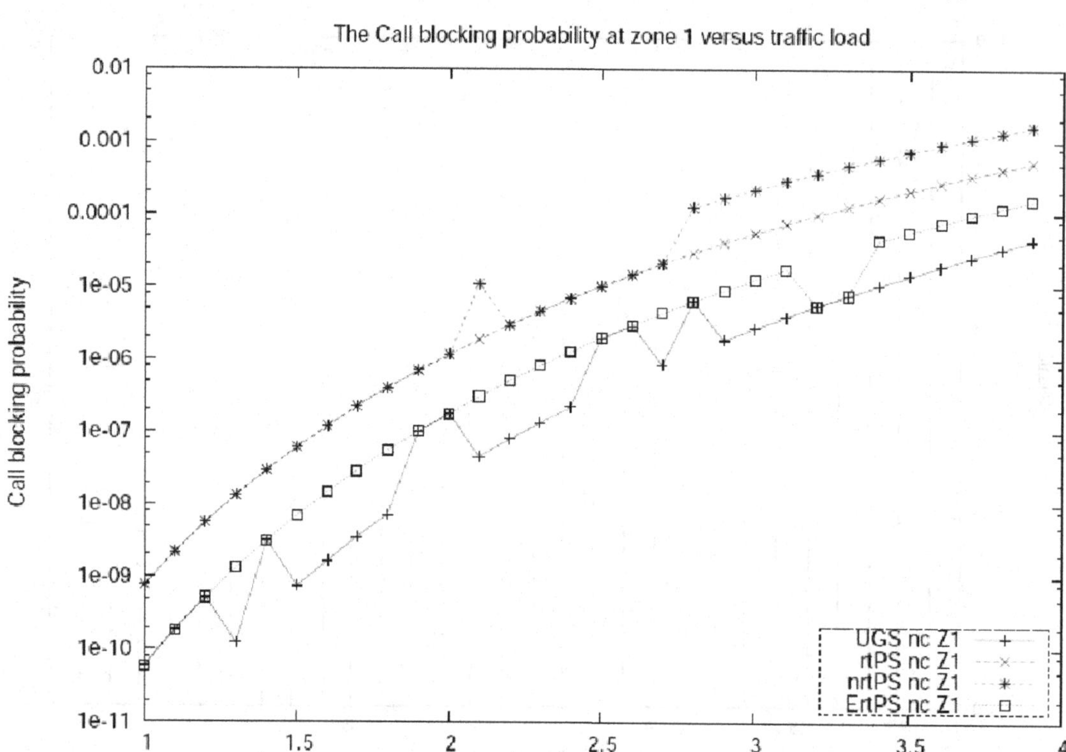

Figure 13. UGS rtPS nrtPS ErtPS Call blocking probability in the area 1 for handoff call type

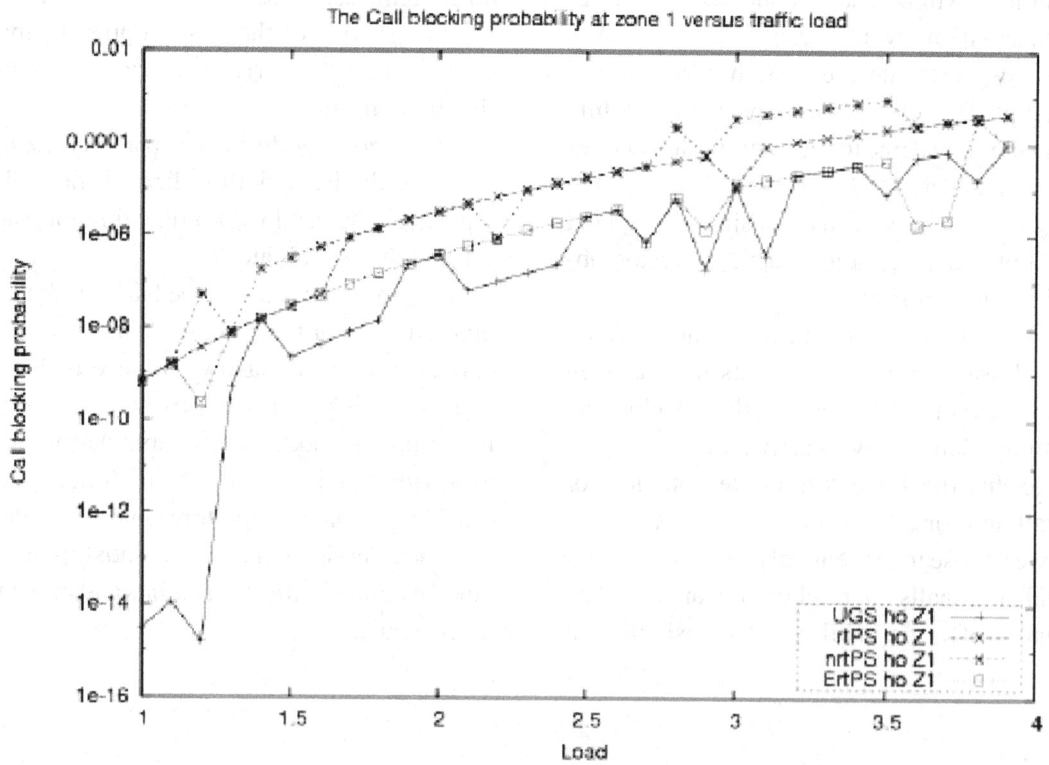

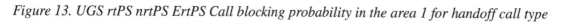

Table 5. Bandwidth utilization ratio

		New call				Handoff			
λ_{ij}^{nc}	Area	UGS	rtPS	nrtPS	ErtPS	UGS	rtPS	nrtPS	ErtPS
1	1	47.7	48.7	48.7	48.5	40.6	31.6	31.7	29.2
1	2	48.5	48.7	48.7	48.6	55.4	56	56	55.4
1	3	48.4	48.7	48.7	48.6	53.8	50.1	50.1	50.0
2	1	50	51.4	51.4	50.5	39.5	38.9	38.9	39.5
2	2	50.5	51.4	51.4	50.9	53.0	54.1	50.78	53
2	3	50.7	51.4	52.4	51.0	53.2	50.30	54.9	53.2
3	1	42.6	45.4	47.1	43.9	38.5	45	41.1	41.5
3	2	43.4	45.4	47.1	44.3	40.3	43.8	45.7	42.2
3	3	43.7	45.4	47.1	44.5	41.5	43.5	47.4	44.6
3.9	1	36.9	40.5	42.8	38.6	37.6	45.0	41.1	40.5
3.9	2	37.6	42.7	42.7	38.9	34.2	43.8	45.7	36.2
3.9	3	38.9	42.7	42.7	39.1	39.8	43.5	47.4	38.6

We can choose CAC strategy near the optimal CAC with which we obtain a bandwidth utilization ratio near 100% but we increase the difficulty of the implementation of such a system. In other word, it is a compromise between the complexity of the system model and the performance of the system.

The final results concerns the system Gain, we note that the gain is near the optimal solution one.

It increases linearly with the increase of the arrival rate. This could be explained by the the fact that thresholds are very low, so if we look at the System Revenue equation we could make an approximation of the system revenue as follows:

$$R = \sum_{i,j=1}^{3,4} R_j \left[p_{ij}^{nc} + p_{ij}^{ho} \right] \quad (5)$$

CONCLUSION

In this work we have demonstrate that QoS management is fundamental in mobile WiMAX network. We have proposed a fast tool of bandwidth partitioning for mobile WiMAX service provisioning. This tools depends on the allocation BW strategy used by the system. It is based on a double constraint point of view.

- **Subscribers Point of View:** Satisfies statistical constraints and call blocking probabilities constraints
- **Providers Point of View:** Maximizing the system Gain

The purpose of the paper is to determine, using an analytical model and a heuristic approach, the nearly-optimal sizes of the partition sizes dedicated to each type of connection, which is characterized by its service class, type of request and modulation and coding scheme. The basic idea is a simple 2 steps algorithm that we can summarize as follows:

- Compute an Initial capacity vector that satisfies Subscribers constraints.
- Compute a final capacity vector that satisfies providers constraints.

This tool presents the advantage to be Fast, efficient, scalable and simple. Concerning future work we will focus on the comparison of this work with other strategies like those described in Kim, Lee, Choi, Chung, and Lee (2011), Hwang, Hwang, and Su (2011), and Tung, Tsang, Lee, and Ko (2008).

REFERENCES

Altman, E. (2002). Applications of Markov decision processes in communication networks. In Feinberg, A. E., & Schwartz, A. (Eds.), *Handbook of Markov decision processes* (pp. 489–536). New York, NY: Springer. doi:10.1007/978-1-4615-0805-2_16.

Baynat, B., Nogueira, G., Maqbool, M., & Coupechoux, M. (2009). An efficient analytical model for the dimensioning of WiMAX networks. In *Proceedings of the International Conference on Mobile Technology, Applications, and Systems* (pp. 521-534).

Choudhury, G., Leung, K., & Whitt, W. (1995). An algorithm to compute blocking probabilities in multi-rate multi-class multi-resource loss models. *Advances in Applied Probability*, *27*, 1104–1143. doi:10.2307/1427936.

Dusit, N., & Ekram, H. (2006). A queuing-theoretic and optimization-based model for radio resource management in IEEE 802.16 broadband wireless networks. *IEEE Transactions on Computers*, *55*, 1473–1488. doi:10.1109/TC.2006.172.

Eklund, C., Marks, B., Stanwood, K. L., & Wang, S. (2002). IEEE Standard 802.16: A technical overview of the WirelessMAN air interface for broadband wireless access. *IEEE Communications Magazine*, *40*(6), 98–107. doi:10.1109/MCOM.2002.1007415.

Elayoubi, S., & Fourestié, B. (2008). Performance evaluation of admission control and adaptive modulation in OFDMA WiMAX systems. *IEEE/ACM Transactions on Networking*, *16*, 1200–1211. doi:10.1109/TNET.2007.911426.

Hou, F. P., Ho, H., & Shen, X. (2006, November). Performance analysis of a reservation-based connection admission scheme in 802.16 networks. In *Proceedings of the IEEE Global Telecommunications Conference* (pp. 1-5).

Hwang, I., Hwang, B., & Su, R. (2011). Maximizing downlink bandwidth allocation method based on SVC in mobile WiMAX networks for generic broadband services. *International Scholarly Research Network Communications and Networking*, *2011*, 5.

IEEE. Computer Society. (2005). IEEE Standard 802.16e-2005: Amendment to IEEE standard for local and metropolitan area networks - Part 16: Air interface for fixed broadband wireless access systems- Physical and medium access control layers for combined fixed and mobile operation in licensed bands. Washington, DC: IEEE Computer Society.

Khemiri, S., Boussetta, K., Achir, N., & Pujolle, G. (2007a). Wimax bandwidth provisioning service to residential customers. In *Proceedings of the International Conference on Mobile Wireless Communications Networks* (pp. 116-120).

Khemiri, S., Boussetta, K., Achir, N., & Pujolle, G. (2007b). Optimal call admission control for an IEEE802.16 wireless metropolitan area network. In *Proceedings of the International Conference on Network Control and Optimization*.

Kim, H., Lee, J., Choi, Y., Chung, Y., & Lee, H. (2011). Dynamic bandwidth provisioning using ARIMA-based traffic forecasting for mobile WiMAX. *Computer Communications*, *34*(1), 99–106. doi:10.1016/j.comcom.2010.08.008.

Liping, W., Fuqiang, L., Yusheng, J., & Nararat, R. (2007). Admission control for non-preprovisioned service flow in wireless metropolitan area networks. In *Proceedings of the Fourth European Conference on Universal Multiservice Networks* (pp. 243-249).

Rong, B., Qian, Y., Lu, K., Hsiao-Hwa, C., & Mohsen, G. (2008). Call admission control optimization in WiMAX networks. *IEEE Transactions on Vehicular Technology*, *57*, 621–632.

Shafaq, B., & Ratan, K. (2007). Adaptive connection admission control and packet scheduling for QoS provisioning in mobile WiMAX. In *Proceedings of the IEEE International Conference on Signal Processing and Communications*, Dubai, UAE (p. 1355).

Tarhini, C., & Chahed, T. (2008). Density-based admission control in IEEE802.16e mobile WiMAX. In *Proceedings of the 1st IFIP Wireless Days Conference* (pp. 1-5).

Tung, H. Y., Tsang, K., Lee, L., & Ko, K. (2008). QoS for mobile wimax networks: Call admission control and bandwidth allocation. In *Proceedings of the IEEE Consumer Communications and Networking Conference* (pp. 576-580).

Wang, H., & Li, W. (2005). Dynamic admission control and QoS for 802.16 wireless man. In *Proceedings of the Wireless Telecommunications Symposium* (pp. 60-66).

This work was previously published in the International Journal of Wireless Networks and Broadband Technologies, Volume 1, Issue 2, edited by Naveen Chilamkurti, pp. 42-61, copyright 2011 by IGI Publishing (an imprint of IGI Global).

Chapter 12
IP Paging for Mobile Hosts in Distributed and Fixed Hierarchical Mobile IP

Paramesh C. Upadhyay
Sant Longowal Institute of Engineering & Technology, India

Sudarshan Tiwari
Motilal Nehru National Institute of Technology, India

ABSTRACT

The concept of Paging has been found useful in existing cellular networks for mobile users with low call-to-mobility ratio (CMR). It is necessary for fast mobility users to minimize the signaling burden on the network. Reduced signaling, also, conserves scarce wireless resources and provides power savings at user terminals. However, Mobile IP (MIP), a base protocol for IP mobility, does not support paging concept in its original form. Several paging schemes and micro-mobility protocols, centralized and distributed, have been proposed in literature to alleviate the inherent limitations of Mobile IP. In this paper, the authors propose three paging schemes for Distributed and Fixed Hierarchical Mobile IP (DFHMIP) and develop analytical models for them. Performance evaluations of these schemes have been carried out and results have been compared with DFHMIP without paging and with Dynamic Hierarchical Mobile IP (DHMIP) for low CMR values.

1. INTRODUCTION

Paging is one of the key concepts being used since first generation cellular networks. Most of the cellular network standards are using paging in one form or the other. In cellular networks, mobility management involves two operations. First is tracking the mobile users during their movements. This is called location update. In this operation, a user is required to update its location whenever it moves from one cell to another. Second operation is called as paging. This operation is performed by the network to locate a user on the basis of user's current location update information for accurate call delivery. Paging process is primarily used for dormant users that cross the

DOI: 10.4018/978-1-4666-3902-7.ch012

cell boundaries frequently without any ongoing session. The benefit of paging lies in the fact that a dormant user is allowed to move freely within a group of cells without updating its location to the network. This group of cells is called a paging area. The paging area consists of a group of base stations, which are under the same mobile switching center (MSC). Thus, the network has only a coarse knowledge of the user location. When a call arrives, MSC sends a paging message to all the base stations in the paging area to locate the user. The base stations broadcast the paging message in their own cells. The system determines the mobile station's accurate location after receiving a paging response message from the paged mobile device. The precise location information is then used to establish the call.

Mobile IP (MIP) (Perkins, 1997, 2002) a base protocol for IP mobility, uses only location updates or registrations for the mobility management, and does not support paging concept in its original form. Several paging schemes and micro-mobility protocols, centralized (Ramjee, Varadhan, Salgarelli, Thuel, Wang, & La Porta, 2002; Valko, 1999; Gustafsson, Jonsson, & Perkins, 2005) and distributed (Xie & Akylidiz, 2002; Ma & Fang, 2004; Bejerano & Cidon, 2003) have been proposed in literature to alleviate the inherent limitations of Mobile IP. The paging is beneficial for dormant users that cross the subnet boundaries frequently without an ongoing session. Therefore, the researchers have identified IP paging altogether a different issue. Distributed and Fixed Hierarchical Mobile IP (DFHMIP) also does not distinguish between active and dormant mobility users. Therefore, signaling costs at lower CMR values become significantly high.

This paper endeavors to develop paging schemes for DFHMIP so as to take account of the dormant users differently than the active users. Following this section, the paper is alienated in four major sections. Section 2 provides state-of-art scenario in the area of IP paging. Section 3 explains the system description and three paging mechanisms for DFHMIP. Analytical models have been developed to compute signaling costs for the proposed paging schemes in Section 4. Section 5 evaluates the performance of the paging schemes, in terms of signaling costs, and compares them with DFHMIP without paging and DHMIP for low CMR values. Finally, the conclusion of paper is presented in Section 6.

2. RELATED WORKS

IP paging enables a common infrastructure and protocol to support different wireless interfaces. According to Mobile IP regional paging (Haverinen, 2000), each host can stay in either of the two states: active, or idle, also called as dormant mode. For active hosts, it acts similar to Mobile IP whereas in dormant mode, the hosts can freely move within a paging area, a group of subnets, without any location registration with the network. Thus, the system has only coarse knowledge of whereabouts of the host. A mobile host performs registration only when it changes paging areas. The packets intended for a dormant host are terminated at a paging initiator. The paging initiator buffers the packets and sends the IP paging messages within the paging area. The subnet where the host is currently residing responds to this message. Then, the packets are forwarded to the recipient.

P-MIP (Zhang, Gomez Castellanos, & Campbell, 2002) proposes each subnet to have a base station, similar to a cell in cellular networks, which can act as an FA in its own subnet. P-MIP considers both, overlapping and non-overlapping paging areas, and uses fluid flow model for host mobility. Analytical and simulation studies show that with growing paging area size, the increase in signaling cost is considerably less with P-MIP than with MIP. In Yun, Sung, and Aghvami (2003) authors have shown suitability of P-MIP over MIP for small values of session-to-mobility ratio (SMR), same as CMR. On the contrary, performance of MIP is better for high values of SMR. The authors in

Kwon, Nam, Hwang, and Sung (2004) have similar paging area as for P-MIP, but they have analyzed their results using additional mobility states for the hosts for battery power conservation.

The authors in Choi, Kim, Nah, and Song (2004) suggest a paging scheme that uses hierarchical network architecture similar to HMIP. However, in order to reduce signaling cost, a dormant mobile host, after every FA crossing, registers with its local FA only, not with its GFA. During call delivery, IP paging messages are sent to all the FAs in the domain. These FAs check their visitor lists, and the FA which finds an entry of the host in its visitor list responds back to the GFA.

Micro-Mobile IP (μMIP) (de Silva & Sirisena, 2002) also implements the same hierarchical structure as HMIP or Regional Registration and uses two-level hierarchy. At the highest level are gateway mobility agents (GMA) that perform the role of a border router, filtering between intra- and inter-domain signaling. At the second level is the Subnet Agent (SA) which inter-works with a particular RAN consisting of access points or base stations that cover a specific geographic area. The GMA in this scheme corresponds to GFA of HMIP, and the SA is same as the FA. In addition to the flags used in RFC 2002, this scheme uses two new flags N (new registration) and P (paging update) to determine the next step in processing. GMA associates each SA with a preconfigured multicast paging group. During packet delivery, the GMA buffers the packets and uses multicast to send a page solicitation message, which is broadcast over access points to locate the dormant users.

Ramjee et al. (2002) proposed three IP paging protocols namely home agent paging, foreign agent paging, and domain paging. The authors have shown that domain paging is able to support the highest call load among the three protocols by utilizing available processing resources (routers) in the domain efficiently, and has the least number of updates to the home agent. A paging area size of six subnets is an optimal value for low paging latency.

Kempf (2001) and Kempf et al. (2001) have proposed a paging functional architecture that contains four entities: host, paging agent, tracking agent, and dormant monitoring agent. The jobs of a paging agent are two folds; one, it alerts the dormant host on a packet arrival, and second, it maintains paging areas by periodically wide casting information over the host's link to identify the paging area. The tracking agent keeps the location records of a dormant or active host by receiving updates from them. The dormant monitoring agent is used to coordinate all the activities related with paging of a dormant host. The packets arriving for a host are put in a buffer at dormant monitoring agent, which queries the tracking agent about the last reported paging agent of the host and then, it informs the paging agent to page the host in its paging area. After receiving a response from the paging agent, dormant monitoring agent forwards the packets to the intended mobile host.

Castelluccia (2001) proposed an individual adaptive paging scheme for Mobile IP to reduce signaling burden on the network. Each mobile host, in this scheme, computes its optimal paging area according to its mobility and incoming call parameters. The authors have used hexagonal network architecture, in which each base station is assumed to act as either an FA or a paging agent. This concept has been further extended to HMIP to reduce the signaling burden on the network. In Castelluccia and Mutaf (2001) authors have combined the two schemes proposed in Kempf et al. (2001) and Castelluccia (2001) to extend the IP paging architecture defined in RFC3154 to support adaptive and per-host paging areas dependent on individual user's call and mobility pattern. The adaptive paging schemes are flexible and provide optimal paging areas to reduce the signaling cost, but they lack in ease of their implementation. Also, a regular hexagonal architecture cannot be a valid assumption for Internet environment. This issue has been attempted in Xie (2006), where the author has proposed a user-independent multi-step paging scheme. Instead of individual user information, this scheme uses mobility rate of each subnet as the paging criterion. The paging process is a combination of last-location-first paging and highest-mobility-first paging. The information regarding the mobility rate of each subnet is gathered using a new semi-idle mode for a mobile host. A host, in semi-idle mode, periodically sends its home address and CoA in response to current FA advertisement message. Based on this information, the FA decides whether the mobile is in idle or semi-idle state. This scheme avoids the broadcasting of IP messages for paging purposes, and employs a paging initiator to page the dormant user in smaller areas sequentially. The partitioning of these smaller areas is based on up-to-date subnet mobility rates, and is similar to uniform paging concept.

In Lee, Lee, and Cho (2005) authors proposed an IP paging method, which is based on dormant registration threshold (DRT). This scheme is based on conventional Mobile IP and IP paging standardized in IETF. It uses an FA counter in each MH, which counts the number of FAs managing subnets through which the MN has passed. For an active MH, this scheme works similar to MIP.

3. SYSTEM DESCRIPTION

3.1. An Overview of DFHMIP

DFHMIP (Upadhyay & Tiwari, in press) like MIP, uses two mobility agents, viz. Home Agent (HA) and Foreign Agent (FA), to make user mobility possible from one subnet to another in the network. Each mobile host is permanently registered with a subnet where it started its mobility services first time. This subnet is called the home subnet of the host. Any other subnet in the network is a foreign subnet for the host. The HA, located at the home subnet, assigns a permanent IP address to the MH, called its home address. Similarly, when a mobile host leaves its home subnet and enters a foreign subnet, the FA, located at the foreign subnet assigns a temporary address, also known as care-of-address (CoA) of the host. The proposed scheme is based on having neighbors' information, wherein each FA is assumed to be aware of the IP addresses of its neighboring FAs.

The FA_x is said to be the neighbor of FA_y if a mobile host moves from area covered by FA_x to another area covered by FA_y. This means that the neighbors are located at a single hop-distance only. The neighboring information of each FA is assumed to have been embedded in the FA advertisement messages. This assumption does not have significant impact on data rate in the network (Zhang, Jaehnert, & Dolzer, 2003). Also, unlike the scheme proposed in Chang, Li, and Morikawa (2005), wherein the entire domain list is broadcast in the coverage area of an AR through the advertisement messages, in our proposed scheme, each FA broadcasts the list of neighboring FAs only in its coverage area. An MH, residing in the coverage area of the FA, listens the advertisement messages, and uses it solely to decide if a registration is necessary.

An FA, in this proposal, can concurrently work in three modes of operation: a Gateway FA (GFA), a Regional FA (RFA), or simply as an FA. The FA acts as a GFA whenever a mobile host requests it to perform a registration with its HA. This registration is termed as a macro-registration. Having performed a macro-registration, a mobile host can freely roam from one subnet to another until the next macro-registration is requested with the visiting FA of the mobile host. At this moment, the visiting FA of the mobile host becomes its new GFA. The group of the subnets, wherein a mobile host can move without a macro-registration, constitutes a registration area, called as macro-registration area (MacRA). The GFA resides at the center of a MacRA and any two consecutive MacRAs overlap with each other. The size of a MacRA is subnet or FA-specific rather than the user-specific.

A MacRA consists of a number of FA-specific smaller regions. These FA-specific regions are called micro-registration areas (MicRA). In fact, each FA is surrounded by its neighboring FAs, with the FA being at the center. This FA together with its neighboring FAs constitutes MicRA. The center FA acts as an RFA for itself as well as for the neighboring FAs. Thus, the size of MicRA depends upon the number of neighboring FAs only. When a mobile host leaves a MicRA, the visiting FA becomes its new RFA, and a new MicRA is formed. This makes two successive MicRAs overlapping with each other. When a mobile host changes its MicRA, it registers with its GFA via its RFA. This registration is called a micro-registration. A host can move in its MicRA without any registration with the GFA. Thus, the movement of a mobile host within MicRA is transparent to its GFA. Based on the algorithm given in Upadhyay and Tiwari (in press), a mobile host can decide to perform a macro, micro, or a local registration.

When a mobile host registers an FA as an RFA, the registered FA and its neighboring FAs simply act as FAs for the host. While moving from one FA to another within the MicRA, the host registers with its RFA locally. This registration is referred as a local registration. Note that when a host registers an FA as a GFA, at that moment, the registered FA behaves like the GFA, RFA, and the FA for the mobile host. Similarly, if the host registers an FA, other than the GFA, as its RFA, the registered FA acts as an FA for the host, as well.

Thus, a macro-registration is performed at MacRA level, a micro-registration at MicRA level, and a local registration at subnet level. Thus, the scheme uses three level hierarchies in the network, the GFA being at the highest level, RFA at the intermediate level, and the FA at the lowest level. The size of a MacRA or a MicRA for all the mobile hosts using same FA as a GFA or a RFA, respectively, is fixed. The traffic load in the network is evenly distributed at each FA. Therefore, the proposed scheme has been given a name: Distributed and Fixed Hierarchical Mobile IP, abbreviated as DFHMIP.

Each FA maintains three visitor lists namely, GFA visitor list, RFA visitor list and FA visitor list to keep location records of each mobile host at its GFA, RFA and an FA, respectively. The packets intended for a mobile host are first intercepted by the HA, which then tunnels them to the GFA. The GFA forwards these packets to the RFA, which forwards them to the mobile host via itself or through one of its neighboring FAs, if necessary. The formation of overlapping MacRAs and MicRAs with host movement is shown in Figure 1.

3.2. Proposed Paging Schemes

In DFHMIP, paging can be used either at MacRA level, or at MicRA level. Accordingly, proposed paging schemes have been categorized as below:

1. If paging is used at MacRA level, the GFA acts as a paging initiator for the mobile hosts residing in its MacRA. Since, in hierarchy the RFA comes next to the GFA, one possibility is that a dormant host could move freely from one MicRA to the other without any micro-registrations with the GFA. The local registrations with RFA are necessary; otherwise the hosts will be required to be paged at RFAs in their respective MacRA as well as at FAs in the MicRA. This will contribute to high paging cost and may not serve the intended purpose. Therefore, in this scheme, the local registrations are performed same way as in DFHMIP so that the RFA knows the exact location of the current FA of the mobile hosts in its MicRA. However, the GFA is unaware of the RFAs of the mobile hosts registered at the GFA, i.e., a dormant

Figure 1. Formation of overlapping macro- and micro-registration areas

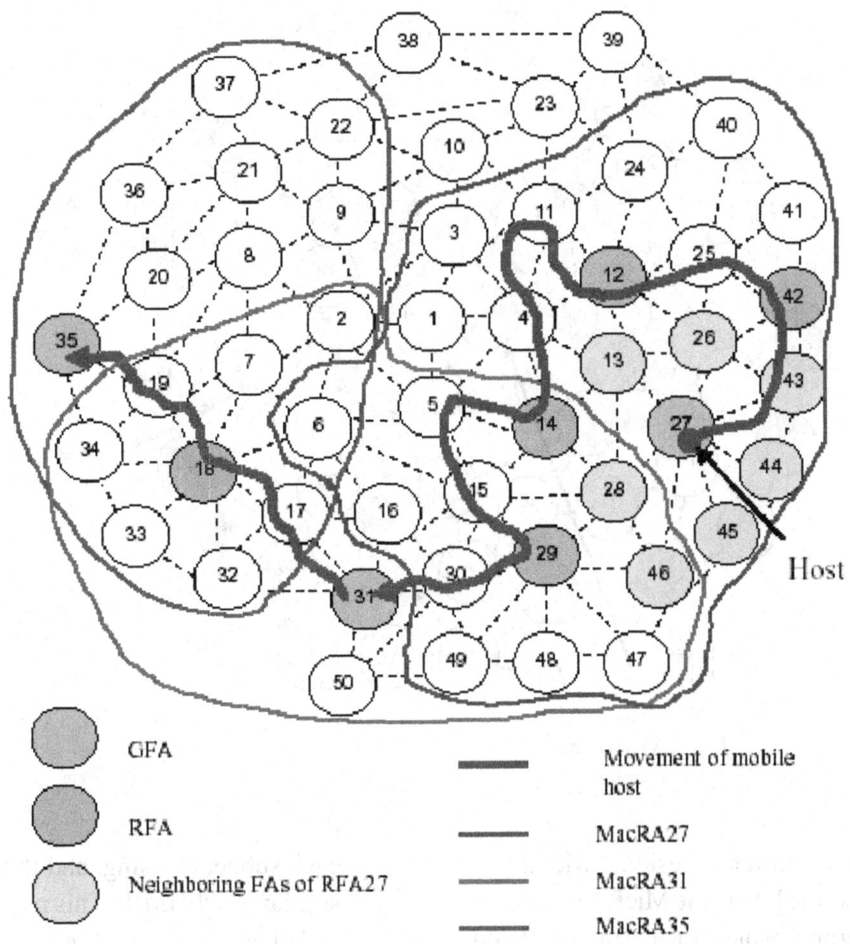

host does not request for micro-registrations while on move. Micro-registrations take place during session only. The paging area for this scheme is shown in Figure 2.

The packets destined for a mobile host arrive at its HA, which are tunneled to the registered GFA of the host. The GFA, which acts as a paging initiator for the mobile hosts in its MAcRA, buffers the packets and sends the IP paging messages to all the RFAs in the MacRA. The RFA, where the host is currently registered sends the paging response to the GFA. After receiving the paging response, the GFA forwards the packets to the RFA of the host, which, in turn, forwards them to the FA of the host. Then, the FA delivers these packets to intended recipient. When the host enters into active mode, the micro-registrations are performed same way as in DFHMIP. This paging scheme for DFHMIP is referred as paging scheme I.

2. In this proposed scheme, paging is employed at MicRA level and the corresponding RFA acts as a paging initiator. A dormant mobile host registered at an RFA in a MicRA may be allowed to move from one FA to the other without any local registration with its registered RFA. Thus, the FAs in the MicRA do not have the location information of the

Figure 2. Paging area of a host in paging scheme I for GFA 1 acting as a paging initiator

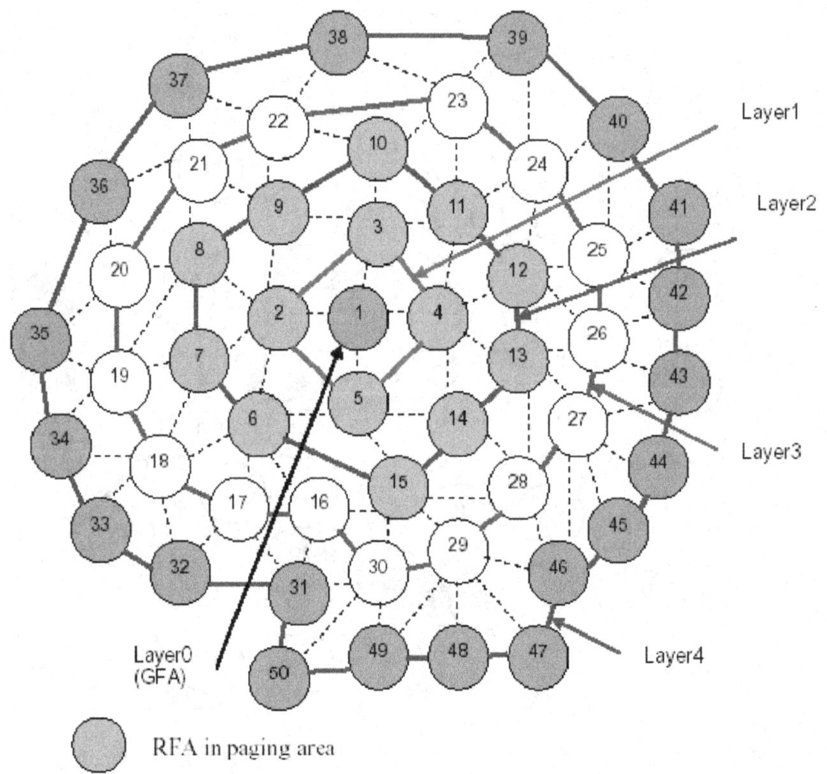

hosts that are currently residing with them. In this case, the FAs in the MicRA constitute the paging area of the host, and the registered RFA acts as a paging initiator. When a host changes its paging area, the new RFA i.e., the visiting FA becomes its new paging initiator, and a new paging area with RFA and its neighboring FAs is constituted. In this case, packets received for the mobile host at GFA, from the HA of the host, are forwarded to the RFA of the host. The RFA, the paging initiator, stores these packets in its buffer and sends a paging message in its paging area. Then, each FA in the paging area broadcasts the paging messages in its own coverage area. The FA, where the host is currently residing, responds to the paging message, and the packets are delivered to the host via this FA. When session is in progress, the active host registers with its RFA after each subnet crossing, and this scheme acts similar to DFHMIP. This paging scheme is called as paging scheme II.

3. In this approach, the paging areas are same as in (2). However, when a dormant user moves in a MicRA from one FA to the other, it registers its location with its current FA only as in Choi, Kim, Nah, and Song (2004). This registration is termed as FA-registration. Thus, the FAs in a paging area are aware of the hosts that are currently residing with them, but RFA knows nothing about the host in its paging area. The RFA acts as a paging initiator for the hosts in the paging area. Figure 3 shows a paging area for paging schemes II and III.

The packets reaching at the HA are directed to the RFA, via GFA, of the host, which buffers the packets and then, sends paging messages to

Figure 3. Paging area in paging scheme II and III for RFA 11 acting as a paging initiator

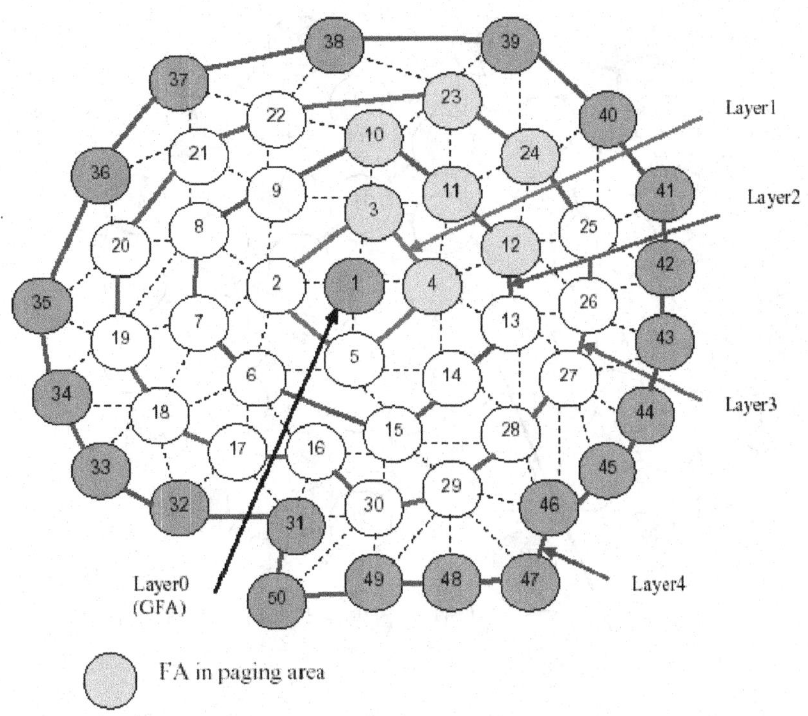

the underlying FAs in the MicRA. The FA which has exact location information of the host responds to this message. Then the packets from RFA are forwarded to current FA of the host for their correct delivery. After the host has entered in active mode, local-registrations are resumed, as in DFHMIP. This paging scheme is termed as paging scheme III.

4. SIGNALING COSTS

For the analysis of the proposed paging schemes, we use random-walk model for the host mobility. In random-walk mobility model, a mobile host can move to its neighboring subnets with equal probability (Akyildiz & Ho, 1995). The analytical model for micro-registration and macro-registration are developed using continuous-time Markov chains. It is assumed that the subnet-residence time of a mobile host follows an exponential distribution with mean:

$$\frac{1}{\lambda_m}$$

and the packet arrival rate for a host follows Poisson distribution with mean λ_a. A MacRA, in DFHMIP, can be considered as consisting of four layers namely, layer 0, layer 1, layer 2, and layer 3, as shown in Figure 4.

Layer 0 consists of a single FA only, and lies in the interior most vicinity of a MacRA. This FA is referred as a GFA. A mobile host performs a macro-registration in two situations. First, when it moves to an FA at layer 4 and second, when it moves to an FA at layer 3 and finds that it has changed its MicRA. Under both the circumstances, the visiting FA becomes the new GFA of the host, and a new MacRA is formed. A mobile host experiences a change of MicRA or RFA at an FA in layer 3 only when it's current RFA is located at either layer 1 or layer 2. This is due to the fact that, in proposed DFHMIP, the layer 3 FAs in a MacRA can act as general FAs only, and not as RFAs.

Figure 4. Layered network architecture for DFHMIP

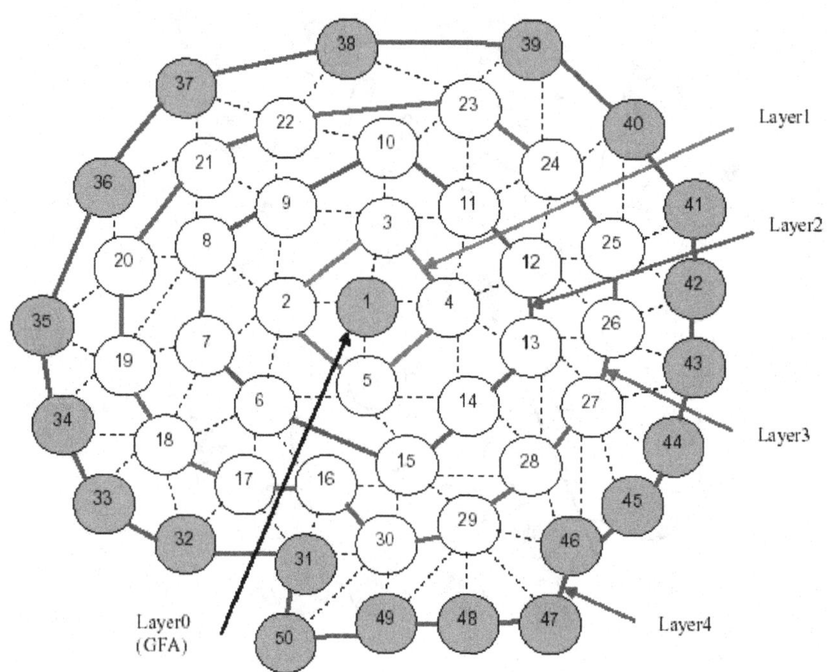

The total signaling cost, in proposed paging schemes, comprises of local registration cost, micro-registration cost, macro-registration cost, and packet delivery cost. These costs have been computed separately for the three paging schemes as follows:

4.1. Signaling Cost for Paging Scheme I

Each local registration, in MicRA, causes a registration message to travel from the host's current FA to its RFA, resulting in a local registration cost. This registration takes place after each subnet boundary crossing in a MicRA. The average local registration cost per unit time, $Cost_{Local}$, can be expressed as:

$$Cost_{Local} = \lambda_m C_{local} \tag{1}$$

where, C_{local} is the cost of unit local registration.

A dormant host, in paging scheme I, performs micro-registrations during sessions only. Therefore, the micro-registration cost depends on the average packet arrival rate of the host, and can be written as:

$$Cost_{Micro}^{(l,k)} = \lambda_a C_{micro}^{(l,k)} \tag{2}$$

Here, $C_{(micro)}^{(l,k)}$ is the cost of unit micro-registration in l^{th} MicRA of kth MacRA.

The macro-registration cost for paging scheme I is same as for DFHMIP, and is given as:

$$Cost_{Macro} = \lambda_m \pi_2^{(k)} (p_{2,3}^{(k)} + p_{2,4}^{(k)}) C_{macro} \tag{3}$$

where $\pi_2^{(k)}$ is steady state or stationary probability vector of an MH having its RFA at layer 2 in k^{th} MacRA. $p_{2,3}^{(k)}$ and $p_{2,4}^{(k)}$ are the transition probabilities of a mobile host that it has changed its current MicRA with RFA at layer 2 to new MicRAs with RFAs at layers 3 and 4, respectively, in k^{th} MacRA. C_{macro} is the cost of unit macro-registration.

The packet delivery cost, denoted as $Cost_{Packet\_Delivery\_I}$ is dependent upon the packet arrival rate, and is given below:

$$Cost_{Packet\_Delivery\_I} = \lambda_a C^{(k)}_{packet\_delivery\_I} \quad (4)$$

Here, $C_{packet\_delivery\_I}$ is unit packet delivery cost for paging scheme I.

The total signaling cost per packet arrival for paging scheme I is given as:

$$Cost_{Paging\_Scheme\_I} = \frac{1}{\rho}[\pi_2^{(k)}(p_{2,3}^{(k)} + p_{2,4}^{(k)})C_{macro} + C_{local}] + C^{(l,k)}_{micro} + C_{packet\_delivery\_I} \quad (5)$$

Where, ρ is call-to-mobility ratio.

4.2. Signaling Cost for Paging Scheme II

In this scheme, the macro-registrations and micro-registrations are performed same way as in DFHMIP. But, a local-registration takes place only when packets arrive for a host. The micro and local-registration costs for this scheme are given below:

$$Cost^{(l,k)}_{Micro} = \lambda_m \pi_1^{(l,k)} p_{1,2}^{(l,k)} C^{(l,k)}_{micro} \quad (6)$$

and,

$$Cost_{Local} = \lambda_a C_{local} \quad (7)$$

where $\pi_1^{(l,k)}$ is steady state or stationary probability vector of layer 1 of lth MicRA in k^{th} MacRA. $p_{1,2}^{(l,k)}$ is the transition probability of a mobile host that it has changed its current MicRA to a new MicRA in k^{th} MacRA.

The packet delivery cost, denoted as $Cost_{Packet\_Delivery\_II}$, can be written as:

$$Cost_{Packet\_Delivery\_II} = \lambda_a C^{(k)}_{packet\_delivery\_II} \quad (8)$$

where, $C_{packet\_delivery\_II}$ being unit packet delivery cost for paging scheme II.

Total signaling cost per packet arrival, for paging scheme II, is given below:

$$Cost_{Paging\_Scheme\_II} = \frac{1}{\rho}[\pi_2^{(k)}(p_{2,3}^{(k)} + p_{2,4}^{(k)})C_{macro} + \pi_1^{(l,k)} p_{1,2}^{(l,k)} C^{(l,k)}_{micro}] + C_{local} + C^{(k)}_{packet\_delivery\_II} \quad (9)$$

4.3. Signaling Cost for Paging Scheme III

In this scheme also, the macro and micro-registrations are same as in DFHMIP. Since, FA-registrations is performed at corresponding FA only, the FA-registration cost, denoted as $Cost_{FA}$, can be written as:

$$Cost_{FA} = \lambda_m C_{fa} \quad (10)$$

where, C_{fa} is unit FA-registration cost.

Local registrations are performed during sessions only; therefore, the local cost is same as in paging scheme II. The packet delivery cost for scheme III, denoted as $Cost_{Packet\_Delivery\_III}$, can be written as:

$$Cost_{Packet\_Delivery\_III} = \lambda_a C^{(k)}_{packet\_delivery\_III} \quad (11)$$

where, $C_{packet\_delivery\_III}$ is unit packet delivery cost for paging scheme III.

The total signaling cost per call arrival for paging scheme III, denoted as $Cost_{Total\_Paging III}$, is given as follows:

$$Cost_{Paging\_Scheme\_III} = \frac{1}{\rho}[\pi_2^{(k)}(p_{2,3}^{(k)} + p_{2,4}^{(k)})C_{macro} + \pi_1^{(l,k)}p_{1,2}^{(l,k)}C_{micro}^{(l,k)} + C_{fa}] + C_{local} + C_{packet\_delivery\_III}^{(k)} \quad (12)$$

5. PERFORMANCE EVALUATION AND COMPARISONS

When a signaling message flows across the network, two types of costs are involved. First, the communication cost between the two entities, and the processing cost at each entity through which the message travels in the network. These costs, for example, can be in terms of delays. To evaluate the performance of proposed DFHMIP, the unit registration costs and unit packet delivery cost need to be defined. Following parameters are used for this purpose:

- $\mathbf{c_{GFA,HA}}$: Communication cost of registration messages between a GFA and the HA
- $\mathbf{c_{RFA,GFA}}$: Communication cost of registration messages between an RFA and a GFA
- $\mathbf{c_{FA,RFA}}$: Communication cost of registration messages between an FA and an RFA
- $\mathbf{c_{MH,FA}}$: Communication cost of registration messages between a Host and an FA
- $\mathbf{p_{GFA,HA}}$: Processing cost of registration messages between a GFA and the HA
- $\mathbf{p_{RFA,GFA}}$: Processing cost of registration messages between an RFA and a GFA
- $\mathbf{p_{FA,RFA}}$: Processing cost of registration messages between an RFA and a GFA
- $\mathbf{p_{MH,FA}}$: Processing cost of registration messages between a Host and an FA
- $\mathbf{c_{HA,GFA}}$: Communication cost of packet delivery between the HA and a GFA
- $\mathbf{c_{GFA,RFA}}$: Communication cost of packet delivery between a GFA and an RFA
- $\mathbf{c_{RFA,FA}}$: Communication cost of packet delivery between an RFA and an FA
- $\mathbf{c_{FA,MH}}$: Communication cost of packet delivery between an FA and a Host
- $\mathbf{p_{HA,GFA}}$: Processing cost of packet delivery between the HA and a GFA
- $\mathbf{p_{GFA,RFA}}$: Processing cost of packet delivery between a GFA and an RFA
- $\mathbf{p_{RFA,FA}}$: Processing cost of packet delivery between an RFA and an FA
- $\mathbf{p_{FA,MH}}$: Processing cost of packet delivery between an FA and a Host

Using these parameters, C_{macro}, $C_{micro}^{(l,k)}$, C_{local}, and $C_{packet\_delivery}^{(k)}$ can be expressed as:

$$C_{macro} = 2c_{MH,GFA} + 2p_{MH,GFA} + 2dc_{GFA,HA} + p_{GFA,HA} \quad (13)$$

$$C_{micro}^{(l,k)} = 2c_{MH,RFA} + 2p_{MH,RFA} + 2\alpha\, c_{RFA,GFA} + \log(N_{RFA}^{(k)})p_{RFA,GFA} \quad (14)$$

$$C_{local} = 2c_{MH,FA} + 2p_{MH,FA} + 2c_{FA,RFA} + \log(N_{FA}^{(l)})p_{FA,RFA} \quad (15)$$

$$C_{fa} = 2c_{MH,FA} + p_{MH,FA} \quad (16)$$

$$C_{packet\_delivery\_I}^{(k)} = dc_{HA,GFA} + p_{HA,GFA} + N_{RFAs}^{(k)}c_{GFA,RFA} + \log_{10} N_{RFAs}^{k}p_{GFA,RFA} + c_{RFA,FA} + \log_{10} N_{FAs}^{(l,k)}p_{RFA,FA} + c_{FA,MH} + p_{FA,MH} \quad (17)$$

$$C^{k}_{packet\_delivery\_II} = dc_{HA,GFA} + p_{HA,GFA} + \alpha c_{GFA,RFA} + \log_{10} N^{k}_{RFAs} p_{GFA,RFA} + (c_{RFA,FA} + C_{paging}) N^{(l,k)}_{FAs} + c_{FA,MH} + p_{FA,MH} \quad (18)$$

$$C^{(k)}_{packet\_delivery\_III} = dc_{HA,GFA} + p_{HA,GFA} + \alpha c_{GFA,RFA} + \log_{10} N^{k}_{RFAs} p_{GFA,RFA} + c_{RFA,FA} \phi^{(i,k)}_{j} + C_{paging} + c_{FA,MH} + p_{FA,MH} \quad (19)$$

Here, C_{paging} is the unit cost for paging the coverage area of an FA, $N^{(l,k)}_{FAs}$ is the number of FAs in l^{th} MicRA, and $N^{(k)}_{RFAs}$ is the number of RFAs in a MacRA.

The parameter values used for performance evaluation are same as in Ma and Fang (2004) and are listed in Table 1. Unit macro-registration and packet delivery costs depend upon the distance, d, between GFA and HA. The average hop-distance between GFA and RFA, in proposed DFHMIP, is fixed. Therefore, following relationships between the parameters are used in this analysis:

$$c_{HA,GFA} = dc_{RFA,FA}; \quad c_{GFA,HA} = dc_{FA,RFA} \quad (20)$$

$$c_{GFA,RFA} = 1.5c_{RFA,FA;} \quad c_{RFA,GFA} = 1.5c_{FA,RFA} \quad (21)$$

For analysis purpose, 50 subnets in the network have been considered. The cost analysis uses a MacRA and MicRA with FA1 acting as GFA as well as RFA. For performance evaluation of the proposed paging schemes, the unit paging cost, C_{paging}, of broadcasting the polling messages over the air is set equal to the processing cost at each FA, as in Xie (2006). The network architecture is assumed to have 50 FAs. A mobile host is assumed to have registered FA 1 as its GFA, and at a particular time its RFA is RFA 1. The DFHMIP without paging scheme is compared with DFHMIP with paging schemes I, II and III. As shown in Figure 5, introducing paging concept improves the performance of DFHMIP significantly for CMRs ≤ 0.5 i.e., for highly mobile users. It is also observed that the paging scheme II and III perform better than the paging schemes I. This is due to the fact that the number of RFAs in a MacRA that constitute the paging area for paging scheme I is much higher than the number of FAs in MicRA that form paging area for scheme II and III. This is true since, during packet delivery for paging scheme I, the network will search all the RFAs in MacRA to locate the mobile host,

Table 1. Parameter values for analysis

Communication costs of packet delivery from the HA to a Mobile Host	$c_{HA,GFA}$ d*25	$c_{GFA,RFA}$ α*25	$c_{RFA,FA}$ 25	$c_{FA,MH}$ 5
Communication costs of registration messages from a Mobile Host to the HA	$c_{MH,FA}$ 10	$c_{FA,RFA}$ 50	$c_{RFA,GFA}$ α*50	$c_{GFA,HA}$ d*50
Processing costs of packet delivery from the HA to a Mobile Host	$p_{HA,GFA}$ 10	$p_{GFA,RFA}$ 1	$p_{RFA,FA}$ 0.5	$p_{FA,MH}$ 1
Processing costs of registration messages from a Mobile Host to the HA	$p_{MH,FA}$ 2	$p_{FA,RFA}$ 1	$p_{RFA,GFA}$ 0.5	$p_{GFA,HA}$ 20

Figure 5. Comparison of signaling costs of DFHMIP with and without paging schemes; d=2

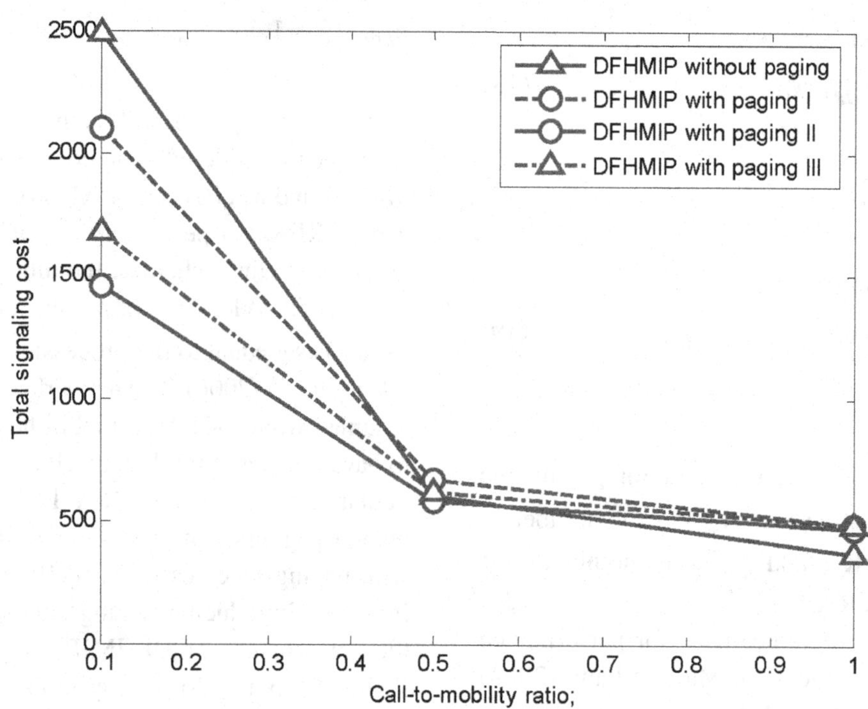

Figure 6. Comparison of signaling costs of DFHMIP with and without paging schemes; d=5

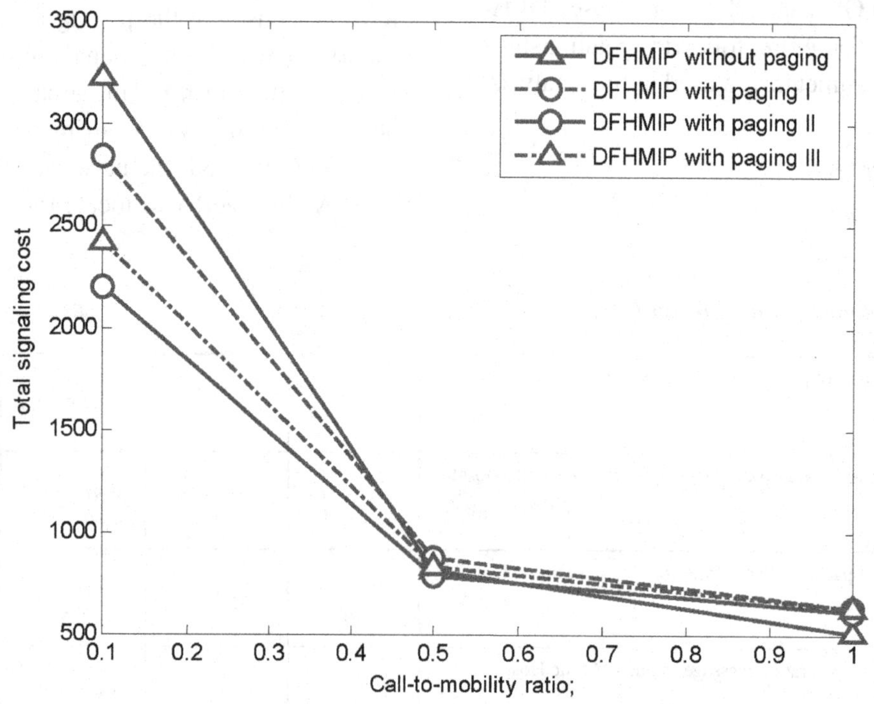

Figure 7. Comparison of signaling costs of DFHMIP with and without paging schemes; d=10

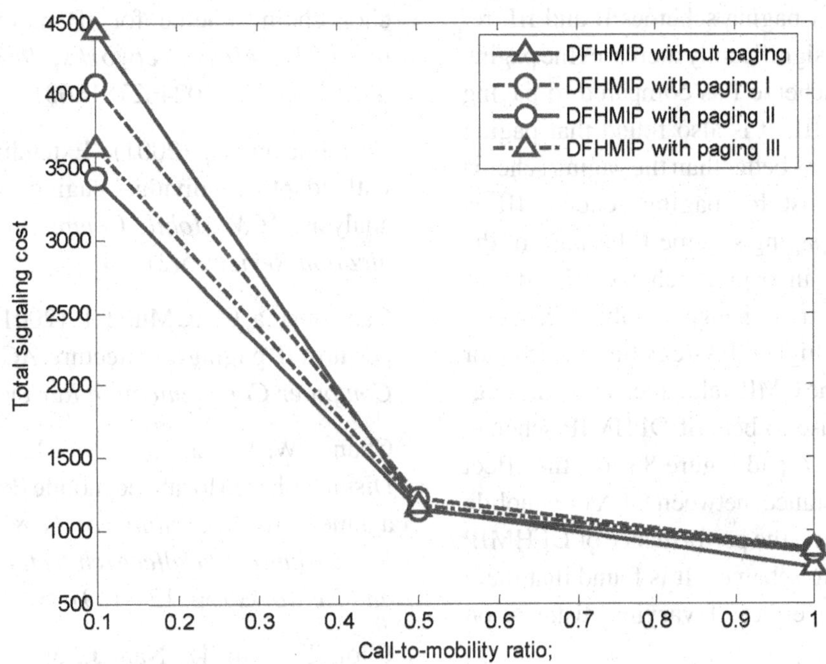

Figure 8. Comparison of signaling costs of DFHMIP with and without paging schemes; d=15

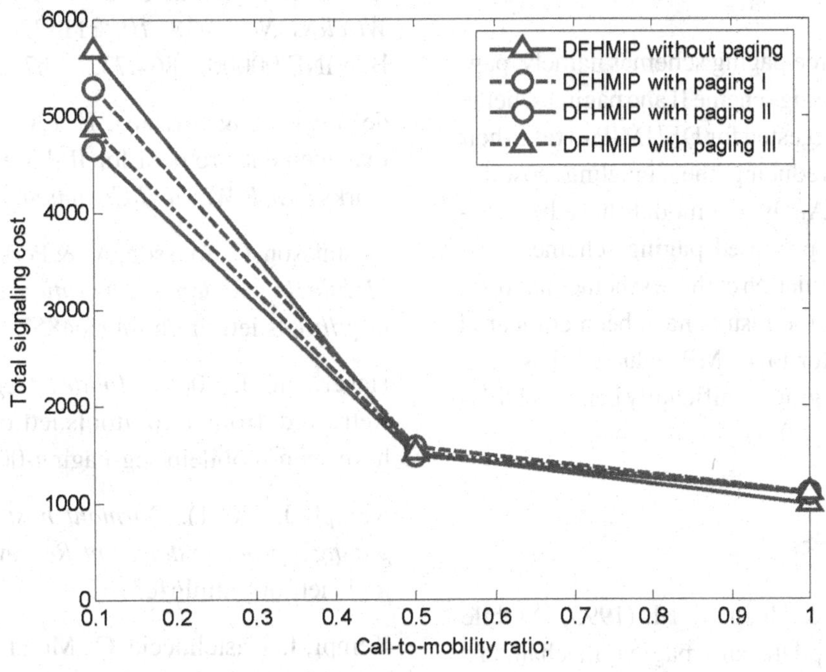

whereas it pages only smaller MicRA and a single FA in the case of paging schemes II and III, respectively.. This significantly increases the paging cost in paging scheme I as compared to paging schemes II and III. It is also found that paging scheme II performs better than the paging scheme III. Signaling cost for paging scheme III is higher than the paging scheme II because of the FA-registrations in paging scheme III. At low CMRs, a mobile may change its subnet frequently, resulting in higher FA-registration cost for scheme III. As the CMR value increases, the paging schemes cease to benefit DFHMIP scheme. Figure 6, Figure 7, and Figure 8 show the effect of increasing distance, between GFA of a mobile host and its HA, on the performance of DFHMIP with three paging schemes. It is found that there is no significant effect of varying distance on performance.

6. CONCLUSION

In this paper, three paging schemes namely, paging scheme I, paging scheme II and paging scheme III have been suggested for DFHMIP to take their advantages in reducing the signaling cost for dormant users. Analytical models have been developed for the proposed paging schemes. The performance evaluation of these schemes has been carried out, and the results have been compared with DFHMIP for low CMR values. It has been observed that paging significantly benefits DFHMIP for CMRs ≤ 0.5.

REFERENCES

Akyildiz, I. F., & Ho, J. S. M. (1995). Mobile user location update and paging mechanisms under delay constraints. *ACM-Baltzar Journal on Wireless Networks*, *1*(4), 413–425. doi:10.1007/BF01985754.

Bejerano, Y., & Cidon, I. (2003). An anchor chain scheme for IP mobility management. *Wireless Networks*, *9*(5), 409–420. doi:10.1023/A:1024627814601.

Castelluccia, C. (2001). Extending mobile IP with adaptive individual paging: A performance analysis. *ACM Mobile Computing and Communication Review, 5*(2).

Castelluccia, C., & Mutaf, P. (2001). An adaptive per-host IP paging architecture. *ACM SIGCOMM Computer Communication Review*, 48-56.

Chang, W. P., Li, J., & Morikawa, H. (2005). Distance-based localized mobile IP mobility management. In *Proceedings of the 8th International Symposium on Parallel Architectures, Algorithms and Networks* (pp. 154-159).

Choi, T., Kim, L., Nah, J., & Song, J. (2004). Combinatorial mobile IP: A new efficient mobility management using minimized paging and local registration in mobile environments. *Wireless Networks*, *10*, 311–321. doi:10.1023/B:WINE.0000023864.73253.b7.

de Silva, P., & Sirisena, H. (2002). A mobility management protocol for IP-based cellular networks. *IEEE Wireless Communications*, 31-37.

Gustafsson, E., Jonsson, A., & Perkins, C. (2005). *Mobile IPv4 regional registration.* Retrieved from http://tools.ietf.org/html/rfc4857

Haverinen, H. (2000). *Mobile IP regional paging.* Retrieved from http://tools.ietf.org/html/draft-haverinen-mobileip-reg-paging-00

Kempf, J. (2001). *Dormant host alerting ("IP paging") problem statement. Retrieved* from http://tools.ietf.org/html/rfc3132

Kempf, J., Castelluccia, C., Mutaf, P., Nakajima, N., Ohba, Y., Ramjee, R., et al. (2001). *Requirements and functional architecture for an IP host alerting protocol.* Retrieved from http://tools.ietf.org/html/rfc3154

Kwon, S.-J., Nam, S. Y., Hwang, H. Y., & Sung, D. K. (2004). Analysis of a mobility management scheme considering battery power conservation in IP-based mobile networks. *IEEE Transactions on Vehicular Technology*, *53*(6), 1882–1890. doi:10.1109/TVT.2004.836964.

Lee, J. W., Lee, H.-J., & Cho, D.-H. (2005). Effect of dormant registration on performance of mobility management based on IP paging in wireless data networks. *International Journal of Electronics and Communications*, *59*, 319–323. doi:10.1016/j.aeue.2004.11.009.

Ma, W., & Fang, Y. (2004). Dynamic hierarchical mobility management strategy for mobile IP networks. *IEEE Journal on Selected Areas in Communications*, *22*(4), 664–676. doi:10.1109/JSAC.2004.825968.

Perkins, C. (1997). Mobile IP. *IEEE Communications Magazine*, 84–99. doi:10.1109/35.592101.

Perkins, C. (2002). *IP mobility support for IPv4*. Retrieved from http://www.ietf.org/rfc/rfc3220.txt

Ramjee, R., Li, L., La Porta, T., & Kasera, S. (2002). IP paging service for mobile hosts. *Wireless Networks*, *8*, 427–441. doi:10.1023/A:1016534027402.

Ramjee, R., Varadhan, K., Salgarelli, L., Thuel, S. R., Wang, S.-Y., & La Porta, T. (2002). HAWAII: A domain-based approach for supporting mobility in wide-area wireless networks. *IEEE/ACM Transactions on Networking*, *10*(3), 396–410. doi:10.1109/TNET.2002.1012370.

Upadhyay, P. C., & Tiwari, S. (in press). Distributed and fixed mobility management strategy for IP-based mobile networks. *International Journal of Business Data Communications and Networking*.

Valko, A. G. (1999). Cellular IP: A new approach to internet host mobility. *Computer Communication Review*, *29*(1), 50–65.

Xie, J. (2006). User independent paging scheme for mobile IP. *Wireless Networks*, *12*, 145–158. doi:10.1007/s11276-005-5262-2.

Xie, J., & Akylidiz, I. F. (2002). A novel distributed dynamic location management scheme for minimizing signaling costs in mobile IP. *IEEE Transactions on Mobile Computing*, *1*(3), 163–175. doi:10.1109/TMC.2002.1081753.

Yun, C. W., Sung, D. K., & Aghvami, A. H. (2003). Steady state analysis of P-MIP mobility management. *IEEE Communications Letters*, *7*(6), 278–280. doi:10.1109/LCOMM.2003.813797.

Zhang, W., Jaehnert, J., & Dolzer, K. (2003, April 22-25). Design and evaluation of a handover decision strategy for 4th generation mobile networks. In *Proceedings of the 57th IEEE Semiannual Vehicular Technology Conference* (Vol. 3, pp. 1969-1973).

Zhang, X., Gomez Castellanos, J., & Campbell, A. T. (2002). P-MIP: Paging extensions for mobile IP. *Mobile Networks and Applications*, *7*(2), 127–141. doi:10.1023/A:1013774805067.

This work was previously published in the International Journal of Wireless Networks and Broadband Technologies, Volume 1, Issue 2, edited by Naveen Chilamkurti, pp. 62-76, copyright 2011 by IGI Publishing (an imprint of IGI Global).

Chapter 13
Cooperation Among Members of Online Communities:
Profitable Mechanisms to Better Distribute Near-Real-Time Services

M. L. Merani
University of Modena and Reggio Emilia, Italy

M. Capetta
University of Modena and Reggio Emilia, Italy

D. Saladino
University of Modena and Reggio Emilia, Italy

ABSTRACT

Today some of the most popular and successful applications over the Internet are based on Peer-to-Peer (P2P) solutions. Online Social Networks (OSN) represent a stunning phenomenon too, involving communities of unprecedented size, whose members organize their relationships on the basis of social or professional friendship. This work deals with a P2P video streaming platform and focuses on the performance improvements that can be granted to those P2P nodes that are also members of a social network. The underpinning idea is that OSN friends (and friends of friends) might be more willing to help their mates than complete strangers in fetching the desired content within the P2P overlay. Hence, an approach is devised to guarantee that P2P users belonging to an OSN are guaranteed a better service when critical conditions build up, i.e., when bandwidth availability is scarce. Different help strategies are proposed, and their improvements are numerically assessed, showing that the help of direct friends, two-hops away friends and, in the limit, of the entire OSN community brings in considerable advantages. The obtained results demonstrate that the amount of delivered video increases and the delay notably decreases, for those privileged peers that leverage their OSN membership within the P2P overlay.

DOI: 10.4018/978-1-4666-3902-7.ch013

1. INTRODUCTION

Online Social Networks and P2P are both tools contributing to the way people are using and approaching the Internet today. The first pursues the target to bring, inside the network, social relationships like friendship or professional acquaintances that constitute a significant fraction of our everyday life. In this manner people, regardless of being close or far away, may stay in touch and keep alive their social connections. The second, instead, aims at effectively providing networked services -- such as file distribution, video streaming -- via resource sharing, where resource means bandwidth, processing power, memory of the network users.

Although OSN and P2P were born and have evolved independently, only very recently a few proposals have arisen in academia, that aim at merging some of their features, taking advantage of the strengths of both. Indeed, if P2P allows to better spread contents, reducing the server stress, OSNs can, e.g., greatly ease the search for content, taking advantage of the similarities in the personal taste of connected users. TRIBLER, the social-based P2P system proposed by Pouwelse et al. (2006), uses the OSN relationships as the base layer of a P2P system, not only for content discovery or recommendation, but also to improve download performance. The improvement is achieved thanks to the cooperative downloading implemented by the users that join the same OSN groups, where members who trust each other cooperate. Along a parallel path, Graffi, Gross, Stingl, Hartung, Kovacevic, and Steinmetz (2011) suggest that OSNs will be the next main application field for the P2P paradigm: through their prototype, the authors show that a P2P-based online social network is feasible and testify that the distributed approach is indeed profitable.

Other works have recently provided interesting contributions on the mobile side of P2P services: Kubo, Shinkuma, and Takahashi (2010) consider how effective a social networking service is as a platform for mobile P2P multicast; Qureshi, Min, and Kouvatsos (2010) lie a framework to identify trustworthy users and to allow secure transmissions, while isolating untrustworthy nodes from the mobile community of a social network.

However, most of the previous works focus on P2P for file-sharing and there is little work that merges the OSN friendship concept with P2P streaming. The concise contribution of Abboud, Zinner, Lidanski, Pussep, and Steinmetz (2010) indicating how social networks can be used to build new incentive mechanisms, represents a first attempt in this direction. Our work develops a similar idea: we aim at incorporating in a P2P streaming architecture the notion of social relationships, assuming these are preferential links to retrieve content. The delivery mechanisms we devise aim at favoring OSN peers whenever the P2P overlay operates in a critical regime, i.e., when the overall bandwidth is scarce. In such circumstance, an OSN peer requesting video contents and not finding any, asks for the help of direct friends that discard non-OSN peers currently served to make room for their mate. Several variants to this priority concept implemented in favor of OSN friend peers are examined. Priority is extended to friends of friends, a choice motivated by the observation that mutual friendship is a strong bond that can lead to personal information sharing, as shown in Nagle and Singh (2009) and Acquisiti and Gross (2005, 2006). The strategy where priority is granted to all other OSN members, regardless of direct friendship, is also investigated as the limit case.

To validate our proposal, the first step we have taken has been to spot in literature (and properly correct) the graph model of a popular social network for picture and video sharing, Flickr (http://www.flickr.com/). Next, we have implemented and developed a simulative tool that evaluates the performance achieved by the proposed delivery strategies. Both the amount of delivered video and the delay suffered in retrieving it are determined, demonstrating that a clear separation in performance between users belonging to the OSN and those outside is achieved.

The remainder of the paper is organized as follows. Section 1 illustrates the proposal; Section 2 describes: (i) the model used to capture the most salient features of the OSN for the sharing of pictures and videos that we took as a reference; (ii) the amendments introduced to it, to better fit the experimental data; Section 3 details the way the newly proposed P2P system operates; Section 4 numerically quantifies the performance achieved and Section 5 reports our concluding remarks.

2. THE PROPOSAL

The majority of current P2P systems treat their users as anonymous and uncorrelated entities, neglecting any social connection among them. To fill this gap, this work introduces a novel, OSN-based P2P streaming overlay, where social relationships are exploited to develop a privileged video content distribution mechanism among peers that are also OSN members.

The focus is on a specific operating condition of the P2P overlay, namely, that of overloading, where the resource index σ (Wu, Liu, & Ross, 2009) the system displays is lower than one: if N active nodes are present within the system, the i-th node displaying an upload capacity C_i, $i = 1, 2, \ldots, N$, given the video server exhibits an upload capacity S of its own and the desired streaming rate is d, then the resource index σ is defined as the ratio between the overall upload capacity and the overall rate that system nodes would require to successfully download the video, i.e., as

$$\sigma = \frac{\sum_{i=1}^{N} C_i + S}{N \cdot d}. \quad (1)$$

When $\sigma \leq 1$, a critical condition that can often occur, an ordinary P2P overlay cannot provide all of its members an adequate viewing experience. In this circumstance, the functioning of the new system is simple: peers that do not belong to the OSN exploit the video distribution mechanism in the usual manner; hence are subject to a degraded service; on the contrary, peers that are OSN members have an advantage, as they can request their friends' help to retrieve the video.

In detail, in our first proposed strategy an OSN member that newly joins the P2P overlay and meets difficulty in finding portions of the video, is allowed to contact those among its OSN friends that fall in the list of the potential parent peers, and to ask for their help. Upon receiving such request, the contacted peers search among their children, looking for a peer that does not belong to the OSN, and if they find one, its connection is discarded, to make room for the mate in need for content.

For the second proposed strategy, the request for help of an OSN peer unable to fetch the desired video stream is extended to friends of friends within the OSN, i.e., to nodes that are up to two-hops away from it on the OSN graph: any of them, once contacted, tries to fulfill the received request applying the mechanism described.

Finally and in the limit, the third strategy proposes all OSN members can help any other OSN node, when the latter runs into difficulty retrieving the video.

In essence, OSN members that are also active P2P nodes implement a preemptive priority mechanism in serving the new connection requests, favoring a specific subset of other OSN users.

The Numerical Results Section will illustrate what benefits different policies can achieve and what prices are paid by non privileged users. However, before commenting upon these points, it is mandatory to describe in detail the OSN model we resorted to and its interaction with the P2P overlay.

3. THE ONLINE SOCIAL NETWORK

3.1. Network Evolution Model

In order to create the OSN graph, where vertices represent members of the OSN and edges the relationships between them, our starting point is the general model of network evolution developed in Leskovec, Backstrom, Kumar, and Tomkins (2008) that also reports data collected from four distinct OSNs: Flickr, LinkedIn, Delicious and Yahoo Answers. In essence, the model the authors put forth in Leskovec, Backstrom, Kumar, and Tomkins (2008) relies on three distinct processes:

- The node arrival process, that is, the process ruling the arrival of new users joining the social network.
- The process of initiating new edges within the osn, that details the behavior of an active node, statistically specifying the times when the node creates new edges.
- The edge destination selection process, that indicates what node will be the destination of a newly created edge.

As our interest lies in videos, we concentrated on Flickr, a social network that was born to share pictures and that later on was extended to allow the sharing of videos: we therefore resorted to the specific statistics and parameter setting summarized below, that according to Leskovec, Backstrom, Kumar, and Tomkins (2008) tailor the OSN model to the features Flickr exhibits.

Namely, given $N(t)$ indicates the number of nodes within the network at a generic time t of its evolution, its behavior was suggested to be exponential, with

$$N(t) = e^{0.25t}. \tag{2}$$

As for the active lifetime of a node, a, this obeys an exponential probability density function,

$$p_l(a) = \lambda \cdot e^{-\lambda a}, \tag{3}$$

with $\lambda = 0.0092$.

During its active lifetime, a node adds a new edge every δ time steps, where δ is a temporal gap described by a truncated exponential random variable, whose probability density function is

$$p_g(\delta) = \frac{1}{Z} \cdot \delta^{-\alpha} e^{-\beta d\delta}, \tag{4}$$

where Z is the normalization constant,

$$Z = \int_0^\infty \delta^{-\alpha} e^{-\beta d\delta} d\delta = \frac{\Gamma(1-\alpha)}{(\beta d)^{1-\alpha}}, \tag{5}$$

$\alpha = 0.84$ and $\beta = 0.002$.

Last, the authors of Leskovec, Backstrom, Kumar, and Tomkins (2008) found that a satisfying way to choose the destination edge (hence setting up a mutual relationship between two OSN members) was the simple random-random triangle closing model: based on it, during the active lifetime of a node, when δ expires, first the originating node s picks at random one of its neighbors, say u, next it randomly picks one of u's neighboring nodes, say v, and finally a new edge is initiated between the origin node s and the last chosen node v, therefore closing -- at random -- the (s, u, v) triangle.

3.2. Amendments to the Model

We actually had to introduce a few amendments to the model, to have it satisfyingly fit the experimental data.

The first modification dealt with $N(t)$ behavior: data show that Flickr would count up to approximately 580.000 nodes in a relatively short time span of observation, 25 months. However, it is impossible to reach such significant size exclusively relying upon (2) (even Leskovec, Backstrom, Kumar, & Tomkins, 2008, indicates (2) fits well $N(t)$ only in the long term). We therefore decided to rapidly and forcedly increase the number of OSN users at the beginning of the simulation, to later converge to the steady trend described by the exponential function given in (2). Only in this manner the artificial data fit the experiments, as Figure 1 shows.

The exponentially distributed lifetime of the nodes has also been modified: once more, experimental data highlight that there is a relatively high probability that a node, once invited to join Flickr, creates very few edges and then never returns, a condition that the exponential distribution cannot capture. We believe this difference is non negligible and according to the data set reported in Leskovec, Backstrom, Kumar, and Tomkins (2008) we put forth a possible solution to this problem, modifying the exponential p.d.f. in (3) into a mixed one: the latter displays a probability $p(0) = 0.55$ that a newly invited node has a null lifetime, a probability $p(1) = 0.05$ that the node lifetime is 1 day and a probability $p(2) = 0.05$ that the node lifetime is 2 days. As a consequence, (3) is modified into:

$$p_l(a) = p(0) \cdot u(y) + p(1) \cdot u(y-1) + p(2) \cdot u(y-2) + (1 - p(0) - p(1) - p(2)) \cdot \lambda \cdot e^{-\lambda a}, \quad (6)$$

where $u(y)$ is an impulse located at the origin. Figure 2 indicates that the recreated data are now very close to the real ones (actually much closer than those obtained sticking to the suggestion of Leskovec, Backstrom, Kumar, & Tomkins, 2008).

Last, the truncated exponential shape of (4) also had to be amended. When employing the α and β values suggested by the authors of Leskovec, Backstrom, Kumar, and Tomkins (2008) we found that the values of time gaps δ generated in

Figure 1. Experimental and recreated data for $N(t)$

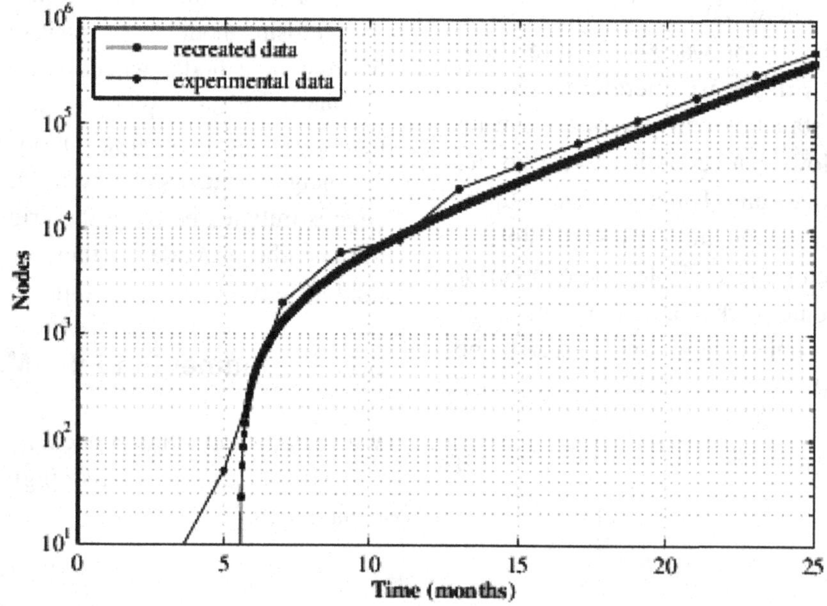

Figure 2. Experimental and recreated data of the node lifetimes

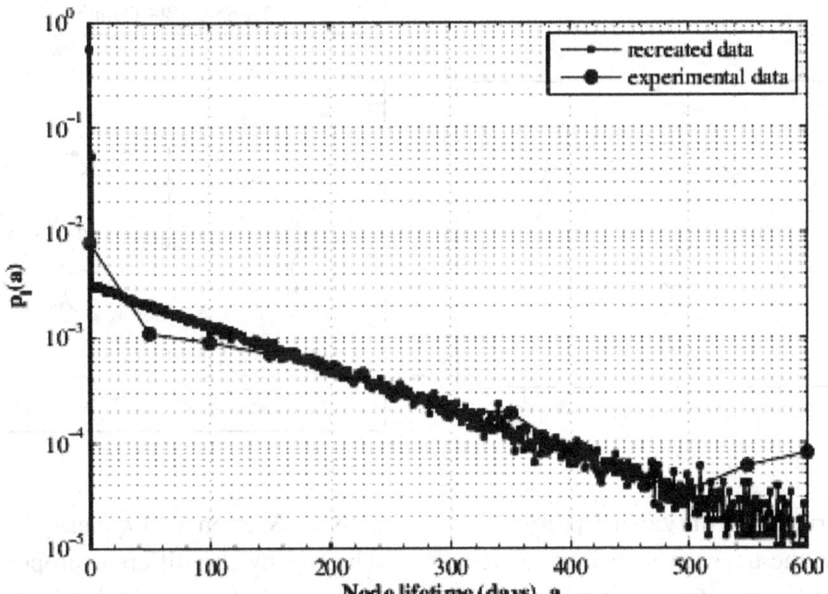

accordance to (4) are much larger than in reality: this condition does not have to be underestimated, as the longer the node time gap, the lower the number of edges the node creates, hence the lower its degree, i.e., the number of edges incident to the vertex representing the node in the graph. After several tests, β was set equal to 0.02, a choice that guarantees a slope for the network degree distribution of -1.76, very close to the real measured value, -1.74.

4. P2P STREAMING OVERLAY AND OSN FRIENDS

Now that all tools to generate the OSN graph have been properly set, it is appropriate to thoroughly describe the P2P streaming system functioning.

We begin by observing that the OSN and the P2P architecture lie on two distinct planes, evolving at a completely different time pace. It takes several days, if not months, to build an OSN with a fairly large size; besides, relationships among its members can be considered stable in the medium to long term. It is definitely not so within the P2P overlay, where peers join and leave much more frequently and unpredictably. We therefore choose to create one sufficiently large instance of the examined OSN, whose graph remains static throughout the entire P2P system evolution; on the contrary, the birth and death of peers within the streaming system will occur much more frequently, obeying two distinct exponential distributions. Then, whenever a new user asks to join the P2P overlay, with probability P it will be tagged as belonging to the OSN and randomly associated to a specific node of the OSN graph (whereas with probability $1 - P$ will exhibit no relationships within it). A visual representation of the way P2P users are associated to OSN members is reported in Figure 3.

As regards the examined P2P architecture, its features reproduce the macroscopic behavior that most current P2P prototypes display. The video to be distributed among peers is divided into m substreams, each with rate d / m, where we recall d is the streaming rate (Figure 4). All m substreams have to be received, in order to guarantee

Figure 3. Mapping inSN peers onto the OSN graph

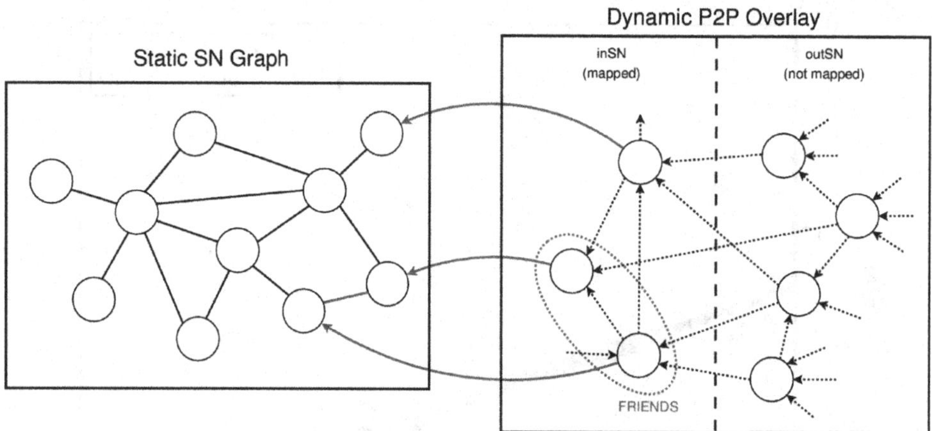

a proper reconstruction of the video. Upon joining the P2P overlay, the tracker server immediately passes the new peer a sufficiently wide list of neighbors, and it is among them that the peer randomly selects its potential parent peers: once these are contacted, if they possess the desired substream and have not exhausted their upload capacity, they start providing the newcomer with video chunks.

We additionally observe that the distribution scheme of the P2P overlay is push-pull: once a parent peer starts delivering video chunks to a child peer, it continues to do so until either the parent leaves the overlay or the child itself departs. Moreover, every peer is forced to provide each of its children with only one substream, to avoid the very likely disruption in video quality that its sudden departure would cause.

When the P2P overlay operates in normal conditions, i.e., it is underloaded so that its resource index σ is greater than 1, the behavior of P2P nodes that are also OSN members (and the quality of service they experience) by no means has to be different from that of ordinary peers. However, when the system happens to be overloaded

$\sigma < 1$

(), they become privileged users and one of the three strategies delineated in Section 2 comes into play.

Next Section will quantify the performance achieved by the different proposed schemes in terms of delivery effectiveness, for both OSN peers and for nodes of the P2P system that do not belong to the OSN.

5. NUMERICAL RESULTS

To evaluate the behavior of the proposed system, we have resorted to simulation. In order to realistically describe the dynamic behavior of the P2P overlay, we resorted to PeerSim (http://peersim.sourceforge.net/), a Java based simulator, properly tailoring and mimicing a hybrid push-pull, mesh-based streaming protocol devised for real-time content distribution.

The results presented hereafter refer to an OSN graph whose size is 50×10^4 nodes, whereas the P2P overlay displays an average of $N = 10 \times 10^4$ peers. The average peers' lifetime coincides with the distributed video duration, while their average interarrival time is set so as to guarantee that, after a short initial transient, the population size steadily takes on the N value. Moreover, we initially fix the probability of a peer to belong to the OSN equal to $P = 0.5$. In the examined framework, the video to be distributed among the peers is divided into $m = 8$ substreams, each

Figure 4. Percentages of peers that receive all substreams as a function of time

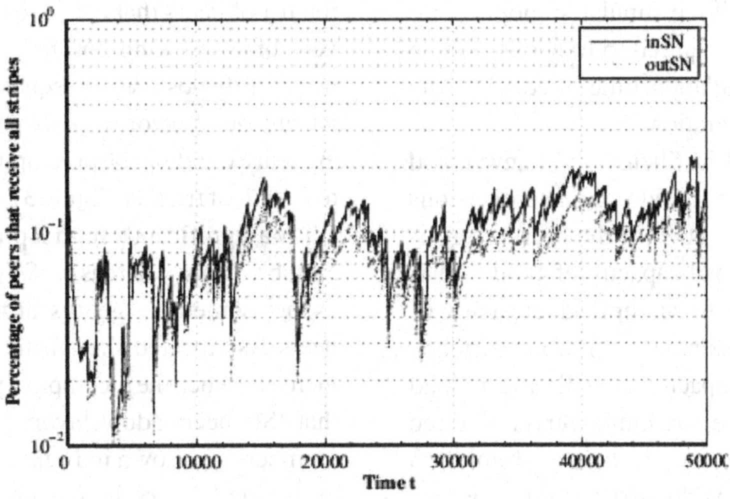

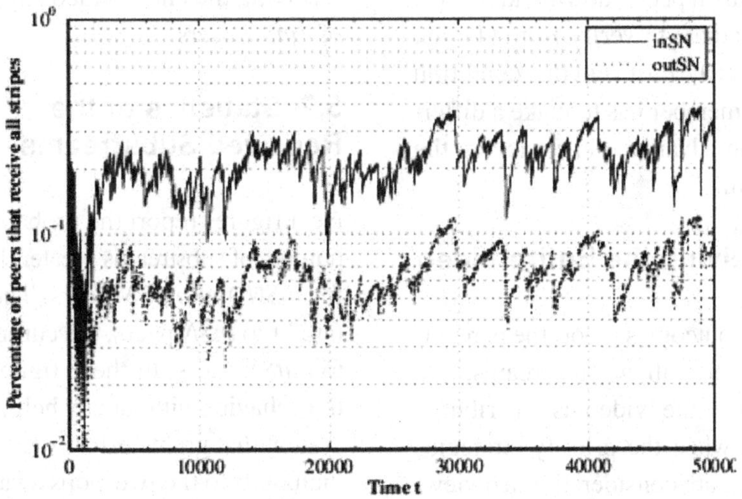

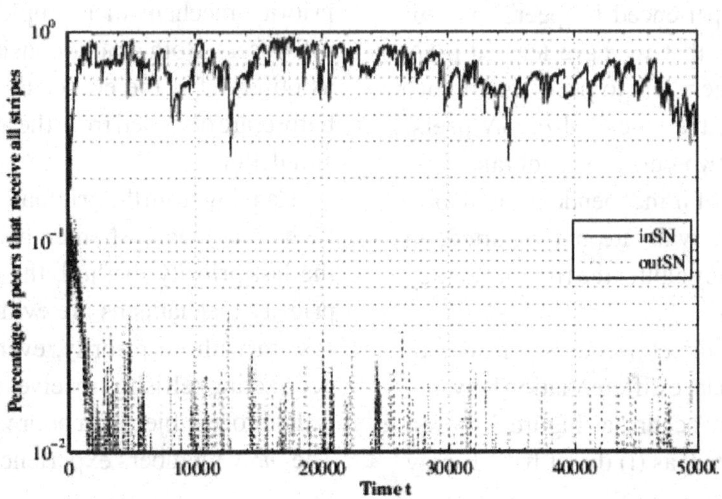

with rate $d / 8$. The video stream is made of 10^4 chunks, the time unit the simulator adopts coincides with the time required to transmit a video chunk, and the simulation time is equal to the total video stream duration.

As suggested in Liu, Shen, Ross, Panwar, and Wang (2009) we have considered a heterogeneous population of nodes: 30% of the peers are residential, with an upload capacity of $0.5d$, 60% are institutional, with an upload capacity of $1.2d$, while the remaining 10% are free riders, i.e., their upload capacity is null; the upload bandwidth S of the streaming server is three times the streaming rate d. This configuration corresponds to an overloaded condition, where the capacity to fulfill all peers' download request is scarce: it is immediate to verify from (1) that the resource index $\sigma;0.87$. It is in this condition that being an OSN member has to make a difference, introducing an advantage in retrieving the desired video stream.

5.1. Peers Receiving the Entire Video

The first simulation outcomes report the percentage of peers that receive all m substreams, as a function of the time the video is distributed within the overlay, when the priority strategies previously described are considered. Each viewgraph displays two curves: the solid line refers to the performance experienced by peers that are also OSN members, that we have termed *inSN* peers; the dashed line refers to peers that do not belong to the OSN, the so-called *outSN* peers. Every point of the two curves was obtained by averaging the results of 5 independent simulation runs; confidence intervals are not reported, in order not to excessively clutter the figure, but they are very tight.

When the help of direct friends is considered, only a slight performance differentiation between *inSN* and *outSN* peers occurs, as Figure 3 shows: this had to be expected, as (i) direct friends may not be that many; (ii) not all of them belong to the list of peers that can be contacted by an *inSN* newcomer looking for help; (iii) they do not necessarily possess the requested substream. The advantage in favor of *inSN* peers becomes more evident extending the preemptive priority concept to friends of friends: Figure 3 points out *inSN* peers enjoy a significantly higher probability to receive all substreams, whereas *outSN* peers are penalized, as they sometimes get discarded in favor of *inSN* peers. As expected, the gap in performance further increases when the preemptive priority mechanism that ISN peers adopt favors the totality of *inSN* members, as shown in Figure 3: now *inSN* nodes experience a very high probability of correctly receiving the entire video stream, at the expense of *outSN* users.

5.2. Statistics of the Received Substreams

Next figures report the probability of receiving a number of substreams greater than k, $0 \leq k \leq 8$, for *inSN* and *outSN* peers. The curves in Figure 6 refer to *inSN* peers, the curves in Figure 6 refer to *outSN* users. In these figures squares refer to the behavior without any help, circles refer to the system that grants priority to direct friends only, diamonds to the two-hops away friends help and triangles to the condition where the preemptive priority mechanism is implemented among all *inSN* users. Both graphs consider the system state as observed at the end of the simulation and as before, are obtained from the average of 5 distinct simulations.

Ranging from the original configuration where no help among OSN members is foreseen, until the last priority method, the effects of different priority mechanisms are evident. In the original system with no priority, every *inSN* and *outSN* peer, acting selfishly, receives few substreams and is therefore subject to a poor viewing experience. The *inSN* members experience a steady increase in the number of received substreams the more OSN users can assist them; on the contrary, *outSN* peers get fewer and fewer substreams. As desired, *inSN* peers aid friends, friends of friends or ultimately all OSN members, always at the expense of *outSN* nodes.

It is worth pointing out that the presented results are subject to a further interpretation. If the m video substreams are obtained via an encoding technique such as Multiple Description Coding (MDC), so that the more substreams are retrieved, the higher the quality of the reconstructed video, then in an overloaded system *inSN* peers would be able to watch a good quality video, whereas *outSN* peers would experience a basic viewing quality, without being locked out.

5.3. Impact of OSN Members Percentage

We have also investigated the effects that a different percentage of *inSN* peers over the total P2P overlay population size has on system performance, considering different values of the probability P introduced at the end of Section 4.

The following results refer to the strategy where the help with preemptive priority is implemented among peers that are either direct friends or lie two-hops away (friends of friends) within the OSN graph. Figure 5 reports the probability an *inSN* peer receives a number of substreams greater than k: the five curves summarize system behavior for different P values, namely, $P = 0$, 0.2, 0.5, 0.8 and 1. A modest degradation of *inSN* performance with the increase of the P value, from $P = 0.2$ to $P = 0.8$, is noticeable. This effect has to be ascribed to the growing fraction of *inSN* peers: if they represent a significant portion of the entire overlay population, then there is a decreasingly lower possibility that either a friend or a friend of a friend can discard *outSN* nodes. The results obtained for $P = 1$ correspond to the original system, where no differentiation among *inSN* and *outSN* users is present (the totality of the peers are also OSN members), whereas for $P = 0$ there is no possibility to determine a CDF, *inSN* peers being absent.

Figure 5. Probability that the number of substreams received by inSN peers is greater than the value k on the abscissa, for different P values, when the help from friends and friends of friends is considered

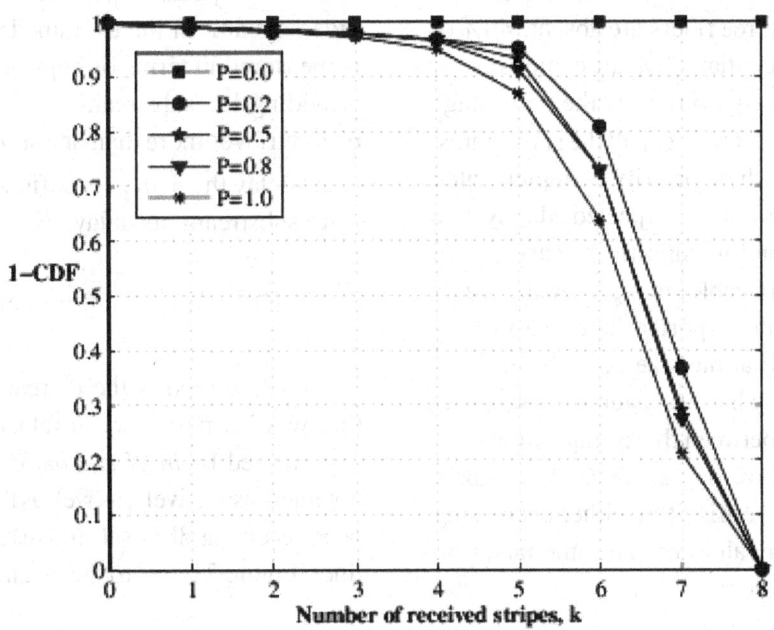

Figure 6. Percentage of peers retrieving the entire video as a function of time, in underloaded conditions, when the help from friends and friends of friends is considered

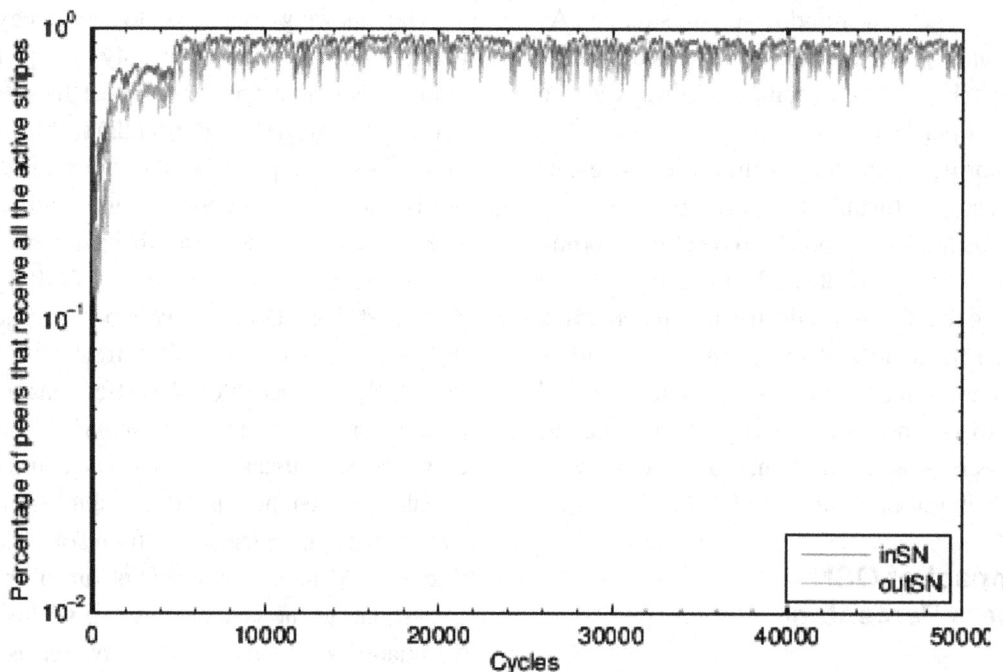

To verify that the introduction of the priority mechanism would not cause any malfunctioning in the P2P system when operating in underloaded conditions, i.e., when its resource index is lower than one, a second scenario was examined, termed underloaded: now free riders are absent, 40% of P2P nodes are residential, 60% are campus users, so that $\sigma;1.22$. Figure 6 reports the percentage of peers that are able to collect all the substreams, for the limit case where priority is implemented among all *inSN* peers: as expected, the system works smoothly and no significant variation in performance is noticeable among *inSN* and *outSN* nodes. Although not reported here, we investigated the effects that the size of both the OSN and the P2P overlay has on system performance. Keeping the ratio between the average number of peers N in the P2P overlay and the OSN members fixed to $5:1$, we verified that either decreasing or increasing both values by the same factor no relevant changes occur.

5.4. Experienced Delay

In parallel, we have determined the effects that different priority strategies have on the delay *inSN* and *outSN* peers experience in retrieving the video stream. In the examined context ``delay'' is measured in terms of hops between the server providing the substreams and the peer: as each peer retrieves more than one substream, given d_{ij} is the delay the i-th peer suffers in retrieving the j-th substream, its delay, D_i, is defined as

$$D_i = max\ D_{ij}, \qquad j = 1, 2, \ldots, m. \tag{7}$$

Figure 8 reports the dicrete density function that was ``a posteriori'' evaluated for the delay experienced by *inSN* and *outSN* peers, solid and dot lines respectively, as well as the delay that *inSN* peers receiving all m substreams encounter (dashed line): Figure 7 refers to the original system, where

Figure 7. Probability that the received number of substreams is greater than the value k on the abscissa

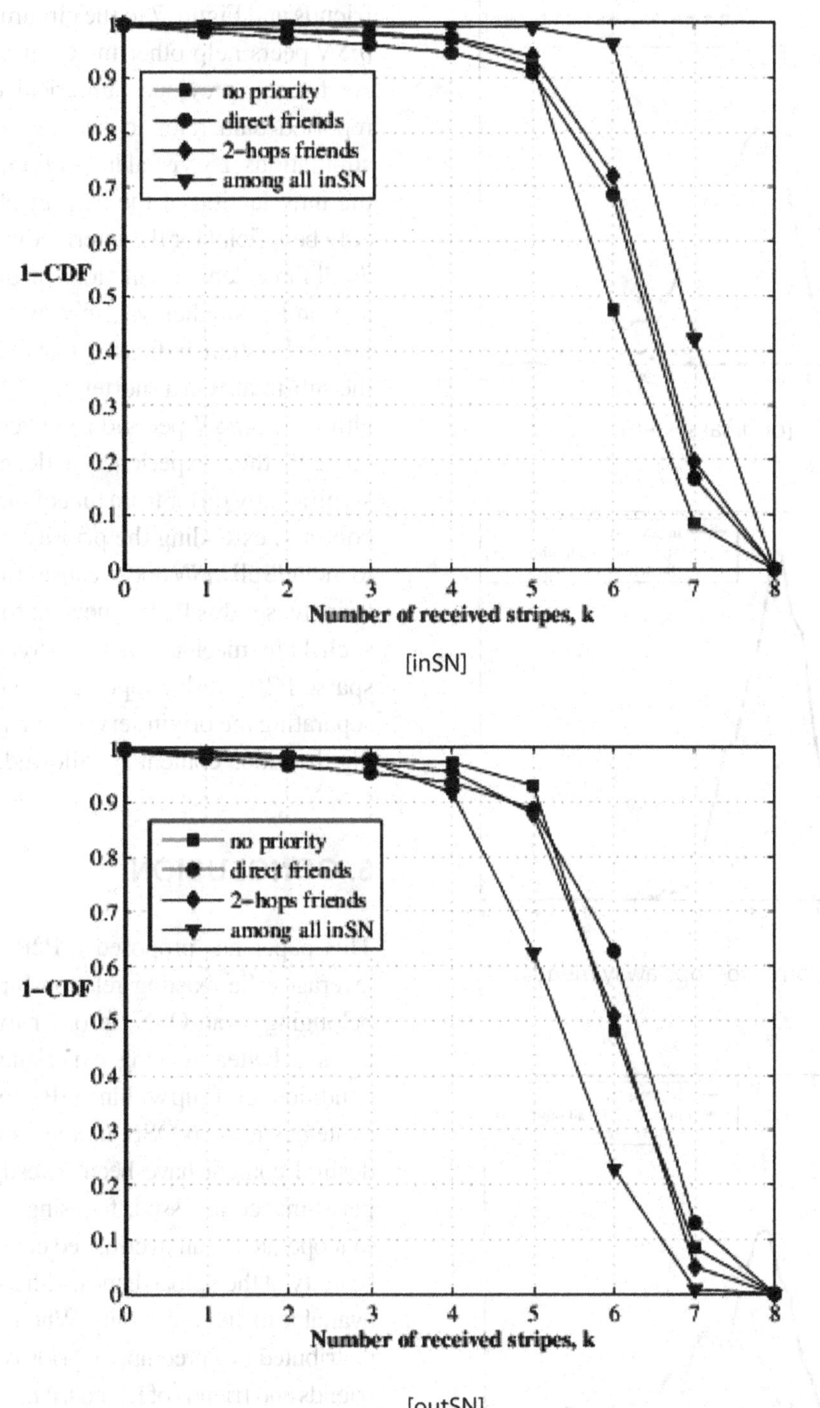

Figure 8. Discrete density function of the delay

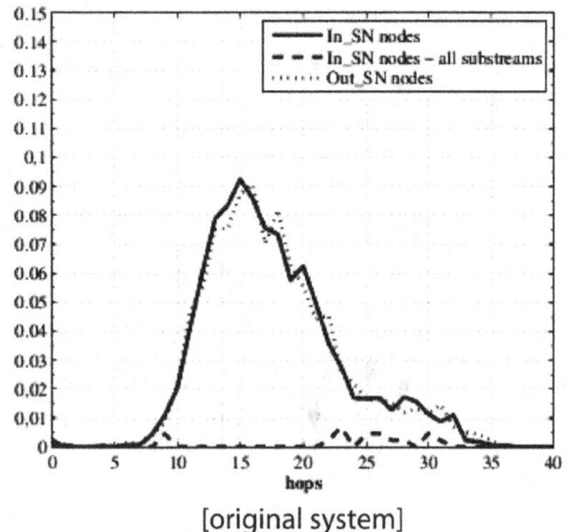

[original system]

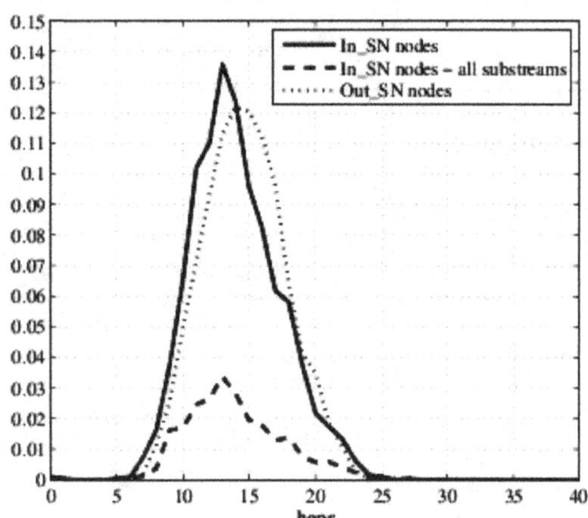

[Help from two-hops away friends]

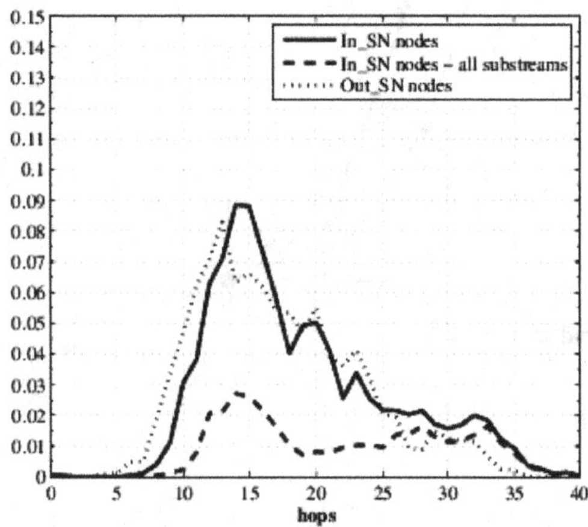

[Help from all inSN peers]

no priority mechanism is implemented; Figure 7 to the case where *inSN* peers favor two-hops away friends and Figure 7 to the circumstance where all *inSN* peers help other mates in need for content. As for the previous numerical evaluations, the reported data refer to an average of 5 distinct simulations. By visual inspection, it is evident that the introduction of the help mechanism is markedly beneficial for the intermediate case of Figure 7: all three density functions display a lower average and a smaller variance with respect to the original system, indicating that *inSN* nodes receive the substreams in a shorter time. At the same time, although *outSN* peers do not receive that much content, they experience a delay that does not significantly differ from that of *inSN* peers. On the contrary, extending the priority mechanism so as to include all *inSN* nodes causes the delay variance to increase: this had somewhat to be expected, as such help mechanism forcedly induces a more sparse P2P overlay topology, where longer paths separating the origin server from the peers retrieving the video content are allowed.

6. CONCLUSION

This paper has proposed a P2P architecture that leverages the existing relationships among nodes belonging to an OSN, in order to guarantee such users a better viewing experience when critical conditions build up within the P2P overlay. Different strategies to help OSN members in retrieving the desired content have been investigated, and their performance assessed, focusing on a P2P system that operates in an overloaded condition, due to the scarcity of the upload bandwidth its members make available to the community. When video streams are distributed in a preemptive priority fashion to OSN friends and friends of friends, it has been shown that OSN peers experience a very good probability of flawlessly receiving the entire video or a significant portion of it, achieving a clear service differentiation with respect to peers that do not belong to the OSN.

REFERENCES

Abboud, O., Zinner, T., Lidanski, E., Pussep, K., & Steinmetz, R. (2010, June). StreamSocial: a P2P Streaming System with Social Incentives. In *Proceedings of the IEEE International Symposium on a World of Wireless Mobile and MultiMedia Networks*, Montreal, QC, Canada.

Acquisiti, A., & Gross, R. (2005, November). Information Revelation and Privacy in Online Social Networks (the Facebook case). In *Proceedings of the ACM Workshop on Privacy in the Electronic Society*.

Acquisiti, A., & Gross, R. (2006, June). Imagined Communities: Awareness, Information Sharing, and Privacy on Facebook. In *Proceedings of the 6th Workshop on Privacy Enhancing Technologies*.

Graffi, K., Gross, C., Stingl, D., Hartung, D., Kovacevic, A., & Steinmetz, R. (2011, January). LifeSocial.KOM: A secure and P2P-based solution for online social networks. In *Proceedings of the IEEE Consumer Communications and Networking Conference* (pp. 554-558).

Kubo, H., Shinkuma, R., & Takahashi, T. (2010, December). Mobile P2P Multicast based on Social Network Reducing Psychological Forwarding Cost. In *Proceedings of the IEEE Conference on Global Communications* (pp. 1-5).

Leskovec, J., Backstrom, L., Kumar, R., & Tomkins, A. (2008). Microscopic Evolution of Social Networks. In *Proceedings of the ACM SIGKDD International Conference on Knowledge Discovery and Data Mining*, Las Vegas, NV (pp. 462-470).

Liu, Z., Shen, Y., Ross, K. W., Panwar, S. S., & Wang, Y. (2009). LayerP2P: Using Layered Video Chunks in P2P Live Streaming. *IEEE Transactions on Multimedia, 11*(7), 1340–1352. doi:10.1109/TMM.2009.2030656.

Nagle, F., & Singh, L. (2009, July 20-22). Can Friends Be Trusted? Exploring Privacy in Online Social Networks. In *Proceedings of the International Conference on Advances in Social Network Analysis and Mining* (pp. 312-315).

Pouwelse, J. A., Garbacki, P., Wang, J., Bakker, A., Yang, J., & Iosup, A. et al. (2006). Tribler: A social-based Peer-to-Peer System. *Concurrency and Computation: Recent Advances in Peer-to-Peer Systems and Security, 20*(2), 127–138.

Qureshi, B., Min, G., & Kouvatsos, D. (2010). A Framework for Building Trust Based Communities in P2P Mobile Social Networks. In *Proceedings of the IEEE Conference on Computer and Information Technology* (pp. 567-574).

Wu, D., Liu, Y., & Ross, K. W. (2009, April). Queuing Network Models for Multi-Channel P2P Live Streaming Systems. In *Proceedings of the IEEE Conference on INFOCOM* (pp.73-81).

This work was previously published in the International Journal of Wireless Networks and Broadband Technologies, Volume 1, Issue 3, edited by Naveen Chilamkurti, pp. 1-14, copyright 2011 by IGI Publishing (an imprint of IGI Global).

Chapter 14
Load Balancing Aware Multiparty Secure Group Communication for Online Services in Wireless Mesh Networks

Neeraj Kumar
Neeraj Kumar, CSED, Thapar Univesity, Patiala (Punjab), India

ABSTRACT

The internet offers services for users which can be accessed in a collaborative shared manner. Users control these services, such as online gaming and social networking sites, with handheld devices. Wireless mesh networks (WMNs) are an emerging technology that can provide these services in an efficient manner. Because services are used by many users simultaneously, security is a paramount concern. Although many security solutions exist, they are not sufficient. None have considered the concept of load balancing with secure communication for online services. In this paper, a load aware multiparty secure group communication for online services in WMNs is proposed. During the registration process of a new client in the network, the Load Balancing Index (LBI) is checked by the router before issuing a certificate/key. The certificate is issued only if the value of LBI is less than a predefined threshold. The authors evaluate the proposed solution against the existing schemes with respect to metrics like storage and computation overhead, packet delivery fraction (PDF), and throughput. The results show that the proposed scheme is better with respect to these metrics.

1. INTRODUCTION

Over the years, wireless mesh network (WMN) has gained lot of attention due to unique features such as multihop nature, easily scalable and maintainable with low network cost applicable to wide range of applications, and self healing and self configurable nature (Akyildiz, Wang, & Wang, 2005). A generalized architecture of the WMN consists of mesh gateways (MG), mesh routers (MRs) and mesh clients (MCs). Every node in WMNs may acts as a router, forwards the packet to other nodes. Some of these routers may act as gateways which are directly connected to

DOI: 10.4018/978-1-4666-3902-7.ch014

internet (1). A WMN combines the fixed network (backbone) and mobile network (backhaul). Every node in WMNs can act as a router and is able to forward the packet to other nodes. A node which didn't have access to the backbone network can establish a connection by routing the packets from a neighbouring node that has a backbone network connection. MCs in WMNs connect to the internet using gateways which act as a relay nodes as shown in Figure 1.

There are number of applications where WMN can be used such as video on demand, Voice over IP (VoIP), online single/multiple player/players multimedia games etc. Because users accessing the different services are located in different domains, security is a major concern for all these service provided by the underlying network. Only the legitimate users are allowed to access the available resources over the network, i.e., before performing any operation in the network these users must use the key provided by the trusted party for a particular duration. The key to these users is given for a particular time interval and may be renewed for the next duration by trusted party so that only the authenticated person can use the available services of the network. Because most of the traffic flows from mesh routers (MRs) to mesh clients (MCs), so key is kept at MRs for the duration of communication between MRs and MCs. Broadly, there are two types of attacks boundaries namely as outside and inside within which an attacker can access the network resources. The attacker from outside the boundary can destroy the MRs or MCs while within the inside boundary it can access the messages that are not meant for it (2).

Although security in wireless networks has been investigated by various researchers from different prospective such as data confidentiality, integrity, trust management etc. The existing solutions for the security are divided into two folds namely as for centralized and distributed

Figure 1. Scenario in wireless mesh network

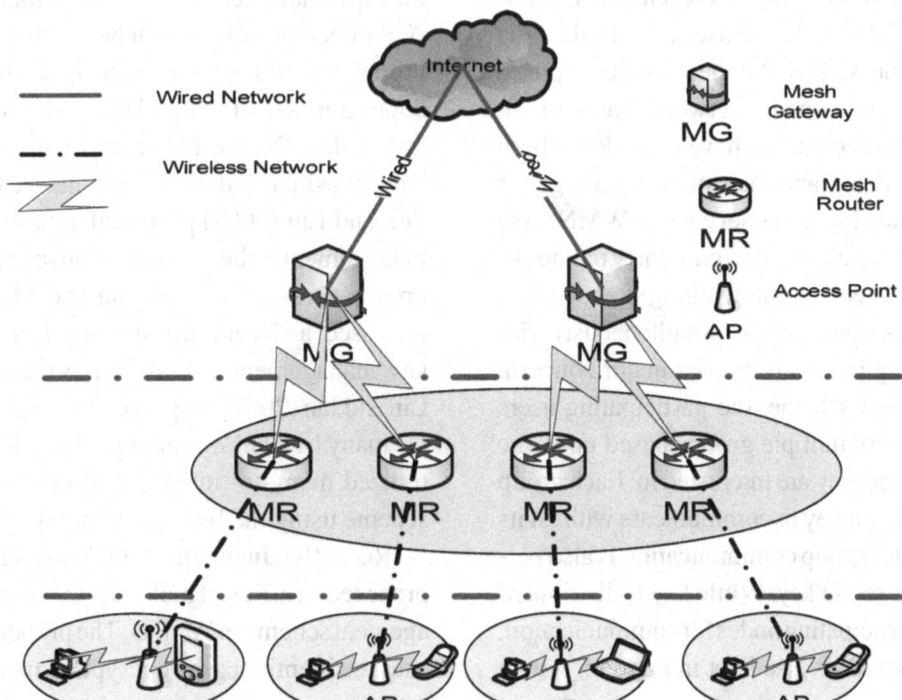

scenarios. In case of centralized scenario, standard encryption/decryption mechanisms are applied but these techniques have single point of failure (Han, Gui, Wu, &Yang, 2011; Rafaeli & Hutchison, 2003). The distributed approaches divide the group of users into several subgroups with each group uses a separate shared key for communication or multicasting the message (Han, Gui, Wu, &Yang, 2011; Han & Gui, 2009; Wu, Mu, Susilo, & Qin, 2009). Han, Gui, Wu, and Yang (2011) have used the proxy based secure multicast group communication mechanism. Both centralized and distributed strategies have their advantages and disadvantages. As the MCs are distributed in different regions which are controlled by the respective mesh points, this may rise security challenges such as data confidentiality, integrity, authentication. Hence any solution for the secure group communication must be scalable and efficient with respect to the available resources of the network.

Keeping in view of the issues, in this paper we propose a new load balancing aware secure multiparty group communication for online services in WMNs. Although recently Zhao, Al-Dubai, and Min (2010) proposed a load balancing mechanism for WMNs. But the solution proposed is applicable in general network scenario. In the proposed scheme, we have considered both the load balancing and multiparty secure group communication for online services in WMNs. We have specifically chosen the multiparty online applications such as social networking sites in which server handles many requests simultaneously with load balancing and security as a major concern. In the proposed scheme, the participating users are divided into multiple groups based upon the type of services they are interested in. Each group uses the common key to communicate with all its members. Inter group communication is also performed using secret key. While key is distributed among the participating nodes for communication, load on the server is also kept in mind for which LBI is calculated. Based upon the value of LBI, the new user is allowed or disallowed to use the network resources.

Rest of the paper is organized as follows. Section 2 describes the related work. Section3 discusses the network model. Section 4 gives the proposed solution and algorithm. Section 5 explore on simulation results and discussion. Section 6 provides the conclusion.

2. RELATED WORK

With the evolution of internet there are number of online services available to the users who may work in different domains to avail these available services. Security and privacy for all these services are a major concern for successful execution in WMNs as these networks are effected to various inside and outside attacks. Anyone can acts as an intruder and can eavesdrop, inject, or disturb the normal traffic pattern in these networks.

As these users use the services located at different domains, security becomes more difficult as they may have centralized or distributed control. A number of solutions have been proposed in literature for this purpose (Zhang &Varadharajan, 2006; Sun & Liu, 2007; Huang & Medhi, 2004; Lin & Jan, 2007), but none of these solutions have considered the dynamic nature of WMNs. Sun and Liu (2007) proposed a multi-group key management scheme that achieves hierarchical group access control. Huang and Medhi (2004) proposed a decentralized many-to-many group key management scheme using the hash chain. Lin and Jan (2007) proposed a tree based structure for many to many communications. It is a decentralized many-to- many group key management scheme using the broadcast encryption.

Recently, Junbeom, and Yoon (2010) have proposed an efficient multi service group key management scheme in WMNs. The proposed scheme uses the identity based encryption mechanism so that key can be distributed to legitimate users in an efficient manner. The scheme has low storage and rekeying cost and also authors have considered the dynamic nature of mesh points. To increase the efficiency and to reduce the high decryption and re-encryption overheads, scalable and adaptive key management (SAKM) (Challal, Bettahar, & Bouabdallah, 2004) has been proposed which depends upon the states of an agent. This work is extended further in which agent calculates the cost to adapt to a particular state (Dong, Ackermann, & Nita-Rotaru, 2009).

To maintain the routing privacy in WMNs, Wu and Li (2006) proposed an onion ring structure so that it would be very difficult for the intruder to get the secret code. The proposed scheme is quite efficient to locate any intruder present in the system. Wu, Xue, and Cui (2006) use multiple paths for data transmission so that attacker could not get any valuable information by getting only partial information. Kong and Hong (2003) proposed an anonymous on demand routing protocol (ANODR) in which each node generate one time password for authenticity. The scheme used by Kong and Hong (2003) is also used in SDAR (Boukerche, El-Khatib, Xu, & Korba, 2004), CAR (Shokri, Yazdani, & Khonsari, 2007), and ODAR (Sy, Chen, & Bao, 2006) in which secure routing is used for communication among the participating nodes. Attack resilient security architecture for WMNs is presented by authors in Zang and Fang (2006). In this proposal, authors eliminate the need for establishing bilateral roaming agreements and having real-time interactions between potentially numerous WMN operators. Recently, Wan, Ren, Zhu, Preneel, and Ming (2010) presented an anonymous user communication for privacy protection in WMNs. In this proposal, a group signatures scheme is used to deliver security and privacy protection. The pairwise secrets between any two users are used to achieve stronger privacy protection and the user is kept anonymous to mesh routers. Sun, Zhang, and Zhang (2011) proposed security architecture for achieving anonymity and traceability in wireless mesh networks. The proposed architecture maintain the privacy, confidentiality and data integrity in WMNs. Zhu, Lin, Rongxing, Pin-Han, and Shen (2008) proposed a secure localized authentication and billing scheme for WMNs. The authors proposed a novel secure system having resilience capability with minimum inter domain handoff authentication latency and workload.

Although there exists above defined solutions for security, but none of these solutions consider the security concern in online service controlled by multiple users. But there are many challenges yet to be addressed which have direct impact on the performance of the system. The most difficult challenge which is yet to be address is load aware secure efficient scheme with respect to various types of existing attacks. Keeping in view of the above drawbacks and challenges, we propose a new load aware multiparty secure group communication for WMNs. Our solution is different from the existing solutions in which we have considered applications which are accessed and controlled by multiple users simultaneously.

3. NETWORK MODEL

The network model in the proposed scheme is shown in Figure 2. It consists of three tier architecture as shown in Figure 2. At the lower layer are the MCs which want to communicate with each others in secure manner using mesh infrastructure. The MCs may be wired or wireless devices. At the second layer are the Access points (APs) and Mesh routers (MRs) which builds the mesh backbone infrastructure. All client requests are handled using this mesh backbone infrastructure, i.e., it is responsible for all communication among MCs both within inter and intra domains. At the last layer are the certification authorities (CAs) that provides the certificates to clients for secure communication whenever the MCs enter into the network. Any new client entering in the network

Figure 2. Network model in the proposed scheme

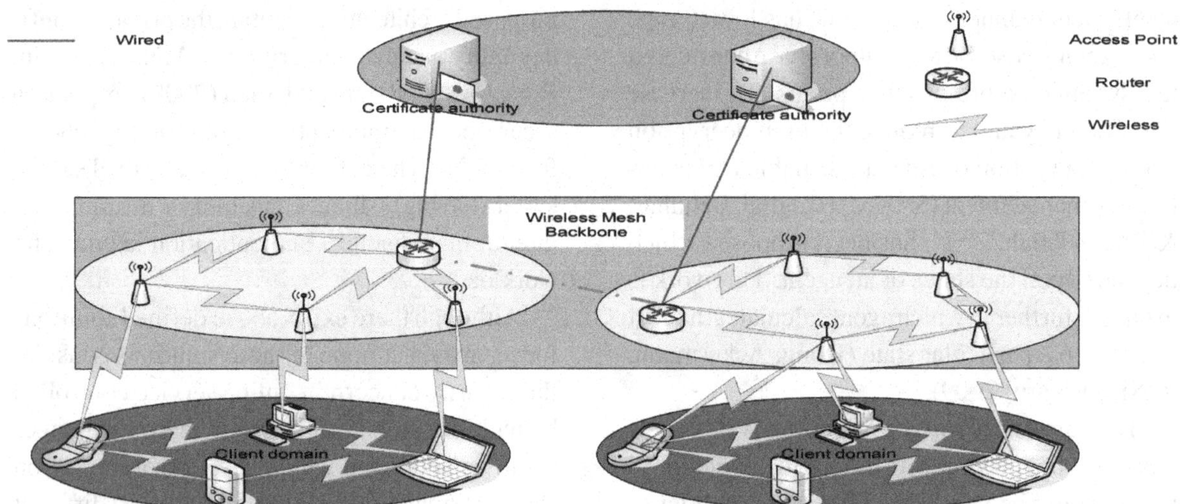

will get issue its certificate from CAs for communication with the other client. CAs issue the certificate with some constraints such as duration of certificate etc.

Wireless security has been the emerging area in the past years as the research in this direction is still in its infancy state. Many new proposals in this direction have come up from the research community to address the new and emerging issues such as attack resilient architecture, detection of intrusion in the network with mobility of the MCs, countermeasures to various types of inside and outside attacks. In particular, Anonymity and privacy issues have gained considerable attention with respect to the context of the application such as internet where a number of social networking websites are being used by the group of users simultaneously. These users may be in the same region or they may be in different regions accessing the internet.

4. PROPOSED APPROACH

As shown in Figure 2, the proposed network model consists of three major components as: mesh clients (MCs), wireless mesh backbone and certification authority. Before performing any operation in the network, all MCs must register themselves in backbone which takes the secure certificate from the trusted certificate authority and pass back this information to MCs. MCs may be mobile such as laptops, personal digital assistance (PDA), smart phones etc. or dedicated such as thin clients, personal computers (PCs) or servers. MCs are divided into different domains so that each domain is handled by their respective router. This router is responsible for the overall operations and traffic control form MCs to router and vice-versa.

As soon as any new MC enter into the network domain for resource sharing, it must register itself to the respective router and get the session key from the router for communication. This request for session key is passed through APs of the respective domain. Once the MCs of each domain get the key from router, they can start communication with the MCs in other domain. The key is maintained at MRs which distribute the key for both inter domain and intra domain communication

The proposed approach consists of following steps:

- Registration of MCs.
- Intra and inter domain data dissemination.
- Data dissemination.

Figure 3 describes the operation performed during the MCs communication in different domains. As shown in figure, MCs pass their information to MRs using APs. The sequences of operations are described as labelled arrow which shows the flow of information. For the sake of simplicity, we have considered two domains with MCs, APs, and MRs. The communication can be started with any MC within a domain. Before starting any operation, a MC sends the request for registration to nearest MR using the AP in the domain. For registration purpose each MCs send its ID as identification which is used as the public key for it. In return, MR returns the session key with time stamp for MCs to communicate with other MCs in different / same domain. The complete sequences of operations are described in Figure 3.

The proposed algorithm consists of following steps. In step 1, MCs sends the registration requests $\mathrm{Re}\,q$ from respective domains with their IDs, access point IDs, and mesh routers ID to MRs through APs in respective domains. In the second operation, MRs may or may not acceptthe registration requests for new MCs depending upon the current load. The value of LBI is calculated which consists of multiplication of the total number of requests (N) to the ratio of total number of request served (req_served) to the total number of request pending $(req\_pending)$ with in a particular time interval $(t_{i+1} - t_i)$. This procedure is repeated for each time interval having total n intervals. If the value of LBI is less than predefined threshold thr, then reply to MCs corresponding to its request for registration is sent with mesh IDs, access point IDs, with time stamp based key for communication. This key is used for communication with the other MCs. In the next step, a session key is generated along with time dependent nonce t^{non} and expiry time of the same is set using the value of TTL. This is sent to the demanding MCs so that they can start communication with intra and inter domain MCs. In step 2, 3, intra and inter domain communication with data dissemination is performed. We assume that MCs, A and B are in the same domain. Then these two MCs communicate with each other by sending AP, their own ID and message authentication code (MAC). The MAC is generated same as by the authors in Sun, Zhang,

Figure 3. Sequence of operations in the proposed scheme

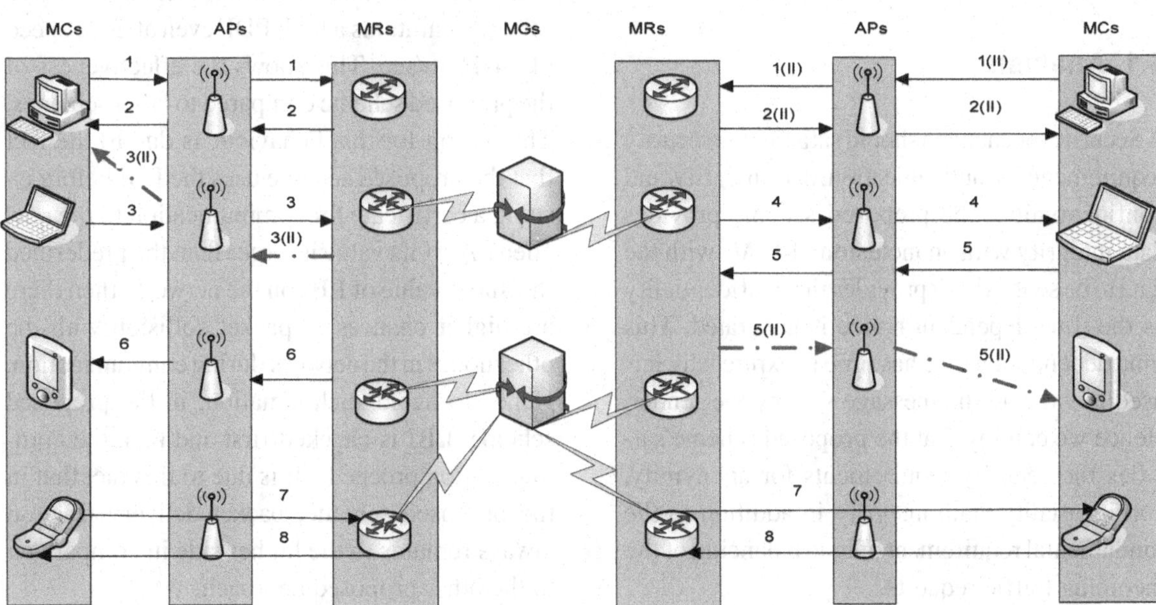

and Zhang (2011). The message sent, nonce generated and Id of the destination is protected by the MAC. MAC is calculated both from sender and receiver. If value of both the MAC is same, then the message is not captured and original and received messages are same else message has been changed in between the sender and receiver. In the proposed scheme, the identity of the destination is also protected using MAC, as it may happen that intruder may change the identity of the destination and message destination for which it was originally sent. In case of inter domain communication, instead of AP identity, MR identity is sent. Identity of MC, MR, and time dependant nonce are protected by MAC. Again both at receiver and sender the value of MAC is tested and verified. If both the values are matched, then the message is legitimate else it is not legitimate. In the final step, once both intra and inter domain communication is secured, then trust is established and using this trust mechanism, a secure multiparty communication among the participating MCs for accessing online web services is performed. The key feature of the proposed scheme is time dependent nonce which is generated as a large random number.

4.1. Analysis

A Security mechanism should satisfies the security requirements for authentication, data integrity, and confidentiality. The proposed scheme provides data integrity with an inclusion of MAC with the data to be sent. Also it provides the confidentiality as the time dependant nonce is generated. This time dependant nonce has time to expire which is used to validate the message sent by the sender. Hence we can say that the proposed scheme satisfies the security requirements for anonymity, confidentiality, data integrity in addition to the fundamental requirement of load balancing of the incoming traffic requests.

5. RESULTS AND DISCUSSION

We have implemented the proposed scheme for WMNs on ns2 (Fall & Varadhan, 2000) and compare its performance with well known Ad hoc On-Demand Distance Vector (AODV) and SeGroM (Dong, Ackermann, & Nita-Rotaru, 2009) protocols. The network topology consists of 8×8 grid (Figure 4) with source S communicates with destination D. There can be multiple paths from sources S to destination D shown in dotted line but the most efficient path is that one that has highest PDF with minimum delay and is tolerant to various types of attacks in the network. The metrics used for comparison are: Packet delivery fraction (PDF), packet delivery latency, computational time, Computational overhead, and network throughput. The simulation parameters are listed in Table 1.

Impact of the Proposed Scheme on Packet Delivery Fraction

Figure 5 shows the impact of the proposed scheme on PDF with other two schemes such as AoDV and SeGroM. With an increase in node speed PDF falls in all three schemes. But the proposed scheme maintains a high PDF even at node speed of 14-16 m/sec. This shows the effectiveness of the proposed scheme compared to other schemes. The reason for this behaviour is due to the fact that the proposed scheme uses the LBI before issuing a certificate for communication to the new client. As if its value is larger than the predefined threshold value of LBI on the network, then there are higher chances of packet collision with the other nodes in the network during communication. Hence to avoid such situation, in the proposed scheme, LBI is checked first and then the communication proceeds. It is due to this fact that in the proposed scheme, packet delivery fraction always remains to the higher side in comparison to the other proposed approach.

Figure 4. A 8×8 grid network topology in the proposed scheme

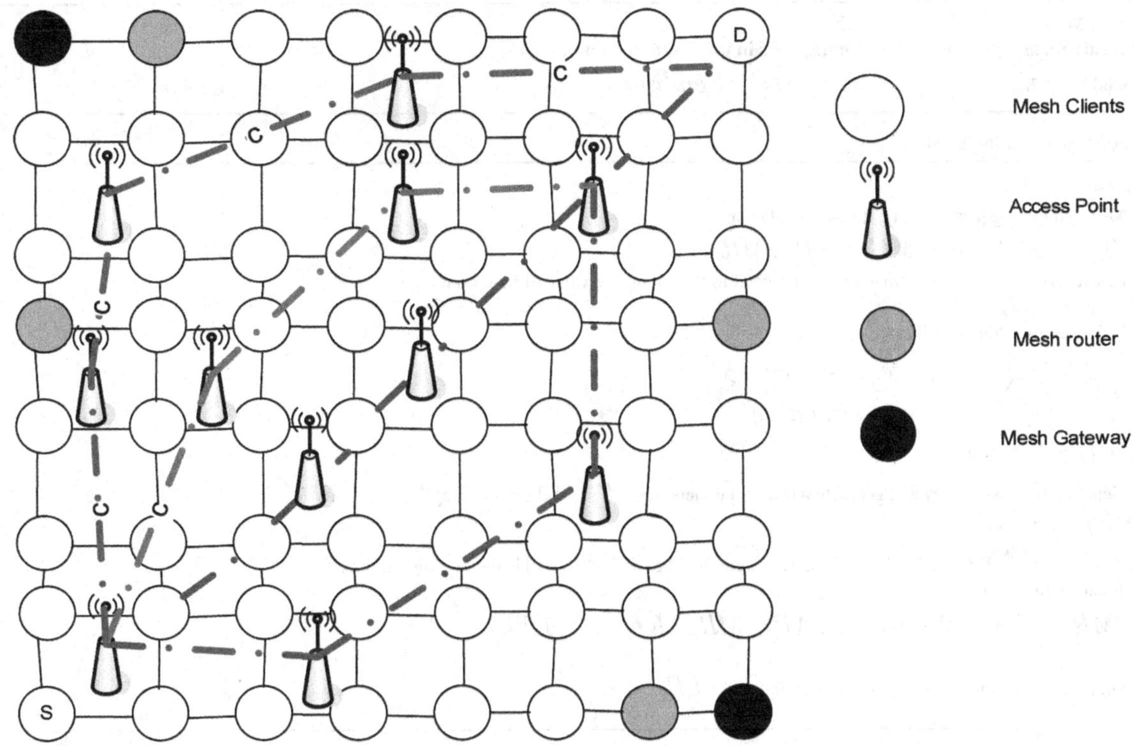

Table 1. Simulation parameters

Simulation Time	200s
Simulation Area	500×500
Transmission Range	200m
Mobility Model	Random waypoint
Packet Size	512 bytes
Number of Traffic Flows	50

Impact of the Proposed Scheme on Packet Delivery Latency

Figure 6 shows the impact of the proposed scheme on packet delivery latency with node speed. The results obtained show that the proposed scheme has lowest latency which is comparable to SeGroM protocol. This shows that the proposed scheme is quite effective in delivering the packets to the desired destination compared to other schemes. Although it may take some time during the registration process in the proposed scheme but once the registration process is completed then there is a secure communication among the various parties involved in the conversation. Hence the packet delivery latency reduces in the proposed scheme compared to other schemes as the packet transmitted from the source would have less chances of being captured in the middle of its path and have fewer chances to be captured or stolen. As shown in figure, even with an increase in node speed packet delivery latency is less in the proposed scheme in comparison to other schemes.

Impact on Computation Overhead

Figure 7 shows the impact of the proposed scheme on computation overhead. The computation overhead is mainly due to the number of encryption/ decryption operations. Because we are using

Box 1.

Input: MCs requests, IDs of MCs, N
Output: Secure communication among MCs in inter and intra domains
Variables: LBI, $t^{non}, req\_served, req\_Pending$

Step 1: Registration of MCs

Operation 1, 1(II)
1. MCs send the request $(\mathrm{Re}\,q)$ for session key as:
$MC \rightarrow MR : \mathrm{Re}\,q(MC_{ID}, AP_{ID}, MR_{ID})$
2. MR receive the requests from MCs and proceed for checking the value of load on it
Operation 2, 2(II)
3. Calculate the value of LBI as:
4. $LBI = \sum_{i=1}^{n} N * \frac{req\_served}{req\_Pending} * (t_{i+1} - t_i)$
5. $if(LBI < thr)$
6. Generate the session key and generate a time dependent none t^{non} and send the reply $(\mathrm{Re}\,p)$ as follows:
7. $t^{non} = K * rand();$ where K is very large number and $rand()$ is the random number Generation function.
8. $MR \rightarrow MC : \mathrm{Re}\,p(MC_{ID}, AP_{ID}, MR_{ID}, Key, t^{non}, TTL)$
Else
9. Drop the request for registration temporarily until ($LBI < thr$)

Step2, and 3: Intra and inter domain communication with data dissemination

Operation 3,3(II), and 5, 5(II)
10. Send the message from MC, A to B in the same domain as follows:
11. Generate the MAC address same as in (24).
12. $A \rightarrow B : (AP_{ID}, {MC_{ID}}^{A}, MAC(msg, {MC_{ID}}^{B}, t^{non}), TTL))$
13. On receiving the message from MC A, B calculates the MAC of the message
14. **If** $(new(MAC) == old(MAC))$ **then**
15. Message sent by MC A is legitimate
16. **Else**
17. Message is not legitimate
Operation 4, 6
18. Send the message from MC, A to B in different domain as follows:
19. $A \rightarrow B : ({MR_{ID}}^{A}, {MC_{ID}}^{A}, MAC({MR_{ID}}^{B}, msg, {MC_{ID}}^{B}, t^{non}), TTL))$
20. On receiving the message from MC A, B which lies in different domain calculates the MAC of the message and matches the same as follows:
21. **If** $(new(MAC) == old(MAC))$ **then**
22. Message sent by MC, A is legitimate
23. **Else**
24. Message is not legitimate
Operation 7, 8
25. Trust is established between different MCs which are in different domains
26. Communication proceeds among the MCs as the trust is already established

Figure 5. Impact of the proposed scheme on packet delivery fraction with node speed

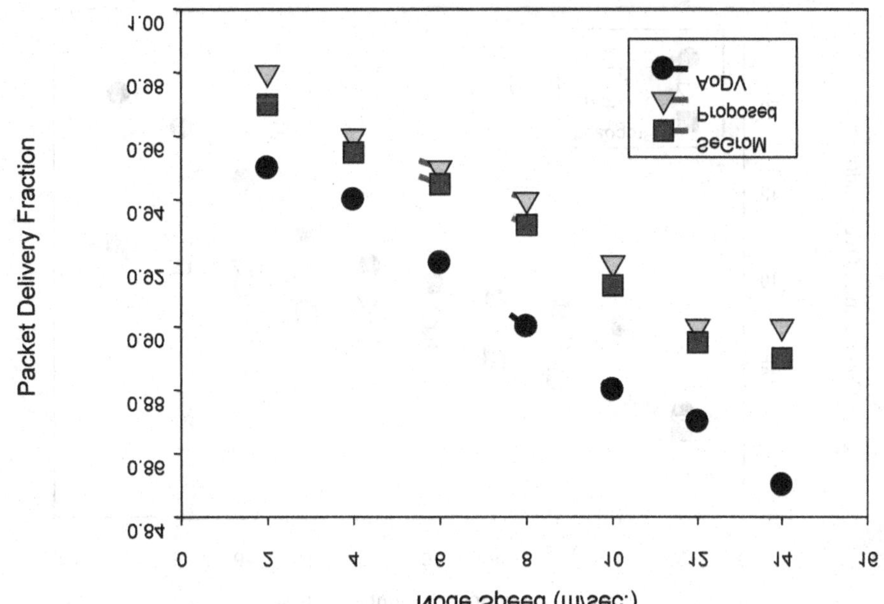

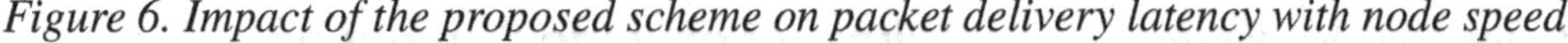

Figure 6. Impact of the proposed scheme on packet delivery latency with node speed

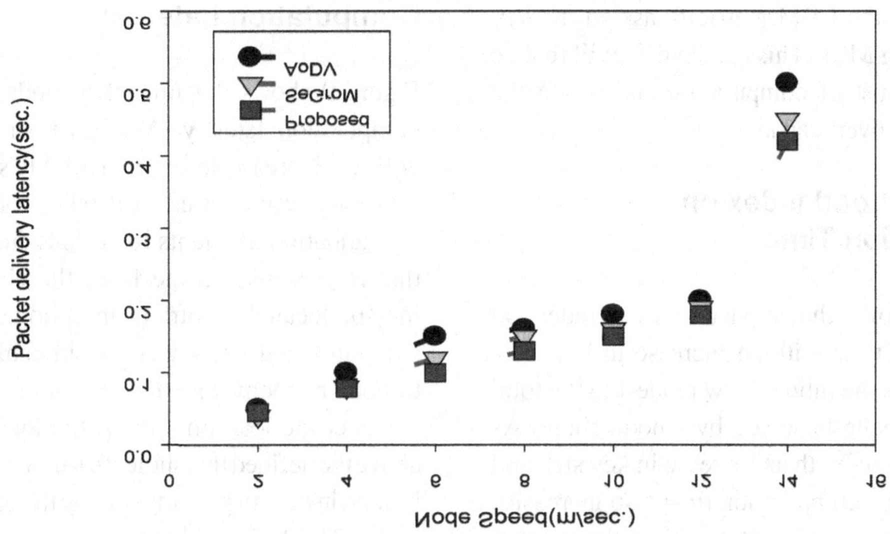

the identity of the node as the public key for the identification, hence the computation overhead is smaller in the proposed scheme than its counterpart. The computation overhead increases with an increase in the size of the group. As the group size is increased, it will take more time to distribute the key among the members and to perform encryption/decryption mechanisms. But as the proposed scheme is using the identity of the node as the public key, lesser number of encryption/decryption operations is required and hence the computation overhead is reduced to a great extent compared to other schemes. The results obtained show that the proposed scheme has comparable computation overhead with SeGroM and better than AoDV scheme. The reason for this type of

Figure 7. Computation overhead in the proposed scheme

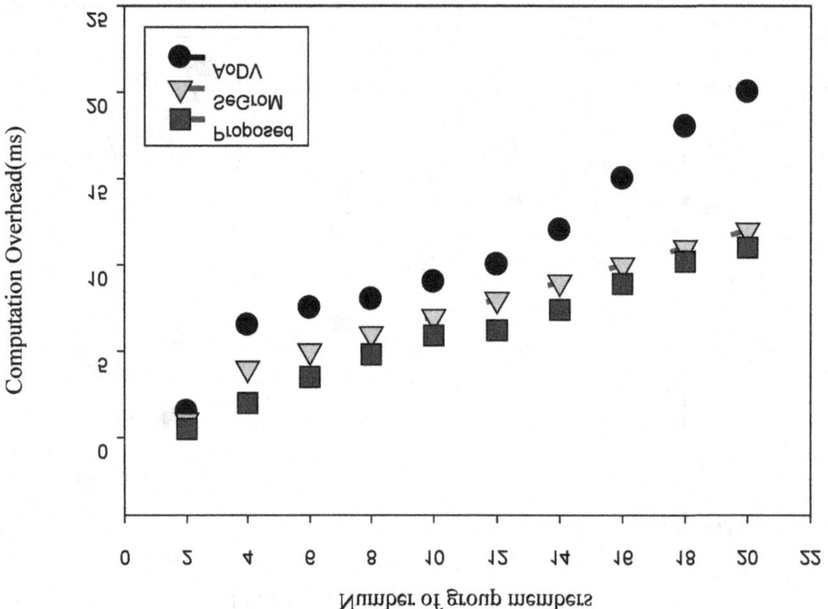

behaviour is due to the fact that the proposed scheme uses the LBM metric to assign the key to demanding MCs. This mechanism will reduce the overall cost of computation and hence the computation overhead also.

Impact of Load Index on Computation Time

Figure 8 shows the impact of load index on computation time with an increase in key size. Load index is the ratio of new request to the total number of requests served by a node/client. As shown in figure, with an increase in key size and load index, the computation time also increases. With an increase in computation time, the security of the system also increases. But in this case the network resource consumption also increases. Hence some tradeoff between network security with increased in load index is always required. As shown in Figure 7, with an increase in load index, the computation time increases with the size of the key.

Impact of Node Density on Computation Latency

Figure 9 shows the impact of node density on computation latency. As shown in the figure, with an increase in number of MCs, computation latency increases as it takes a longer time for validation of clients Id. This is due to the fact that request may come from the clients which may be located in intra or inter domain. But the computational latency is considerably less than the other schemes as the proposed scheme also consider the load on APs. If the load on APs is above the defined threshold then the authentication is done in peer to peer manner by the participating nodes. The decision that which participating node authenticates the newly arrived MC is taken by considering the past behaviour of the most trusted node. Although, the computational complexity increases with an increase in number of MCs, but still the latency is less in the proposed scheme compared to other schemes of its category. Hence in the proposed scheme, with an increase in node

Figure 8. Computation time with load index

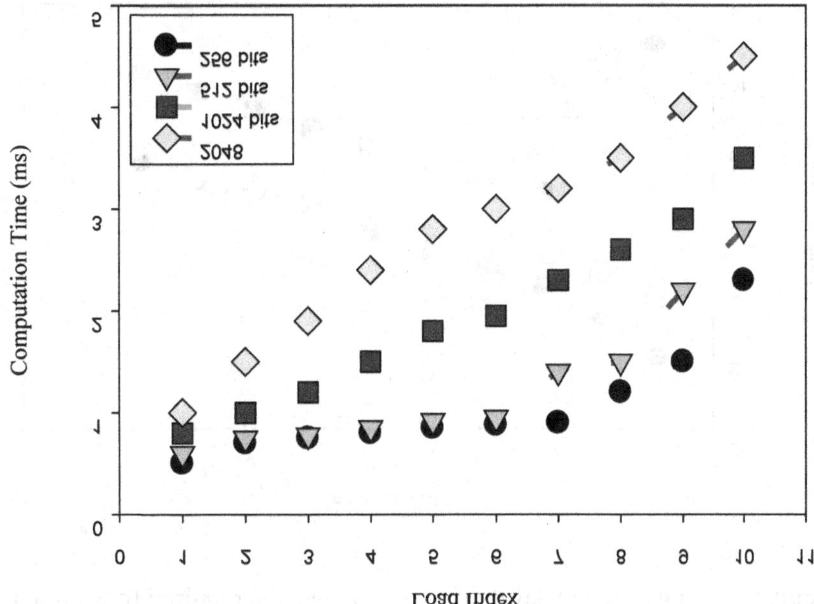

Figure 9. Impact of node density on computational latency

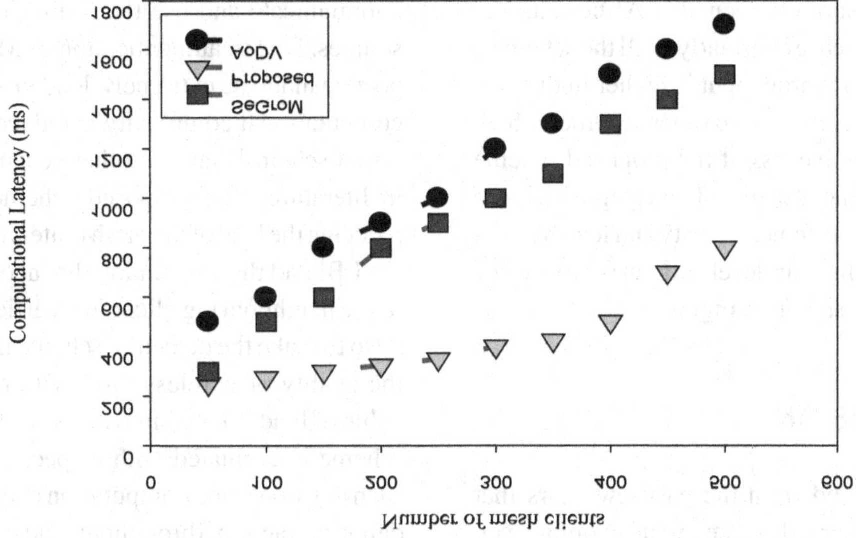

density, computational latency decreases to considerable amount compared to other approaches as shown in Figure 9.

Figure 10 shows the impact of the proposed scheme on the network throughput. As shown in the figure with an increase in the number of MCs, throughput increases to a considerable amount compared to other schemes. This is primarily due to the increase in packet delivery fraction as discussed above because in this case, load balancing on the network along with the security is considered. Although with an increase in number of

Figure 10. Impact of the proposed scheme on throughput

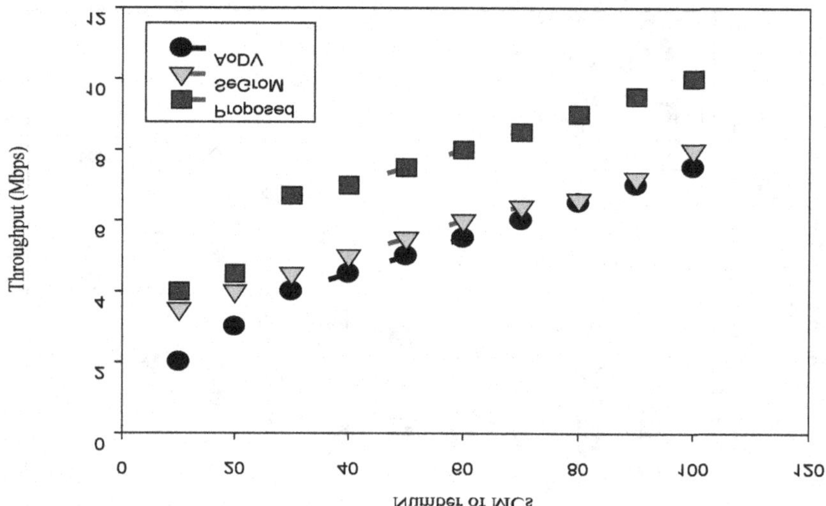

MCs, it is difficult to maintain same level of performance of the system, but with an increase in the MCs, the proposed scheme intelligently manage the load index of the system and also provide the enough security. Although, the throughput increase constantly in all the schemes, but the % gain in throughput is higher in the proposed scheme compared to other schemes which shows the effectiveness of the proposed scheme compared to other scheme of its category. Hence in the proposed scheme, security and load balancing are kept at the same level without performance degradation as shown in Figure 10.

6. CONCLUSION

It has been found from the past few years that security has emerged as a key issue in online web services due to the exponential growth of internet and its associated applications. In this paper, we propose a new load aware secure multiparty group communication for wireless mesh networks. For assigning a certificate and perform authentication to the new clients, load balancing index (LBI) is checked. The key feature of the proposed scheme is peer to peer authentication if LBI of the router is above the required threshold. On that case the authentication is done by the neighbouring node which has the maximum time stay in the network. Once this authentication process is over, the MCs communicate and use the available network resources. Due to authentication of MCs in peer to peer manner in extremely loaded situation, the computational complexity is reduced in the proposed scheme than the other schemes proposed in literature. The proposed scheme consists of reducing the load of the mesh routers by calculating the LBI and then allocating the incoming request to the neighbouring client in an efficient manner. Also to make the detection scheme more accurate the quality of wireless links with respect to the value of load is taken into account. The proposed scheme is evaluated with respect to the metrics such as storage and computation overhead, packet delivery fraction, throughput, and computational complexity. The results obtained show that the proposed scheme is quite effective than the other proposed scheme with respect to the above metrics. Hence, we have shown through simulation that the proposed scheme have performance improvement over an existing scheme with respect to the metrics. Due to the use LBI metric, the scheme gives accurate load status of the network so that quick action can be taken if the incoming requests have to be sent to the other nodes in the network. In near future, we will like to extend the proposed scheme to support detection of multiple inside and outside attacks patterns. Also, the proposed scheme will be tested with soft and hard handover scheme scenario for security in inter and intra domain clients.

REFERENCES

Akyildiz, F., Wang, X., & Wang, W. (2005). Wireless Mesh Networks: a survey. *Journal of Computer Networks*, *47*(4), 445–487. doi:10.1016/j.comnet.2004.12.001.

Boukerche, A., El-Khatib, K., Xu, L., & Korba, L. (2004). SDAR: A secure distributed anonymous routing protocol for wireless and mobile ad hoc networks. In *Proceedings of the 29th Annual IEEE International Conference on Local Computer Networks* (pp. 618-624).

Challal, Y., Bettahar, H., & Bouabdallah, A. (2004). SAKM: A scalable and adaptive key management approach for multicast communications. *ACM Computer Communication Review*, *34*(2), 55–70. doi:10.1145/997150.997157.

Dong, J., Ackermann, K., & Nita-Rotaru, C. (2009). Secure group communication in wireless mesh networks. *Ad Hoc Networks*, *7*(8), 1563–1576. doi:10.1016/j.adhoc.2009.03.004.

Fall, K., & Varadhan, K. (2000). *NS notes and documentation: The VINT project*. Retrieved from http://www.isi.edu/nsnam/ns/

Han, Y., & Gui, X. (2009). Adaptive secure multicast in wireless networks. *International Journal of Communication Systems*, *22*(9), 1213–1239. doi:10.1002/dac.1023.

Han, Y., Gui, X., Wu, X., & Yang, X. (2011). Proxy encryption based secure multicast in wireless mesh networks. *Journal of Network and Computer Applications*, *34*(2), 469–477. doi:10.1016/j.jnca.2010.05.002.

Huang, D., & Medhi, D. (2004). A key-chain based keying scheme for many-to-many secure group communication. *ACM Transactions on Information and System Security*, *7*(4), 1–30. doi:10.1145/1042031.1042033.

Junbeom, H., & Yoon, H. (2010). A Multi-service Group Key Management Scheme for Stateless Receivers in Wireless Mesh Networks. *Mobile Networks and Applications*, *15*(5), 680–692. doi:10.1007/s11036-009-0191-4.

Kong, J., & Hong, X. (2003). ANODR: Anonymous on demand routing with untraceable routes for mobile ad-hoc networks. In *Proceedings of the 4th ACM International Symposium on Mobile and Ad Hoc Networking and Computing* (pp. 291-302).

Lin, R., & Jan, J. (2007). A tree-based scheme for security of many to many communications. *Journal of High Speed Networking*, *16*(1), 69–79.

Rafaeli, S., & Hutchison, D. A. (2003). Survey of key management for secure group communication. *ACM Computing Surveys*, *35*(3), 309–329. doi:10.1145/937503.937506.

Shokri, R., Yazdani, N., & Khonsari, A. (2007). Chain-based anonymous routing for wireless ad hoc networks. In *Proceedings of the 4th IEEE Consumer Communications and Networking Conference* (pp. 297-302).

Sun, J., Zhang, C., & Zhang, Y. (2011). SAT: A security Architecture Achieving Anonymity and Traceability in Wireless Mesh Networks. *IEEE Transactions on Dependable and Secure Computing*, *8*(2), 295–307. doi:10.1109/TDSC.2009.50.

Sun, Y., & Liu, K. J. R. (2007). Hierarchical group access control for secure multicast communications. *IEEE/ACM Transactions on Networking*, *15*(6), 1514–1526. doi:10.1109/TNET.2007.897955.

Sy, D., Chen, R., & Bao, L. (2006). ODAR: On-demand anonymous routing in ad hoc networks. In *Proceedings of the IEEE Conference on Mobile Adhoc and Sensor Systems* (pp. 267-276).

Wan, Z., Ren, K., Zhu, B., Preneel, B., & Ming, G. (2010). Anonymous User Communication for Privacy Protection in Wireless Metropolitan Mesh Networks. *IEEE Transactions on Vehicular Technology*, *59*(2), 519–532. doi:10.1109/TVT.2009.2028892.

Wu, Q., Mu, Y., Susilo, W., Qin, B., & Domingo-Ferrer, J. (2009). Asymmetric group key agreement. In A. Joux (Ed.), *Proceedings of the 28th International Conference on the Theory and Applications of Cryptographic Techniques* (LNCS 5479, pp. 153-170).

Wu, T., Xue, Y., & Cui, Y. (2006). Preserving traffic privacy in wireless mesh networks. In *Proceedings of the International Symposium on a World of Wireless, Mobile and Multimedia Networks* (pp. 459-461).

Wu, X., & Li, N. (2006). Achieving privacy in mesh networks. In *Proceedings of the 4th ACM Workshop on Security of Ad Hoc and Sensor Networks* (pp. 13-22).

Zhang, J., & Varadharajan, V. (2006). A scalable multi-service group key management scheme. In *Proceedings of the Advanced International Conference on Telecommunications and International Conference on Internet and Web Applications and Services.*

Zhang, Y., & Fang, Y. (2006). ARSA: An Attack-Resilient Security Architecture for Multihop Wireless Mesh Networks. *IEEE Journal on Selected Areas in Communications*, *24*(10), 1916–1928. doi:10.1109/JSAC.2006.877223.

Zhao, L., Al-Dubai, A. Y., & Min, G. (2010). GLBM: A new QoS aware multicast scheme for wireless mesh networks. *Journal of Systems and Software*, *83*(8), 1318–1326. doi:10.1016/j.jss.2010.01.044.

Zhu, H., Lin, X., Lu, R., Ho, P.-H., & Shen, X. (2008). SLAB: A Secure Localized Authentication and Billing Scheme for Wireless Mesh Networks. *IEEE Transactions on Wireless Communications*, *7*(10), 3858–3868. doi:10.1109/TWC.2008.07418.

This work was previously published in the International Journal of Wireless Networks and Broadband Technologies, Volume 1, Issue 3, edited by Naveen Chilamkurti, pp. 15-29, copyright 2011 by IGI Publishing (an imprint of IGI Global).

Chapter 15
Adaptive Sending Rate Over Wireless Mesh Networks Using SNR

Scott Fowler
Linköping University, Sweden

Marc Eberhard
Aston University, UK

Keith Blow
Aston University, UK

Ahmed Shaikh
Aston University, UK

ABSTRACT

Wireless Mesh Networks (WMNs) have emerged as a key technology for the next generation of wireless networking. Instead of being another type of ad-hoc networking, WMNs diversify the capabilities of ad-hoc networks. Several protocols that work over WMNs include IEEE 802.11a/b/g, 802.15, 802.16 and LTE-Advanced. To bring about a high throughput under varying conditions, these protocols have to adapt their transmission rate. This paper proposes a scheme to improve channel conditions by performing rate adaptation along with multiple packet transmission using packet loss and physical layer condition. Dynamic monitoring, multiple packet transmission and adaptation to changes in channel quality by adjusting the packet transmission rates according to certain optimization criteria provided greater throughput. The key feature of the proposed method is the combination of the following two factors: 1) detection of intrinsic channel conditions by measuring the fluctuation of noise to signal ratio via the standard deviation, and 2) the detection of packet loss induced through congestion. The authors show that the use of such techniques in a WMN can significantly improve performance in terms of the packet sending rate. The effectiveness of the proposed method was demonstrated in a simulated wireless network testbed via packet-level simulation.

1. INTRODUCTION

Wireless Mesh Networks (WMNs) provide alternative technologies for last-mile broadband Internet access and high speed connectivity with cost-effectiveness. WMNs have emerged as a key technology for the next generation of wireless networking. Instead of being another type of ad-hoc networking, WMNs diversify the capabilities of ad-hoc networks by integrating additional routing function to support wireless networks such as cellular wireless sensors (Akyildiz & Wang, 2005; Pathak & Dutta, 2011). WMNs have the advantages of being self-organizing, self-config-

DOI: 10.4018/978-1-4666-3902-7.ch015

uring, and offering increased reliability. Nodes in WMNs automatically form an ad-hoc network and maintain mesh connectivity. These features bring further advantages to WMNs, such as low up-front cost, easy network maintenance, robustness, reliable service coverage, etc (Akyildiz & Wang, 2005; Pathak & Dutta, 2011). WMNs are comprised of two types of nodes: mesh routers and mesh clients. In the WMN architecture, mesh routers form an infrastructure for various types of client (Figure 1), where dashes indicate wireless and solid lines indicate wired links on the WMN. Various wireless devices (such as laptops, PDAs, cellular networks) equipped with wireless cards can connect to a WMN through a mesh router with gateway/bridge capabilities. The gateway/bridge integrates various exiting wireless networks such as cellular, wireless sensors, Wireless-Fidelity (Wi-Fi), LTE-Advanced and Worldwide Interoperability for Microwave access (WiMAX) (Akyildiz, Wang, & Wang, 2005). As WMNs are self-organized, self-configured with wireless mesh routers, and automatically establish and maintain wireless mesh connectivity (effectively, creating an ad-hoc network), they can provide wireless transport services to data travelling from other users, access points or base stations (access points/base stations are special wireless routers with a high-bandwidth wired connection to the Internet backbone) (Figure 1).

Currently, WMNs are going through rapid commercialization in several application scenarios such as broadband home networking, community networking, building automation, high speed metropolitan area networks, and enterprise networking (Akyildiz & Wang, 2005; Pathak & Dutta, 2011). This is due to the fact that WMNs can be relatively easily established because all the required components are already available in the form of ad-hoc network routing protocols, IEEE 802.11 MAC protocols, Wired Equivalent Privacy (WEP) security, and so on.

Although WMNs could be established straightforwardly, obtaining high data rates in WMNs is still a big challenge since bandwidth in wireless is limited. One way to achieve higher data rates

Figure 1. Wireless mesh infrastructure

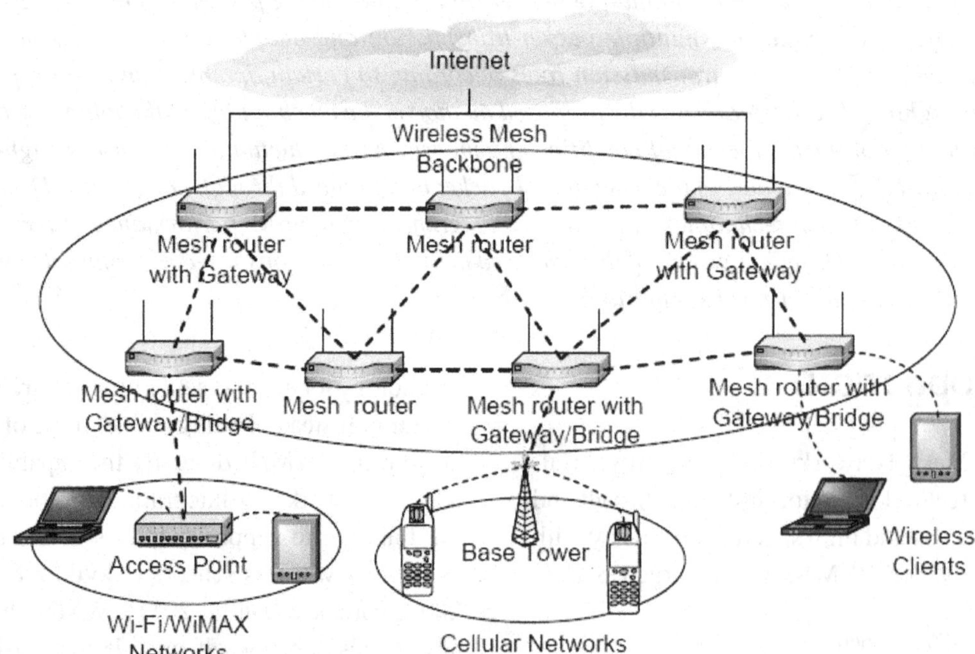

is the use of a well-designed modulation format. The performance of each modulation scheme is measured by its capability to precisely maintain the encoded data, which is represented by the low Bit Error Rate (BER). Variation in the BER is directly related to the received Signal-to-Noise Ratio (SNR) (Figure 2). BER and SNR are inversely related that when the SNR is decreased the BER increases. Therefore, when the SNR is lowered, it is more difficult for the modulation scheme to decode the received signal as the BER is too high. The relationship between the BER and the SNR for various modulation schemes is illustrated in the references (Holland, Vaidya, & Bahl, 2001, 2000). Generally, when data rate is increased, the BER also increases consequently. A logical question arises would be "Are there any ways to attain high data rates while maintaining lower BER?" One of the effective approaches for that is adaptation of different modulation schemes (Holland et al., 2001, 2000). This as a result, will lead to improvements in the performance of a wireless device.

Indeed WMNs have the advantage of being self-organizing and self-configuring by sharing information with their neighbours to establish communication pathways. However, a new method needs to be developed that is capable of adapting to the dynamic nature of WMNs. Therefore in this paper, we propose a mechanism that

Figure 2. General performance of BER versus SNR

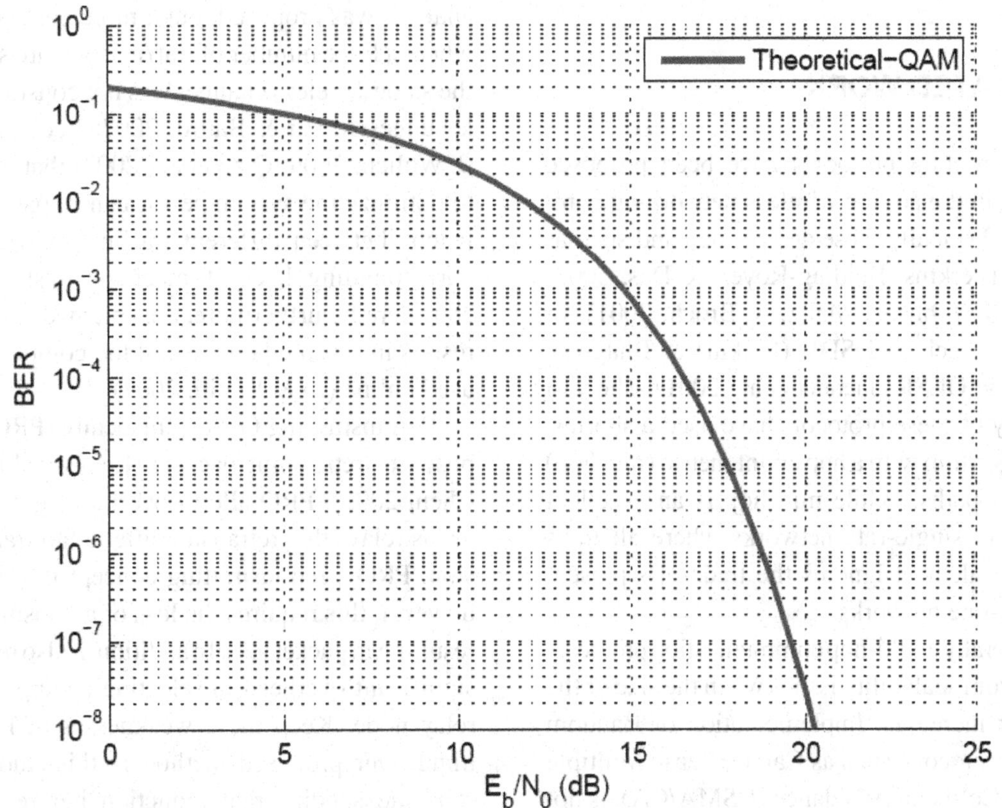

BER: Bit Error Rate

E_b: Energy-per-bit

N_o: Noise Power Spectrum Density

SNR: Signal-to-Noise Ratio (which is E_b/N_o)

does dynamic monitoring and adapts to changes in channel quality, thereby maximally taking advantages of the dynamic nature of WMNs. Our proposal consists of the followings:

- Exploiting the channel conditions by adapting the transmission rate accordingly, thereby utilizing the multi-rate ability of WMNs.
- Maximizing the throughput using multiple packets algorithm when the channel quality is optimal.
- Utilizing the standard deviation of the recent background noise for the receiver to be able to adapt the sending rate accordingly.
- Determining the network status using packet loss before the physical channel degrades.

2. RELATED WORK

Several routing protocols have been proposed to work in the ad-hoc wireless networking community. Typically these are: on-demand such as AODV (Perkins, Belding-Royer, & Das, 2003) and DSR (Johnson, Maltz, & Broch, 2001), or proactive such as DSDV (Perkins & Bhagwat, 1994) and OLSR (Jacquet et al., 2001). The vast majority of these protocols have used a shortest path algorithm with a hop count metric (min hop) to select paths. While min hop is an excellent criteria in single-rate networks where all links are equivalent, it ignores the trade-offs present in multi-rate networks.

Generally, multi-hop communication protocols suffer from scalability issues when the size of the network increases. Implementation of Random access protocols, such as Carrier Sense Multiple Access/Collision Avoidance (CSMA/CA), is not an efficient solution to improve scalability. Instead, one might consider the use of access schemes such as Time Division Multiple Access (TDMA) and Code Division Multiple Access (CDMA). However, access schemes are too complicated to implement in WMNs, due to its design associated with the following two factors (Akyildiz & Wang, 2005; Pathak & Dutta, 2011). The first is the complexity of developing a distributed and cooperative MAC with TDMA or CDMA. The second is the compatibility of TDMA or CDMA with the existing MAC protocols as they do not scale well in relation to the physical layer channel conditions.

A MPDU (MAC Protocol Data Unit)-based link adaptation scheme was proposed (Qiao, Choi, & Shin, 2002), which is based on good- put analysis. In MPDU-based link adaptation, the wireless node computes a table of physical modes indexed by the system status offline, where each entry of the table is the best physical mode for maximizing the expected goodput. In the paper, a detailed analysis was proposed, but a more general case, where a large number of traffic flows are sharing the same wireless channel was not considered. It is suggested (Li & Battiti, 2007; Redlich, Ezri, & Wulich, 2009; Qiao et al., 2002) that in order to optimize the system performance, packet collisions between different traffic flows and the corresponding backoff mechanism should be considered. Furthermore, the obtained analytical results in this method are still too complex to be used online (Li & Battiti, 2007).

Retransmitting Opportunistically (PRO) proposed to retransmit packets (Lu, Steenkiste, & Chen, 2009). PRO allows overhearing nodes to act as relays that retransmit after a fail transmission. PRO's retransmitting concept is valuable; however, this requires the loss of a transmission, resulting in packet loss. In addition, it also requires significant overhead in selecting an appropriate relay node. Keep these weaknesses of PRO in mind. Our proposed method in this study will start the sending-rate reduction before packet loss occurs at the link-layer. (Chowdhury, Felice, & Bononi, 2009) proposed a Cross-layer Channel adaptive Routing protocol (XCHARM) for Multi-Radio Multi-Channel WMNs. This protocol jointly performs route discovery and channel assignment. The Simple Opportunistic Adaptive Routing protocol (SOAR) is combination of the adaptive-rate protocol and the routing-adjusting protocol (Rozner, Seshadri, Mehta, & Qiu, 2009). SOAR combine these techniques by selecting forwarding nodes and employing priority-based timers. The strength of SOAR is that this protocol further incorporates adaptive rate control to dynamically adjusted sending rates according to network conditions. Furthermore it is capable of recovering lost packets using efficient local feedback and recovery. However, challenges in integrating a new routing protocol into the existing protocol are that duplicate retransmission or incurring significant coordination overhead caused by channel assignments during set up

In Luo, Rosenberg, and Girard (2010) and Hayajneh and Abdallah (2004) considered a jointly adapt power and adoptive rate scheme in cellular networks. However, these schemes may have detrimental performance in ad hoc networks since nodes able to choose their transmission parameter based on their own preferences and observations. The possibility of greedy behaviours in which nodes choose their transmit power and data rates without considering others users resources. In Lu, Sun, Ge, Dutkiewicz, and Zhou (2010), joint power and adoptive rate problem is addressed as a tradeoff to be balanced where each node determines its transmission power and rate as a fictitious game to attain global utility. This approach is similar to utility based adjustable joint power and rate adaptive algorithm investigated (Zheng & Ma, 2009) in which variation in channel conditions in considered. In both algorithms, nodes are aware of the highest utility attainable within their domain depending on SNR measurements. However, the schemes assume that the power choices would always be optimal in terms of interference reduction.

Other approaches such as resource-allocation to improve throughput (Kodialam & Nandagopal, 2005; Brzezinski, Zussman, & Modiano, 2006) can provide guidelines for initial network resource planning. However, they often need global configuration changes, which are undesirable in case of frequent local link failures. Fault-tolerant routing protocols like (Nelakuditi et al., 2005) or multipath routing (Chen & Nahrstedt, 1999) can be adopted to use network-level path diversity for avoiding the faulty links, However, they rely on detour paths or redundant transmissions, which may require more network resources than link-level network reconfigurations (Kim & Shin, 2007).

3. BACKGROUND

The conventional procedure for a narrow-band wireless device can be summarized in brief as following. Once signal arrives as a wireless (analogue) signal, it will be converted into a digital signal by a digital demodulator. When the wireless device send a message, digital modulator then converts the digital signal into a wireless (analogue) signal. Subsequently, this modulated signal is amplified and transmitted. Since the modulation process significantly impacts on the performance of the transmission, we will describe it in detail in the following section.

3.1. Modulation

BER, by definition, is the ratio of the number of bits incorrectly received to the total number of bits sent during a certain time interval. Thus, there is a general tendency that when the data rate is increased the BER is also increased. The logical extension will be "Are there any way to obtain higher data rates while maintaining minimal BER?" One way to achieve this is adaptation of various modulation schemes to optimize the performance of a wireless device (Holland et al., 2001, 2000).

BER and SNR are in inverse relationship, i.e., when SNR increases the BER decreases, whereas when the SNR decreased then the BER increases.

Therefore, when the SNR is lowered, it is more difficult for the modulation scheme to decode the received signal as the BER is too high.

Effective modulation methods enable higher data rates while maintaining lower BER. One of the factors which determine the effectiveness of modulation method is the tolerance of the modulation scheme to the impairments introduced by the channel. In other words, channel quality has a strong impact on the effectiveness of modulation scheme.

The key elements of our modulation methods in this paper are as following: In our proposed method, multiple packets will be sent when the physical channel condition is good. It is enabled by the implementation of various modulation methods correspondingly to the varying sending rates. In other words, depending on the high or low sending rate, our proposed protocol will select the most appropriate modulation method out of the several choices. We also will demonstrate how the performance of the physical layer channel condition can be detected before channel condition diminishes completely. It is enabled by observing the packet loss in relation to SNR by exploiting the Interface Queue (IFQ). We employed the SNR as it is commonly used measure for the quality of a received signal. SNR is the signal power over the background noise (i.e., channel noise) power as shown in (1)

$$SNR = \frac{Psignal}{Pnoise} \quad (1)$$

where P_{signal} is the signal power and P_{noise} represents the channel noise.

The SNR may be varied by path loss, interference, and/or fading in wireless devices. (Please note, as our focus is on stationary wireless devices, fading issues will not be addressed in this paper.).

3.2. Modifying the 802.11 for Adaptive Rate

In our method, the Source (Src) chooses a data rate based on some heuristic algorithm, such as the most recent rate that was successful for transmission to the Destination (Dst), and then stores the rate and the size of the data packet into the RTS. Other nodes which overhear the RTS will calculate the duration of the requested reservation using the rate and packet size carried in the RTS. The other node(s) then will update its NAV to reflect the reservation.

While receiving the RTS, the Dst node generates an estimate of the conditions for the impending data packet transmission using the available information about the channel conditions. The Dst node then selects the appropriate rate based on that estimate, and transmits the information it in the CTS packet back to the sender.

In our method, the Dst node adapts the sending rate using the standard deviation of the recent background noise. To accomplish this, we have modified the SNR for the received signal as we recognized that the received SNR can be utilized as the measure of the packet level performance on the physical layer in relation to the channel quality. More specifically, we incorporated the standard deviation (σ) from the signal noise, the packet loss at the MAC layer, and the physical layer.

Since a node overhears the many different RTS or CTS in 802.11, an efficient means for managing these reserved requests is required. This is accomplished by the NAV. Once the receiver and sender select a different sending rate, NAV reservation will no longer be valid. Hence, a provisional reservation of the NAV needs to be addressed, with some extension for sending of multiple packets. The provisional reservation is only temporal, and any nodes that overhear a provisional reservation do not conflict with any requests received because the sending rate and the number of packets sent will be changed. It should be noted that when the provisional reservation is not correct, the Src will send a data packet with the special MAC header containing Reservation SubHeader (The final reservations are confirmed by the presence or absence of a special subheader named Reservation SubHeader). When other node(s) hear the Reservation SubHeader, they will calculate the final provisional reservation, which informs its NAV of the difference between RTS requested reservation and the provisional reservation. Notice that, for the other node(s) to update the NAV properly, they must know what input the RTS had made to its NAV. This can be done by maintaining a list of the end times of each tentative reservation, which is catalogued by the Src and Dst. A node can use this list to determine if the NAV has to be updated when an update is involved.

We have used the multi-rate ability which accordingly adapts the transmission rate by exploiting the channel conditions. We have also maximized the throughput using a multiple packet algorithm when the channel quality is good. This is done by obtaining the number of packets to be set using the following calculation: Transmission rate divided by Base rate, where the base rate is assumed 2 Mbps. For example, if the transmission rate is to be 5.5 Mbps, the Src will allow [5.5/2] = 3 packets to be sent. More detailed explanation will be provided in the next section.

4. PROPOSED METHOD

Our proposed a method has been implemented in NS-2 (Information Sciences Institute, n. d.) to evaluate the multi sending rate and maximize the throughput by multiple back-to-back packets over a WMN as a function of channel quality conditions and packet loss. These extensions provide a higher 802.11 transmission rate thereby achieving higher throughput, but usually with the cost of higher BERs. Hence our goal is to achieve higher throughput while maintaining lower BERs.

In order to maximize the gain in throughput, it is critical to balance the tradeoff between the MAC conditions and channel quality. In the next part, we will describe how it is achieved in our proposed method.

4.1. Calculating Sending Rate

The sending rate should be determined by the signal strength and the packet loss as detailed previously. Selection of the rate is accomplished by matching the channel quality against the various modulation schemes. The modulation schemes increase the throughput according to the calculated channel quality. The modulation schemes were selected as follows:

M_1 if SNR $< \theta_1$

$$M_i \text{ if } \theta_i \leq \text{SNR} < \theta_{i+1} \quad (2)$$

M_N otherwise

where $M_1, \cdots, M_N$ correspond to the selected modulation methods for increasing the sending rate and θ_i corresponds to the SNR threshold at which $BER(M_i) = 10^{-5} = 1E - 5$.

For the BER calculation, we used the threshold parameters presented in Table 1 which follows the traditional Quadrature Amplitude Modulation (QAM) based radio model. Examples of the models which calculate the probability of bit-error (P_e) include Binary Phase Shift Keying (BPSK) and Quadrature Phase Shift Keying (QPSK) (Rappaport, 2002; Stallings, 2005). Based on these models, our calculation is expressed as following:

Table 1. Physical layer parameters

Parameter:	Description
Frequency range	2.400 GHz
Transmitter Power	15 dBm
11.0 Mbps (CCK) Sensitivity	-82 dBm
5.5 Mbps (CCK) Sensitivity	-87 dBm
2.0 Mbps (DQPSK) Sensitivity	-91 dBm
1.0 Mbps (DBPSK) Sensitivity	-94 dBm
Carrier Sense Threshold	-108 dBm
Capture Threshold	10

$$P_e = Q\left(\sqrt{\frac{2E_b}{N_o}}\right) \tag{3}$$

where E_b is the energy-per-bit, No represents the noise power spectrum density and *Q()* is the Gaussian noise channel disputation.

The error probability for M-ary QAM (Rappaport, 2002; Stallings, 2005) was calculated as following:

$$P_e = 4\left(\frac{1}{\sqrt{M}}\right)Q\left(\sqrt{\frac{3\log_2(M)E_b}{(M-1)N_o}}\right) \tag{4}$$

where *M* is the number of possible signals, $s_1(t)$, $s_2(t), \cdots, s_M(t)$ transmitted during each bit period (Rappaport, 2002).

The ratio E_b/N_o is important because the BER decreases as the signal strength relative to the noise increases (hence, increasing E_b/N_o ratio). We can relate the E_b/N_o to the bit-energy-to-noise ratio of the received signal and the SNR as following:

$$\frac{E_b}{N_o} = SNR\left(\frac{B_t}{R_b}\right) \tag{5}$$

where B_t is the maximum bit rate of the modulation scheme and R_b is the bandwidth of the noise at the receiver.

The received SNR can be exploited for capturing the packet level performance on the physical layer in relation to the channel quality. We incorporated the standard deviation (σ) of the received signal power as an approximation to the background noise in the original SNR Equation (1), which now can be expressed as:

$$SNR = \frac{Psignal}{\sigma^2} \tag{6}$$

where σ is the Root Mean Square (RMS) deviation of values from the mean value representing the collection of P_{signal}. However, in NS-2, the correlation between BER and SNR is not distinctly defined. Therefore, we cannot use the traditional modulation schemes (Rappaport, 2002; Stallings, 2005) directly in our simulations. Here, we compared the received power to the receiver power thresholds, where the thresholds of the receiver power are defined for the each modulation scheme individually so that they can implicitly define the receiver SNRs.

For the received signal in relation to previous channel conditions, we implemented our estimate of σ^2. However, another consideration needs to be given to the possibility that there will also be changes to the perceived wireless channel quality resulted from congestion. A highly congested MAC layer will lead to increased packet loss which can appear to be a reduction in channel quality. Thus, our approach is that if packet loss occurs on the MAC layer, then we will employ a reasonable history of SNR from Equation (6) together with the current packet loss to derive an *Effective SNR*(SNR_E) as follows:

$$SNR_E = \frac{Psignal}{\sigma^2}\left(1 - \frac{P_{loss}}{Q_{size}}\right) \tag{7}$$

where P_{loss} is the packet loss of the Interface Queue (IFQ) and Q_{size} represents the queue length or maximum number of packets of the IFQ. The

MAC module is responsible for channel access. If a packet is not successful after several attempts, this packet is then placed in the MAC buffer which is the IFQ. A part of the Equation (7), the P_{loss}/Q_{size}, plays an important role in detecting perceived channel quality. When congestion occurs at the MAC layer there will be considerable packet loss due to collisions. Treating the packet loss as an equivalent to SNR penalty, we could use the same transmission rate back-off mechanism for responding to channel quality as well as congestion.

We utilized the multi-rate ability by accordingly adapting the transmission rate to the channel conditions. The receiver adapts the sending rate using the effective signal to noise ratio as presented in Equation (7). This way, we could maximize the throughput via implementation of a multiple packet algorithm when the channel quality is good. This was achieved by the number of packets to be sent is determined by the sending rate divided by an assumed base rate of 2Mbps. For example, if the transmission rate is determined to be 5.5 Mbps, the Src will allow [5.5/2] = 3 packets to be sent. Should the sending rate be 2 Mbps or less, only a single packet will be sent rather than transmitting multiple packets back-to-back.

5. SIMULATION

To evaluate the proposed adaptive sending rate over a WMN all experiments were simulated either with TCP, UDP or TCP and UDP connections as competing traffic, dependent on the scenario, for 5 minutes. The results are averaged over 5 simulation runs with different random seeds.

In all simulations presented here the packet size is set to 1000 bytes. We have also performed simulations with different packet sizes, and obtaining very similar results. Thus the results with different packet sizes were not presented in this paper because it would not have any contribution to the quality of the results.

The network link bandwidth for all simulations is fixed to one megabit. All the WMN queue sizes are set to 50 packets. The WMN routing proto- col selected is AD-hoc On-demand Distance Vector (AODV) (Perkins et al., 2003). The net- work covers an area of 2000 x 2000 meters, where the nodes are at least 200m apart from each other. Each node is configured to a transmission range of 250m.

We have also performed simulations with different chain lengths, obtaining very similar results. Again, these results with different chain length would not contribute to the paper. The propagation models used in our simulations were the two- ray ground model (which we refer to as the two- ray model). The two-ray model is the most widely used simulation model, thus provides some level of comparison of our results to previous work. The two-array model is used with the typical settings of 250m radio range and 550m interference range.

In this paper it is assumed the wireless nodes are fixed. Even with the quality of the link changes over the life of the connection, the wireless nodes are not added or removed from the WMN topologies. In other words, there is no need to re- establish the connection due to the link being stable enough. The stability of wireless routers is assumed since wireless mesh routers are implemented for backhaul due to their highly reliable in WMN.

5.1. Topologies

In our simulations with deterministic topologies and fixed packet streams, we considered the following scenario as shown in Figure 3, Figure 4, and Figure 5 which represent the testbeds used.

The first topology (Figure 3, "Chain") is based on the use of the Bidirectional-Chain. We used two connections along this chain. In the Bidirectional- Chain, the one connection travels from the first to the last node in the chain. The second connection running in the opposite direction, from the last towards the first node, provides competing traffic which is carried by UDP or TCP.

Figure 3. Chain wireless mesh topology

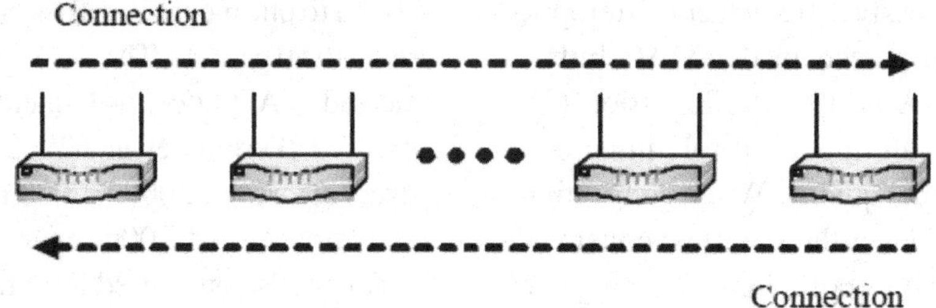

Figure 4. Cross-chain wireless mesh topology wireless mesh

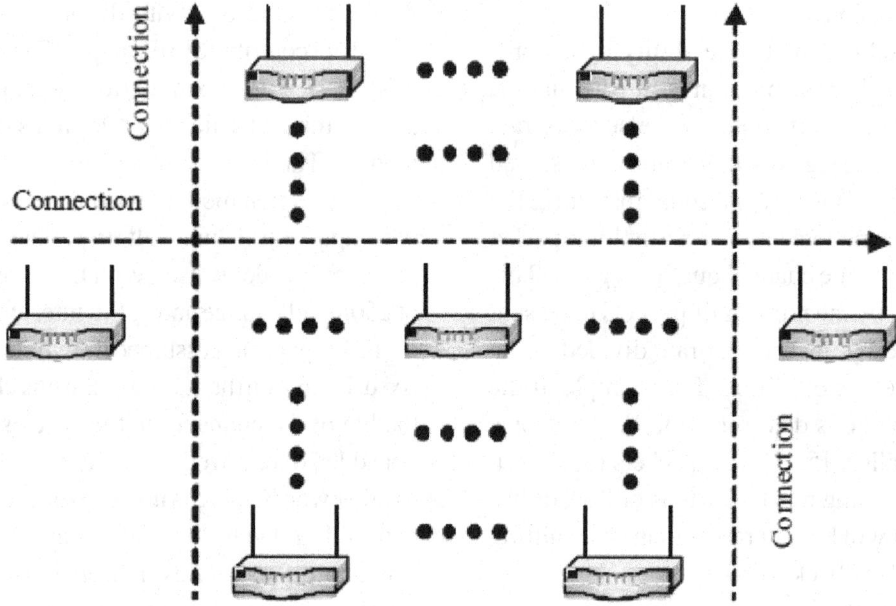

Unlike Chain topology, both "Cross-Chain" (Figure 4) and "Bidirectional Cross-Chain" (Figure 5) consist of more than two chains. One of the chains runs in the x-direction of the plane and the others run in the y-direction of the plane. The chains cross each other in the middle, where they share one common node. While Cross-Chain (Figure 4) has traffic streams travelling in one direction of the y-plane, Bidirectional Cross has traffic streams travelling in both directions of the y- plane chain toward the opposite end of the respective chain. In the Cross-Chain (Figure 4) the vertical chains are either UDP or TCP, whereas in the Bidirectional Cross-Chain (Figure 5), the vertical chains are UDP and TCP. The vertical chains signify connections with varying transmission rates which represent competing background traffic.

6. PERFORMANCE EVALUATION

The focus of this research is to improve the traffic performance, in other words, to refine the average rate of successful message delivery over a wireless communication channel. We evaluated the performance by means of throughput over a given period of time (expressed as average Kbit per second).

Figure 5. Bidirectional cross-chain wireless mesh topology

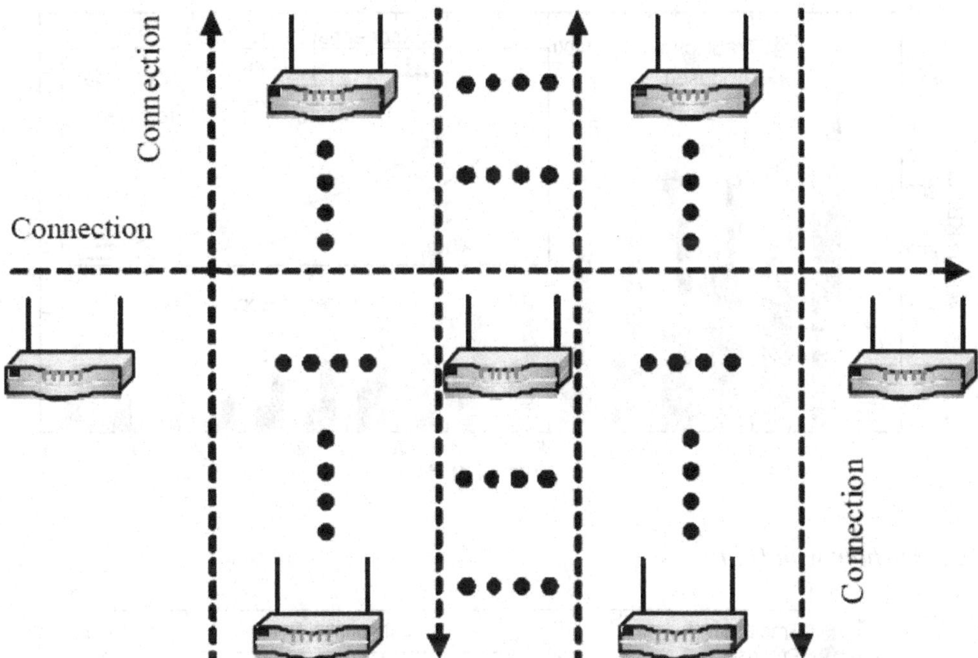

Figures 6, 7, 8, 9. and 10 show the changes of average throughput rate as the number of hops increases over the WMN. Figures 6 and 7 represent the data from Chain topology for UDP or TCP, respectively, while Figures 8, 9, and 10 depict the data from Cross-Chain topology for UDP, TCP, or UDP+TCP combined, respectively. In all cases, we simulated both the traditional method (labeled "Traditional") and our proposed method (labeled "Proposed") and their performance was compared.

Regardless of the topologies (Chain or Cross-Chain), type of traffic (UDP, TCP, or UDP+TCP), or methods (Traditional or Proposed), there was a general tendency in all cases that as the number

Figure 6. Chain with UDP

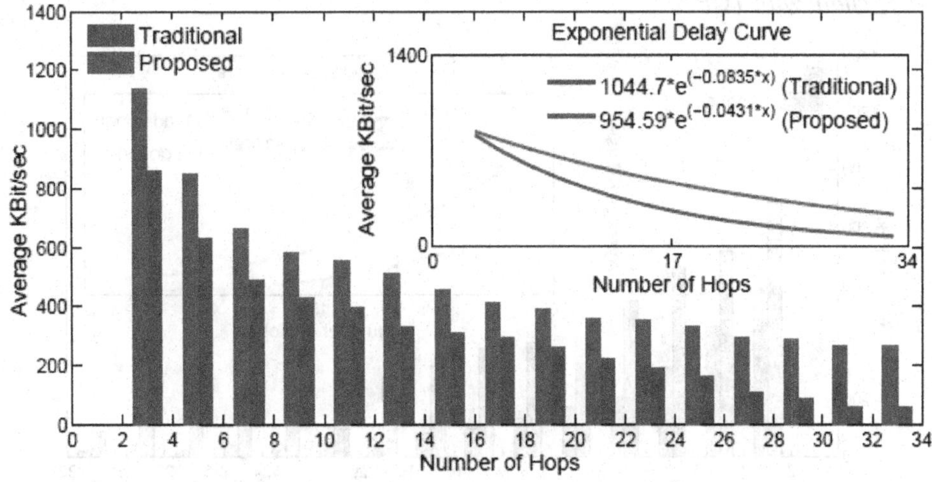

Figure 7. Chain with TCP

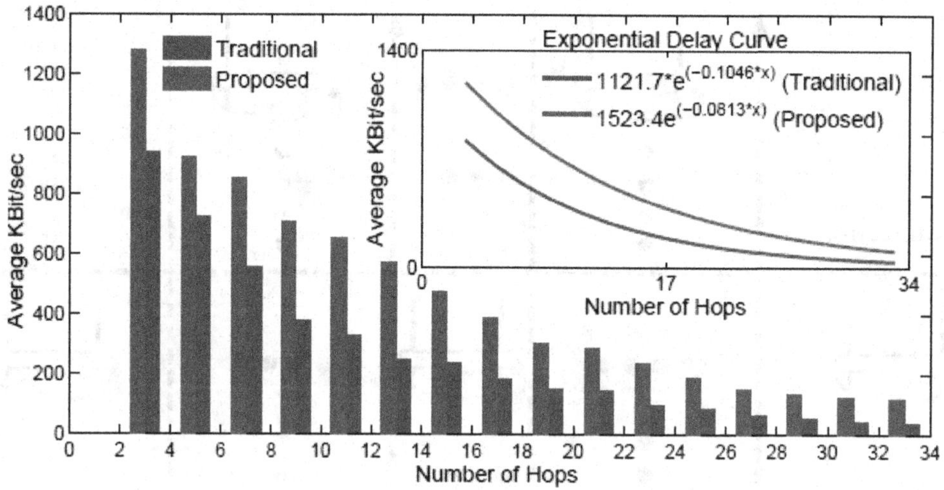

Figure 8. Cross-chain with UDP

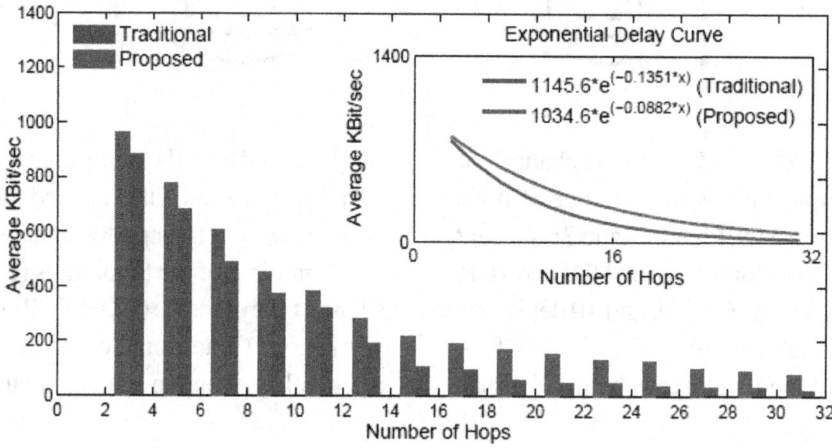

Figure 9. Cross-chain with TCP

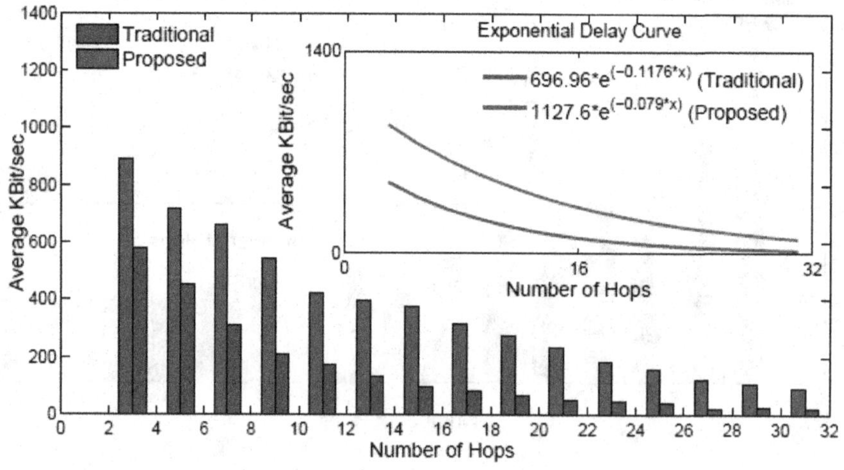

Figure 10. Bidirectional cross-chain with TCP and UDP

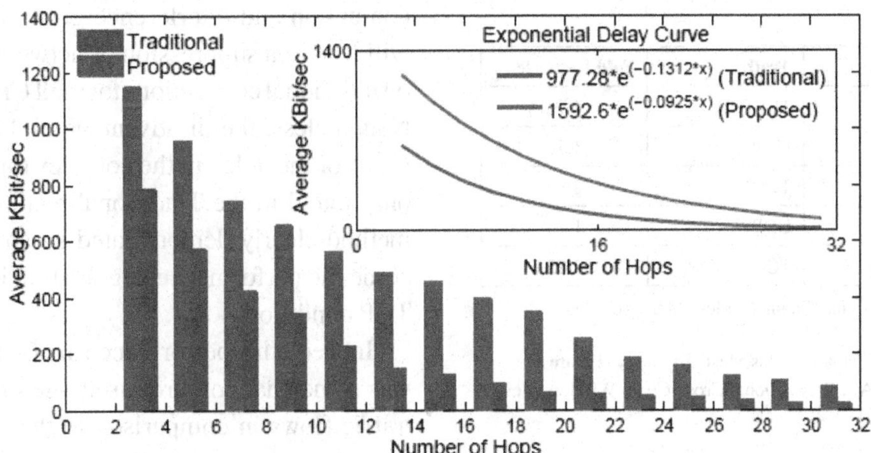

of nodes in a chain increases, the data contention grows. This effect, in turn, will cause throughput degradation. Our observation is in accordance with the report from Heusse, Rousseau, Berger-Sabbatel, and Duda (2003) that in IEEE 802.11, performance degrades considerably as all flows match their throughput to that of the flow with the lowest quality channel.

The decay of throughput rate was exponential manner in all cases. Nonetheless, the degree of decay was significantly different between the Traditional and Proposed methods, regardless of the topology or the type of traffic. In all conditions, the performance degradation was more pronounced in Traditional methods than that in Proposed methods. In order to numerically compare the degree of performance degradation, the data was fitted to exponential decay plot (Figures 6 through 10 small windows).

The formula is expressed as:

$N(t) = N_0 e^{-\lambda t}$(8)

where $N(t)$ is the average throughput rate at time t, N_0 is the initial average throughput rate at time zero, and λ is the decay constant. The larger value of decay constant λ indicates abrupt drop of performance whereas the smaller λ value represents better performance sustainability. The decay constant of Traditional method verses Proposed method were in Chain/UDP (Figure 6) 0.0835 vs. 0.0431; Chain/TCP (Figure 7) 0.1046 vs. 0.0813; Cross/UDP (Figure 8) 0.1351 vs. 0.0882; Cross/TCP (Figure 9) 0.1176 vs. 0.079; and Cross/TCP+UDP (Figure 10) 0.1312 vs. 0.0925, respectively. In other words, performance sustain ability with increasing number of hops was enhanced in our Proposed method regardless of the topology or type of traffic tested.

In addition to the performance sustainability, the overall average throughput rate was significantly higher in the Proposed method than that of Traditional method when they were compared in the same conditions (topology/traffic). Table 2 summarizes fold difference of overall average throughput rate for numerical comparison. The overall average throughput rates were 1.73-, 1.58-, 2.37-, 1.75-, or 2.23-fold higher with Proposed method than those with Traditional method, in either Chain/TCP, Chain/UDP, Cross/TCP, Cross/UDP, or Bidirectional/TCP+UDP, respectively.

The main feature of the current Proposed method is the ability of adjusting its sending rate before packet losses occur. This was clearly demonstrated in the Chain/UDP (Figure 6), where UDP by itself has any congestion avoidance or

Table 2. Performance of proposed against traditional

Topologies	Traffic	Fold Increase
CHAIN	TCPTCP	1.73
	UDP	1.58
CROSS	TCP	2.37
	UDP	1.75
BIDIRECTIONAL	TCP & UDP	2.23

CHAIN: Bidirectional-Chain Wireless Mesh Topology (Figure 3)
CROSS: Cross-Chain Wireless Mesh Topology (Figure 4)
BIDIRECTIONAL: Bidirectional Cross-Chain Wireless Mesh Topology (Figure 5)

control mechanisms. TCP is capable of adjusting its sending rate, whereas UDP traffic has a constant rate. Since the Traditional method lacks the sending rate adjustability, when the hop number increases the resulting congestion is inevitable. This in turn leads to the performance degradation. Our adaptive sending rate mechanism, on the other hand, compensated for the limitation of UDP in terms of congestion control in the Chain topology.

The performance decay curve of Proposed method in Chain/TCP (Figure 7) represents synergistic effect of sending rate adjusting capacity of our method plus TCP's own flexible sending rate mechanism. In simple topologies, such as our current Chain topology, there was no palpable difference in throughput rates of Traditional method, whether it is sent with constant sending rate (Figure 6, UDP) or with adaptable sending rate (Figure 7, TCP). However, when Proposed method was applied, the throughput rate in TCP was remarkably improved, indicating synergistic interaction of the adaptive sending rate mechanisms between TCP traffic and Proposed method.

TCP sometimes incur delays as its priority is accuracy. Unlike in the case with Chain topology, with the more complex Cross topology (Figures 8 and 9) there was a severe throughput decline of Traditional method with TCP than that with UDP traffic. This observation could be explained as follows: Packet congestion occurs when data contention and interference grows in TCP. This will lead to a suppression of network throughput on individual connections for the TCP connection. Nonetheless, the disadvantage of TCP traffic in terms of the delay in the complex topologies was only found in the Traditional method. Proposed method clearly demonstrated its ability to overcome the performance degradation in the Chain/TCP condition.

Indeed, the performance in Cross topology was somewhat compromised due to its complex traffic flows in comparison to that in Chain topology. Regardless of the traffic (TCP, UDP, or TCP+UDP), the throughput rate of Traditional method was strongly suffered from the topology-derived contention, but at slightly different degree, where the impact was most severely on TCP, then on TCP+UDP, and the least on UDP traffic. The proposed adaptive sending rate mechanism demonstrated its capacity to overcome the contention in the complex traffic flow in all the traffic (TCP, UDP, or TCP+UDP), but at the different magnitude, where the improvement was greatest in TCP, then in TCP+UDP, and the least in UDP traffic. This phenomenon is also a good indication of the synergistic effect of the adaptive sending rate mechanism of TCP and the Proposed method.

Unfairness in multi-hop networks is also a well-known problem in IEEE 802.11. In multi-hop wireless networks, fairness is derived from contention of neighbourhood nodes and variable rate channels. Thus, the use of current media access and transport protocols could result in severe unfairness. Our observations that the performance degradation with an increasing number of hops in our study could be explained by the starvation of flow caused by sever unfairness.

Taken together, the proposed method was able to maximize the sending rate over a shorter burst. In other words, the performance improvement was achieved without sacrificing the fairness of traffic distribution by the adaptive sending rate mechanism regardless of the topology or type of traffic.

7. RANDOM TOPOLOGIES

In the next set of simulations, we examined the steady-state throughput of our proposed method. The simulated networks cover an area of 1500 x 1500 square meters, where 40 nodes are placed randomly. Five random connections are set up in each scenario, that continuously try to deliver as much data as possible. All experiments were simulated with TCP connections as competing traffic since majority of traffic over the internet is TCP, and TCP has an adjustable sending rate for the proposed method to be evaluated against.

Figures 11 and 12 shows the results of these measurements for the five random network setups. In these figures, each bar represents throughput rate of a topology, while each segment of the bar stands for the throughput of one stream. An identical colour fill pattern indicates that the respective segment belongs to the same pair of communicating nodes. The chosen representation thus allows not only a comparison of the total throughput, but also of its distribution to the five streams.

Figure 11 represent the results of traditional approach, whereas Figure 12 is that of our proposed approach. Please note that the scale of y- axis in the Figure 11 (traditional method) is only up to 20 Mbps, whereas that in the Figure 12 (proposed method) is up to 80 Mbps. Therefore, although the visual presentation somewhat magnified the throughput rate of traditional method, the actual results were nearly 25% of the data range of the proposed method. More precisely, the total throughput rates of our pro-

Figure 11. Random topology with traditional traffic flows

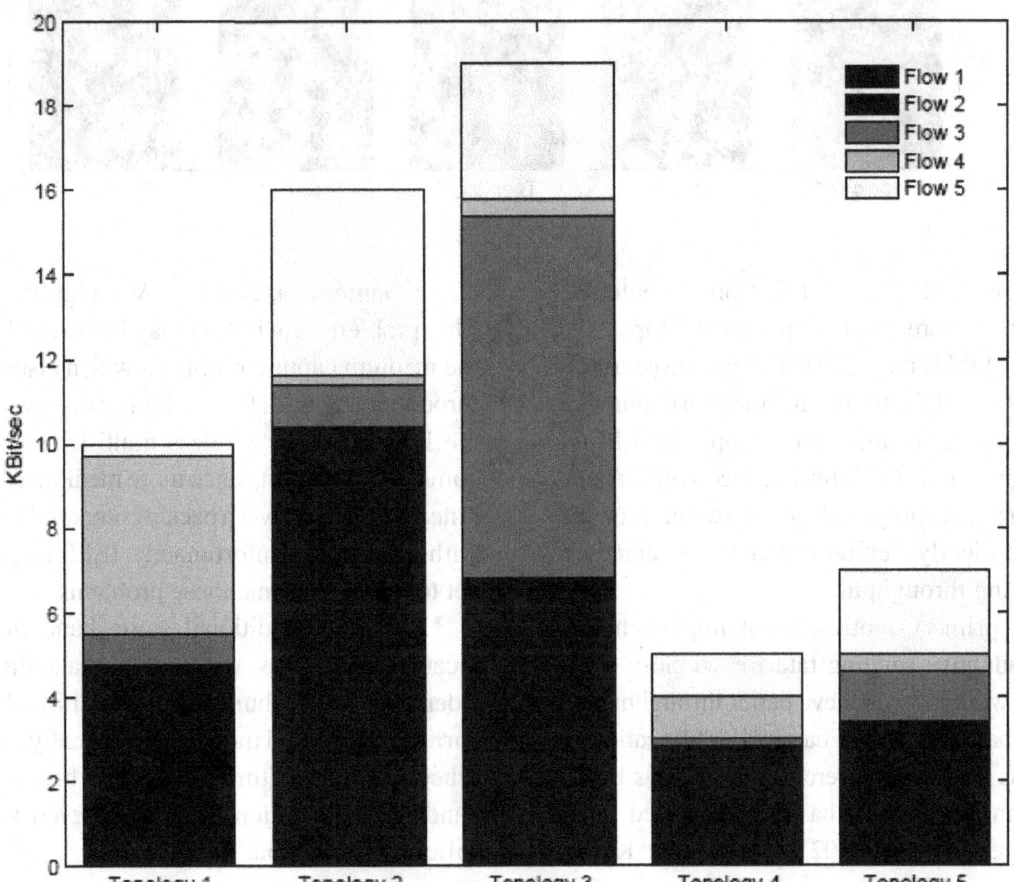

Figure 12. Random topology with proposed traffic flows

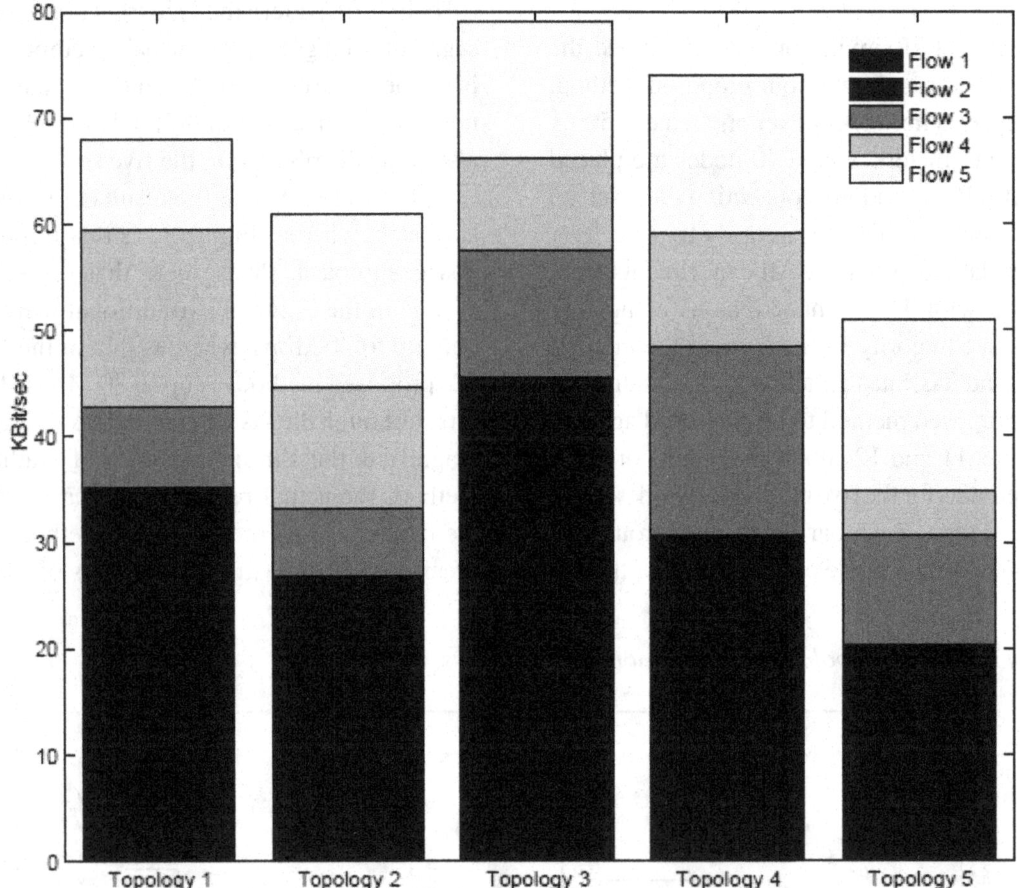

posed method (Figure 12) from Topology 1 through 5 were 68.0 Mbps, 61.0 Mbps, 79.0 Mbps, 74.0 Mbps, and 51.0 Mbps, respectively, while those of the traditional method (Figure 11) were only 10.0 Mbps, 16.0 Mbps, 19.0 Mbps, 5.0 Mbps, and 7.0 Mbps, respectively. Again, the currently proposed adoptive sending rate scheme clearly demonstrated its efficiency in enhancing throughput.

Our primary motivation of implementation of an adaptive sending rate for wireless multi-hop networks is to achieve better throughput rate while maintaining fair bandwidth allocation. In wireless networks, severe unfairness has been a persistent problem as has been reported in the literatures on IEEE 802.11 (Gupta & Kumar, 2000; Gambiroza, Sadeghi, & Knightly, 2004). This problem in wireless may be traced back to the medium capture problems which also induce throughput degradation. A higher throughput for the long flows in wireless multi-hop networks comes at a high cost, since more medium capacity is necessary to deliver a packet compared to a flow with a few hops. Unfortunately, IEEE 802.11 has yet to fully overcome these problems.

Unlike the traditional static bandwidth allocation algorithms, which do not take into consideration for the burstiness of multimedia, our currently proposed method dynamically changes scheduling for multimedia traffic. Therefore, fair bandwidth allocation can be achieved without suffering throughput rate.

The results clearly demonstrated the impact of the adaptive sending rate mechanism on fair bandwidth allocation for wireless multi-hop networks. For the traditional method (Figure 11), Coefficient of Variance (CV) as a measure of throughput dispersion, among the flow 1 through 5 were calculated as 91%, 100%, 80%, 65%, and 88% from Topology 1 through 5, respectively, which signify large fluctuation of bandwidth. On the other hand, the CV of proposed method (Figure 12) were 40% 35%, 45%, 28%, and 50% from Topology 1 through 5, which are less fluctuated as depicted in the graph as more evenly distributed segments of colour fill in the each bar.

8. CONCLUSION

In this paper, we have proposed a scheme to improve channel conditions by performing rate adaptation along with multiple packet transmission using MAC, packet loss and physical layer condition. The proposed solution can be applied to other mesh architectures such as LTE-Advanced.

Greater throughput was achieved by the combination of dynamic monitoring, multiple packet transmission, and adaptation to changes in channel quality by adjusting the packet transmission rates according to certain optimization criteria.

The key feature of the proposed method is the combination of the following two factors: 1) detection of intrinsic channel conditions by measuring the fluctuation of noise to signal ratio via the standard deviation, and 2) the detection of packet loss induced through congestion.

We have shown that the use of such techniques in a WMN can significantly improve performance in terms of the packet sending rate. The effectiveness of the proposed method was demonstrated in a simulated wireless network testbed via packet-level simulation.

Our simulation results showed that regardless of the channel condition, there is an improved performance as shown by the increase in throughput.

In future work, we will examine the performance of adaptive rate over routing protocols, which use different routing metrics. This will allow a better understanding of the complex interactions between TCP, adaptive rate and lower layer protocols.

Another study needs to be done in the future is to investigate the performance of adaptive rate in relation to end-to-end Quality of Service. Different multi-media have different forms of delay sensitiveness and traffic bursts for self-similar traffic. Finally, to further improve network performance and also increase network capacity of adaptive rate for WMNs, we will test the effects of the usage of multiple channels in comparison to a single fixed channel.

ACKNOWLEDGMENT

This work was supported by the Heterogeneous IP Networks (HIPNet) project funded by the Technology Strategy Board (TSB) (formerly Department of Trade and Industry (DTI)), UK. Scott Fowler was also partially supported by the Swedish Excellence Center at Linköping - Lund in information Technology (ELLIIT), Sweden and EC-FP7 Marie Curie CIG grant, EU, Proposal number: 294182.

REFERENCES

Akyildiz, I., & Wang, X. (2005). A survey on wireless mesh networks. *IEEE Communications Magazine*, *43*(9), 23–30. doi:10.1109/MCOM.2005.1509968.

Akyildiz, I. F., Wang, X., & Wang, W. (2005). Wireless mesh networks: a survey. *Computer Networks*, *47*, 445–487. doi:10.1016/j.comnet.2004.12.001.

Brzezinski, A., Zussman, G., & Modiano, E. (2006). Enabling distributed throughput maximization in wireless mesh networks: a partitioning approach. In *Proceedings of the 12th Annual International Conference on Mobile Computing and Networking* (pp. 26-37).

Chen, S., & Nahrstedt, K. (1999). Distributed quality-of-service routing in ad-hoc net- works. *IEEE Journal on Selected Areas in Communications*, *17*, 1488–1505. doi:10.1109/49.780354.

Chowdhury, K. R., Felice, M. D., & Bononi, L. (2009). A Fading and Interference Aware Routing Protocol for Multi-Channel Multi-Radio Wireless Mesh Networks. In *Proceedings of the ACM Symposium on Performance Evaluation of Wireless Ad Hoc, Sensor, and Ubiquitous Networks* (pp. 1-8).

Gambiroza, V., Sadeghi, B., & Knightly, E. W. (2004). End-to-end performance and fairness in multihop wireless backhaul networks. In *Proceedings of the ACM SIGCOMM Mobile Computing and Networking* (pp. 287-301).

Gupta, P., & Kumar, P. (2000). The capacity of wireless networks. *IEEE Transactions on Information Theory*, 388–404. doi:10.1109/18.825799.

Hayajneh, M., & Abdallah, C. T. (2004). Distributed joint rate and power control game- theoretic algorithms for wireless data. *IEEE Communications Letters*, 8.

Heusse, M., Rousseau, F., Berger-Sabbatel, G., & Duda, A. (2003). Performance anomaly of 802.11b. In *Proceedings of the IEEE International Conference INFOCOM* (Vol. 2, pp. 836-843).

Holland, G., Vaidya, N. H., & Bahl, P. (2000). *A Rate-Adaptive MAC Protocol for Multi-Hop Wireless Networks (Tech. Rep. No. UMI TR00-019)*. College Station, TX: Texas A & M University.

Holland, G., Vaidya, N. H., & Bahl, P. (2001, July). A Rate-Adaptive MAC Protocol for Multi-Hop Wireless Networks. In *Proceedings of the ACM International Conference on Mobile Computing and Networking*.

Information Sciences Institute. (n. d.). *NS-2-the network simulation.* Retrieved from http://www.isi.edu/nsnam/ns/index.html

Jacquet, P., Muhlethaler, P., Clausen, T., Laouiti, A., Qayyum, A., & Viennot, L. (2001). Optimized link state routing protocol for ad hoc networks. In *Proceedings of the 5th IEEE Multi Topic Conference.*

Johnson, D. B., Maltz, D. A., & Broch, J. (2001). DSR: The Dynamic Source Routing Protocol for Multi-Hop Wireless Ad Hoc Net- works. In Johnson, D. B. (Ed.), *Ad Hoc Networking* (pp. 139–172). Reading, MA: Addison-Wesley.

Kim, H., Hou, J. C., Hu, C., & Ge, Y. (2006). QoS provisioning in IEEE 802.11-compliant networks: Past, present, and future. *Computer Networks. The International Journal of Computer and Telecommunications Networking*, *51*(8).

Kim, K.-H., & Shin, K. G. (2007). Self-healing multi-radio wireless mesh networks. In *Proceedings of the 13th Annual ACM International Conference on Mobile Computing and Networking* (pp. 326-329).

Kodialam, M., & Nandagopal, T. (2005). Characterizing the capacity region in multi-radio multi-channel wireless mesh networks. In *Proceedings of the Annual ACM International Conference on Mobile Computing and Networking* (pp. 73-87).

Li, B., & Battiti, R. (2007). Achieving optimal performance in IEEE 802.11 wireless LANs with the combination of link adaptation and adaptive backoff. *Computer Networks: The International Journal of Computer and Telecommunications Networking*, *51*(6), 1574–1600.

Lu, M.-H., Steenkiste, P., & Chen, T. (2009). Design, Implementation and Evaluation of an Efficient Opportunistic Retransmission Protocol. In *Proceedings of the ACM International Conference on Mobile Computing and Networking* (pp. 73-84).

Lu, S., Sun, Y., Ge, Y., Dutkiewicz, E., & Zhou, J. (2010). Joint power and rate control in ad hoc networks using a supermodular game approach. In *Proceedings of the IEEE Conference on Wireless Communications and Networking* (p. 1-6).

Luo, J., Rosenberg, C., & Girard, A. (2010). Engineering Wireless Mesh Networks: Joint Scheduling, Routing, Power Control and Rate Adaptation. *IEEE/ACM Transactions on Networking*, *18*(5), 1387–1400. doi:10.1109/TNET.2010.2041788.

Nelakuditi, S., Lee, S., Yu, Y., Wang, J., Zhong, Z., Lu, G. H., et al. (2005). Blacklist-aided forwarding in static multihop wireless networks. In *Proceedings of the Second Annual Conference on Sensor and Ad Hoc Communications and Networks* (pp. 252-262).

Ord, J. K. (1972). *Families of Frequency Distributions*. New York, NY: Hafner.

Pathak, P., & Dutta, R. (2011). A survey of network design problems and joint de- sign approaches in wireless mesh networks. *IEEE Communications Surveys and Tutorials, 13*(3).

Perkins, C. E., Belding-Royer, E., & Das, S. (2003). *RFC 3561: Ad hoc On-Demand Distance Vector (AODV) Routing*. Retrieved from http://www.ietf.org/rfc/rfc3561.txt

Perkins, C. E., & Bhagwat, P. (1994). Highly dynamic destination sequenced distance-vector routing (DSDV) for mobile computers. In *Proceedings of the ACM SIGCOMM Conference on Communications Architectures, Protocols and Applications*.

Qiao, D., Choi, S., & Shin, K. G. (2002). Goodput Analysis and Link Adaptation for IEEE 802.11a Wireless LANs. *IEEE Transactions on Mobile Computing*, 278–292. doi:10.1109/TMC.2002.1175541.

Rappaport, T. S. (2002). *Wireless Communications: Principles and Practice* (2nd ed.). Upper Saddle River, NJ: Prentice Hall.

Redlich, O., Ezri, D., & Wulich, D. (2009). Snr estimation in maximum likelihood decoded spatial multiplexing. Retrieved from http://arxiv.org/abs/0909.1209

Rozner, E., Seshadri, J., Mehta, Y. A., & Qiu, L. (2009). SOAR: Simple Opportunistic Adaptive Routing Protocol for Wireless Mesh Networks. *IEEE Transactions on Mobile Computing*, *8*(2), 1622–1635. doi:10.1109/TMC.2009.82.

Stallings, W. (2005). *Wireless Communications and Networks* (2nd ed.). Upper Saddle River, NJ: Prentice Hall.

Zheng, J., & Ma, M. (2009). A utility-based joint power and rate adaptive algorithm in wireless ad hoc networks. *IEEE Transactions on Communications*, *57*(1), 134–140. doi:10.1109/TCOMM.2009.0901.060524.

This work was previously published in the International Journal of Wireless Networks and Broadband Technologies, Volume 1, Issue 3, edited by Naveen Chilamkurti, pp. 30-48, copyright 2011 by IGI Publishing (an imprint of IGI Global).

Chapter 16
A Source Based On-Demand Data Forwarding Scheme for Wireless Sensor Networks

Martin Brandl
Danube University Krems, Austria

Christian Mayerhofer
St. Pölten University of Applied Sciences, Austria

Andreas Kos
St. Pölten University of Applied Sciences, Austria

Thomas Posnicek
Danube University Krems, Austria

Karlheinz Kellner
Danube University Krems, Austria

Christian Fabian
St. Pölten University of Applied Sciences, Austria

ABSTRACT

Wireless Sensor Networks (WSNs) are becoming more important in the medical and environmental field. The authors propose an on-demand routing protocol using sensor attractiveness-metric (P_a) gradients for data forwarding decisions within the network. Attractiveness-based routing provides an efficient concept for data-centric routing in wireless sensor networks. The protocol works on-demand, is source-initiated, has a flat hierarchy and has its origin in the idea of pheromone-based routing. The algorithm supports node-to-sink data traffic and is therefore a lightweight approach to generalized multihop routing algorithms in WSNs. The performance evaluation of the proposed protocol is done by extensive simulation using a multi-agent based simulation environment called NetLogo. The efficiency of the attractiveness-based routing algorithm is compared in simulations with the well known Dynamic Source Routing algorithm (DSR). The authors conclude that the P_a based routing algorithm is well suited for easy to set up WSNs because of its simplicity of implementation and its adaptability to different scenarios by adjustable weighting factors for the node's attractiveness metric.

DOI: 10.4018/978-1-4666-3902-7.ch016

INTRODUCTION

Wireless Sensor Networks (WSNs) are special forms of ad-hoc networks, built by small, cheap and robust devices, so called sensor nodes. A collection of these nodes, which combine sensing, computation and communication abilities, forms the swarm intelligence. Normally, this term is referred to as the common intelligence of insect colonies, e.g., ant colonies. Each individual in an insect colony comprises of only a very low intelligence. Insects follow their intuitive rules, and show different behavioral patterns according to the current situation. As an example, an ant excretes a chemical called pheromone during foraging, and always follows the path of highest pheromone concentration. Sensor nodes also follow a simple set of rules. In insect colonies, there are a variety of other tasks besides foraging, including guarding, procreation, etc. Nevertheless, all individuals in the entire colony work towards a common goal. The same case occurs with member nodes of a WSN.

Classical approaches to ant colony optimization algorithms are given in Dorigo (1999) and Dorigo and Caro (1991), and their application in packet switching networks are presented (Di Caro & Dorigo, 1997, 1998a, 1998b). The AntNet architecture described was designed for large scale networks where each node can communicate with every other node over the network. The approach to realize a routing protocol for ad-hoc networks by imitating ants has also already been examined in other works (Heissenbuettel, 2005; Liu & Feng, 2005). In contrast to Heissenbuettel (2005) and Liu and Feng (2005) we propose a clear data centric routing concept for WSNs from each node to one sink.

The routing mechanism in Heissenbuettel (2005) is based on vectors containing the forwarding direction (FWD), in which data is broadcasted. The FWD is calculated from the neighbors' forwarding directions and is re-aligned. The authors show an analogy between polar bonds of atoms and chains of nodes formed by running this protocol. The setting of the FWD only takes place mathematically, which means that the x- and y- components of the FWD vector, or the equivalent norm and angle, have to be maintained. Furthermore, every node maintains a tuple containing the x, y coordinates, buffer fill level and the induced charge, a model parameter. Out of this tuple, a parameter called 'Sending Decision' is calculated. This parameter has to exceed a threshold to start a data transmission. The protocol is suitable for highly mobile networks and only uses the information of its nearest neighbors for the routing decisions. It does not consider the energy status of the nodes and requires knowledge of the correct position. The mechanism does not provide an inherent prevention of routing loops.

In Liu and Feng (2005), an ant colony based multi path routing protocol named AMR is proposed, which combines swarm intelligence and node disjoint multi-path routing. The method establishes and utilizes multiple routes of node-disjoint path to send data packets concurrently and adopts the use of pheromone to disperse communication traffic using forward and backward ants for route discovery and maintenance.

Multihop routing protocols for wireless sensor networks have been proposed in several works. Table driven reactive (on-demand) multihop protocols are AODV (Perkins & Royer, 1999), DSR (Johnson & Maltz, 1994; Johnson, 1994), TORA (Park & Corson, 1997), etc. The path discovery procedure terminates either when a route has been found or no route is available after examination of all route permutations. In a mobile ad hoc network, active routes may be disconnected due to node mobility. Therefore, route maintenance is an important operation of reactive routing protocols. Compared to the proactive routing protocols DSDV (Perkins & Bhagwat, 1994), WRP (Murthy & Garcia, 1996), OLSR (Jacquet, 2003) (in proactive routing protocols the routing information from each node to every other node in the network is maintained at all times) for mobile ad-hoc networks, a lower control overhead is a distinct advantage of reactive routing protocols. Thus, reactive routing protocols have better scalability than proactive routing protocols in mobile ad-hoc networks. However, when using reactive routing protocols, source nodes may suffer from long delays for route searching before they can forward data packets.

An additional on-demand routing protocol, the MINTRoute protocol, is based on selecting optimal paths between neighboring nodes by dynamically capturing link connectivity statistics using an adaptive link estimator. The calculated statistics are hosted and managed in a node's neighborhood table for link status and routing information (Woo, 2003).

In Singh (2010) a power aware based routing protocol is given using an algorithm where the next hop node is chosen by optimizing a link metric. The link metric is calculated from the energy level of the next hop node and the transmission success rate. The proposed protocol was compared with AODV. From simulations a better delivery ratio and less energy consumption and delay in comparison with AODV was found.

Our developed attractiveness-metric gradient based routing strategy (P_a) (Kos, 2006), provides a new concept for a data-centric WSN routing protocol. The attractiveness-metric considers the actual energy states of neighboring nodes for the routing decision and therefore ensures energy aware operation. In detail, the routing costs consider the sensor nodes' energy status, as well as the received signal strength and current buffer (on board memory) fill level. It has a flat hierarchy, works on-demand, is source-initiated and has its origin in the concept of ant-based routing. However, the following description of P_a-routing shows that this concept is more similar to a conventional WSN routing protocol, and only the basic idea comes from ants. During development of the algorithm, the attractiveness-metric was turned into a special factor for link costs and is the sole determinant that remains similar to the ant-based commencements.

To properly introduce the system, it is necessary to mention that P_a-routing only supports node-to-sink data traffic and is therefore a lightweight approach to multihop routing in WSNs. One important component of P_a-routing is the concept of automatic synchronization, which must be reconciled with the routing protocol. The majority of the time, the whole network is in a recovery phase to save battery power.

DESCRIPTION OF PA BASED-ROUTING

P_a-routing is a source-initiated on-demand routing mechanism which considers the actual energy states of neighbor nodes for the routing decision and therefore ensures energy aware operation. Before we explain the mode of operation, we will identify the most important definitions:

In this paper, range refers to the transmission range, the maximum distance between two communicating nodes. All nodes within range are the neighbors and are the only affected nodes of a broadcast. Broadcast always means a local broadcast in this paper. Broadcasts are not forwarded in P_a-routing. Furthermore, in P_a-routing we only examine homogenous networks, where all nodes have the same hardware. The hierarchy is flat. Some other protocols use a hierarchical topology, which means they form groups for the purpose of communication. P_a-routing is an on-demand source-initiated mechanism. On-demand routing creates routes only when desired by the source node. When a node requires a route to a destination, it initiates a route discovery process within the network by choosing the next hop node by the highest node attractiveness-level. To start a route discovery, or any other communication, all neighbors must first be synchronized. This means that all communicating nodes need one common active time, which can be used for communication. The rest of the time the nodes are in a low power mode (sleep mode). Every node periodically wakes up to communicate with its neighbors, and then goes back to sleep until the next frame. Meanwhile, new messages are queued. The principle used for synchronizing the time schedules for all nodes is similar to the S-MAC protocol (Heidemann & Estrin, 2002). P_a-routing presumes that the hardware used is able to provide the node's energy state, their memory fill level as well as the received signal strength and these factors are used for its attractiveness-metric calculation.

The logical topology of a network running P_a-routing is always starlike. All nodes share a data sink as a common destination for data packets. A sink is a central data collecting node. There is no node-to-node or sink-to-node communication possible. The sink node has to be better equipped than normal sensor nodes regarding energy resources.

A. Synchronization

The listen/sleep scheme requires synchronization among all neighboring nodes. Frame synchronization is inspired by virtual clustering, as described by the authors of the S-MAC protocol (Heidemann & Estrin, 2002). When a node comes to life, it starts by waiting and listening. If it hears nothing for a certain amount of time, it chooses a frame schedule and transmits a SYNC packet, which contains the time until the next frame starts and the node's attractiveness-level. If the node, during startup, hears a SYNC packet from another node, it follows the schedule in that SYNC packet and transmits its own SYNC accordingly. A node gets its initial attractiveness level from its neighbors, weighted by its own system parameters (energy status, buffer fill level) and multiplied by the network hop to hop decrease rate (Equation 4). The initial attractiveness-level for all nodes is therefore spread by the base station over the whole network during synchronization (flooding). The synchronizing node receives the transmitted SYNC messages from its neighboring nodes and can calculate its attractiveness-level (only if the SYNC sending node is synchronized to the base station otherwise the transmitted attractiveness-level is zero) and the synchronization time. If the attractiveness-level of all received SYNC messages is zero (all neighboring nodes are not synchronized to the base station), the node starts the synchronization procedure once again. After the synchronization process, the node updates its attractiveness-level by choosing the highest received attractiveness-level from the neighbor nodes and calculates its own attractiveness-level by the P_a-routing algorithm Equation 3.

If the synchronizing node does not hear any neighbors during the listening period, the next attempt will commence after a certain time period, $T_{RECOVER}$. In Figure 1, the synchronization procedure for $R_{decr} = 0.3$ is depicted. Nodes with direct access to the base station get an attractiveness-value of (1-0.3) = 0.7. All nodes within the 1st hop area get an attractiveness-value of $(1\text{-}0.3)^2 = 0.49$, in the 2nd hop area $(1\text{-}0.3)^3 = 0.34$ and so on. The attractiveness-value decreases by the number of hop distances from the base station and therefore ensures loop free data centric routing.

B. Sending Decision

Once a node has synchronized with its neighbors', it is ready to transmit data. Due to energy saving behavior, sending actions should only be carried out when really necessary. It becomes necessary to transmit data when the fill level of the buffer memory exceeds a certain threshold. It is not recommended to completely fill the buffer. If a mobile node loses its connection to the network during data collection from its sensor interfaces or the traffic in the surrounding network inhibits a broadcast, it has no chance to dispose its collected data. It is important to have some buffer space conserved to proceed with the sensing task without a memory overflow, and thus resulting in a loss of sensing data (see section "Simulations").

Figure 1. Spreading of the attractiveness-value within the network during the synchronization process with a per-hop decrease rate of R_{decr}=0.3

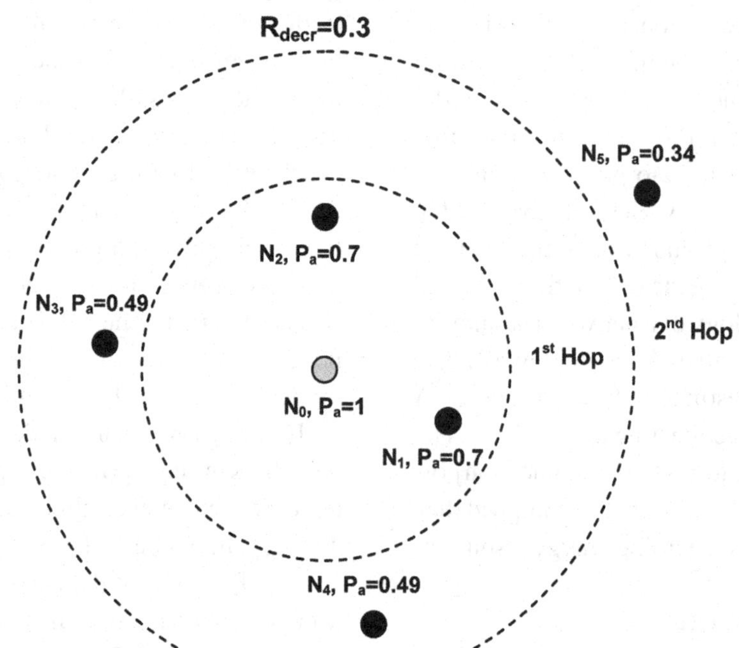

C. Choosing the Route

Once the sending decision is positive, a decision has to be made to determine which neighbor node should become the next hop for the collected data, this decision is called routing decision. Therefore, a node sends out a Route Discovery Query (RDQ) to its neighboring nodes. Before the neighbors give an answer, they make a preprocessing of their attractiveness-value P_a. The highest last received attractiveness-level from the neighbors is weighted with the current energy status and sent to the requesting node, called Route Discovery Reply (RDR), where it is also post-processed. The node which offers the highest remaining attractiveness-metric becomes the next hop.

D. Propagation of Attractiveness

A node obtains attractiveness from its neighbors in the course of a route discovery. As already mentioned, a potential next-hop node (neighbor) preprocesses its attractiveness-value $P_{a,n}$ first, before answering a route discovery query of an asking node (Figure 2). The result of this preprocessing is $P'_{a,n}$ where n is element of the numbers of neighbors $N_{neighbor}$. The number $N_{neighbor}$ is defined as a subset of all nodes N within the transmission range of the asking node.

$$P'_{a,n} = P_{a,n} \cdot \left(F_{energy,\,n} \cdot W_{energy,\,n}\right) \cdot \left(1 - F_{buffer,\,n}\right) \cdot W_{buffer,\,n} \qquad n \in N_{neighbor} \tag{1}$$

After receiving all values $P'_{a,n} \left(n \in N_{neighbor}\right)$, the following post-processing is carried out by the asking node for all received attractiveness-values:

$$P''_{a,n} = P'_{a,n} \cdot \left(F_{rss} \cdot W_{rss}\right) \qquad n \in N_{neighbor} \tag{2}$$

Figure 2. Schematic of a route discovery broadcast (RDQ, RDR) for two neighbor nodes

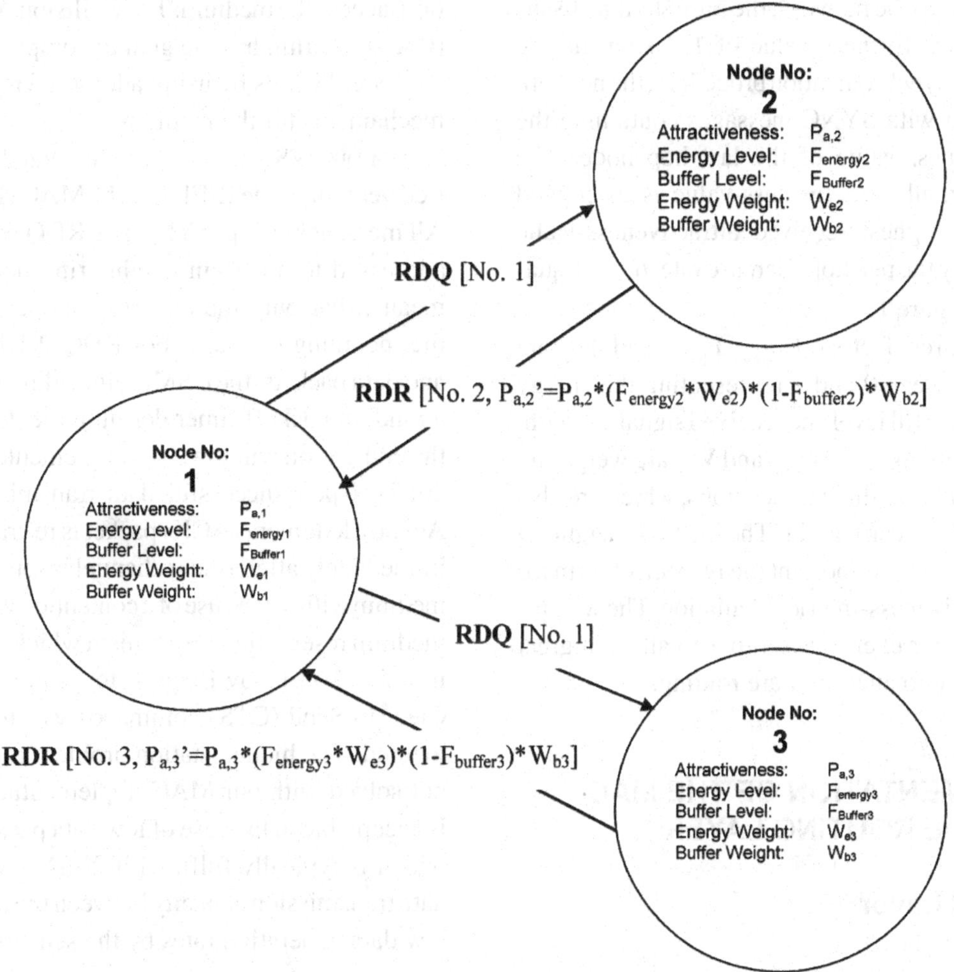

This step is only necessary for the route deciding procedure. The node's attractiveness-value, resulting in the highest remaining $P^{''}_{a,n}$ becomes the next hop node.

Next-hop node decision:

$$\rightarrow \max\left(P^{''}_{a,n}\right)$$

$$n \in N_{neighbor} \qquad (3)$$

The requesting node's attractiveness-metric is then calculated (updated) as follows:

$$P_{own} = \max_{n \in N_{neighbor}}\left(P^{''}_{a,n}\right)\cdot\left(1 - R_{decr}\right) \qquad (4)$$

Where P_{own} is the new attractiveness-value of the broadcasting node and R_{decr} is the per-hop decrease rate which is a predefined network parameter. The data routing is therefore always directed to the node with the highest attractiveness. By setting up the network, the base station always has an attractiveness value of 1. Using the pre described synchronization procedure, the network is flooded with SYNC messages containing the attractiveness-value of the last hop node. The node's actual attractiveness-value is calculated from the highest received attractiveness-value reduced by the per-hop decrease rate R_{decr} (Equation 3), Figure 1.

The three factors F_{energy}, F_{buffer} and F_{rss} are values between 0 and 1 representing the energy level, buffer fill level and received signal strength, respectively. W_{energy}, W_{buffer} and W_{rss} are weighting factors for these three dimensions, which are also numbers between 0 and 1. The higher a weighting factor, the more important the respective term for the attractiveness-metric calculation. The weighting factor for energy is set to 1 to attach a great importance to energy aware routing.

IMPLEMENTATION OF THE MAC AND THE ROUTING LAYER

A. MAC Layer

The main task of the MAC layer is the coordinated utilization of the air interface within the WSN. The data transmission between the nodes must be regulated with each other so that as few as possible concurrent requests take place. Otherwise packet collisions take place reducing the data throughput, causing packet replications and leading to delayed data transmission.

In the wireless network standard IEEE 802.11 (IEEE, 2007) the distributed coordination function (DCF) builds the general basic principle for access methods and implements the Carrier Sense Multiple Access with Collision Avoidance (CSMA/CA) algorithm. The CSMA algorithm defines the time during which every ready to send node must listen to the channel (carrier sense) before the node can, in case of no carrier detection, access the medium. The Collision Avoidance (CA) algorithm has the goal to completely avoid packet collisions by using additional reservation mechanisms for the medium.

For our WSN we have implemented a simplified version of the IEEE 802.11 MAC (Figure 3). All messages except SYNC and RDQ packets are addressed to a certain neighboring node which means, that only the addressed node processes the incoming message. For RDQ, RDR, SYNC and data packets, the CSMA algorithm, including a random backoff timer defining the duration of the contention window, was implemented (IEEE, 2007). Upon successful data transmission, the Acknowledgment (ACK) packet is re-transmitted immediately after a clear channel is sensed on the medium without the use of a contention window. A medium reservation mechanism which is defined in IEEE 802.11 by Request to Send (RTS) and Clear to Send (CTS) commands was not implemented. The hidden station problem is therefore not solved with our MAC implementation. This is acceptable in the case of low data packet traffic which is typically fulfilled in WSN's with high data transmission capacity between the nodes and low data generation rates by the sensors.

1. IEEE 802.15.4 Frame Format

The frame format and error detection/correction is part of the MAC layer, frame synchronization is carried out in the physical layer. The frame formats are given by control bytes in the header which defines the type of frame and contains the frame control information (Figure 4). Frame synchronization is done within the synchronization header (SHR) containing a preamble sequence of 4 bytes followed by a 'start of frame' byte. In the one byte 'frame length field', 7 bits are used for the frame length description and one bit is reserved. The MAC frame length (MHR+MSDU+MFR) can therefore have a maximum of 127 bytes. For

Figure 3. Implemented MAC layer

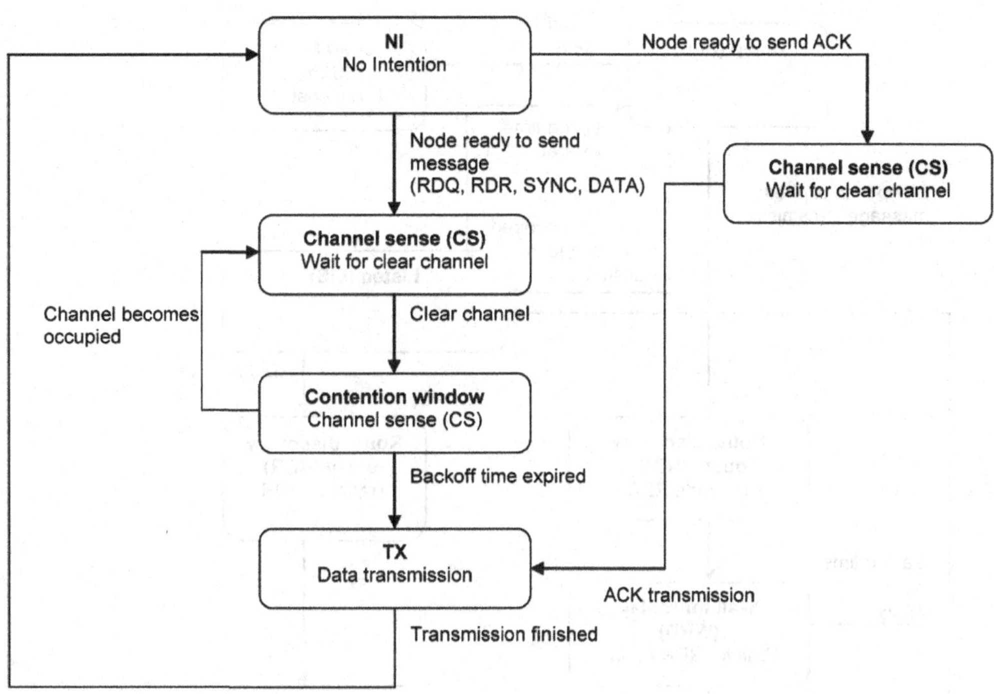

Figure 4. IEEE 802.15.4 frame format

Bytes	4	1	1	2	1	4 to 20	n Bytes	2
Content	Preamble Sequence	Start of Frame Delimiter	Frame Length	Frame Control	Data Sequence Number	Address Information	Data Payload	FCS
Description	SHR		PHR	MHR			MSDU	MFR

MAC Frame: 5+(4 to 20)+n ≤ 127 Bytes

the MAC layer, the content of the MAC header (MHR), of the MAC service data unit (MSDU) and of the MAC footer (MFR) are for interest. All messages within the WSN use a data frame format which is defined by the bit constellation in the frame control field. For addressed messages (RDR, DATA, ACK) the node source and the packet destination address is delivered in the address information field. Unaddressed messages (RDQ, SYNC) contain only the source node's address.

By using the full available 20 bytes for address information, the data payload has a maximum length of 102 bytes which can be transmitted in one frame. The MAC frame is protected by a two bytes frame check sequence (FCS) located at the end of the frame called MAC footer (MFR). In our setup we have used 4 bytes for the address information (2 bytes for the source, and 2 bytes for the destination address) which therefore allows a maximum remaining data payload of 118 bytes.

B. Routing Layer

In Figure 5 the state machine for transmission of data packets is depicted. In general, the nodes are mainly in the sleep phase (SLP) and if the network is synchronized, all nodes wake up at a defined schedule. From the sleep state, the nodes typically switch to the listen state except when

Figure 5. State diagram for P_a based routing

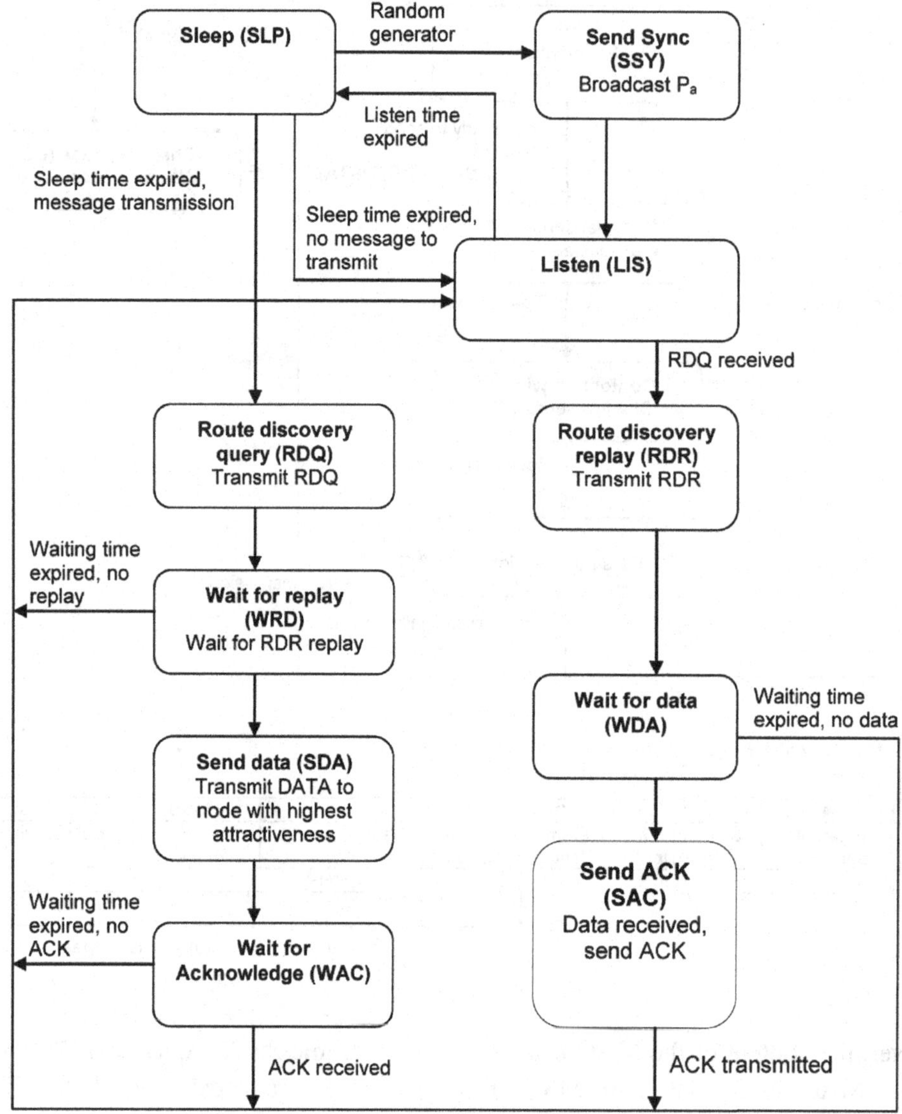

a data transmission is induced. For maintaining synchronization in the network, from time to time, triggered by a random generator, the node switches at wake-up not into the listen state or the data broadcast chain, but into a transmission state where a SYNC message is broadcasted which includes the time until the next wake-up and the node's attractiveness level P_a.

Data transmission is only triggered if a certain threshold of the buffer memory is exceeded. To avoid a group of nodes reaching this at the same time, a fixed buffer memory threshold and starting simultaneously a data transmission, the buffer memory threshold is individually set to a certain range by a random generator for each node.

If a data transmission is triggered, a Route Discovery Query (RDQ) is firstly sent to all neighbors and afterwards the broadcasting node waits for Route Discovery Replays (RDR). In the case of no replays, the transmission task is stopped and the node returns to the listen state. Otherwise the broadcasting node chooses the neighbor with the highest post processed P_a (Equation 2 and Equation 3) for the next-hop node. After a successful receipt of an Acknowledgment (ACK) message, the data transmission finishes and the node returns to the listen state.

A node only leaves the listen state if a RDQ message is received. This means that a neighboring node starts a route discovering process for their next hop decision. The node therefore replays a RDR message back to the asking node including its actual pre-processed attractiveness value P_a' (Equation 1) as well as the source and the destination address. If a data packet is received, the data is stored in the memory and in the case of correct transmission, an ACK message is relayed back. Otherwise if no data is received, this means, another node is addressed for data transmission and when the waiting time has expired, the node returns to the listen state.

In general, for energy saving reasons the nodes change synchronously and periodically from the listen to the sleep state, where the active state is typically shorter than the sleep state. In the case of data transmission, the active state can be extended by the transmitting and the receiving node until the complete data transmission is finished and acknowledged. This provides a higher data transmission rate within the network because the number of transmitted data frames is not limited by the node's wake up duration.

A NETLOGO BASED SIMULATION TOOL

NetLogo (Tisue & Wilenski, 2004) is a multi-agent cross-platform modeling- and simulation-environment for simulating complex systems over time. It is designed for both research and education and is used in a wide range of disciplines and education levels. Modelers can give instructions to hundreds or thousands of independent agents, all operating concurrently. This makes it possible to explore connections between micro-level behaviors of individuals and the macro-level patterns that emerge from their interactions. NetLogo enables users to open simulations and play with them, exploring their behavior under various conditions. NetLogo is also an authoring environment that is simple enough to enable researchers to create their own models, even if they are not professional programmers. Historically, NetLogo is the next generation of the series of multi-agent modeling languages which includes StarLogo.

NetLogo is a standalone application written in Java so it can run on all major computing platforms (Wilensky & Stroup, 1999). As a language, NetLogo is a member of the Lisp family that supports agent and concurrency. Mobile agents called turtles move over a grid of patches, which are also programmable agents. All of the agents can interact with each other and perform multiple tasks in parallel. Niazi and Hussain (2009) showed that NetLogo is a highly flexible agent-based simulation tool and very effective in modeling self-organizing and complex systems, e.g., the simulation of communication networks.

A big advantage is the ability of NetLogo to communicate with other applications, such as other simulation tools. It provides commands to simply write and read any kind of text file. Thus, data generated by NetLogo can be easily used by a visualization program to print an attractive graphic. Furthermore, NetLogo includes a still evolving tool called 'Behavior Space' that allows parameter sweeping. That is, the systematic testing of the behavior of a model across a range of parameter settings.

C. System Modeling with NetLogo

As already mentioned, the P_a based routing protocol was developed with the help of the tool NetLogo. In the model, there are two different breeds of turtles, each describing a group of nodes. One group is the sensor nodes, portrayed by the breed nodes.

The others are the data sinks, described by the breed sinks. Thus, all communicating devices are turtles in the NetLogo model. Each member turtle of the breed nodes has to carry out two different kinds of code for each timestep. One part is the code describing the influences from the outside, like movement (in the case of passive moving nodes) and disturbances. The second part is the code responsible for modeling nodes running P_a based routing. This part comprises the communication and sensing task. In a realization of P_a based routing with real devices, this code has to be implemented in the sensor nodes' main microcontroller. Every sensor node has a set of parameters, which are the attractiveness P_a, the per-hop decrease rate, the maximum buffer size, the buffer fill level threshold, three weighting factors for R_{ss}, energy and buffer fill level, communication range, wake and sleep time and number of collected data sets per wakeup (Figure 6). Global settings are the number of sensor nodes, the number of sinks, the simulation run time and the simulation area. Table 1 shows a complete list of all the relevant parameters and their typical settings in the NetLogo model of P_a based routing.

SIMULATIONS

The simulation tool used is a multiagent programming language and modeling environment. It uses a patch for the unit of length and timesteps for the

Figure 6. Example of the NetLogo based simulation environment

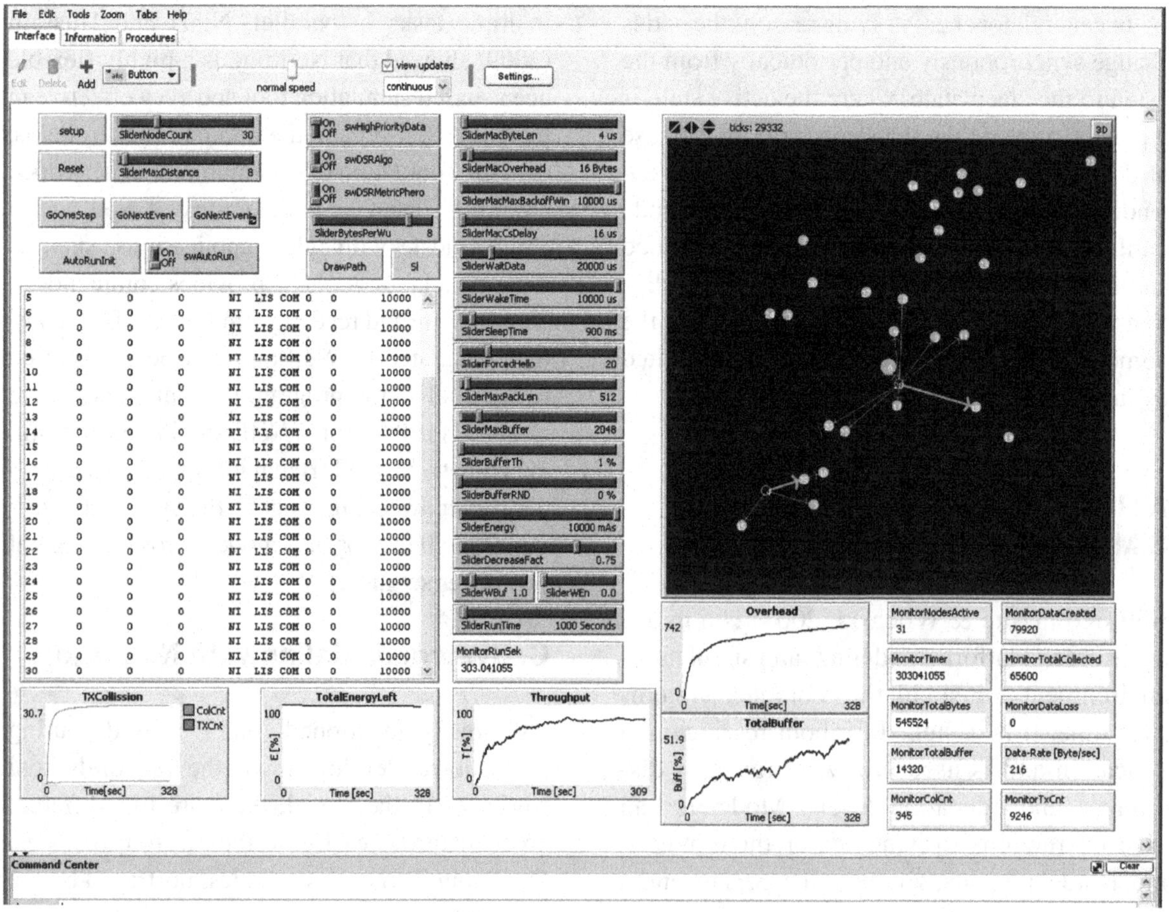

Table 1. Simulation parameters

Node Parameters	Value	Unit
Per-hop decrease-rate	0.25	1
Buffer size	2048	Bytes
Buffer fill level threshold	410 (20% of buffer size)	Bytes
Weighting factor for buffer fill level	1	1
Weighting factor for R_{ss}	1	1
Weighting factor for energy	1	1
Wakeup time	10	ms
Sleep time	1000	ms
Communication range	8	Patches
Data collection rate	1-8	Bytes/wake up (Bytes/WU)
Bit duration	4	μs
Data bytes per frame	118	Bytes
Overhead bytes per frame	15	Bytes
Max frames per TX	4	1
Contention window	5	ms
R_{ss} delay	16	μs
Synchronization maintenance	20	Wake up periods
Global Parameters		
Number of nodes	10-90	1
Number of sinks	1	1
Simulation area	40x40	Patches
Simulation run time	1000	Seconds

unit of time. All simulations are run in a random network using simulation parameters according to Table 1. The simulation tool offers an easy to design graphical user interface comprising action buttons, sliders for the adjustment of variables, a simulation window where the turtles (nodes) are randomly placed and numerical and graphical windows representing the output variables. Communication paths between the nodes are depicted by colored arrows, where the direction of the arrow displays the communication direction and the color of the type of message sent (RDR, RDQ, DATA, ACK, SYNC). In case of a message collision at the receiver, the links to the corresponding node are depicted in light grey and the collision counter is incremented (Figure 7).

In the standard IEEE 802.15.4, the bit duration is defined as 4μs for data rates of 250 kBit/s and is therefore chosen for the time resolution of the simulation software. To speed up the simulation time, the simulation is triggered by events. This means, the time periods where nothing happens (e.g., when the node's are sleeping) are skipped, but if an event occurs (e.g., wake up, sending decision) the simulation runs in timesteps of one bit duration. By using this event triggered simulation principle, the overall simulation time can be dramatically reduced.

At the beginning of the simulation, the base station is positioned at the center and the nodes are randomly placed within the simulation area. A node can only communicate with neighboring nodes within a defined range of patches which is defined by the Communication range. At startup, all nodes are in listening mode waiting for a synchronization message which is firstly spread by the base station and then distributed by flooding over the network. After synchronization, the nodes start to work and collect data sets represented by bytes per wakeup (Bytes/WU). For energy saving reasons, the nodes change periodically and synchronize from wake up into the sleep state where energy consumption is significantly reduced to allow a longer battery life. To maintain the synchronization between the nodes, a SYNC message sent from the base station is flooded from time to time throughout the network. A synchronization message is only forwarded by a node if the

Figure 7. Example of two data transmissions (red arrows) from node 7 to the base station (node 0) and from node 16 to node 12. All other nodes with incoming arrows are listeners, whereas at node 14 a collision event occurs. The data message is an addressed transmission and is indicated by a bold arrow.

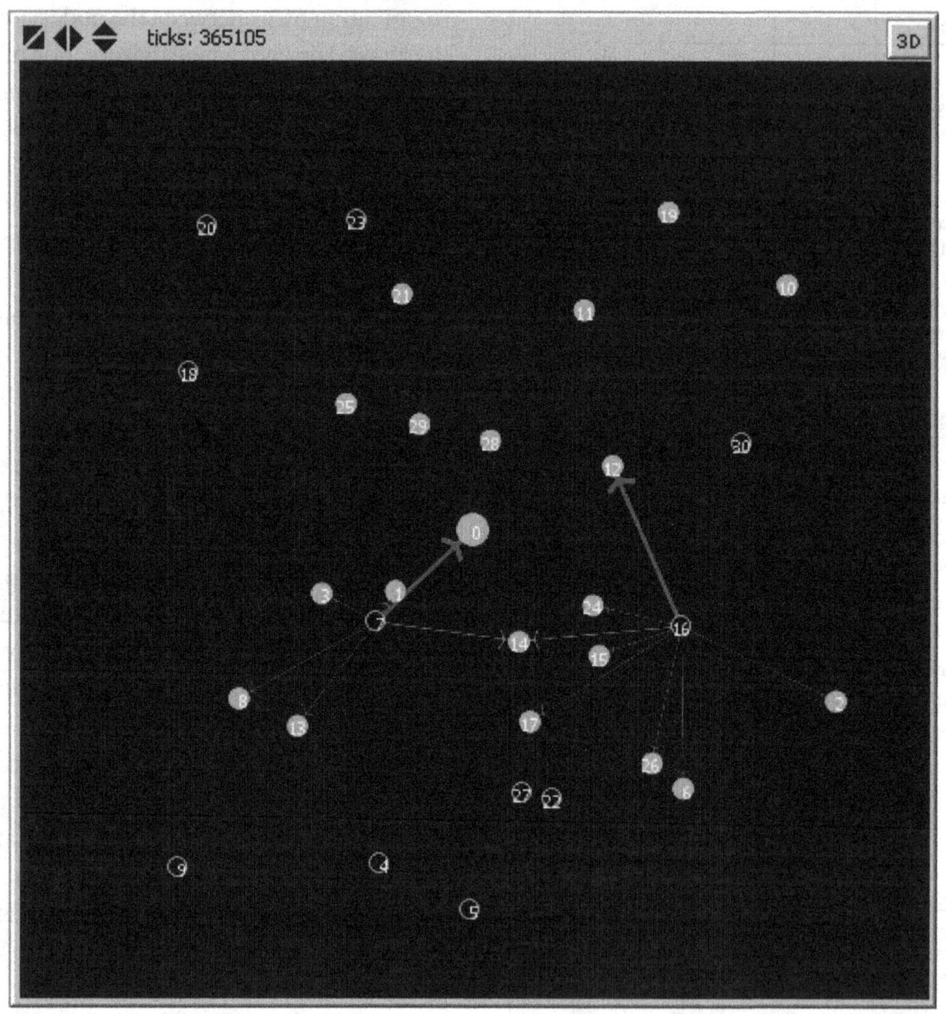

SYNC message is "newer" than the last received SYNC message. After a certain number of wake ups, chosen by a random generator between 1 and Synchronization maintenance (Table 1), the actual SYNC message is forwarded to the neighboring nodes. This procedure avoids collisions due to SYNC messages being forwarded from neighboring nodes with the same synchronization status.

The data transmission between the nodes is defined by the P_a based routing algorithm. For a higher data throughput a burst mode transmission is used where the complete data memory of the sensor node is transmitted at once by a set of sequenced frames.

Example of data forwarding: The node's buffer fill-level = 1000 bytes (~ 49% of the buffer memory) and all node parameters are set to the values in Table 1. In this example the node's buffer fill-level threshold is exceeded (threshold = 20%) and therefore a data sending decision is triggered. If the next best hop node is successfully chosen by P_a based routing, the node's memory content is split into 9 IEEE 802.15.4 frames where 8 frames contain 118 bytes of data and the 9th frame contains the remaining 56 bytes. In addition, each frame comes with SHR, PHR, MHR and MFR blocks which in our setup contain a total of 15 bytes per frame. After a successful transmission, the received data sets are acknowledged by the receiving node.

D. Dynamic Source Routing (DSR)

To evaluate the effectiveness of the proposed P_a based routing algorithm the simulation results are compared with the Dynamic Source Routing (DSR) algorithm (Johnson & Maltz, 1994; Johnson, 1994). The DSR algorithm is popular and widely used in WSNs and therefore we have used it as a reference. DSR is an on-demand algorithm based on routing tables stored at each node which ideally contains the routing paths to all other nodes within the network. At startup or if a routing path is not known, a Route Discovery Message is flooded from the requesting node through the network. When the Route Discovery Message arrives at the designated node, the message will be reversed to the source node as a Replay Message on the cached pass, which is stored in the message header or within a path which is located in the destination node's routing table. If the Replay Messages reach the source node, the message header contains all addresses of the intermediate nodes which defines the routing path and will be stored in the node's routing table. Also in DSR the intermediate nodes within a path discovery process can read out the routing header information and store it in its own path table for later use. If a message is transmitted on a broken path (e.g., a node is damaged) the last working node on the routing path relays a Route Error Packet to the source node which proceeds with a new path discovery query. The node synchronization, the MAC and physical layer, and the node's and global simulation parameters are in DSR, the same as in P_a based routing.

RESULTS

All simulations for P_a based routing and DSR are done with a self written simulation tool running within the NetLogo environment. First a random network topology was chosen with an increasing number of nodes (10 to 90 nodes). The simulation run time for each topology was 1000 seconds which represents about 990 sleep and wake up cycles. All simulation results, e.g., data throughput and data rate, are evaluated at the end of the simulation time.

The P_a based routing algorithm and DSR were simulated together for each randomly chosen network topology. Each topology was simulated for different node data generation rates (1 to 8 bytes per wake up and node). These simulations were conducted three times with a fixed number of nodes but with different network topologies to get representative mean values of the simulation results. After the simulation runs for a certain number of nodes within the simulation area, the number of nodes was increased by 20 and started again until the maximum number of nodes, which was set to 90, was reached. The total simulation time for a complete run on a standard PC was about 2 hours.

For a better comparison of the simulation results, we referenced the simulation outputs to the node density within the communication range of one node and not to the absolute number of nodes. The node density therefore refers to the mean number of nodes within a given node's communication range. For given number of 50 nodes within the simulation area of 40x40 patches (simulation area is 1600 patches^2) and a node's communication range with a radius of 8 patches (node's communication area is about 200 patches^2), the node density within the node's communication area is 6.25.

The mean data throughput from the nodes to the base station is defined by the ratio total data bytes generated at the nodes / total data bytes received at the base station at the end of a simulation run

(1000s) and for different node data generation rates (bytes per wake up: Bytes/WU). The simulated network throughput for P_a based routing and DSR is depicted in Figure 8.

It can be clearly seen that the data throughput for low data generation rates is mainly constant for different node densities but away from 100%. This is caused by the internal memories of the nodes where a part of the collected data sets is buffered and not immediately transmitted. This aberration becomes lower for increasing data collection rates because the relation of collected data sets per wake up and the nodes memory size becomes lower. For low data generation rates, the network operates in a stable and efficient manner up to node densities of 10 nodes per communication area. For higher data collection rates, as well as for higher node densities within the network, the data throughput starts dropping in DSR as well as in P_a based routing. At these points, the limit of data transmission capacity within the network is reached. Beyond this limit the data transmission activity of the nodes becomes very high resulting in too many collisions and therefore in a drop in data throughput. Additionally this can result in a node memory overflow and data loss occurs. The simulated mean values and statistics for the data throughputs are given in Table 2.

In the case of data loss within the network (Figure 9), the limit of data transmission capacity is reached. The data transmission capacity is mainly limited by two factors: First, by the access duration to the base station in the wake up state. Only one node can access the base station at any time without collisions and can transmit data. The transmission of the complete node memory (2048 Bytes) needs about 74 ms (roughly: 17 frames each containing 118 data bytes and 15 bytes overhead with a single frame duration of 4.26 ms and an additional frame with 42 bytes payload and 15 bytes overhead with a frame duration of 1.82 ms). After transmission of a complete node memory, the wake up cycle of 10 ms is left where the network goes synchronous into the sleep mode

Figure 8. Data throughput for different data generation rates and node densities (mean values, N=3)

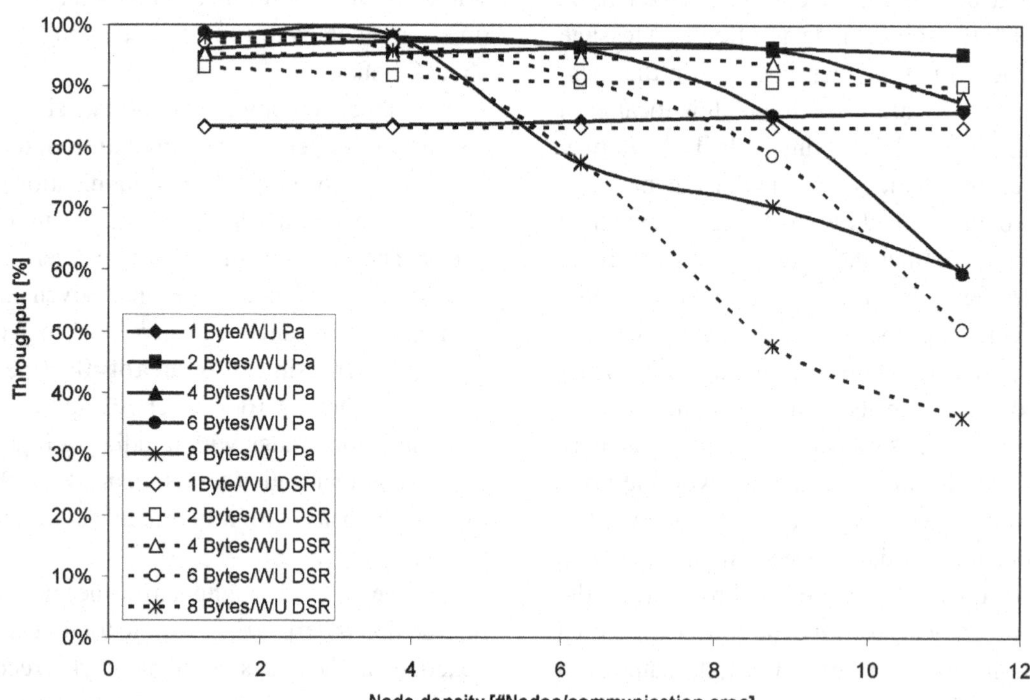

Table 2. Simulation values for the data throughput

a)

Pa [%]	Bytes/WU									
Node density	1		2		4		6		8	
	mean	std	mean	std	mean	std	mean	std	mean	std
1,25	83,45	0,13	94,51	0,50	96,06	0,57	98,72	0,83	98,12	0,14
3,75	83,79	0,53	95,50	0,56	97,45	0,33	98,07	0,16	98,17	0,59
6,25	84,48	0,13	96,28	0,60	96,92	0,35	96,33	2,17	77,51	18,08
8,75	85,23	0,32	96,33	0,49	96,10	0,16	85,24	4,24	70,60	14,95
11,25	85,91	0,33	95,18	2,10	87,38	5,31	59,36	22,02	60,04	3,84

b)

DSR [%]	Bytes/WU									
Node density	1		2		4		6		8	
	mean	std	mean	std	mean	std	mean	std	mean	std
1,25	83,34	0,00	93,07	0,34	95,25	0,00	97,24	0,00	98,13	0,17
3,75	83,35	0,00	91,75	0,30	95,26	0,00	96,93	0,31	95,92	2,05
6,25	83,35	0,00	90,61	0,83	94,90	0,62	91,30	4,01	77,60	6,88
8,75	83,35	0,00	90,72	0,25	93,81	1,67	78,78	6,27	47,67	4,00
11,25	83,26	0,15	90,12	0,10	88,17	2,05	50,36	5,57	36,00	2,23

Figure 9. Data loss within the network for P_a based routing and DSR (mean values, N=3)

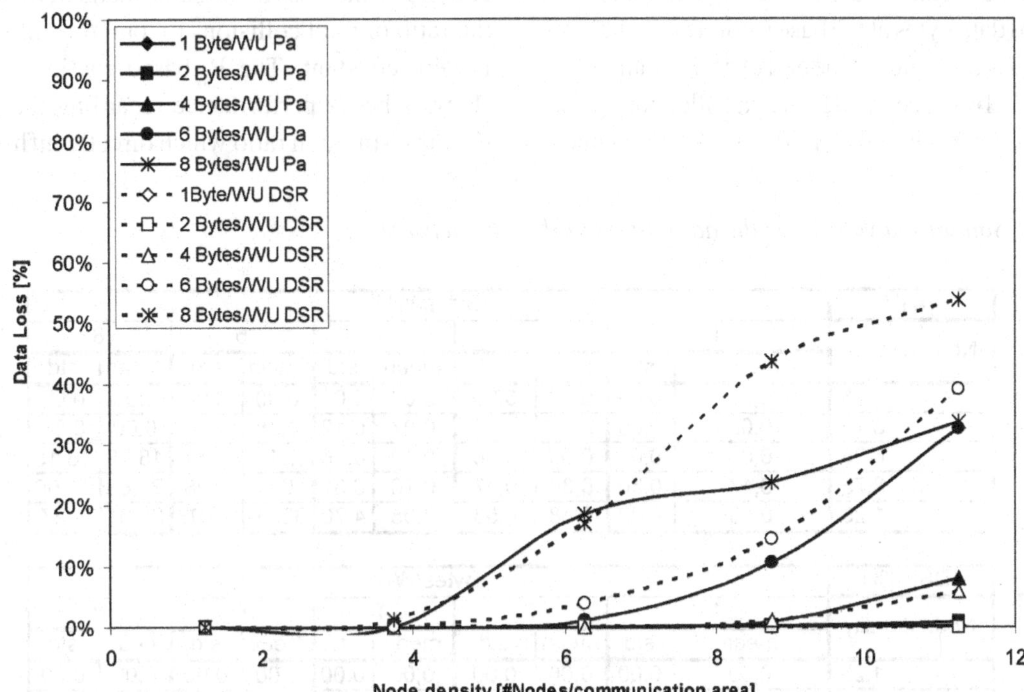

and no further transmissions from neighboring nodes can start. The second main limit for the data transmission capacity is given by collisions during routing requests or data transmissions because no channel reservation mechanism is implemented in our routing algorithms. This limitation becomes more and more important for higher node densities (Figure 9). For low node densities (< 4 nodes/communication area) collisions occurring have not influenced the system's performance. The simulated mean values and statistics for the data losses within the network are given in Table 3.

The average data rate within the network is theoretically given by the amount of data sets collected per wake up cycle and the number of network nodes. The higher the number of network nodes and the number of collected data sets the higher must be the average data rate for a successful data transmission to the base station. The average data rate is defined by the number of received data bytes at the base station / simulation time and is evaluated at the end of each simulation run at t = 1000 seconds. For data collection rates below 4 bytes per wake up, P_a based routing and DSR exhibit the same performance, where the data rate is linear increasing with the node density (Figure 10). The DSR routing algorithm reaches it maximum in data transfer with average data rates of about 300 Bytes/second where P_a based routing performs slightly better with average data rates of up to 400 Bytes/second.

The main limiting factor for the maximum achievable average data rate in both routing algorithms is given by collisions during high traffic within the network. The ratio number of total collisions by number of total transmissions during the simulation runs is depicted in Figure 11. For both routing algorithms, the collision to transmission ratio is approximately linear increasing with the node density and is mainly independent from the number of collected data sets per wake up. These results clearly show that for a higher number of collected data sets per wake up the number of data transmissions, as well as the number of collisions, increase in the same manner, therefore the ratio of total collisions to total transmissions is kept constant. The P_a based routing clearly shows a better performance regarding the collision/transmission ratio which directly influences

Table 3. Simulation values for the data losses within the network

a)

Pa [%]	Bytes/WU									
Node density	1		2		4		6		8	
	mean	std	mean	std	mean	std	mean	std	mean	std
1,25	0,00	0,00	0,00	0,00	0,00	0,00	0,00	0,00	0,00	0,00
3,75	0,00	0,00	0,00	0,00	0,07	0,12	0,08	0,14	0,00	0,00
6,25	0,00	0,00	0,00	0,00	0,10	0,16	1,19	1,61	18,60	16,13
8,75	0,17	0,30	0,30	0,27	0,76	0,31	10,55	3,98	23,83	12,60
11,25	0,13	0,22	0,88	0,53	8,08	4,76	32,66	19,03	33,61	4,27

b)

DSR [%]	Bytes/WU									
Node density	1		2		4		6		8	
	mean	std	mean	std	mean	std	mean	std	mean	std
1,25	0,00	0,00	0,00	0,00	0,00	0,00	0,00	0,00	0,00	0,00
3,75	0,00	0,00	0,00	0,00	0,00	0,00	0,00	0,00	1,24	1,56
6,25	0,00	0,00	0,11	0,20	0,04	0,06	3,99	3,57	17,08	5,74
8,75	0,00	0,00	0,00	0,00	1,18	1,35	14,57	5,91	43,66	3,78
11,25	0,14	0,24	0,00	0,00	5,75	1,67	39,09	4,53	53,86	2,78

Figure 10. Average data transmission rate (mean values, N=3)

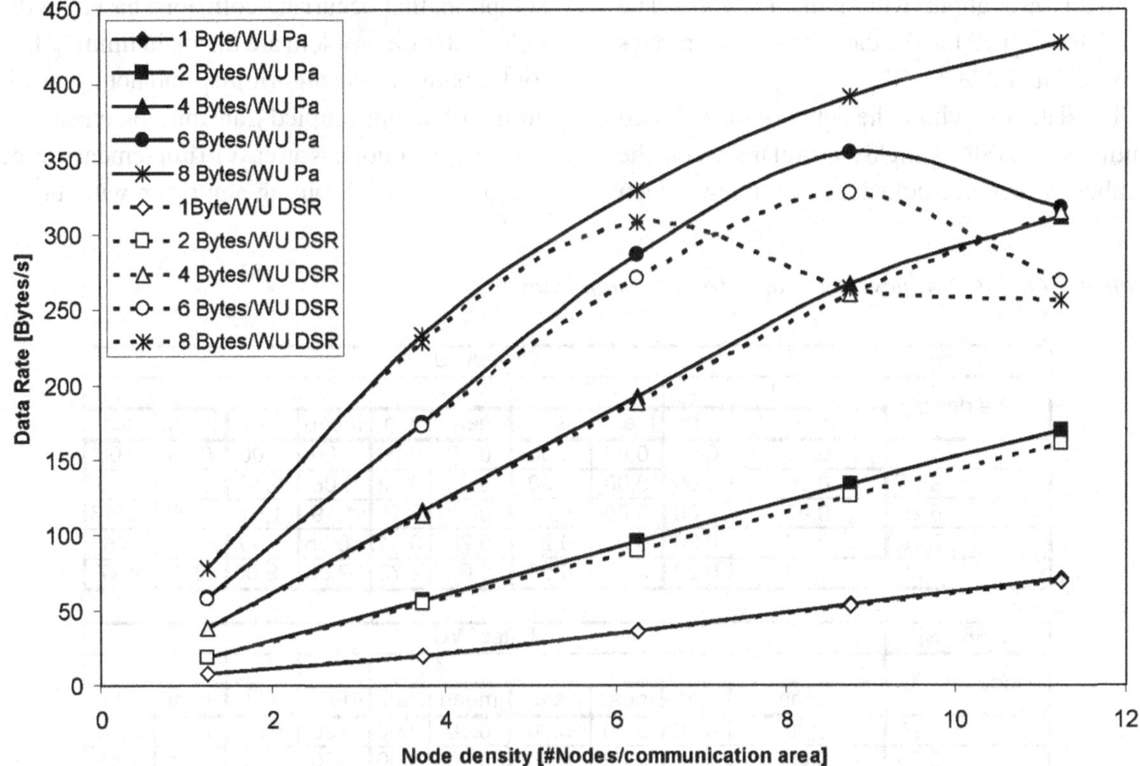

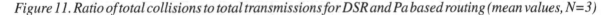

Figure 11. Ratio of total collisions to total transmissions for DSR and Pa based routing (mean values, N=3)

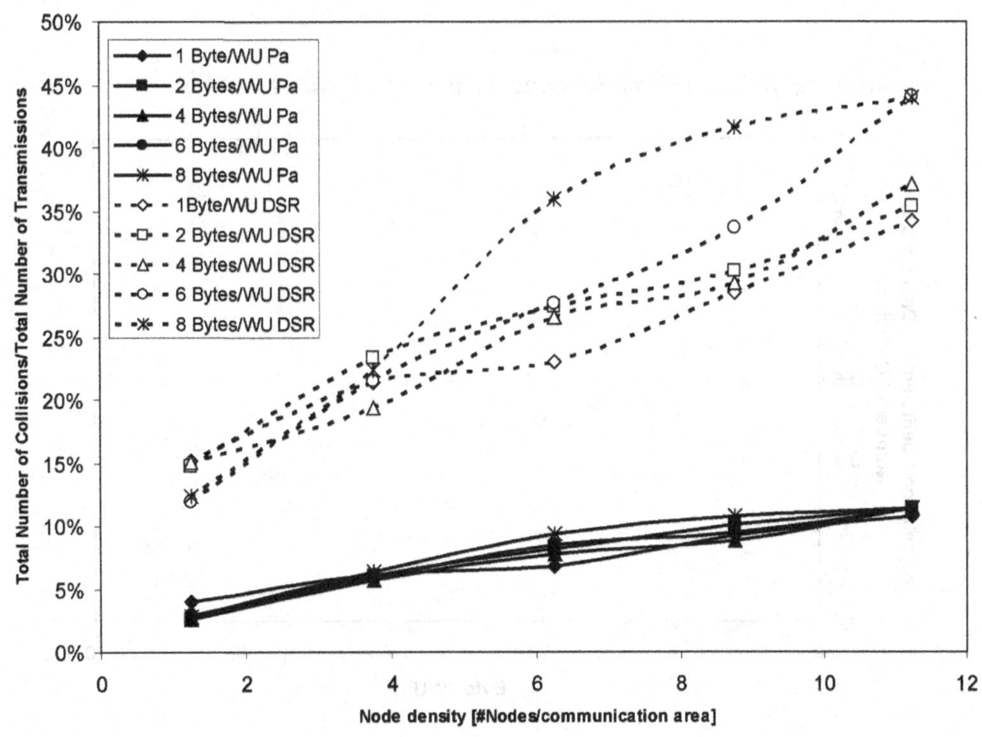

the achievable average data transmission rate and the data throughput within the network. The simulated values for the data transmission rates are given in Table 4.

The data loss within the network for P_a based routing and DSR is highly correlated with the number of message collisions occurring, which is depicted in Figure 12. This validates the assumption, that occurring collisions based on the hidden station problem are the main limiting factor for both routing algorithms, and not the availability of an unoccupied transmission channel.

P_a based routing is an easy to implement source based on-demand routing algorithm without the

Table 4. Simulated values for the data transmission rates

a)

Pa [%]	Bytes/WU									
Node density	1		2		4		6		8	
	mean	std	mean	std	mean	std	mean	std	mean	std
1,25	0,00	0,00	0,00	0,00	0,00	0,00	0,00	0,00	0,00	0,00
3,75	0,00	0,00	0,00	0,00	0,07	0,12	0,08	0,14	0,00	0,00
6,25	0,00	0,00	0,00	0,00	0,10	0,16	1,19	1,61	18,60	16,13
8,75	0,17	0,30	0,30	0,27	0,76	0,31	10,55	3,98	23,83	12,60
11,25	0,13	0,22	0,88	0,53	8,08	4,76	32,66	19,03	33,61	4,27

b)

DSR [%]	Bytes/WU									
Node density	1		2		4		6		8	
	mean	std	mean	std	mean	std	mean	std	mean	std
1,25	0,00	0,00	0,00	0,00	0,00	0,00	0,00	0,00	0,00	0,00
3,75	0,00	0,00	0,00	0,00	0,00	0,00	0,00	0,00	1,24	1,56
6,25	0,00	0,00	0,11	0,20	0,04	0,06	3,99	3,57	17,08	5,74
8,75	0,00	0,00	0,00	0,00	1,18	1,35	14,57	5,91	43,66	3,78
11,25	0,14	0,24	0,00	0,00	5,75	1,67	39,09	4,53	53,86	2,78

Figure 12. Correlation coefficient of data loss and number of collisions

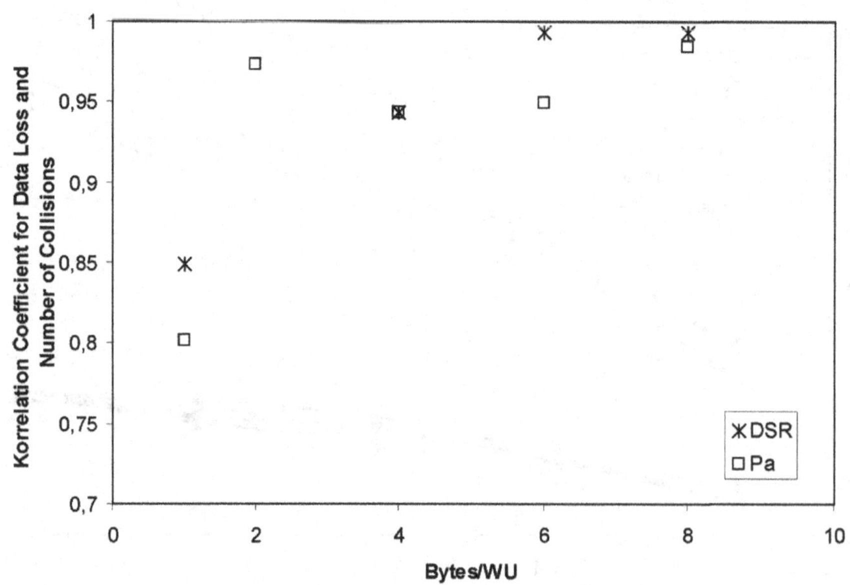

need for any stored routing tables or routing maintenance, and under same preconditions yields a better network performance than DSR. A clear limitation for P_a based routing is given by the data packet forwarding along a dynamic attractiveness gradient which only has its source by the base station. Therefore data packets can only be routed from the data source to the base station and not to other sensor nodes which is generally possible in DSR. Both routing algorithms showed a good data transmission performance for low and medium node densities within the communication range. At high node densities, P_a based routing achieved a higher data throughput as well as a higher average data transmission rate and for all node densities a lower collision to transmission ratio. In DSR, the entries in the node routing tables are dynamic and were optimized by listening to neighboring nodes for possible new routes with a lower hop count to the base station. In a standard DSR implementation, which we have used, the route optimization factor is given by minimizing the number of hops from the actual node to the base station. This can result in routing trees, where the data traffic concentrates at a set of root nodes, allocated with very high traffic. If the root nodes of two or more trees are within its communication range, the hidden station problem will occur, frequently resulting in a substantially higher number of data collisions. In P_a based routing, the route from the actual node to the base station is not given by a routing cache entry. The next hop for a data transmission toward the base station is chosen by asking all neighboring nodes about their attractiveness value which is weighted by several node dependent factors (buffer fill level, radio receive strength and energy level). Nodes which underlay high data traffic typically have a higher buffer memory fill level and a lower energy capacity of the battery resulting in a reduced node attractiveness level. Therefore, P_a based routing does not strictly direct data packets along routes with the lowest hop count to the base station but also chooses longer routes where the data traffic at the participating nodes is moderate. In P_a based routing, the data traffic is better distributed within the network resulting in a more balanced traffic load to each node and a lower number of data collisions.

HARDWARE DESIGN OF THE SENSOR NODES

We have designed the sensor nodes for environmental monitoring, for measuring the environmental temperature and humidity. Therefore, there are special requirements of the hardware:

- Waterproof housing in order to protect the node from environmental influences.
- Lightweight and small to ensure simple attachment to the sensors.
- Low production cost.
- Energy efficiency for a long battery lifetime.

The heart of the node is an ultra low power microcontroller MSP430F149 (Texas Instruments, USA) which has 60K bytes of programmable flash and 2K bytes of data memory. The microcontroller is designed to be battery operated for use in extended-time applications. There are different operation modes where the power consumption can be controlled and optimized (e.g., in stand-by mode: I_{cc}=1.6μA). The radio transceiver on the node is a model CC2420 (Texas Instruments, USA), a single-chip 2.4 GHz IEEE 802.15.4 (ZigBee) (IEEE, 2003) compliant RF transceiver designed for low-power and low-voltage wireless

Figure 13. Functional block diagram of a sensor node

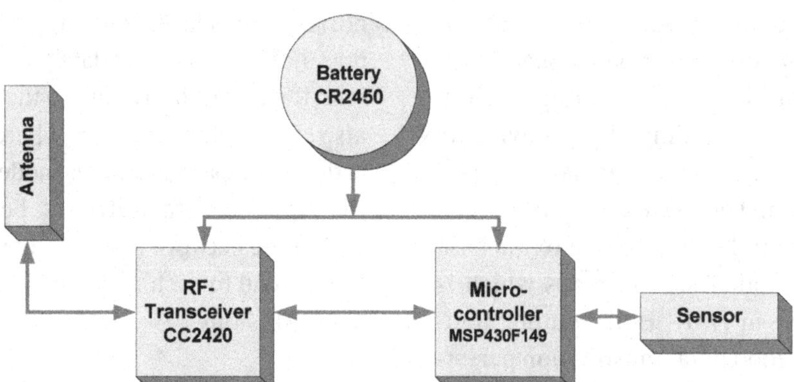

applications. The CC2420 includes a digital direct sequence spread spectrum baseband modem providing a spreading gain of 9 dB and an effective data transmission rate of 250 kbps (Texas Instruments, 2006).

The nodes have three working modes: receiving, transmitting and sleep, drawing an input current of about 18mA, 19mA and 1.6μA respectively. The very low idle current consumption in the sleep mode is obtained by switching off the power supply of the CC2420 RF-transceiver and leaving only the microcontroller running in a low power mode (Figure 13).

E. Technical specifications of a sensor node

The described sensor nodes have a base area of 14 cm^2 and an overall high of 4.5 cm (including SMA socket, but excluding the antenna). The node is powered by a lithium coin battery CR2450 with a typical energy capacity of 620 mAh. The nodes are designed to be suitable for different kinds of external sensors. Therefore a set of analog and digital I/O ports of the microcontroller are routed to the border of the PCB for an easy connection of several sensors via contact pads (Figure 14).

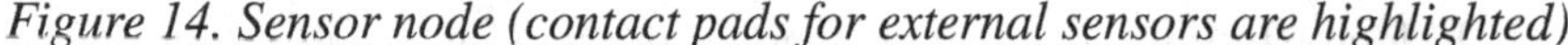

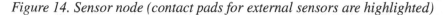

Figure 14. Sensor node (contact pads for external sensors are highlighted)

CONCLUSION

In this paper, a routing protocol for wireless sensor networks is examined. We observed that a WSN is a network for the surveillance of a geographic area with small, robust sensing devices, called sensor nodes. The nodes have restrictions regarding energy, memory, computing power, range and size. A simplified on-demand routing protocol is introduced, especially for WSNs called attractiveness-based routing, using an attractiveness-metric gradient for data forwarding in the WSN. After a detailed explanation of the P_a algorithm and simulations, a practical implementation on a MSP430F149 microcontroller together with a CC2420 RF-transceiver is demonstrated. The P_a based routing algorithm is well suited for easy to set up WSNs because of its simplicity of implementation and its adaptability to different scenarios by adjustable weighting factors for the node's attractively metric. The proposed routing algorithm is mainly investigated for WSNs, with a limited number of neighboring nodes (number of nodes within the communication range) and is typically applicable for environmental monitoring, medical homecare and machine to sensor interfaces.

ACKNOWLEDGMENT

The authors would like to thank the government of Lower Austria and the European Commission (EFRE) for their support of the project (Project ID: WST3-T-91/004-2006).

REFERENCES

Di Caro, G., & Dorigo, M. (1997). *AntNet: A mobile agents approach to adaptive routing* (Tech. Rep. No. IRIDIA/97-12). Brussels, Belgium: Universite Libre de Bruxelles.

Di Caro, G., & Dorigo, M. (1998a). Ant colonies for adaptive routing in packet switched communications networks. In *Proceedings of the 5th International Conference on Parallel Problem Solving from Nature* (pp. 673-682).

Di Caro, G., & Dorigo, M. (1998b). AntNet: Distributed stigmergetic control for communications networks. *Journal of Artificial Intelligence Research*, 317–365.

Dorigo, M., & Di Caro, G. (1991). The ant colony optimization meta-heuristic. In Corne, D., Dorigo, M., Glover, F., Dasgupta, D., Moscato, P., Poli, R., & Price, K. V. (Eds.), *New Ideas in Optimization*. London, UK: McGraw Hill.

Dorigo, M., Maniezzo, V., & Colorni, A. (1991). *The Ant System: An Autocatalytic Optimizing Process* (Tech. Rep. No. 91-016). Milan, Italy: Politecnico di Milano.

Heissenbuettel, M. (2005). *Routing and Broadcasting in Ad Hoc Networks* (Unpublished doctoral dissertation). University Bern, Bern, Switzerland.

IEEE Standards Association. (2003). *IEEE Standard 802.15.4*. Washington, DC: Author.

IEEE Standards Association. (2007). *IEEE 802.11: Wireless LAN Medium Access Control (MAC) and Physical Layer (PHY) Specifications* (Rev. ed.). Washington, DC: Author.

Jacquet, P., Muhlethaler, P., & Qayyum, A. (2003). *Optimized link state routing (OLSR) protocol.* Retrieved from http://www.ietf.org/rfc/rfc3626.txt

Johnson, D., & Maltz, D. (1994). Dynamic Source Routing in Ad Hoc Wireless Networks. In *Proceedings of the Conference on Mobile Computing Systems and Applications*, Santa Cruz, NM (pp. 158-163).

Johnson, D. B. (1994). Routing in Ad Hoc Networks of Mobile Hosts. In *Proceedings of the Workshop on Mobile Computing Systems and Applications*, Santa Cruz, NM (pp. 158-163).

Kos, A. (2006). *Management of Selforganizing Wireless Sensor Networks* (Unpublished master's thesis). Danube University Krems, University of Applied Sciences St. Poelten, Krems, Austria.

Liu, L., & Feng, G. (2005). Swarm Intelligence Based Node disjoint Multipath Routing Protocol for Mobile Ad Hoc Networks. In *Proceedings of the 5th International Conference on Information, Communications and Signal Processing*, Bangkok, Thailand.

Murthy, S., & Garcia-Luna-Aceves, J. (1996). An Efficient Routing Protocol for Wireless Networks. *ACM Mobile Networks and Applications, 1*(2), 183–197. doi:10.1007/BF01193336.

Niazi, M., & Hussain, A. (2009). Agent-based tools for modeling and simulation of self-organization in peer-to-peer, ad hoc, and other complex networks. *IEEE Communications Magazine, 47*(3), 166–173. doi:10.1109/MCOM.2009.4804403.

Park, V., & Corson, M. (1997). A Highly Adaptive Distributed Routing Algorithm for Mobile Wireless Networks. In *Proceedings of the IEEE International Conference INFOCOM.*

Perkins, C., & Bhagwat, P. (1994). Highly dynamic Destination-Sequenced Distance-Vector routing (DSDV) for mobile computers. In *Proceedings of the SIGCOM Conference on Communications Architecture, Protocols and Applications* (pp. 234-244).

Perkins, C., & Royer, E. (1999). Ad-Hoc On-Demand Distance Vector Routing. In *Proceedings of the 2nd IEEE Workshop Mobile Computing Systems and Applications*, New Orleans, LA.

Singh, R., Singh, H., & Kaler, R. S. (2010). An Adaptive Energy Saving and Reliable Routing Protocol for Limited Power Sensor Networks. In *Proceedings of the International Conference on Advances in Computer Engineering* (pp. 79-85).

Texas Instruments. (2006). *Datasheet CC2420, 2.4 GHz IEEE 802.15.4 / ZigBee-ready RF Transceiver*. Dallas, TX: Author.

Tisue, S., & Wilensky, U. (2004). NetLogo: A Simple Environment for Modeling Complexity. In *Proceedings of the International Conference on Complex Systems*, Boston, MA.

Wilensky, U., & Stroup, W. (1999). Learning through Participatory Simulations: Network based Design for Systems Learning in Classrooms. In *Proceedings of the Computer Supported Collaborative Learning Conference*, Stanford, CA.

Woo, A., Tong, T., & Culler, D. (2003). Taming the Underlying Challenges of Reliable Multihop Routing in Sensor Networks. In *Proceedings of the 1st International Conference on Embedded Networked Sensor Systems* (pp. 14-27).

Ye, W., Heidemann, J., & Estrin, D. (2002). An energy-efficient MAC protocol for wireless sensor networks. In *Proceedings of the 21st IEEE Conference on Computer and Communications Societies* (Vol. 3, pp. 1567-1576).

This work was previously published in the International Journal of Wireless Networks and Broadband Technologies, Volume 1, Issue 3, edited by Naveen Chilamkurti, pp. 49-70, copyright 2011 by IGI Publishing (an imprint of IGI Global).

Chapter 17
Wireless Transport Layer Congestion Control Evaluation

Sanjay P. Ahuja
University of North Florida, USA

W. Russell Shore
University of North Florida, USA

ABSTRACT

The performance of transport layer protocols can be affected differently due to wireless congestion, as opposed to network congestion. Using an active network evaluation strategy in a real world test-bed experiment, the Transport Control Protocol (TCP), Datagram Congestion Control Protocol (DCCP), and Stream Control Transport Protocol (SCTP) were evaluated to determine their effectiveness in terms of throughput, fairness, and smoothness. Though TCP's fairness was shown to suffer in wireless congestion, the results showed that it still outperforms the alternative protocols in both wireless congestion, and network congestion. In terms of smoothness, the TCP-like congestion control algorithm of DCCP did outperform TCP in wireless congestion, but at the expense of throughput and ensuing fairness. SCTP's congestion control algorithm was also found to provide better smoothness in wireless congestion. In fact, it provided smoother throughput performance than in the network congestion.

1. INTRODUCTION

With the rise of wireless networking, performance expectations have steadily increased as technology has consistently improved. Similarly, internet applications that depend on streaming timeliness centric data in the midst of congestion, e.g., streaming movies, internet radio, online gaming, etc., have taken the forefront in pushing the limits of wireless and other networking technologies. Congestion based limitations can result from the finite bandwidth and erratic loss due to the characteristics of the physical medium. The protocol layer that deals with these limitations and transporting data in the midst of these challenges is the transport layer.

The most popular protocol in this layer, TCP, has been thoroughly investigated in this context. As a result, several new protocols have been

DOI: 10.4018/978-1-4666-3902-7.ch017

developed and standardized to implement new strategies and features for dealing with the changing demands of these internet applications. The protocols considered in this article include DCCP and SCTP which approach this task in different ways, but which both attempt to provide the necessary functionality to facilitate further growth in performance.

DCCP and SCTP

The Datagram Congestion Control Protocol (DCCP) was designed for applications that need connection based congestion control without reliability (Kohler et al., 2006). The most significant feature of DCCP is its modular configurability of the congestion control algorithm. Depending on the network environment and the requirements of the application, it can handle congestion differently depending on which standardized congestion control algorithm it uses. These algorithms are denoted by their Congestion Control Identifiers (CCID). The CCID's standardized so far include CCID2, which is designed to behave TCP-like with several of the same mechanisms as TCP, and CCID3, which is designed to be TCP-friendly but to control congestion based on an equation so that traffic is affected less rapidly but more smoothly (Kohler et al., 2006). The Stream Control Transport Protocol (SCTP) has similar congestion control mechanisms as TCP including the SACK extension, slow start, congestion avoidance, and fast retransmit (Ye et al., 2002). Though the congestion control mechanisms of these protocols are all related to TCP and focus on interaction with TCP traffic, they each vary in specific ways that can affect how they interact with wireless congestion. It is these variations that will provide different behaviors and performance in different congestion environments.

2. MOTIVATIONS AND RELATED WORKS

Several authors have investigated the individual performance of the transport layer protocols in question. Holland and Vaidya (1999) focused on fairness, a congestion control metric considered in this article. The fairness of TCP in particular was proven to be lacking in wireless networks. Takeuchi et al. (2005), however, points out DCCP's own limitations in fairness when sharing a bottleneck with TCP. Simulation has also been performed using DCCP focusing on throughput smoothness, another metric adopted by this article, where DCCP was shown to be strong in this arena by Navaratnam et al. (2006). Rhee and Xu (2007) analyzes the TFRC congestion control algorithm to determine its limitations. They then verify their findings using ns2 simulations. Active wireless experimentation has been performed with TCP (Bruno et al., 2007) and DCCP (De Sales et al., 2008) where these transport protocols are evaluated in wireless networks similarly to the methodology adopted in this article. SCTP has also been experimented with using heterogeneous network environments using a network evaluation tool designed for this purpose by Emma et al. (2006).

At the time of this writing, the literature was lacking consistent experimental evaluation that included all the aforementioned protocols. This is what inspired the authors to utilize the evaluation techniques used in the related articles and go one step further to perform an active evaluation of the currently available transport layer protocols in a single consistent experimental test-bed environment. This goal was pursued in the master's thesis by Shore (2010) and the results were a baseline comparison of the characteristics of these protocols and their behavior in response to wireless and network congestion. The goal for this article is to now publish the thesis results and lay the groundwork for evaluation of congestion control algorithms in the varying network environments.

3. EXPERIMENTATION

To get the most information about the real-world characteristics of the experimentation protocols and the effects of wireless congestion, the chosen method of experimentation was a real-time active network performance evaluation. In this type of evaluation strategy, consistency in the hardware and software environment is essential for the accuracy and credibility of the experimental results. All experimental factors other than the ones under investigation needed to be taken into consideration to prevent skewing of the experimental data. The methodology of the experimentation in the congestion environment is listed in detail to provide a basis for future experimentation in this area.

Software Environment

To facilitate experimentation in the software side of the experimental environment, certain minimum functionality was necessary. The operating system needed to support network communications with implementations of all the experimentation protocols in a consistent manner. With the popularity of the protocols, several operating systems with implementations were available. Software utilities were also necessary for several critical functionalities to produce the evaluation metric data for analysis including traffic generation, performance tracking, experimental timing automation, and sequence scripting.

The Netbook remix of Ubuntu 9.10, a distribution of the Debian Linux kernel, was selected for its consistent support of all the experimentation protocols. Iperf, an open source command line utility capable of traffic generation and measurement of throughput, delay, and jitter, was chosen as the active network evaluation tool due to the availability of distributions that incorporate the experimentation protocols (Linux Foundation, 2009; Open SS7, 2008). The chosen software environment also provided the Cron utility that was used in the experimentation to automate the test sequences by executing test scripts at regular intervals. For a consistent environment configuration, the experimentation system was prepared on a bootable flash drive. The flash drive was then cloned to several other flash drives for deployment across all of the other experimentation machines. This provided as consistent a software environment configuration as possible.

Hardware Environment

In order to observe the effects of congestion, two competing streams were necessary. This required the use of three machines, one to act as server, and the others to act as competing clients. By manipulating the traffic on these two competing streams, congestion could be introduced in a controlled manner so that measurements could be taken as the two streams competed for a single bottleneck throughput. Performing these sequences across the two physical layer technologies, a differential analysis can be performed on the resulting data to extract the effect wireless congestion had on the protocols undergoing the experiment. The specifications for the equipment used in the experimentation and the two physical layer configurations are listed in Table 1 and Table 2.

Table 1. Wireless experimental device specifications

Device Type	Model	Network Card
Laptop	Acer Aspire 6920	Intel Wireless WiFi Link 4965 AGN
Laptop	HP dv4t 1300	Intel Wireless WiFi Link 5100 AGN
Laptop	Asus EeePC 1005HA	Atheros Wireless WiFi Link AR9285
Router	Belkin F5D8236	N/A

Table 2. Wired experimental device specifications

Device Type	Model	Network Card
Laptop	Acer Aspire 6920	Atheros Gigabit Ethernet
Laptop	HP dv4t 1300	Realtek RTL8102E/8111C Family Gigabit Ethernet
Laptop	Asus EeePC 1005HA	Atheros AR8132 Fast Ethernet
Router	Belkin F5D8236	Fast Ethernet

Isolation

Since the effect of wireless congestion on protocol performance is the focus of this article, the isolation and control of wireless and network congestion are paramount to the success of the experimentation and any conclusions drawn thereof. A means of preventing external sources of interference from affecting the measurements and their derived results was necessary. For the Ethernet networking configuration, isolation was attained electrically by not connecting the system to any other networks. For the wireless networking configuration, the task was a little more complicated. A Faraday cage was constructed to isolate the wireless test environment from possible external wireless interference. This isolation enabled the controlled application of congestion in both environments.

Sequencing

The test sequences were performed in the following fashion. First, one stream was started and allowed to stabilize in the current test environment for a period of sixty seconds. At that point, the second competing stream was started to illustrate the shift in performance of the two protocols and their associated interactions. This stream was then allowed to continue for another sixty seconds until the two protocols had stabilized. Then, the second stream was closed and the original stream was allowed to stabilize in isolation once again. This test sequence provided insight into not only the interaction of the two streams, but also the ability of the protocols to compensate for the creation and removal of competing streams.

Experimental Variables

The sequences were conducted in four differing environments to produce evidence across two experimental variables. The first variable was the physical medium across which the sequences were performed. It was assumed the performance during these sequences would not only provide evidence of the effect the physical medium had on the performance of the test protocols, but they would also give insight into the capability of the test protocols to adapt to the characteristics of the experimental medium.

By beginning a TCP stream of traffic ahead of another stream, it was assumed it would have an advantage over any potential competitor streams started subsequently. The original TCP stream would be well established before another competing stream was started. Since this research focuses on the interaction of the test protocols with the established standard of TCP, thus the second variable was the order in which the streams of traffic were established. By varying whether TCP or the test protocol was the established protocol, the characteristics of the test protocols would be better highlighted.

The first two sequence sets consist of the establishment of a TCP stream in the background and after it was established, a second competing stream was started of the protocol under examination, as illustrated in Figure 1. This provided a baseline for the available bandwidth that TCP was able to utilize, as well as the smoothness of the TCP stream when isolated in the physical medium. Once the test protocol stream was begun, this provided insight into how the two protocols interacted. In addition, once the test stream com-

Figure 1. Test protocol on background TCP

pleted, the remainder of the sequence illustrated how well the background TCP protocol was able to readjust back to its previous steady state.

The next two sequence sets consisted of the establishment of a test protocol stream in the background, followed by a second competing TCP stream, as illustrated in Figure 2. This gave a baseline for the bandwidth the test protocol was able to utilize, as well as the smoothness of the test protocol stream when isolated in the physical medium. Further processing of these sequences of data allowed the extraction of a mean value over the sequences during certain periods of interest.

Figure 2. TCP on background test protocol

The control for the experiment was a set of TCP vs. TCP sequences using the exact same tools and environments as the test protocols for each of the four sequence types. This was particularly effective since TCP is well known as the standard for internet applications. This sequence also served a dual purpose. It had the potential to uncover any experimental differences in the hardware used to conduct the experimentation.

Metrics

The selected metrics by which the protocols in question were evaluated consisted of raw throughput as measured and recorded by Iperf at set consistent intervals throughout the test sequences of generated competing traffic. From the concurrent raw throughput measurements, the metrics were derived to index the performance of the competing protocols in terms of fairness and smoothness.

Fairness as defined by Jain (1991) can be found through analysis of concurrent throughput measurements as

$$f\left(x_1, x_2, \ldots, x_n\right) = \frac{\left(\sum_{i=1}^{n} x_i\right)^2}{n\sum_{i=1}^{n} x_i^2}$$

where n is the number of throughput measurements, $x_1, x_2, \ldots, x_n$ are the concurrent throughput measurements, and $f(x_1, x_2, \ldots, x_n)$ is the fairness index for that set of concurrent throughput measurements. Smoothness as discussed in Navaratnam et al. (2006) can be found through the analysis of subsequent throughput measurements as

$$s\left(x_i,\ x_{i-1}\right) = \min\left(\frac{x_i}{x_{i-1}}, \frac{x_{i-1}}{x_i}\right)$$

where x_i and x_{i-1} are subsequent throughput measurements and $s\ (x_i, x_{i-1})$ is the smoothness index.

4. EXPERIMENTAL RESULTS AND DISCUSSION

By averaging the indexes of fairness and smoothness during the competition period, the sequences could be compared across the experimental variables. Several interesting conclusions as to the effect of wireless congestion and stream establishment on the test protocols in question could then be drawn. The conclusions in each measurement dimension are as follows.

Wireless vs. Wired Fairness

The fairness of the alternative protocols, DCCP using CCID2, DCCP using CCID3, and SCTP were already lower than TCP's when subjected to network congestion. When subjected to wireless congestion, their fairness was not affected. TCP's fairness when subjected to wireless congestion, however, decreased to the point that it became comparable to the fairness of the alternative protocols. This is in accordance with the fairness research mentioned in the literature review, which concludes that TCP's fairness suffers in a wirelessly congested environment. The fairness comparisons are illustrated in Figure 3 and Figure 4.

Figure 3. Jain fairness comparison with established tcp stream

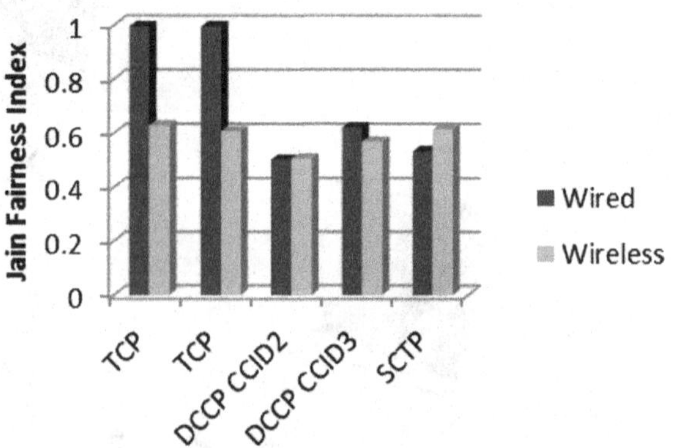

Figure 4. Jain fairness comparison with established test stream

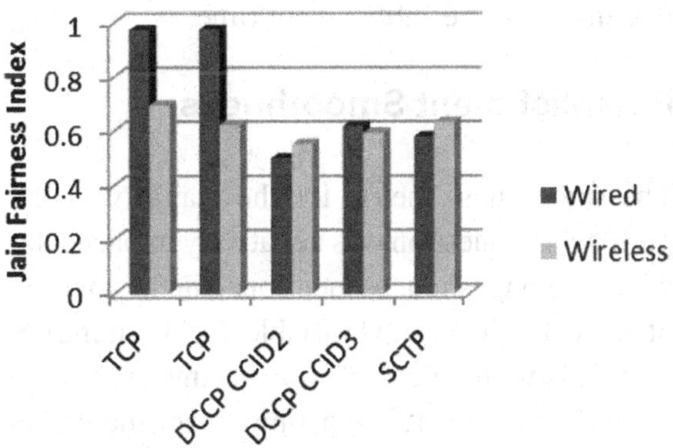

Wireless vs. Wired Smoothness

Smoothness was negatively impacted by wireless congestion for almost all of the protocols in question, as seen in Figure 5, Figure 6, Figure 7, and Figure 8. The two protocols that stood out, DCCP using CCID2 and SCTP, both behaved a little differently. The smoothness of DCCP using CCID2 was not as impacted by wireless congestion as the other protocols, as seen in Figure 6 and Figure 7. This is peculiar, as its congestion control algorithm should be TCP-like, and as such have the same algorithmic deficiencies as TCP. Since DCCP using CCID3 is TCP-friendly and equation based, it was surprising that this protocol was affected as much as it was.

Figure 5. Background smoothness comparison with established tcp stream

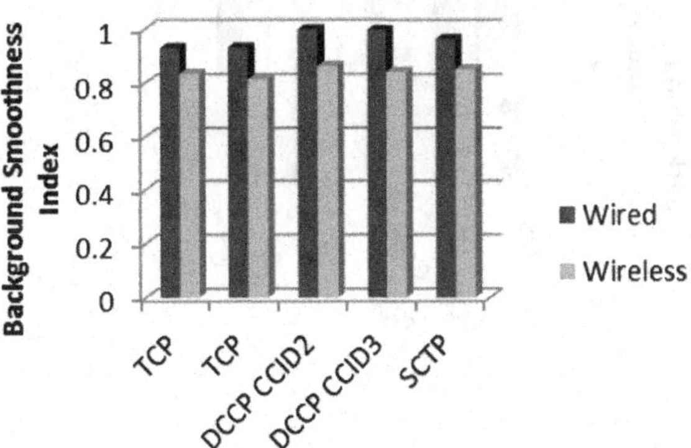

Figure 6. Background smoothness comparison with established test stream

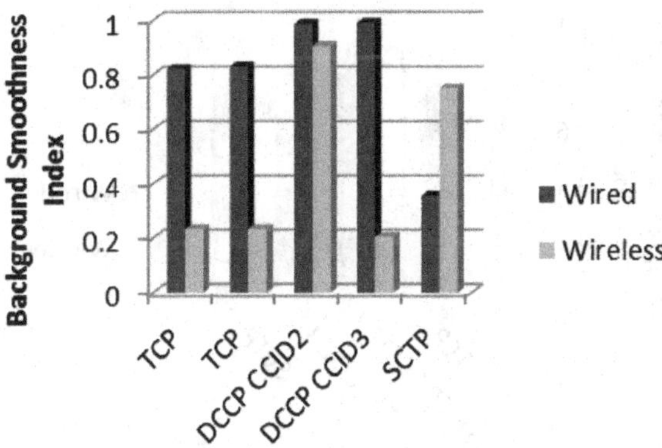

Figure 7. Foreground smoothness comparison with established tcp stream

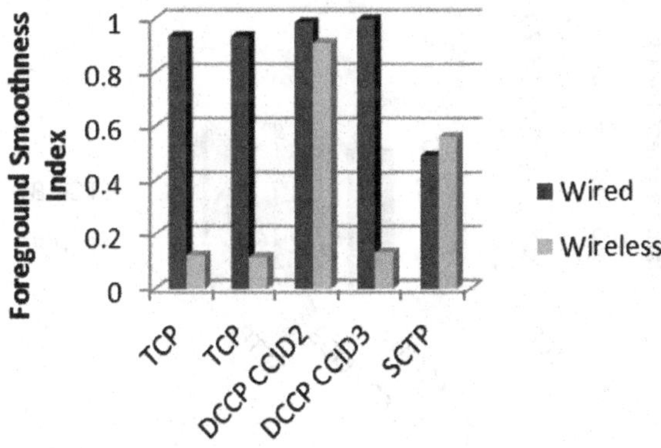

The most surprising results came from the SCTP smoothness analysis. SCTP actually performed smoother when subjected to wireless congestion.

Establishment Fairness

Fairness was not impacted by establishment time for all of the protocols in question. This comparison is illustrated in Figure 9 and Figure 10. This demonstrates the time invariance of the congestion control algorithms used by the protocols in question. This is significant as this characteristic is what allows competing streams to start at any point in time and still be treated with the same

Figure 8. Foreground Smoothness comparison with established test stream

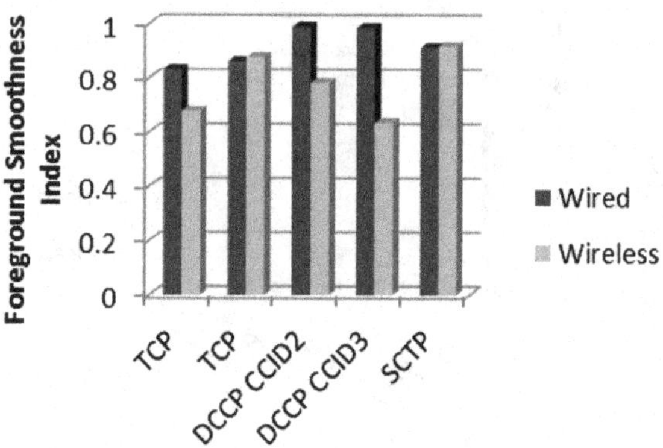

Figure 9. Jain fairness comparison with wired congestion

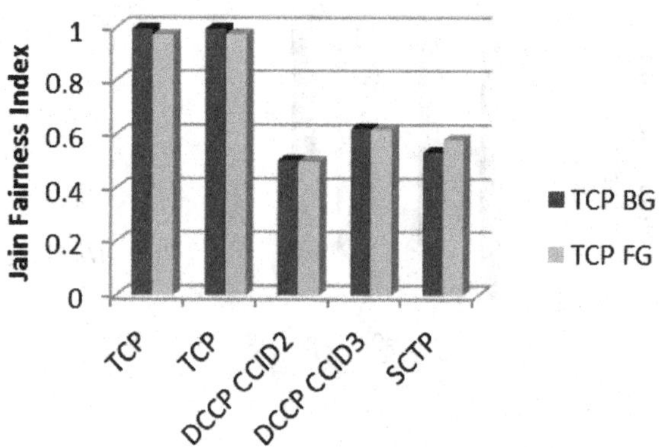

Figure 10. Jain fairness comparison with wireless congestion

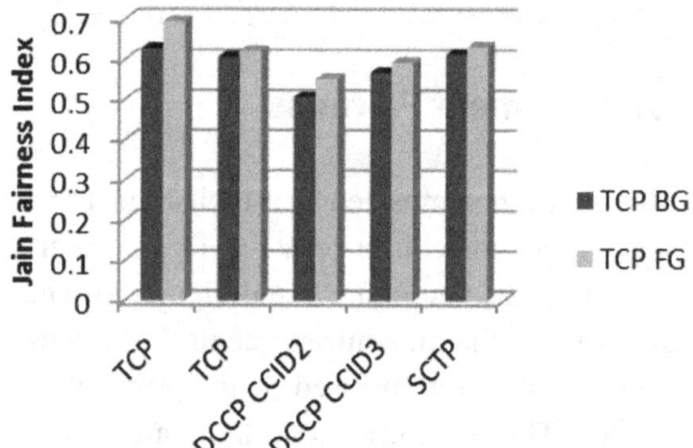

fairness. It also verifies that the experiment does not need to have competing streams start at varying amounts of establishment time.

Establishment Smoothness

The smoothness metric for the majority of the protocols in question was negatively impacted by wireless congestion, as mentioned in the analysis of the first experimental variable. By providing the protocols with an establishment time advantage before beginning the competing stream, the effects of the wireless congestion on the smoothness performance were amplified. The two protocols that stood out before, DCCP using CCID2 and SCTP, both made an impact in this experimental variable. The smoothness of DCCP using CCID2 was also not as impacted by establishment as the other protocols in all of the sequences, as shown in Figures 11, Figure 12, Figure 13, and Figure 14. The SCTP comparison in the wired environment was the only series affected by an establishment advantage, as seen in Figure 11 and Figure 13. This illustrates the significance of establishment time to the smoothness of the SCTP congestion control algorithm in wired environments.

Figure 11. Background smoothness comparison with wired congestion

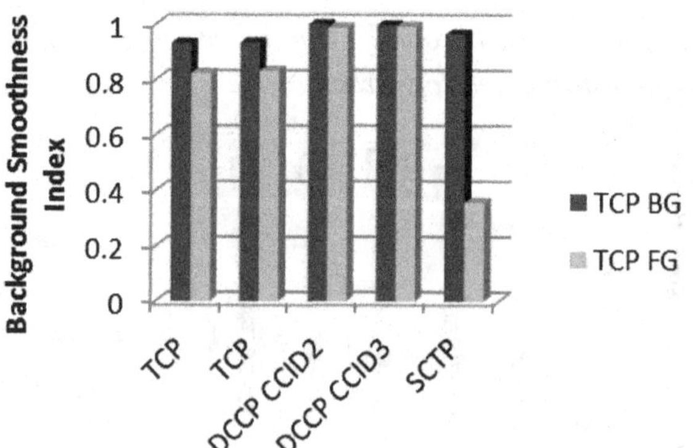

Figure 12. Background smoothness comparison with wireless congestion

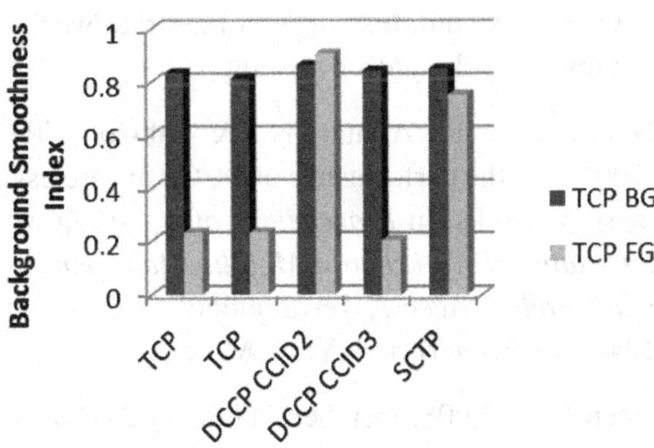

Figure 13. Foreground smoothness comparison with wired congestion

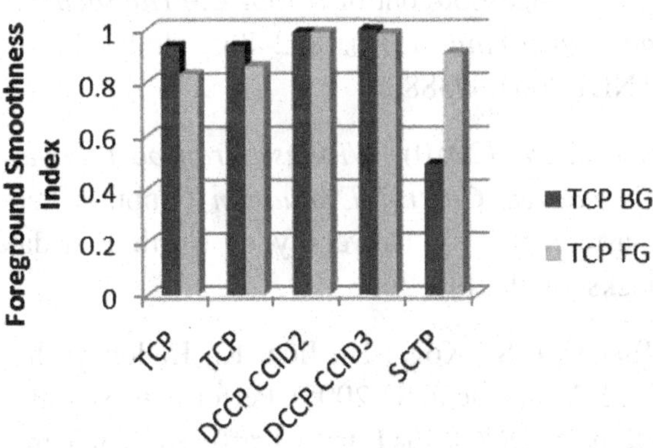

Figure 14. Foreground Smoothness comparison with wireless congestion

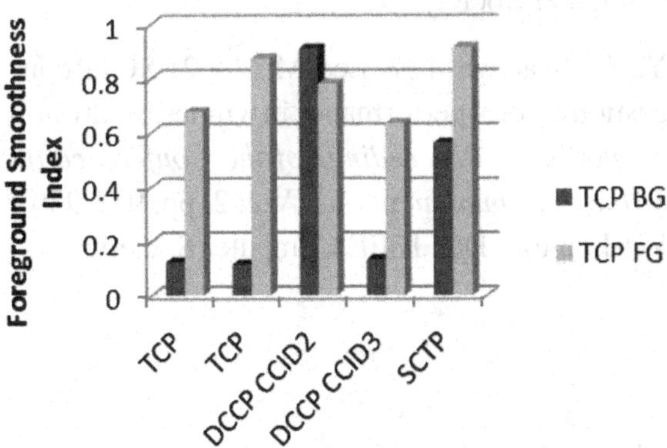

5. CONCLUSION

In this paper we presented an empirical series of experiments comparing the congestion control of several transport layer protocols using their Linux implementations in wireless and wired network environments. TCP's fairness concerns were verified when its performance in this metric was found to suffer in wireless congestion. However, the results showed that it still outperformed the alternative protocols in both experimental scenarios. In actuality, the fact that TCP outperformed the other protocols in fairness when exposed to purely network congestion was of more significance and illustrates the overly submissive behavior of the alternative protocols. For the smoothness metric, the TCP-like congestion control algorithm of DCCP was found to outperform all the others in wireless congestion, but at the expense of throughput and ensuing fairness, again highlighting its submissive behavior. Perhaps this transport protocol/ congestion control algorithm would be an ideal candidate for applications where fairness is not essential, but smoothness is paramount. Another interesting discovery was that SCTP's congestion control algorithm was found to provide better smoothness in wireless congestion than in network congestion. This illustrates SCTP's focus on wireless applications. The result of this study could enable network protocol performance profiling to discern the best algorithms and strategies for the individual requirements of innovative internet applications. The strategic development and usage of these protocols and their congestion control algorithms will hopefully stimulate the future growth of the internet and its ever-expanding world of applications, even in the wireless world ahead.

REFERENCES

Bruno, R., Conti, M., & Gregori, E. (2007). Throughput Analysis and Measurements in IEEE 802.11 WLANs with TCP and UDP Traffic Flows. *IEEE Transactions on Mobile Computing*, 171–186.

De Sales, L. M., Almeida, H. O., & Perkusich, A. (2008). On the performance of TCP, UDP and DCCP over 802.11 g networks. In *Proceedings of the ACM Symposium on Applied Computing*, Fortaleza, Ceara, Brazil (pp. 2074-2078). New York, NY: ACM.

Emma, D., Loreto, S., Pescape, A., & Ventre, G. (2006). Measuring SCTP Throughput and Jitter over Heterogeneous Networks. In *Proceedings of the 20th International Conference on Advanced Information Networking and Applications* (Vol. 2, pp. 395-399). Washington, DC: IEEE Computer Society.

Holland, G., & Vaidya, N. (1999, August). Analysis of TCP performance over mobile ad hoc networks. In *Proceedings of the 5th Annual ACM/IEEE International Conference on Mobile Computing and Networking* (pp. 219-230). New York, NY: ACM.

Jain, R. (1991). *The Art of Computer Systems Performance Analysis: Techniques for Experimental Design, Measurement, Simulation, and Modeling*. New York, NY: John Wiley & Sons.

Kohler, E., Handley, M., & Floyd, S. (2006). Designing DCCP: congestion control without reliability. In *Proceedings of the Conference on Applications, Technologies, Architectures, and Protocols for Computer Communication* (pp. 27-38). New York, NY: ACM.

Linux Foundation. (2009, November 19). *DCCP Testing*. Retrieved August 8, 2010, from http://www.linuxfoundation.org/ collaborate/workgroups/networking/dccp_testing

Navaratnam, P., Akhtar, N., & Tafazolli, R. (2006). On the performance of DCCP in wireless mesh networks. In *Proceedings of the 4th ACM international Workshop on Mobility Management and Wireless Access*, Terromolinos, Spain (pp. 144-147). New York, NY: ACM.

Open SS7. (2008, October 31). *Iperf 2.0.8 Released*. Retrieved August 8, 2010, from http://www.openss7.org/ rel20081029_1.html

Rhee, I., & Xu, L. (2007). Limitations of equation-based congestion control. *IEEE/ACM Transactions on Networking*, *15*(4), 852–865. doi:10.1109/TNET.2007.893883.

Shore, W. (2010). *Wireless Transport Layer Congestion Control Evaluation* (Unpublished master's thesis). University of North Florida, Jacksonville, FL.

Takeuchi, S., Koga, H., Iida, K., Kadobayashi, Y., & Yamaguchi, S. (2005). Performance Evaluations of DCCP for Bursty Traffic in Real-Time Applications. In *Proceedings of the IEEE/IPSJ International Symposium on Applications and the Internet* (pp. 142-149). Washington, DC: IEEE Computer Society.

Ye, G., Saadawi, T., & Lee, M. (2002). SCTP congestion control performance in wireless multi-hop networks. In *Proceedings of the Conference on Military Communications* (Vol. 2, pp. 934-939). Washington, DC: IEEE Computer Society.

This work was previously published in the International Journal of Wireless Networks and Broadband Technologies, Volume 1, Issue 3, edited by Naveen Chilamkurti, pp. 71-81, copyright 2011 by IGI Publishing (an imprint of IGI Global).

Chapter 18
Co-Operative Load Balancing in Vehicular Ad Hoc Networks (VANETs)

G. G. Md. Nawaz Ali
City University of Hong Kong, Hong Kong

Edward Chan
City University of Hong Kong, Hong Kong

ABSTRACT

Recently data dissemination using Road Side Units (RSUs) in Vehicular Ad Hoc Networks (VANETs) received considerable attention for overcoming the vehicle to vehicle frequent disconnection problem. An RSU becomes overloaded due to its mounting location and/or during rush hour overload. As an RSU has short wireless transmission coverage range and vehicles are mobile, a heavily overloaded RSU may experience high deadline miss rate in effect of serving too many requests beyond its capacity. In this work, the authors propose a co-operative multiple-RSU model, which offers the opportunity to the RSUs with high volume workload to transfer some of its overloaded requests to other RSUs that have light workload and located in the direction in which the vehicle is heading. Moreover, for performing the load balancing, the authors propose three different heuristic load transfer approaches. By a series of simulation experiments, the authors demonstrate the proposed co-operative multiple-RSU based load balancing model significantly outperforms the non-load balancing multiple-RSU based VANETs model against a number of performance metrics.

INTRODUCTION

The Federal Communications Commission (FCC) allocates the 5.850 to 5.925 GHz frequency band for vehicle to vehicle and vehicle to Road Side Units (RSUs) communication. A number of applications has been envisioned in Vehicular Ad Hoc Networks (VANETs) such as road safety, driving assistance, emergency public service, business, entertainment etc. (Schoch, Kargl, & Leinmüller, 2008), as well as internet service from on board vehicles (Lee, Ernst, & Chilamkurti, 2011). In VANETs, as vehicles are normally on the move, vehicle to vehicle frequent disconnection is very common. Recently a number of researchers have proposed RSU-based VANETs to handle this

DOI: 10.4018/978-1-4666-3902-7.ch018

connectivity problem (Zhang, Zhao, & Cao, 2010; Schoch, Kargl, & Leinmüller, 2008; Lochert, Scheuermann, Caliskan, & Mauve, 2007). In this kind of model, RSUs are placed along roadside to provide vehicle to infrastructure (V2I) connectivity (Liu & Lee, 2010; Chang, Cheng, & Shih, 2007). RSU is a stationary infrastructure which acts like a server, with memory storage, substantial computational capacity and short wireless range transmission system. RSUs are usually placed at locations with high vehicle density such as at the intersection of roads, market places, gas stations, bus terminals etc. The RSU-based VANETs model is very useful during unfriendly VANETs environment (off peak hour, night time etc.) and on highways where vehicle density is usually low (Lochert, Scheuermann, Caliskan, & Mauve, 2007). In our proposed co-operative multiple-RSU model, an RSU acts as a buffer point (Zhang, Zhao, & Cao, 2010) which stores the information that is useful for the vehicles, RSUs are interconnected, know the workload information of each other and can exchange their workloads.

A given RSU may experience heavy workload during the peak hours and some requests may not get serviced due to overload. To alleviate this problem, we propose transferring workload to less heavily loaded RSUs, where each RSU use the DSIN on-demand scheduling algorithm (Zhang, Zhao, & Cao, 2010; Ali, Chan, & Li, 2011) to give priority to delay sensitive and smaller sized popular data items for scheduling the RSU received queue requests and we propose a number of heuristic load transferring approaches for performing the load balancing.

RELATED WORK

Considerable researches have been carried out to find a stable data dissemination infrastructure in highly mobile and sparsely connected VANETs environments (Chen, Kung, & Vlah, 2001; Wu, Fujimoto, Guensler, & Hunte, 2004; Zhao, Zhang, & Cao, 2007; Zhao & Cao, 2008). Nadeem, Shankar, and Iftode (2006) formulate a data push communication model for vehicle to vehicle communication without any road side infrastructure support. Zhang, Zhao, and Cao (2010) provide a single RSU-based VANETs model which maintains both upload and download queues and try to get the service balance among them. Yi, Bin, Tong, and Wei (2008) provide a mesh RSUs infrastructure based VANETs on both space and time dimensions and formulate reliability and fairness based algorithms. Ali, Chan, and Li (2011) analyze the performance of different on-demand scheduling algorithms for incorporating upload queue with download queue in RSU-based VANETs. Chen, Cao, Zhang, Xu, and Sun (2009) study the performance of the effectiveness of certificate revocation distribution in VANETs with and without enabling vehicle to vehicle communication. Liu and Lee (2010) analyze dynamic traffic characteristics in RSU-based VANETs. They propose to use different channels to disseminate different types of data and apply push and pull data dissemination technique based on the volume of requests at RSU server.

Some researchers study the performance of on-demand scheduling algorithms for data dissemination in real-time environments (Lee, Wu, & Ng, 2006; Liu & Lee, 2010; Xu, Tang, & Lee, 2006; Wu & Cao, 2001).

However, none of the above works considers co-operative load balancing among multiple RSUs to share the load among all RSUs in the networks and improve overall system performances.

SYSTEM MODEL

RSU Architecture

We assume VANETs services are provided to the vehicles at hot spots such as gas stations or intersections of the roads, where the density of vehicles is typically higher than other areas. RSUs are deployed along roadsides as shown in Figure 2. When a vehicle is in the transmission range of an RSU, it can generate requests and receive responses.

Each RSU support two channels, one for user requests and another for responses. Through the request channel vehicles can upload their requests which are inserted into the RSU's received queue. Before making the serving decision the RSU invokes the underlying scheduling algorithm to select the request for the next service cycle and responds through the response channel by broadcasting. The architecture of an RSU is shown in Figure 1.

Notations and Assumptions

- **Request:** We assume when a vehicle submits a request R_i, it submits the following tuples:

$$R_i = (V_{ID}, ReqID_i, ID_i, T_i^{in}, T_i^{out}, T_i^{r}, T_i^{deadline}, T_i^{DTolerance}, HD_i).$$

- V_{ID}: Vehicle ID; $ReqID_i$: Request ID; ID_i : The ID of the requested data item; T_i^{in} : The time the vehicle enters into the communication range of the RSU; T_i^{out} : The time the vehicle leaves the communication range of the RSU. A vehicle can estimate this value from its own driving speed and the RSU transmission range; T_i^{r} : The time the request is generated; $T_i^{deadline}$: The deadline assigned by a request, beyond this time the request R_i will be dropped; $T_i^{DTolerance}$: The maximum delay tolerance of a request. Beyond this time request R_i will neither be served nor transferred; HD_i : Vehicle's intended heading direction.

Figure 1. An RSU architecture

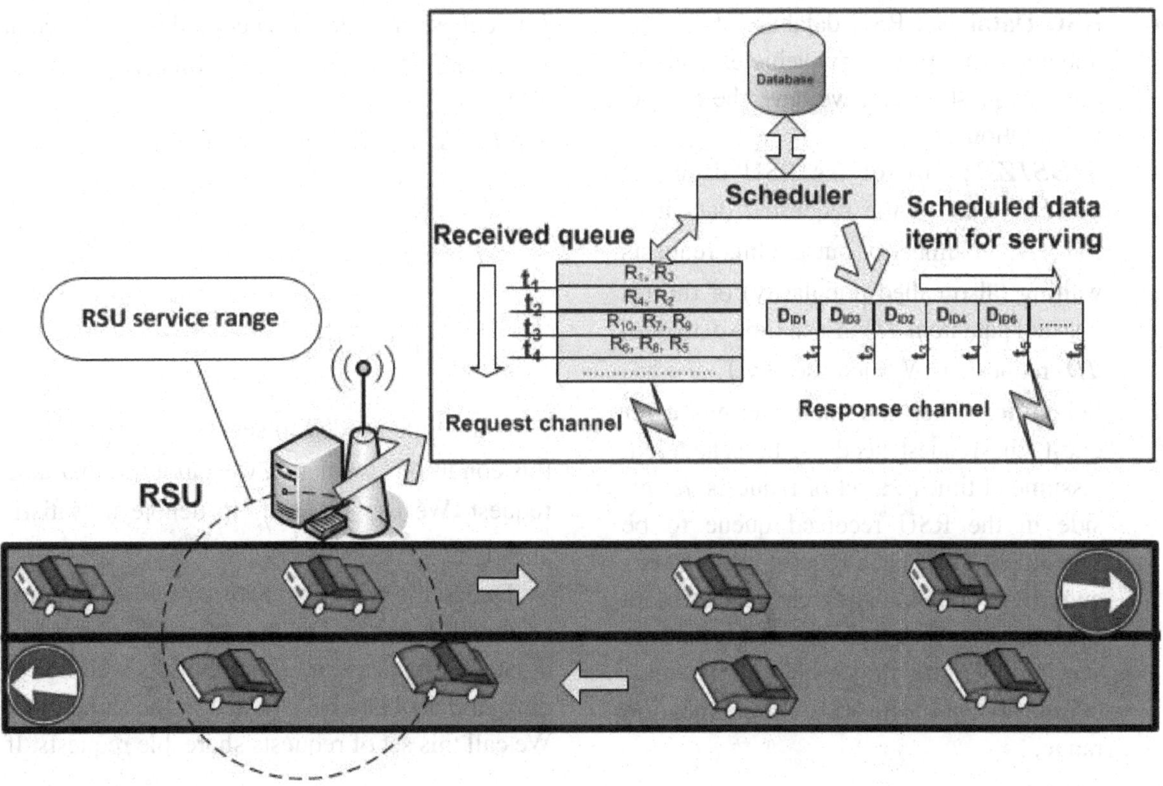

Figure 2. Different numbers of roads conjugate/pass by different RSUs

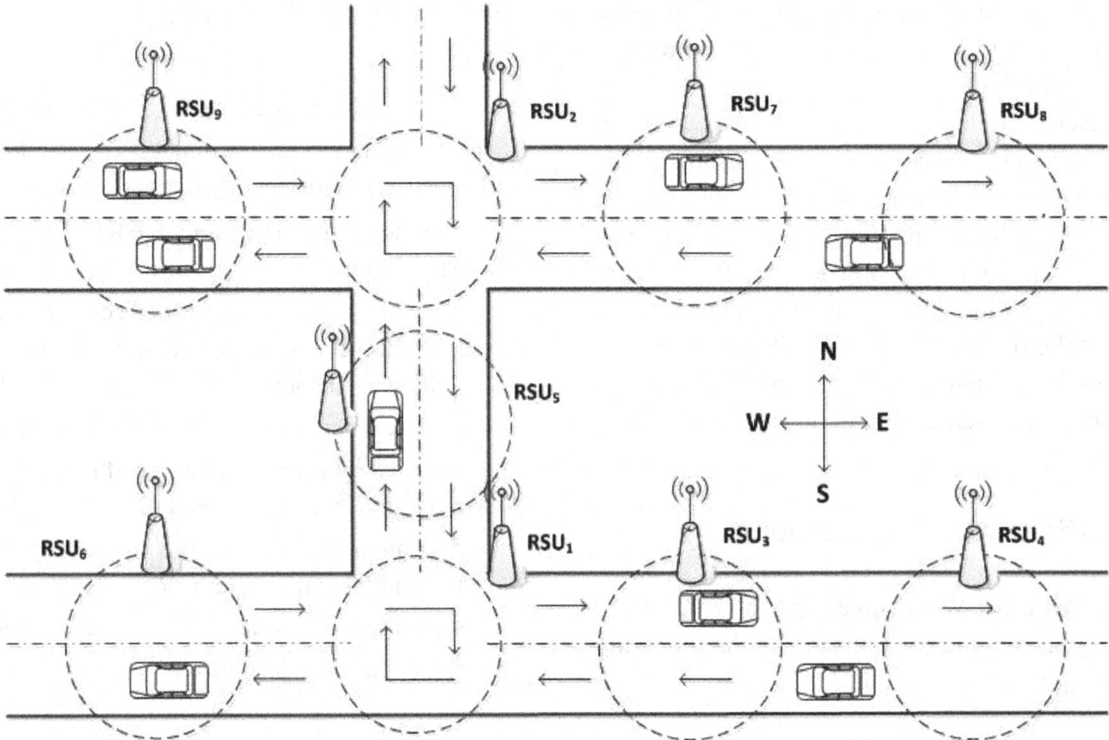

- **RSU Database:** RSU database stores the updated data to satisfy vehicles' on-demand requests. Here, we have the following notations:
- $\boldsymbol{DBSIZE}$: Size of the RSU database; $SIZE_i$: Size of the requested data item ID_i; N_i : Number of outstanding requests waiting (also called popularity) of the requested data item ID_i. Each time data item ID_i requested, N_i increased by 1.
- **Schedule:** When a vehicle submits a request, the request needs to be scheduled. Assume at time t, a set of requests R^t reside in the RSU received queue to be scheduled. Since each request needs to occupy the communication channel for data transmission, it should make sure that the serving operation finishes before the vehicle moves out of the RSU communication range.

If a request $R_i \in R^t$ is scheduled to be served at t, we call R_i is satisfiable if it meets:

$$t \geq T_i^{in} \ \& \ t + T_i^{serv} \leq T_i^{out}$$

where

$$T_i^{serv} = \frac{SIZE_i}{ChannelBW}$$

is the time needed to serve data item ID_i. If this condition is violated, we call it unsatisfiable request. We use $\left(R_i, R^t\right)_*$ to denote its satisfiability.

Due to the broadcast nature of wireless communication, for the downloading operation, when a data is broadcast, a set of requests waiting for the same data can be satisfied at the same time. We call this set of requests shareable requests. If a downloading request $R_i \in R^t$ is scheduled at time t, its shareable requests set is denoted as $SA(R_i,\ R^t)$, which can be defined as:

$$SA\left(R_i,\ R^t\right)=\{R_j \mid \forall j \neq i, R_j \in R^t,$$
$$ID_j = ID_i \ \&\left(R_j,R^t\right)_* = satisfiable\ \}.$$

- **Request's Life Time:** A vehicle can generate requests only within time range $[T_i^{in}, T_i^{out} - T_i^{serv}]$. Assume the radius of the transmission range of an RSU is R meters and average speed of a vehicle within the transmission range of RSU is $V m / s$. If a vehicle reaches the transmission range of an RSU at time $t=0$, and generates the first request at the same time, the average deadline of the first request of the vehicle is, $T_i^{deadline} = \frac{2R}{V}$.

Hence at any time T, the deadline of a request is:

$$T_i^{deadline} = \frac{2R}{V} - T_i^r,$$

where

$$T_i^r = random(0.0,\ (\frac{2R}{V} - T_i^{serv})).$$

Here T_i^r estimates the time when the request R_i is generated within the transmission range of an RSU. As time passes, the deadline value of a request decreases. Usually a request R_i will be discarded from the scheduler, when $T_i^{deadline} < T_i^{serv}$ (called as deadline missed request).

- **Layer Implementation:** Mac layer: For RSU MAC layer implementation, we use the IEEE 802.11 MAC specification. Physical layer: We assume each RSU has a single response channel and it is non-preemptive, i.e., no other data items will be served until the current item has been served. Broadcasting type is assumed to be omnidirectional.

SCHEDULING ALGORITHM

At each RSU, we use the DSIN algorithm as a scheduler. This algorithm is widely considered as an efficient scheduling algorithm (better than *FCFS, MRF, EDF, LWF* etc.) in the VANETs environment (Zhang, Zhao, & Cao, 2010; Ali, Chan, & Li, 2011). DSIN algorithm incorporates deadline (D) of a request, size (S) of the requested data item and number of requests (N) waiting for that requested data item. To make a scheduling decision for the next service cycle, DSIN calculates the DSIN_Value of all the received requests and chooses one which has the lowest DSIN_Value, where DSIN_Value of a request R_i:

$$DSIN\_Value(R_i) = \frac{T_i^{deadline} \times SIZE_i}{N_i}$$

PROBLEM STATEMENT

At any given time t, in RSU-based VANETs not all the RSUs receive the same amount of workload. Moreover as vehicles are mobile so the RSU workload also changes dynamically. An RSU's workload depends on how many vehicles are in the RSU transmission range for generating requests. However, the number of vehicles again depends on the following factors: (1) Vehicle arrival rate on a road, (2) Number of roads within the RSU transmission range, (3) Number of lanes of each road.

An RSU installed at the intersection of the road usually experiences more number vehicles to generate requests than an RSU installed at the non-intersection area or at the edge of a city. In Figure 2, there are 4 and 3 roads conjugate at RSU_2 and RSU_1 respectively, however only 1 road passes by the other RSUs. Hence, it is likely that RSU_2 may have to serve more number of vehicles than RSU_1 and other individual RSUs (RSU_3, RSU_4,RSU_8 etc.) in that topology.

The greater the number of vehicles in the RSU range, the larger the number of requests submitted to the RSU. After entering the RSU service range, a vehicle can generate requests. Before initiating a service cycle to provide service, the RSU scheduler needs to select a request using the selected scheduling algorithm. We use the DSIN schedule here. The scheduler has to find a suitable request among the requests it already has in its received queue and the requests it will receive while it serve the previously scheduled request.

Assume in Figure 3, after T_P^{serv} service cycle there is N_P number requests in the RSU_i received queue. Request generation rate per vehicle is λ_v, $N_{vehicle}$ number of vehicles is in the service range and RSU_i serves request R_i in the current service cycle (T_C^{serv}). Hence in the next service cycle (T_N^{serv}) RSU_i has to schedule total N_T number of requests, where

$$\begin{aligned} &N_T\left(RSU_i\right) \\ &= N_p + N_{vehicle}\left(RSU_i\left(T_c^{serv}\right)\right).\lambda_v.T_c^{serv} \qquad (1) \\ &-\left|SA\left(R_i, R^t\right)\right| - 1 \end{aligned}$$

From Equation (1), if

$$N_{vehicle}\left(RSU_i\left(T_c^{serv}\right)\right)$$

is high, total number of request in the RSU_i received queue $N_T(RSU_i)$ is also high. Following we see, total number of request $N_T\left(RSU_i\right)$ affects the RSU serving performance.

According to the DSIN algorithm, while serving a request R_i having minimum DSIN_Value, a number of requests $\in R^t$ might have smaller deadline than the sum of a request's waiting time and serving time of the request R_i. We call these requests deadline conflicted requests with R_i, because while server serves R_i, these requests miss their deadline. For example, in Figure 4, request $R_1\, to\, R_4$ scheduled according to their DSIN_Value. Request R_1 needs 8 sec of service time, while request $R_3\, and\, R_4$ have 7 and 9 sec deadline values respectively. Hence, while the server will finish serving request R_1, by this time R_3 will miss its deadline. Although after serving R_1, immediately R_4 will not miss its deadline, it does not have enough deadline value (it has 9 sec) to be satisfied, required a total of 15 sec (8 sec waiting time + 7 sec serving time). However, R_2 doesn't conflict with R_1 and hence will be served successfully.

Below we now estimate how many requests are conflicted in an RSU_i at time *t* among N_T requests in the RSU received queue.

Figure 3. Requests generation and serving in the service cycle of an RSU

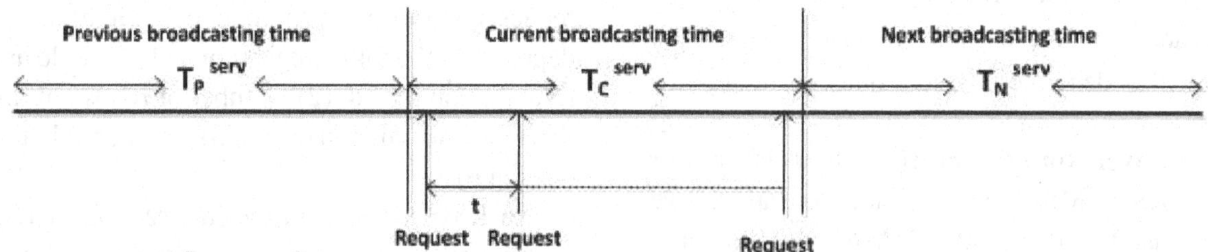

Figure 4. A scenario of request confliction

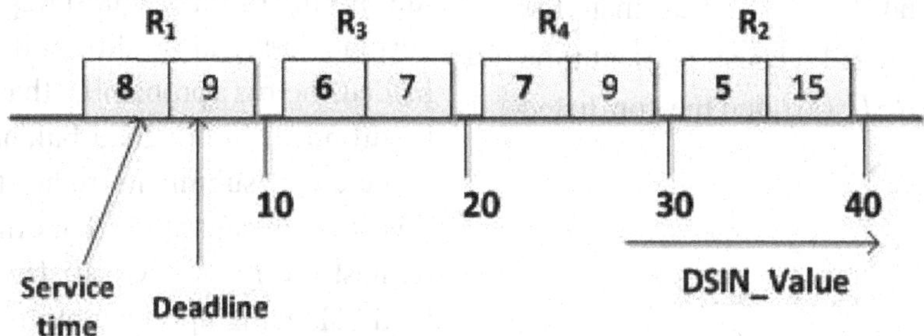

Assume at RSU_i among N_T number requests, *N* number requests are non-shareable in total received queue request set R^t, i.e., each request R_i asks for a separate data item. Using DSIN, if only one request R_i is scheduled from a $SA(R_i,R^t)$, none of the other requests in the same shareable set needs to schedule, as if R_i is served all requests in this set also will be served concurrently. All the requests in the same shareable requests ask for a same data item, hence the size *(S)* and popularity *(N)* of the asked data item are the same, and the only difference is the deadline *(D)* at a given time t. So, we need to schedule only the request R_i from a shareable request set which has minimum deadline, in turns minimum DSIN_Value among them. So, we get a total of *N* number non-shareable request in R^t as follows:

$$N = |\forall R_i \in R^t|,$$

where

$$R_i = min\{DSIN\_Value(R_j) \\ |\forall R_j \in (SA(R_i,R^t)+\{R_i\})\}.$$

After DSIN scheduling, we know the deadline ($T_i^{deadline}$), waiting time (T_i^{wait}) and serving time (T_i^{serv}) of a request R_i. Assume *N* number of non-shareable requests are scheduled, denoted as $R^{t(sch)}$. As scheduled request R_i has the minimum deadline value among its shareable request set $SA(R_i,R^t)$, if R_i doesn't conflict with another scheduled requests, no other requests in $SA(R_i,R^t)$ will be conflicted as well. However, if R_i is conflicted, other requests in $SA(R_i,R^t)$ may or may not be conflicted with other scheduled requests in $R^{t(sch)}$. Hence while scheduling non-shareable requests only, if a request R_i become conflicted we also need to check whether other members in $SA(R_i,R^t)$ conflicted or not. The following procedure shows how to find out a conflicted request from the scheduled request set $R^{t(sch)}$.

$$R_i = \{R_j \mid \forall R_j \in R^{t(sch)} \\ \& T_j^{wait} + T_j^{serv} > T_j^{deadlne}\} \quad /*$$

R_i is conflicted*/

$$R^{t(sch)} \\ = R^{t(sch)} - \{R_i\} + min\{DSIN\_Value(R_j) \mid \\ \forall R_j \in (SA(R_i,R^t) - \{R_i\})\}$$

$$ConflictSet(RSU_i(t)) \\ = ConflictSet(RSU_i(t)) + \{R_i\}$$

and

$$N^{conf}(RSU_i(t)) \\ = |ConflictSet(RSU_i(t))| \quad (2)$$

The above procedure continues until we get $R_i = \varnothing$ in $R^{t(sch)}$ and then we can estimate the total number of conflicted request N^{conf} at time t in RSU_i, $N^{conf}\left(RSU_i\left(t\right)\right)$ and the conflicted requests set denoted as

$$ConflictSet(RSU_i(t))$$

Hence, $N^{conf}(RSU_i(t))$ number of requests may miss their deadline because of their deadline confliction with other scheduled request in RSU_i. If received requests queue R^t is long, (N_T *is higher*) $N^{conf}(RSU_i(t))$ is higher (number of conflictions is higher) which leads to increase deadline miss ratio of RSU_i.

From the above analysis we see that, in VANETs at any given time, different RSUs experience different workload condition (based on their installed locations) and this condition changes dynamically. An RSU having many vehicles in its service range may receive a number of requests which exceeds its servicing capacity and this leads to poor performance (high deadline miss rate and response time). To solve this problem we propose a co-operative load balancing approach where the requests of an overloaded RSU will be shared by its neighbor RSUs. We discuss our approach in the following section.

A CO-OPERATIVE LOAD BALANCING APPROACH

We assume RSUs are networked either wirelessly, i.e., the RSU is a IEEE802.11 wireless access point (Yi, Bin, Tong, & Wei, 2008) or wired, i.e., the RSU is a cell of WiMAX or UMTS network (Lochert, Scheuermann, Caliskan, & Mauve, 2007). We further assume that an RSU knows the current workload situation of its neighboring RSUs. This can be done by having each RSU periodically broadcasts their workload status information to its neighboring RSUs.

After the submission of a request to an RSU, the serving of that request depends on the RSU current workload condition. If that RSU is overloaded, there is a probability that the request might be dropped. In our load balancing approach, a vehicle can submit its request to any RSU. If this RSU is capable of handling the submitted request the request is satisfied. Otherwise our load balancing approach will select a neighbor RSU in whose direction the vehicle is heading with the best chance to satisfy this request. The request is then forwarded to that neighbor RSU. When the vehicle enters into the vicinity of the transferred RSU, it submits an inform request through the request channel. This request is much smaller in comparison of initial submitted request. The inform request may only contain the vehicle ID and transferred request ID. This smaller size request helps to save the scarce bandwidth of the request channel. Upon getting the inform request, the transferred RSU consider that transferred request for scheduling in its next service cycle.

We now discuss in detail the above load balancing procedure with respect to three major decisions that need to make in the load transfer process.

Step 1: When to Transfer

First, an RSU needs to find out whether it needs to transfer its load or not, i.e., determining the workload situation. Assume at time *t*, at RSU_i there are N number non-shareable requests among calculated total received N_T number of requests using Equation (1). Using Equation (2), RSU_i can estimate the total number of requests that are conflicted at time t, $N^{conf}\left(RSU_i\left(t\right)\right)$. If:

$$N^{conf}\left(RSU_i\left(t\right)\right) > 0,$$

RSU finds the request R_i which needs to be transferred among its scheduled request, where

$$R_i = \{R_j \mid \forall R_j \in R^{t(sch)} \\ \& T_j^{wait} + T_j^{serv} > T_j^{deadlne}\}.$$

Step 2: Where to Transfer

In our model, a vehicle's next heading direction is determined by the Manhattan mobility model (Bai, Sadagopan, & Helmy, 2003).

From an RSU to a specific direction, until a vehicle reaches at a new intersection, it drives straight. We call the set of RSUs that a vehicle passes up to reaching a new intersection, the neighboring set of RSUs in that direction of that RSU and denote it as $Neighbor(RSU_i(D))$, here D stands for the direction in which a vehicle is heading. D could be

$$North\ (N), South\ (S), East\ (E) and West\ (W).$$

For example, in Figure 5 the neighboring set of RSU_1 in $N,S,W and E$ directions are $\{RSU_5\ , RSU_2\}$,

Figure 5. Neighboring RSUs

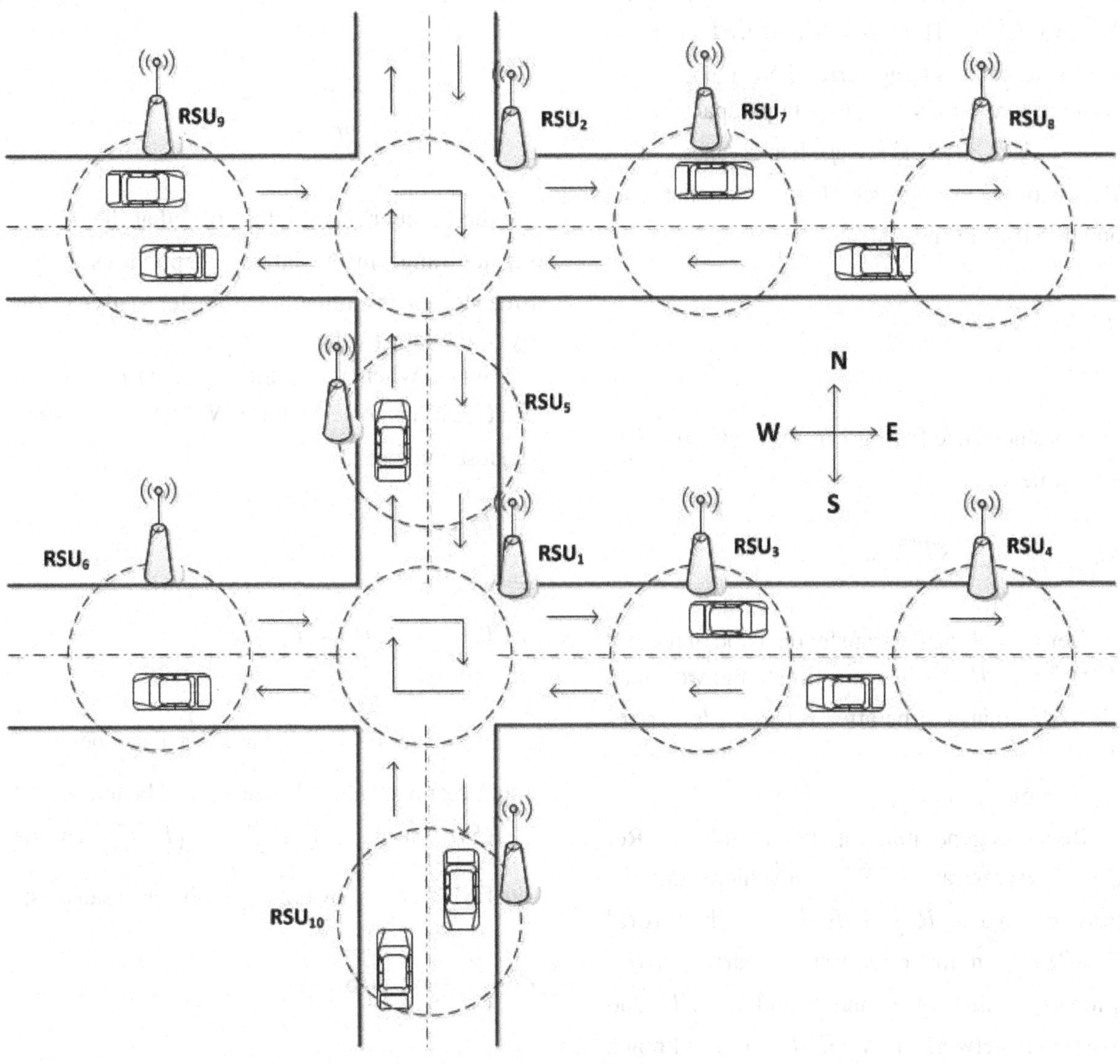

$\{RSU_{10}\}$, $\{RSU_6\}$, *and* $\{RSU_3, RSU_4\}$

respectively. For RSU_3 in the W and E directions, they are $\{RSU_1\}$ and $\{RSU_4\}$ respectively; however there are no neighbors in N and S directions.

After knowing the vehicle heading direction D, the RSU needs to find out a transferred

$$RSU_j \in Neighbor(RSU_i(D))$$

by calculating their serving capability.

Without loss of generality, we consider RSU_i wants to transfer request R_i (submitted by vehicle Vh_i) to RSU_j. Here RSU_i and RSU_j termed as transferring and transferred RSU respectively. Assume two RSUs are n_{hops} hops apart, where $n_{hops} = 1,2,3,\cdots$ and average hop distance is d_{avg}. Hence for V_{ij} average vehicle speed on that road, one hop travel time,

$$T_{hop}^{travel} = \frac{d_{avg}}{V_{ij}}$$

A vehicle needs to go from RSU_i *to* RSU_j total time,

$$T_{ij}^{travel} = n_{hops} \times T_{hop}^{travel}.$$

Before making the transfer decision of request R_i *to* RSU_j, RSU_i has to estimate the workload of RSU_j that will be after vehicle Vh_i arrives there.

Assume at RSU_j,

Requests generating rate per vehicle λ_v; Requests serving rate δ_j; When a vehicle starts its journey from RSU_i *to* RSU_j, it has total $N_p(RSU_j)$ number requests to serve; RSU_i knows the values of λ_v and δ_j and as RSU_i and RSU_j are networked, $N_p(RSU_j)$ is also known to RSU_i. To calculate the estimated total load at RSU_j after T_{ij}^{travel}, now RSU_i needs to know the expected number of vehicles $N_{vehicle}(RSU_j)$ that may arrive by this time. RSU_i calculates this value by the following exponential weighted moving average (EWMA) equation:

$$N_{vehicle}(RSU_j) = (1-\beta).N_{vehicle}(RSU_j(prev)) + \beta.N_{vehicle}(RSU_j(latest))$$

where

$$N_{vehicle}(RSU_j(latest))$$

and

$$N_{vehicle}(RSU_j(prev))$$

are the number of vehicles arrived at the RSU_j service range in the latest and previous T_{hop}^{travel} time slot. β is a tunable parameter to give priority to the latest value.

Hence when the vehicle Vh_i will reach at RSU_j, it has to serve total $N_R(RSU_j)$ number requests, where:

$$N_R(RSU_j) = N_P(RSU_j) + N_{vehicle}(RSU_j).\lambda_j.T_{ij}^{travel} + 1(R_i) - T_{ij}^{travel}.\delta_j$$

As transferred R_i will have $\frac{2R}{V}$ deadline (maximum deadline value), hence R_i will be transferred to RSU_j, if and only if all $N_R(RSU_j)$ can be satisfied at RSU_j by time $\frac{2R}{V}$, that is, it satisfies:

$$\delta_j.\frac{2R}{V} \geq N_R(RSU_j).$$

RSU_i does the same above load calculation and transfer procedure among its neighboring RSUs in different directions based on requests submitted vehicles' heading directions. It continuously do so until

$$ConflictSet(RSU_i(t))$$

is not empty. However, while an RSU_i tries to transfer its conflicted requests to the neighbor RSUs, at the same time it serves the conflict free scheduled requests from its received queue. If RSU_i could transfer all the requests:

$$R_i \in ConflictSet\left(RSU_i\left(t\right)\right)$$

performance will improve (almost all requests would be satisfied).

Step 3: Which to Transfer

It is possible that at any given time, more than one requests in the

$$ConflictSet(RSU_i(t))$$

need to be transferred in a particular direction, because these submitted requests' vehicles are heading in the same direction. In this case, there needs a decision to determine which request to be transferred first. We propose the following heuristic approaches to transfer a request

$$R_i \in ConflictSet(RSU_i(t))$$

to the neighbor RSU_j.

Transferring Request Selection Based on Request Types (SRT)

Transferring RSU_i selects a request from

$$ConflictSet(RSU_i(t))$$

based on request types. Requests for safety applications also called delay intolerant applications (such as accident warning) are delays sensitive and need immediate service. On the other hand requests for non-safety applications (such as entertainment data) are delay tolerant, and so can tolerate some delay for getting service (Mak, Laberteaux, & Sengupta, 2005).

Denote delay-intolerant and delay-tolerant requests as *DIT* and *DT* respectively. RSU tries to transfer the conflicted requests from

$$ConflictSet\left(RSU_i\left(t\right)\right).$$

Neighbor RSU_j is capable to handle a request and located in a particular direction

$$LD\left(RSU_j\right), where\, LD\left(RSU_j\right) \in \left\{N,S,E,W\right\}.$$

If a vehicle heading direction (HD_i) is towards $LD\left(RSU_j\right)$ and this vehicle submitted request is DIT type,

$$(R_i^{DIT} \in ConflictSet\left(RSU_i\left(t\right)\right))$$

this R_i^{DIT} will be chosen to transfer to RSU_j.

$$R_i = \begin{Bmatrix} R_j^{DIT} \mid \exists R_j^{DIT} \in ConflictSet\left(RSU_i\left(t\right)\right) \\ \& HD_j = LD\left(RSU_j\right) \end{Bmatrix}$$

We call this approach "Delay Intolerant Load Transfer (DIT LT)" approach. At a time, if there is more than one DIT type request need to transfer to RSU_j located direction, it chooses one randomly among them. If there is no *DIT* type request in

$ConflictSet\left(RSU_i\left(t\right)\right)$,

it then starts to transfer *DT* type requests.

Transferring Request Selection Greedily

In a greedy approach, transferring request selection done based on specific values. Here, we propose the following 2 greedy approaches.

G-DSIN Approach

The G-DSIN Load Transfer (G-DSIN LT) approach gives the priority to the request

$R_i \in ConflictSet\left(RSU_i\left(t\right)\right)$

having minimum DSIN_Value. The rationale behind this approach for selecting a request for transferring to another RSU is to minimize the deadline misses of the requests which have lower DSIN Value. Hence it transfers a request R_i:

$$R_i = min\{DSIN\_Value\left(R_j\right) | \forall R_j \in ConflictSet \left(RSU_i\left(t\right)\right) \ \&HD_j = LD\left(RSU_j\right)\}.$$

G-DSize Approach

In the G-DSize Load Transfer (G-DSize LT) approach RSU_i transfers the request R_i asking for minimum sized data item (DSIZE) in the conflicts request set

$ConflictSet\left(RSU_i\left(t\right)\right)$,

$$R_i = \min\{DSIZE\left(R_j\right) | \forall R_j \in ConflictSet \left(RSU_i\left(t\right)\right) \ \&HD_j = LD\left(RSU_j\right)\}$$

The rationale behind this approach is to transfer the request which needs less time to serve at the transferred RSU_j. As a smaller sized requested data item takes less time to be served, the transferred RSU can serve the transferred requests more quickly, hence it will not be overloaded easily by transferred requests, can serve its own workload efficiently and be available to serve more transferred request from overloaded neighbor RSUs.

Co-Operative Load Balancing Algorithm

Without loss of generality, Algorithm 1 shows our co-operative load balancing approach, by transferring overloaded request from $RSU_i \, to \, RSU_j$. After scheduling the non-shareable requests set $\left(R^{t(sch)}\right)$, RSU_i removes the conflicted request from it, insert into

$Conflictset(RSU_i(t))$

and serve the 1st non-conflicted request from the scheduled queue (Line 8-12). Then to transfer the conflicted requests according to conflicted requests submitted vehicle heading direction (HD), RSU_i calculates the load handling capacity of the neighbor RSUs in that direction (D) to find a suitable RSU_j (Line 13-18). If there are more than one request

$\in Conflictset(RSU_i(t))$

to transfer to RSU_j to the same direction (D), NP LT approach selects a request R_i randomly (Line 20), DIT LT approach selects R_i based on request type (Line 21), while G-DSIN and G-DSize base their selection on DSIN_Value (Line 22) and requested data item size (Line 23) respectively. After transferring request $R_i \, to \, RSU_j$, RSU_i removes R_i from

$Conflictset(RSU_i(t))$

(Line 24-25). The RSU_j schedules transferred request R_i along with its own load, when transferred vehicle Vh_i reaches in RSU_j's transmission range (Line 32). For scheduling the received requests all the RSUs invoke DSIN_Scheduling (Line 1-5). All the RSUs in the topology perform load transfer like RSU_i and serve the overload (of RSU_i) like RSU_j when they need to do so based on their workload condition.

PERFORMANCE EVALUATION

Experimental Setup

We perform our simulation experiment using CSIM19 (Schwetman, 2001). Other than the CSIM default parameters, the parameters we use for simulation are shown in the Table 1. Our simulation environment is set up based on the multiple-RSU topology shown in Figure 6.

A vehicle can generate requests until it exceeds the RSU transmission range. Different vehicles may generate different number of requests, the number of requests defined by a Poisson process for each individual vehicle. During simulation, for varying workload in the RSU range, we very the Poisson mean request generation, μ (the mean number of requests a vehicle generates while it is within a RSU transmission range) value from 5 to 125. The vehicle request generation interval is (RGIV) exponentially distributed, with a mean value of 0.15 for our simulation. The vehicles request access pattern for accessing the data item in the RSU database follows the Zipf distribution. Here the skewness parameter is θ, we use default θ value is 0.8. The access probability of ith data item in the database is:

$$P(i) = \frac{\frac{1}{i^{\theta}}}{\sum_{n=1}^{N} \frac{1}{n^{\theta}}},$$

where, $N = DBSIZE$.

We generate vehicles from the both west and east end of the road on opposite directed lanes and let the vehicles to move. When a vehicle reaches at the end of a lane, it does not generate any further request and will be removed from the simulation. Vehicle Generation InterVal (VGIV) also follows the exponential distribution.

Vehicle mobility follows the Manhattan mobility model (Bai, Sadagopan, & Helmy, 2003). Each road has 2 lanes and vehicles in the same lane always maintain a safety distance (SD). The simulation completes when data is stable and a 95% confidence interval achieved. We generate the RSU database using increment data item size distribution (INCRT) which is defined in (Xu, Hu, Lee, & Lee, 2004).

While generating traffic, we generate both delay-tolerant (DT) and delay-intolerant (DIT) requests. We calculate the mean DSIN_Value of all the generated requests, and then classify the requests in lower DSIN_Value (LDSIN) and higher DSIN_Value (HDSIN) requests to facilitate the performance analysis of our proposed heuristic G-DSIN approach. Similarly, we calculate the mean data item size from the requested data items to classify them into "Requests for Lower Sized Data item (RLSD)" and "Requests for Higher Sized Data item (RHSD)", which are used to analyze the performance of another proposed heuristic approach G-DSIZE.

Performance Metrics

We evaluate the overall multiple-RSU model performance. To do so, we adopt the following performance metrics:

Algorithm 1. Co-operative load balancing algorithm

DSIN_Scheduling($\boldsymbol{R^t}$) /* Call by all RSUs in the topology*/
$\forall R_i \in R^t\ calculate\ DSIN\_Value(R_i) \leftarrow \dfrac{T_i^{deadline} \times Size_i}{N_i}$
$According\ DSIN\_Value\ scheduling\ of Non-shareable\ requests \in R^t, R^{t(sch)}$
$\boldsymbol{Return}\ R^{t(sch)}$
End DSIN_Scheduling

$\boldsymbol{RSU_{i\_Scheduling}}$ /*Run by RSU_i*/
$\boldsymbol{while}\ R^t(RSU_i) \neq \emptyset$ **do**
    $R^{t(sch)} \leftarrow$DSIN_Scheduling($R^t(RSU_i)$)
    $R_i = \{R_j | \forall R_j \in R^{t(sch)}\ \&\ T_j^{wait} + T_j^{serv} > T_j^{deadlne}\}$ /*R_i is conflicted*/
    $R^{t(sch)} = R^{t(sch)} - \{R_i\} + min\{DSIN\_Value(R_j) | \forall R_j \in (SA(R_i, R^t) - \{R_i\})\}$
    $ConflictSet(RSU_i(t)) = ConflictSet(RSU_i(t)) + \{R_i\}$
    $Serve\ request\ R_1$ and $R^{t(sch)} = R^{t(sch)} - \{R_1\}$ /*Serve scheduled non conflicted first request*/
        for each alternative heading direction (D) from RSU_i **do** $/*\ D \in \{N, S, E, W\}\ */$
            if $\exists R_i \in Conflictset(RSU_i(t))\ having\ HD_i = D$ **then**
            $Find\ an\ RSU_j \in Neighbor(RSU_i(D))$ which satisfies $\delta_j . \frac{2R}{V} \geq N_R(RSU_j)$ where,
            $N_R(RSU_j) = N_P(RSU_j) + N_{vehicle}(RSU_j) . \lambda_j . T_{ij}^{travel} + 1(R_i) - T_{ij}^{travel} . \delta_j$,
            and $N_{vehicle}(RSU_j) = (1-\beta) . N_{vehicle}\left(RSU_j(prev)\right) + \beta . N_{vehicle}\left(RSU_j(latest)\right)$.
            end if

            if there are more than one, $R_i \in Conflictset(RSU_i(t))\ having\ HD_i = D$ **then**
            By NP LT approach: select any $R_i \in Conflictset(RSU_i(t))$
            By DIT LT approach: $R_i = \{R_j^{DIT} | \exists R_j^{DIT} \in ConflictSet(RSU_i(t))\}$
            By G-DSIN LT approach: $R_i = min\{DSIN\_Value(R_j) | \forall R_j \in ConflictSet(RSU_i(t))\}$
            By G-DSize LT approach: $R_i = min\{DSIZE\ (R_j) | \forall R_j \in ConflictSet(RSU_i(t))\}$
            Transfer request $R_i to\ RSU_j$
            $ConflictSet(RSU_i(t)) = ConflictSet(RSU_i(t)) - \{R_i\}$
            end if
        end for
    end while
End $\boldsymbol{RSU_{i\_Scheduling}}$

$\boldsymbol{RSU_{j\_Scheduling}}$ /*Run by RSU_j*/
$\boldsymbol{while}\ (R^t(RSU_j) + R') \neq \emptyset$ **do** /*at time t, RSU_j received R' transferred requests from neighbor */
    $R^{t(sch)} \leftarrow$DSIN_Scheduling$\left(R^t(RSU_j) + R_i\right)$ /*Schedule transferred R_i when transferred vehicle Vh_i in RSU_j range*/
    $Serve\ request\ R_1$ and $R^{t(sch)} = R^{t(sch)} - \{R_1\}$
end while
End $\boldsymbol{RSU_{j\_Scheduling}}$

Figure 6. Simulation topology

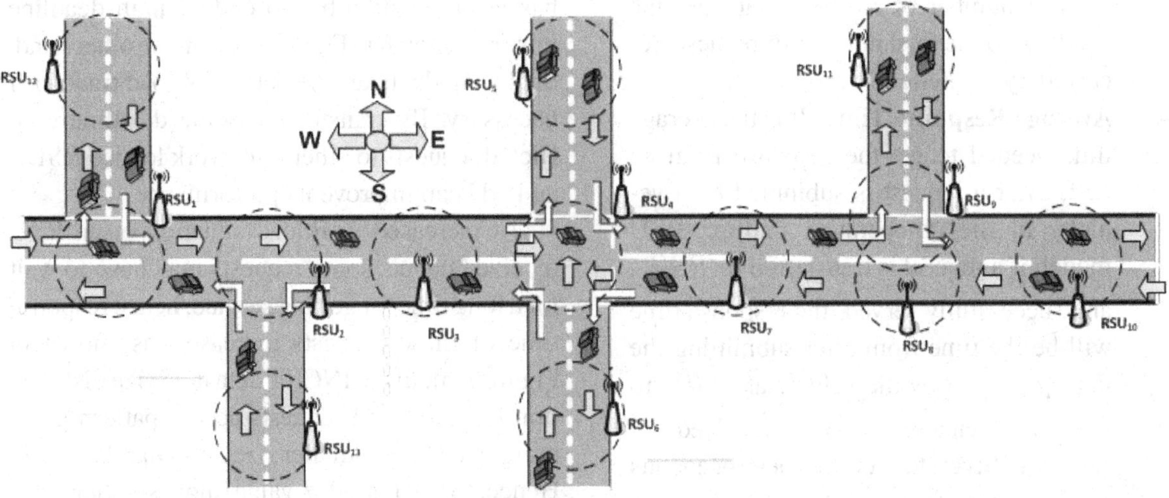

Table 1. Simulation parameters

Parameter	Default	Range	Description
NoOfRSU	12	---	Number of RSUs in the topology
VGIV	0.50	----	Vehicle generation interval
RGIV	0.15	---	Request generation interval by a vehicle
DBSIZE	600	---	Number of data items in the database
DataItemsize	---	10 - 512	Size of each requested data item (K bytes) ($SIZEMIN - SIZEMAX$)
ChannelBW	100	---	Channel broadcast bandwidth K bytes/s
RSU Range (R)	350	----	RSU communication range
VSpeed	40 km/h	---	Vehicle average speed
Distance	1 km	---	Inter RSU distance
THETA (θ)	0.8	---	Zipf distribution parameter
Poisson mean, μ	---	5 - 125	Mean no. of requests generation by a vehicle at each RSU
SD	6 m	---	Safety distance between vehicles
N_l	02	---	Number of lanes of each road
ρ	0.25	---	Vehicle density per meter on a lane of a road
β	0.25	---	EWMA parameter

- **Deadline Miss Ratio (DMR):** It is the ratio of number of requests that miss the deadline to the total number of requests received by the system.
- **Average Response Time:** It is the average time needed to get the response from an RSU after a request is submitted by a vehicle. In the co-operative multiple-RSU model, if a request is transferred to RSU_j, and successfully served, the response time will be the time from after submitting the *inform request* by the vehicle at RSU_j to the time when a response it is received.
- **Average Stretch:** As for heterogeneous size data item response time alone is not a fair performance metric, we adopt average stretch as an additional measure to evaluate efficiency of our system. Average stretch is measured as the ratio of the response time of a submitted request to its service time, where service time is the time taken to serve the request from an RSU when the RSU channel is idle.

Performance Evaluation

In this section we evaluate the performance differences of co-operative load transferring (LT) multiple-RSU model with non-load transferring (NLT) multiple-RSU model. Furthermore, we investigate the performance of the different variants of the LT model.

NLT vs. Variants of LT

Figure 7 exhibits the overall system performance for NLT model and different variants of the LT model in terms of deadline miss rate, average response time and average stretch. Varying the workload (Poisson mean value, μ) has an impact on RSU serving performance. When an RSU receives a number of requests beyond its capacity, deadline miss rate starts to climb. A NLT RSU has no alternative but to endure high deadline miss rate shown in Figure 7(a). On the other hand, the LT model takes the benefit of load transfer if necessary. By transferring some deadline conflicted requests to other light work loaded RSUs, an RSU can improve its performance.

At increased workload situation in an RSU received queue, some requests may have to wait for a long time for getting service; hence response time of these requests increases, as shown in Figure 7(b). In the INCRT data item size distribution more skewed requests access pattern gives preference to the smaller sized popular data item. Hence, for increased θ value, more smaller sized popular data items are requested, serving one request satisfies many shareable requests, and consequently the response time of shareable requests decreases. That is the reason (we use θ value 0.8) why the average response time and average stretch in Figure 7 do not increase sharply after Poisson mean value 75.0. In the LT model when a request is transferred to another RSU, we calculate the response time from submitting the inform request by a transferred vehicle at the transferred RSU to get the response from it. Hence, though increasing workload increases the response time in LT model, it is lower than NLT model for different workload conditions.

Though increasing workload causes both the NLT and LT models performance to decline, regarding different performance metrics the different LT models' performance is clearly better than that the NLT model for both light and heavy workload.

At low workload condition, the LT variants show essentially the same performance, because they only differ in handling conflicted requests. At low workload condition the number of conflicted requests is small, all the transferred requests at the transferred RSU are served successfully and hence deadline miss rate is also same. However, for increased workload condition (after Poisson mean value 75.0), the LT variants have different deadline miss ratio (DMR); G-DSize LT

Figure 7. Performance comparison of non-load transferring (NLT) and different variants of co-operative load-transferring (LT) multiple-RSU model for varying workload in RSU service range: (a) deadline miss ratio, (b) average response time, (c) average stretch

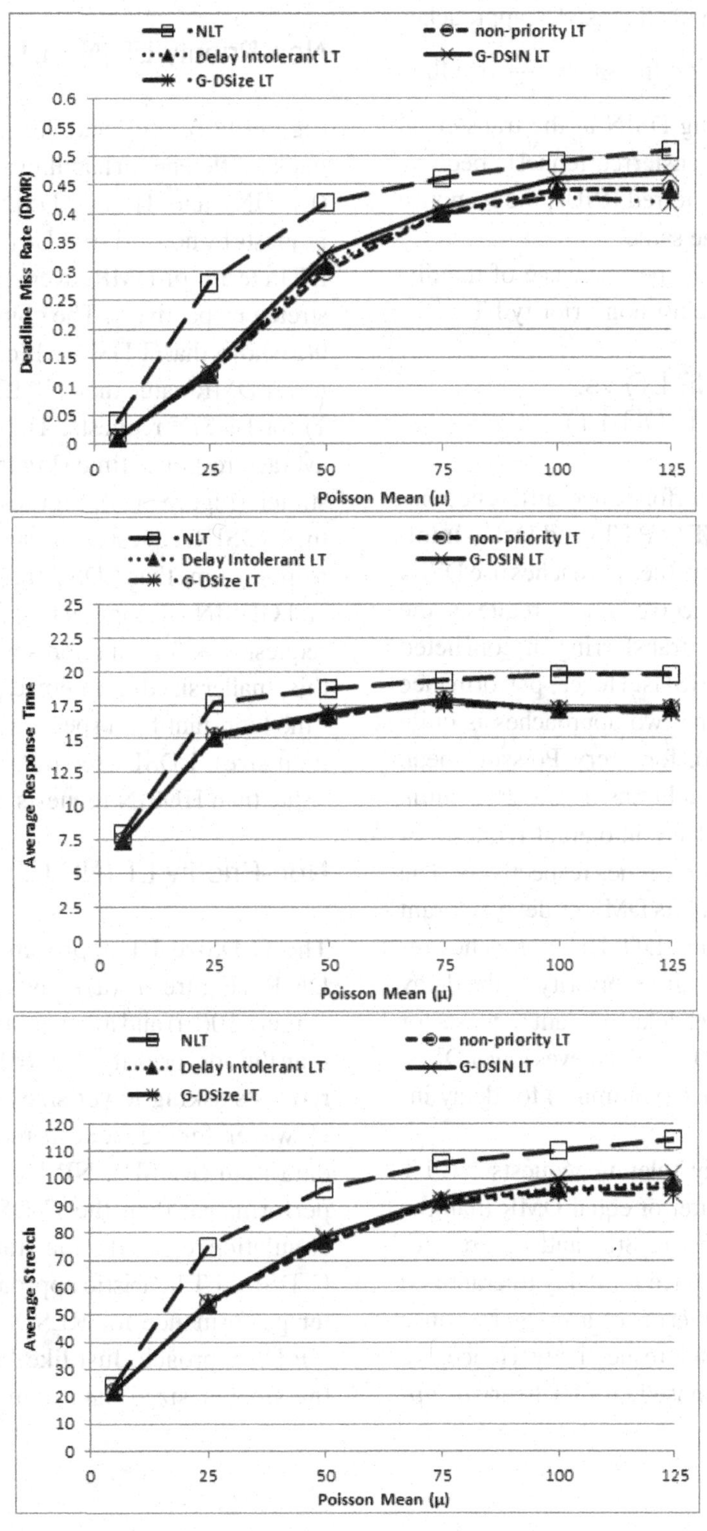

has the lowest and G-DSIN LT has the highest DMR among the LT variants. However, all the LT variants basically have the same average response time, because in all these LT approaches a transferred request get the same $\frac{2R}{V}$ deadline and is scheduled using DSIN at the transferred RSUs, hence if the transferred requests become successful at the transferred RSUs, the response times are basically the same.

We now analyze the performance of the different LT variants against non-priority LT.

Non-Priority LT (NP LT) vs. Delay Intolerant LT (DIT LT)

Figure 8 shows the performance difference between non-priority LT (NP LT) and Delay Intolerant LT (DIT LT). Both the approaches use DSIN for scheduling their received queue requests, the only difference is the transferring of conflicted requests to neighbor RSUs. Hence, performance difference between the two approaches is quite small. In Figure 8(a), for every Poisson mean value (μ), there are 4 columns: 1st and 2nd column show the DMR for delay intolerant requests of NP LT and DIT LT approaches respectively. The 3rd and 4th column exhibits DMR of delay tolerant requests for NP LT and DIT LT approaches respectively. As DIT LT gives priority to the delay intolerant requests over delay tolerant requests for transfer, predictively DIT LT achieves lower DMR (column 2) than NP LT (column 1) for delay intolerant requests.

However for delay tolerant requests NP LT (column 3) shows better or equal DMR than the DIT LT approach. Figure 8(b) and Figure 8(c) show similar performance for delay tolerant and intolerant requests in terms of average response time and average stretch respectively. Hence, we conclude that our proposed DIT LT heuristic approach can successfully minimize the DMR and response time of delay intolerant requests compared to the NP LT approach.

Non-Priority LT (NP LT) vs. G-DSIN LT

Figures 9(a), 9(b), and 9(c) illustrate the performance difference of handling Lower DSIN_Valued (LDSIN) and Higher DSIN_Valued (HDSIN) requests by non-priority LT (NP LT) and G-DSIN LT in terms of DMR, average response time and stretch respectively. The column graph in Figure 9(a) shows that G-DSIN LT (column 2) can achieve lower DMR value than NP LT approach (column 1) for LDSIN requests. The result is similar for average response time (Figure 9(b)) and average stretch (Figure 9(c)). Figures 9(a) and 9(b) show that HDSIN requests experience higher DMR and response time than LDSIN requests for both NP LT and GDSIN LT approaches. However, as LDSIN requests ask for data items with smaller size and this smaller sized data item requires shorter service time (constant for a specific service channel and item size), LDSIN requests have higher stretch value than HDSIN requests in Figure 9(c).

Non-Priority LT (NP LT) vs. G-DSize LT

The G-DSize LT approach can achieve better DMR (Figure 10(a)), average response time (Figure 10(b)) and average stretch (Figure 10(c)) than the non-priority LT (NP LT) approach for the requests asking lower sized data item (RLSD). However, for the requests requiring higher sized data item (RHSD), NP LT has equal or better performance than the G-DSize approach. The simulation results demonstrate that our proposed G-DSize LT heuristic approach can achieve better performance for RLSD compared with the NP LT approach. Just like LDSIN requests, for the smaller sized data item reason, RLSD has

Figure 8. Performance comparison of non-priority LT (NP LT) and Delay Intolerant LT (DIT LT) multiple-RSU model for varying workload in RSU service range: (a) deadline miss ratio, (b) average response time, (c) average stretch

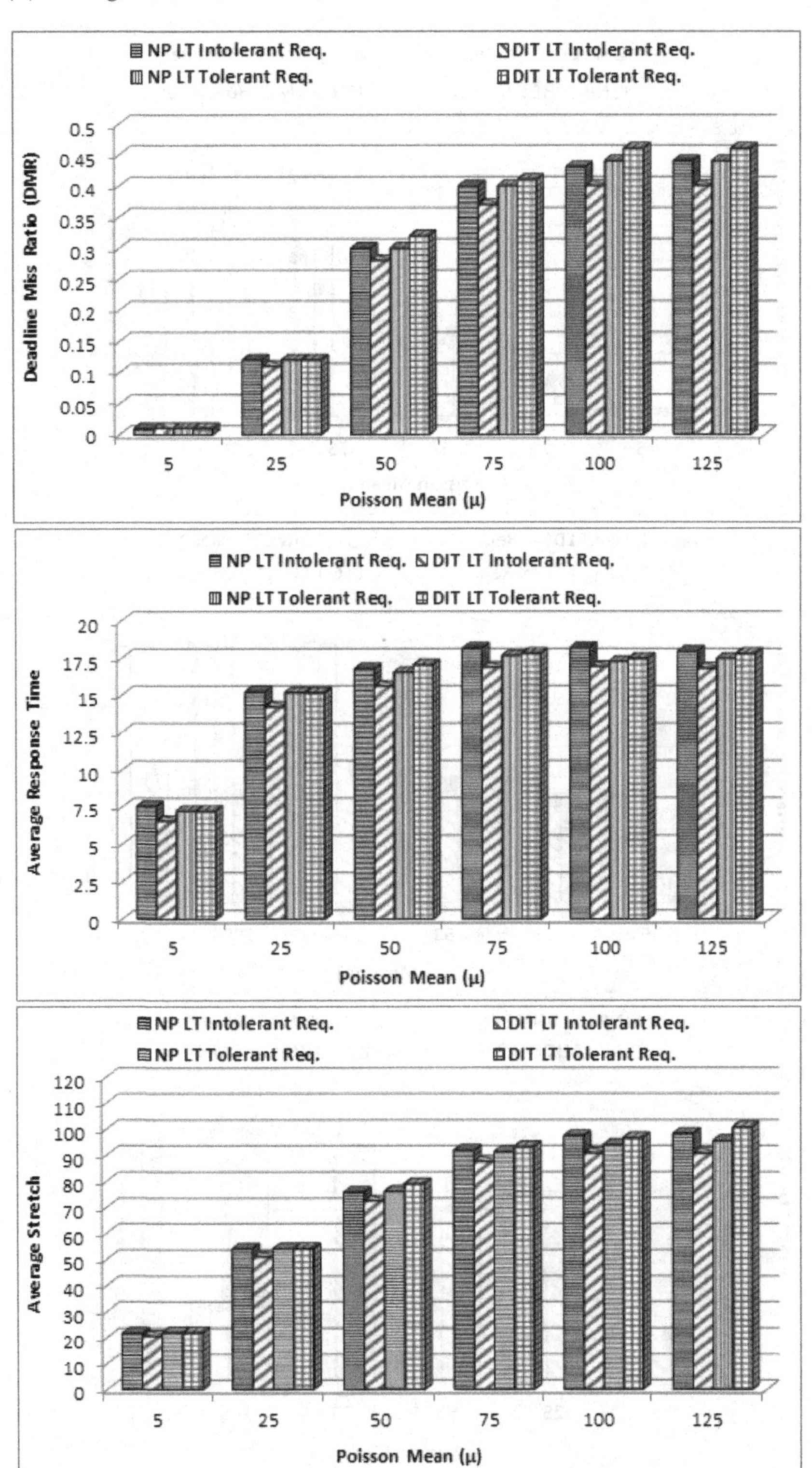

Figure 9. Performance comparison of non-priority LT (NP LT)and G-DSIN LT multiple-RSU model for varying workload in RSU service range: (a) deadline miss ratio, (b) average response time, (c) average stretch

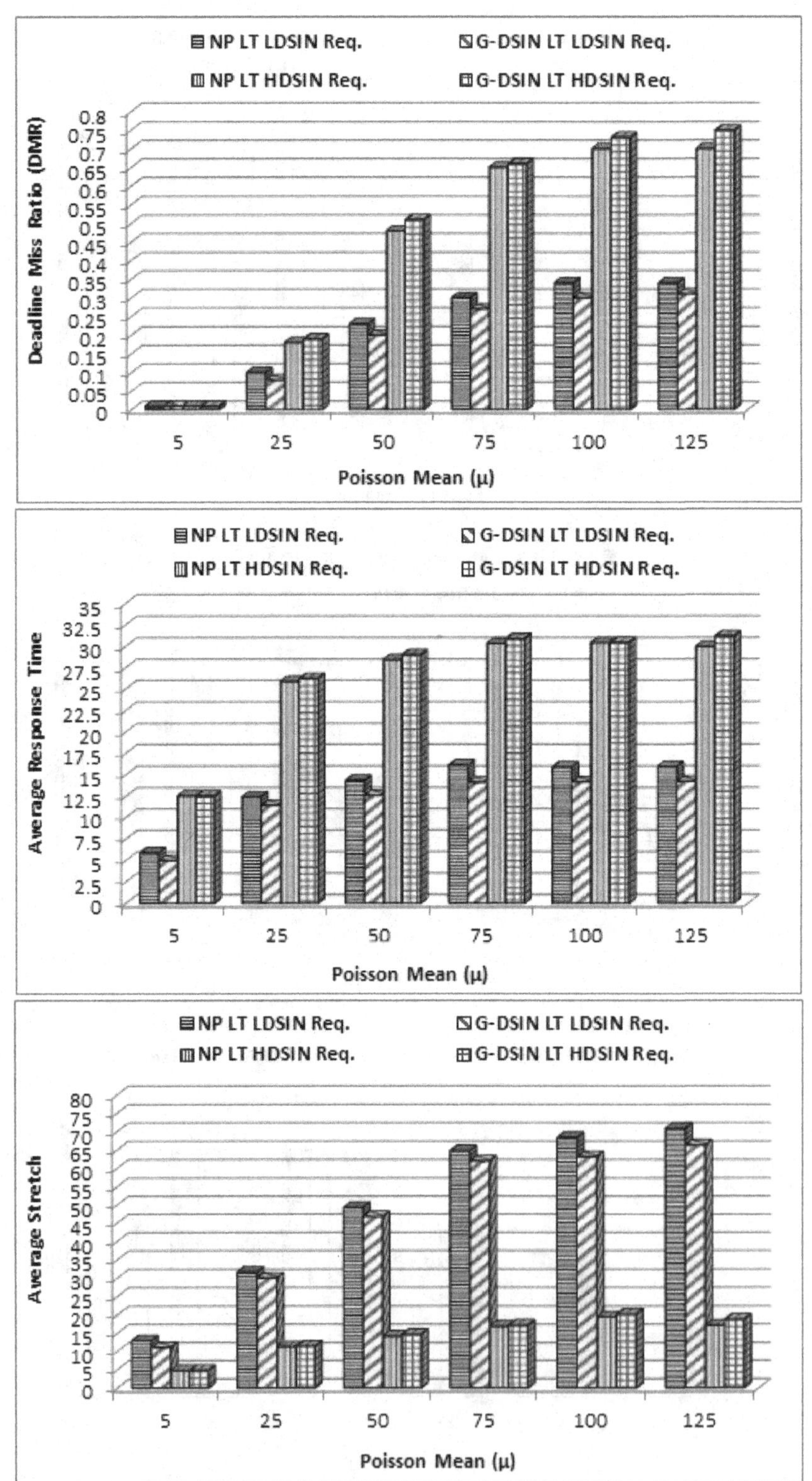

Figure 10. Performance comparison of non-priority LT (NP LT) and G-DSize LT multiple-RSU model for varying workload in RSU service range: (a) deadline miss ratio, (b) average response time, (c) average stretch

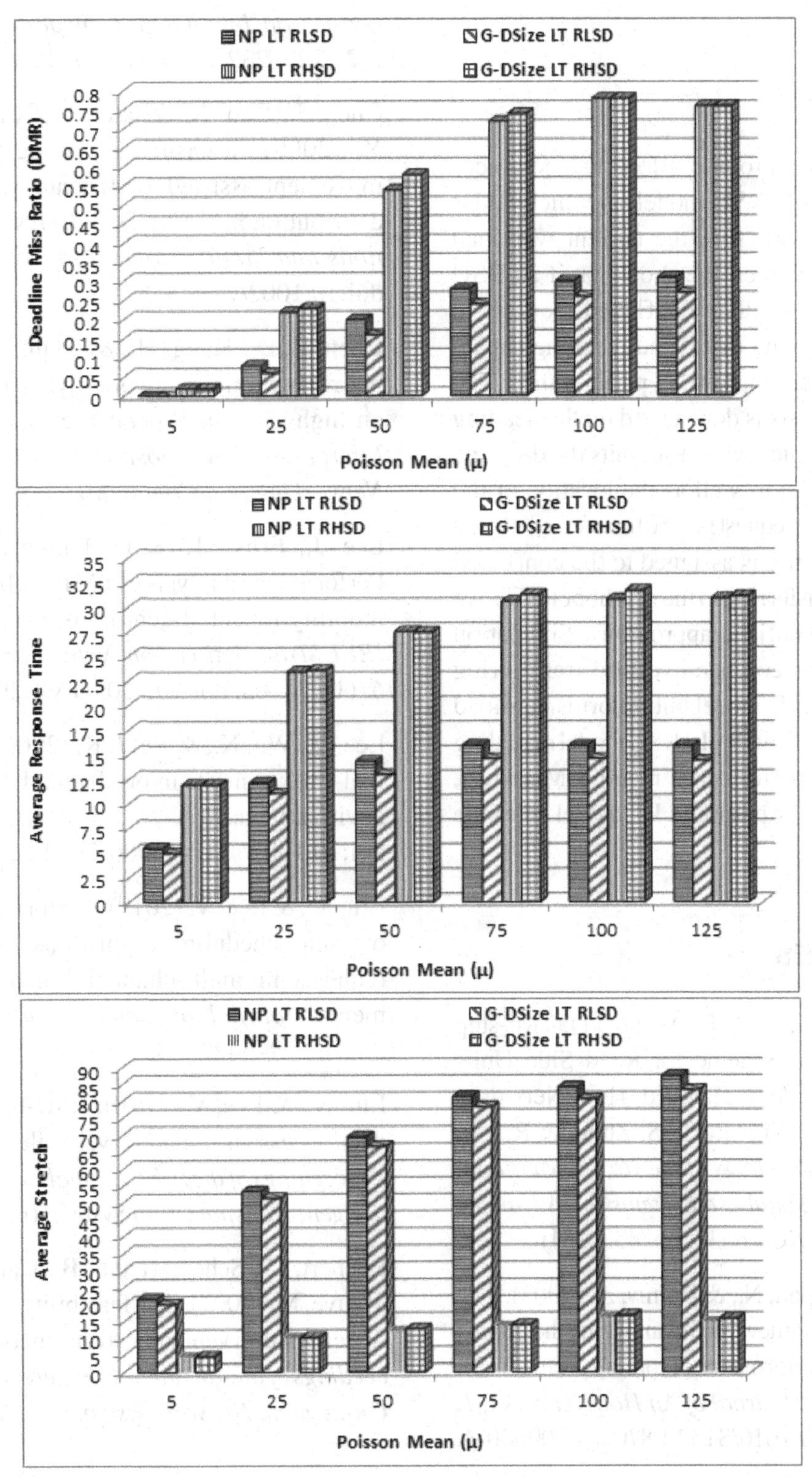

higher stretch (Figure 10(c) column 1, 2) than RHSD (column 3, 4) for different Poisson mean values (5-125).

CONCLUSION

In this paper, we propose a load balancing co-operative multiple-RSU model, in which RSUs are networked and share the current workload status with each other. We show that if an RSU can transfer its deadline conflicted requests to another RSU having extra load handling capacity, that RSU can improve its performance. The transfer of requests is done based on the heading direction of a vehicle which submits the deadline conflicted request as well as the location of the RSU to which the request is to be transferred. Based on the way priority is assigned to the conflicted requests for transferring to the neighbor RSUs, we propose three heuristic approaches. Simulation result shows our co-operative load transferring (LT) multiple-RSU model outperforms non-load transferring (NLT) multiple-RSU model regarding to a number of performance metrics. Moreover, the variants of our proposed LT model offer the expected result.

REFERENCES

Ali, G. G., Chan, E., & Li, W. (2011). Two-step Joint Scheduling Scheme for Road Side Units (RSUs)-based Vehicular Ad Hoc Networks (VANETs). In J. Xu, G. Yu, S. Zhou, & R. Unland (Eds.), *Proceedings of the 16th International Conference on Database Systems for Advanced Applications* (LNCS 6637, pp. 453-464).

Bai, F., Sadagopan, N., & Helmy, A. (2003). The IMPORTANT framework for analyzing the impact of mobility on performance of routing protocols for adhoc networks. *Journal of Ad Hoc Networks, 1*, 383–403. doi:10.1016/S1570-8705(03)00040-4.

Chang, C., Cheng, R., & Shih, H., C. Y. (2007). Maximum freedom last scheduling algorithm for downlinks of DSRC networks. *IEEE Transactions on Intelligent Transportation Systems, 8*(2), 223–232. doi:10.1109/TITS.2006.889440.

Chen, J., Cao, X., Zhang, Y., Xu, W., & Sun, Y. (2009). Measuring the performance of movement-assisted certificate revocation list distribution in VANETs. *Wireless Communications and Mobile Computing, 11*(7), 888–898. doi:10.1002/wcm.858.

Chen, Z. D., Kung, H., & Vlah, D. (2001). Ad hoc relay wireless networks over moving vehicles on highways. In *Proceedings of the 2nd ACM International Symposium on Mobile Ad Hoc Networking and Computing.*

Lee, J., Ernst, T., & Chilamkurti, N. (2011). Performance analysis of PMIPv6 based network mobility for intelligent transportation systems. *IEEE Transactions on Vehicular Technology, 61*(1), 74–85. doi:10.1109/TVT.2011.2157949.

Lee, V., Wu, X., & Ng, J. K. (2006). Scheduling real-time requests in on-demand data broadcast environments. *Real-Time Systems, 34*(2), 83–99. doi:10.1007/s11241-006-7982-5.

Liu, K., & Lee, V. (2010). Performance analysis of data scheduling algorithms for multi-item requests in multi-channel broadcast environments. *Journal of Communications Systems, 23*(4), 529–542.

Liu, K., & Lee, V. (2010). RSU-based real-time data access in dynamic vehicular networks. In *Proceedings of the IEEE Annual Conference on Intelligent Transportation Systems* (pp. 1051-1056).

Lochert, C., Scheuermann, B., Caliskan, M., & Mauve, M. (2007). The feasibility of information dissemination vehicular ad-hoc networks. In *Proceedings of the 4th Annual Conference on Wireless On-demand Network Systems and Services.*

Mak, T., Laberteaux, K., & Sengupta, R. (2005). A multi-channel VANET providing concurrent safety and commercial services. In *Proceedings of the 2nd ACM International Workshop on Vehicular Ad Hoc Networks*, Cologne, Germany (pp. 1-9).

Nadeem, T., Shankar, P., & Iftode, L. (2006). A comparative study of data dissemination models for VANETs. In *Proceedings of the 3rd Annual International Conference on Mobile and Ubiquitous Systems: Networking & Services* (pp. 1-10).

Schoch, E., Kargl, F. M., & Leinmüller, T. (2008). Communication patterns in VANETs. *IEEE Communications Magazine*, *46*(11), 119–125. doi:10.1109/MCOM.2008.4689254.

Schwetman, H. (2001). CSIM19: A powerful tool for building system models. In *Proceedings of the 33th IEEE Winter Simulation Conference*, Arlington, VA (pp. 250-255).

Wu, J., Fujimoto, R., Guensler, R., & Hunte, M. (2004). MDDV: A mobility-centric data dissemination algorithm for vehicular networks. In *Proceedings of the 1st ACM International Workshop on Vehicular Ad Hoc Networks* (pp. 47-56).

Wu, Y., & Cao, G. (2001). Stretch-optimal scheduling for on-demand data broadcasts. In *Proceedings of the Tenth International Conference on Computer Communications and Networks* (pp. 500-504).

Xu, J., Hu, Q., Lee, W., & Lee, D. L. (2004). Performance evaluation of an optimal cache replacement policy for wireless data dissemination. *IEEE Transactions on Knowledge and Data Engineering*, *16*(1), 125–139. doi:10.1109/TKDE.2004.1264827.

Xu, J., Tang, X., & Lee, W. (2006). Time-critical on-demand data broadcast algorithms, analysis and performance evaluation. *IEEE Transactions on Parallel and Distributed Systems*, *17*(1), 3–14. doi:10.1109/TPDS.2006.14.

Yi, L. Z., Bin, L., Tong, Z., & Wei, Y. (2008). On scheduling of data dissemination in vehicular networks with mesh backhaul. In *Proceedings of the IEEE International Conference on Communications Workshops* (pp. 385-392).

Zhang, Y., Zhao, J., & Cao, G. (2010). Service scheduling of vehicle-roadside data access. *Mobile Networks and Applications*, *15*(1), 83–96. doi:10.1007/s11036-009-0170-9.

Zhao, J., & Cao, G. (2008). VADD: Vehicle-assisted data delivery in vehicular ad hoc networks. *IEEE Transactions on Vehicular Technology*, *57*(3), 1910–1922. doi:10.1109/TVT.2007.901869.

Zhao, J., Zhang, Y., & Cao, G. (2007). Data pouring and buffering on the road: A new data dissemination paradigm for vehicular ad hoc networks. *IEEE Transactions on Vehicular Technology*, *56*(6), 3266–3277. doi:10.1109/TVT.2007.906412.

This work was previously published in the International Journal of Wireless Networks and Broadband Technologies, Volume 1, Issue 4, edited by Naveen Chilamkurti, pp. 1-21, copyright 2011 by IGI Publishing (an imprint of IGI Global).

Chapter 19
Cooperative Error Control Mechanism Combining Cognitive Technology for Video Streaming Over Vehicular Networks

Ming-Fong Tsai
Industrial Technology Research Institute, Taiwan

Naveen Chilamkurti
La Trobe University, Australia

Hsia-Hsin Li
Industrial Technology Research Institute, Taiwan

ABSTRACT

Video streaming over vehicular networks is an attractive feature to many applications, such as emergency video transmission and inter-vehicle video transmission. Vehicles accessing road-side units have a few seconds to download information and experience a high packet loss rate. Hence, this paper proposes Cooperative Error Control (CEC) mechanism combining Cognitive Technology (CT) for video streaming over wireless vehicular networks. CEC mechanism combining CT uses a cooperative error recover scheme to recover lost packets not only from a road-side unit which uses the primary channel but also from the other vehicles using a free channel. Hence, CEC mechanism with CT can enhance error recovery performance and quality of video streaming over vehicular networks. Simulation results show the error recover performance of the CEC mechanism combining CT performs better than the other related mechanisms.

DOI: 10.4018/978-1-4666-3902-7.ch019

1. INTRODUCTION

Vehicular networks have emerged as a promising wireless communication technology in recent years (Leinmueller, Schoch, Kargl, & Maihofer, 2010; Coronado & Cherkaoui, 2010; Haas, Hu, & Laberteauz, 2010). Video streaming over vehicular networks is challenging because the video data dependencies within the compressed stream make any data loss damaging. The organization of video streaming over vehicular networks was investigated in Guo, Ammar, and Zegura (2005). The quality of video streaming will decrease with packet loss which errors will propagate in the dependency frames. Chilamkurti, Tsai, Ke, Park, and Kang (2010) proposed a Reliable Transmission Mechanism (RTM) which proposed the roadside unit relies on extra parity bytes adaptively added to a byte-level Forward Error Correction (FEC) to enhance the transmission quality. However, vehicular networks are characterized by intermittent connectivity to the roadside unit. Hence, the lost packets can't recover when the vehicles aren't in range of roadside unit. Therefore, innovative error recover mechanism is needed. Because passing vehicles may not linger sufficiently for a full video sequence to be transferred from a roadside unit, partial storage in any one vehicle may occur. Morillo-Pozo, Trullols, Barcelo. and Garcia-Vidal (2008) proposed a Cooperative Automatic Repeat reQuest (CARQ) mechanism to solve this problem. CARQ mechanism is used in wireless vehicular networks where cars download delay-tolerant information form a road-side unit, suffering an intermittent connectivity. Cooperation among cars is established in the dark areas where connectivity with the road-side units is lost. CARQ can effectively reduce packet losses of transmission from road-side unit to vehicles in a platoon. The main idea is to effectively reduce the packet losses of transmission from road-side unit to vehicles in a platoon. However, the lost packets can't recover when the data packet losses are the same in the different neighbour vehicles. Moreover, the transmission channel of road-side unit to vehicles and transmission channel of vehicles to vehicles are the same which will decrease the transmission performance because of channel collision.

In order to solve previous problems, this paper proposes Cooperative Error Control (CEC) mechanism combining Cognitive Technology (CT) for video streaming over vehicular networks. CEC mechanism combining CT uses packet-level FEC mechanism to recover the lost source packets. The sender side overcomes packet losses by transmitting redundant information, allowing the recovery of a certain amount of missing video data at the receiver without retransmitting the lost packets using the primary channel in the range of road-side unit. When packet loss happens as vehicles leave the range of road-side unit, vehicles will try to receive the lost packet from the other vehicle's free channel which is searched by CT. The lost packets received can not only recover lost packets but can also help the receiver recover the other lost packets again with a packet-level FEC mechanism. Hence, CEC mechanism combining CT uses the error recover scheme to recover the lost packets not only from the road-side unit using the primary channel but also from the other vehicles using the free channel. CEC mechanism combining CT uses an analytical model to decide the number of FEC redundant packets in order to obtain the minimum recovery overhead. Accordingly, CEC mechanism combining CT not only can avoid network collision stemming from unlimited packet retransmission but also can reduce the FEC redundancy to conserve network bandwidth by predicting the effective packet loss rate. Moreover, CEC mechanism combining CT can obtain a higher reduction in the packet losses of transmission. CEC mechanism combining CT is tested to show the benefits of low Effective Packet Loss Rate (EPLR) in improving the Peak Signal to Noise Ratio (PSNR) and the Decodable Frame Rate (DFR) of video streaming over vehicular networks. Compared with two related works, CEC mechanism combining CT shows the promising results when delivering video streaming over vehicular networks.

The remainder of this paper is organized as follows. The background and related works are introduced in Section 2. CEC mechanism combining CT is presented in Section 3. Section 4 discusses the experimental set-up and analyzes the experimental results. Final, the paper is summarized in Section 5.

2. BACKGROUND AND RELATED WORK

2.1. Background

2.1.1. Wireless Access in the Vehicular Environment and Dedicated Short-Range Communication

WAVE/DSRC standards are developed to facilitate the provision of wireless access in vehicular environments. The protocol stacks of the WAVE/DESR architecture are illustrated in Figure 1. The concepts are detailed in Ho, Kang, Hsu, and Lin (2010), Tseng, Jan, Chen, Wang, and Li (2010), Li and Lin (2010), and IEEE (2010a, 2010b, 2010c, 2011) and described as follows. It can be observed that the WAVE physical and MAC layers are based on IEEE 802.11p, which is an amendment to IEEE 802.11a intended to apply in the rapidly varying vehicular environment. In the physical layer, the main modifications in IEEE 802.11p involve the number of channels, the bandwidth of the channels, and the modulation and coding schemes. Seven channels, each with 10 MHz bandwidth, are specified to overcome the extreme multipath environment. With the utilization of enhanced modulation and coding schemes, data transmission rates ranging from 3 to 27 Mbps are supported in high-speed mobile environments. The functionalities of each channel and enhanced priority control for traffic delivery are specified in the MAC layer of IEEE 802.11p. Furthermore, the authentication process for a device can be optionally ignored, and the transmission of data frames with different basic service set identifications is allowed among the communicating devices.

Figure 1. WAVE/DSRC protocol stacks

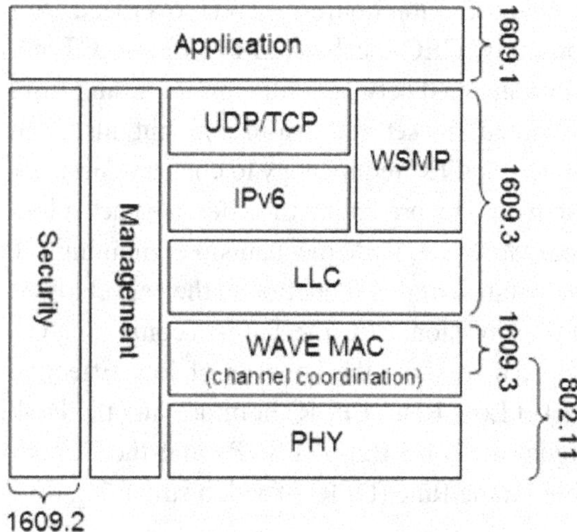

As shown in Figure 1, the upper layers of the WAVE/DSRC architecture are the IEEE 1609-family standards, which extend the specifications of IEEE 802.11p to cover additional layers in the protocol suite. The documents include IEEE 1609.1 (IEEE, 2010a), IEEE 1609.2 (IEEE, 2010b), IEEE 1609.3 (IEEE, 2010c), and IEEE 1609.4 (IEEE, 2011). Following the order of the layers in the Open Systems Interconnection (OSI) model from the bottom up, the purpose of each document is presented. IEEE 1609.4 provides enhancements to the IEEE 802.11p MAC to support multi-channel operation. Specification of the functions associated with the logical link control, network and transport layer of the OSI model are given in IEEE 1609.3. Specifically, IEEE 1609.3 provides addressing and data delivery services within a WAVE/DSRC system for Vehicle-to-Vehicle (V2V) and Vehicle-to-Roadside (V2R) communications. IEEE 1609.2 covers the format of secure messages and their encryption/decryption processes. IEEE 1609.1 defines methods for remote management between a resource manager and a remote device, which is organized into the application layer of the OSI model.

2.1.2. IEEE 1609.3 and IEEE 1609.4

IEEE 1609.3 and IEEE 1609.4 play important roles in the WAVE/DSRC standards. In IEEE 1609.4, one control channel (CCH) and six service channels (SCHs) are defined. Different types of messages need to be transferred on different channels. The concepts are detailed in Ho, Kang, Hsu, and Lin (2010), Tseng, Jan, Chen, Wang, and Li (2010), Li and Lin (2010), and IEEE (2010a, 2010b, 2010c, 2011) and described as follow. Channel coordination is achieved by an interval-based mechanism that utilizes the GPS timer or other methods to synchronize channel intervals among devices. For a WAVE/DSRC device, a series of sync intervals are provided. A sync interval consists of one CCH interval and the following SCH interval. The default channel of operation of WAVE/DSRC devices is the CCH. Three channel access mechanisms, including alternating access, immediate access and extended access, are specified in IEEE 1609.4, as shown in Figure 2. Under alternating access, the antenna of a WAVE/DSRC device will stay on an indicated CCH/SCH during a CCH/SCH interval. On the other hand, under immediate access, the device can access a SCH during a CCH interval. Extended access allows a WAVE/DSRC device to enlarge the duration of SCH access. The details of these mechanisms are described in IEEE (2011). In a WAVE system, a device providing services is named a provider, while the user is defined as a service subscriber. IEEE 1609.3 specifies the operation of providers and users in multi-channel environments. A provider broadcasts WAVE Service Advertisement (WSA) frames on CCH to announce the availability of services. A WSA frame records a detailed list of the services provided by the provider and their corresponding SCH allocation. The user monitors WSA on CCH during CCH intervals. After receiving a WSA with a subscribed service, the user will exchange data frames with the provider on the indicated SCH. If the provider or user is equipped with only one antenna, the aforementioned IEEE 1609.4 channel switching mechanism is required.

2.1.3. Forward Error Correction

Byte-level FEC is done on a per-packet basis (Tsai, Chilamkurti, Shieh, & Vinel, 2011). The concepts are detailed in Tsai, Chilamkurti, Shieh, and Vinel (2011) and described as follow. Tradi-

Figure 2. Channel access mechanisms

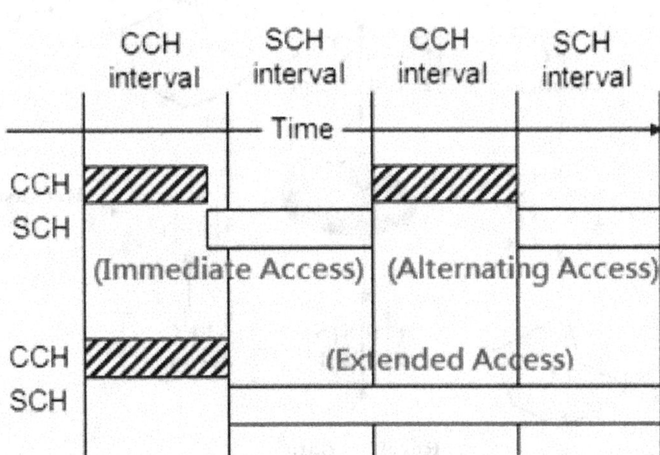

tionally, every packet carries a Cyclic Redundancy Check (CRC) field for error detection only. Because of more powerful processing abilities, this field recently is used for error correction. This technique is adopted by the Reed-Solomon (RS) encoder to process symbols where a symbol is a group of m bits. The maximum encoding data length can be GF(2^m). The RS processes k_b data symbols to generate 2t symbols, where t is the number of symbols that can be corrected by a RS decoder and $k_b + 2t = n_b$. The recovery capability of Byte-level FEC follows equation $t = (n_b - k_b)/2$ bytes, which means that Byte-level FEC can recover from errors if there are less than t bytes of error in a block. GF(2^8) coding is always used in Byte-level FEC, and the maximum coding size of GF(2^8) is 255 bytes. Byte-level FEC uses the byte interleaving technique to achieve robustness in face of burst errors. Byte-level FEC in MAC-layer applies Byte-level FEC to the MAC protocol data unit as the format in Figure 3 (Tsai, Huang, Shieh, & Chu, 2010). The RS block in the frame body can be shorter than 255 octets by using a shortened code such as (40, 24) RS code (which is a shortened version of (255, 239)) for the MAC header and CRC-32 for the Frame Check Sequence (FCS) (Shevtekar, Stille, & Ansari, 2008). A RS block can correct up to 8-byte errors. The outer FCS allows the receiver to skip the RS decoding process if the FCS is correct. The FCS allows the receiver to identify a potential false decoding of the RS decoder since the RS decoder may incorrectly decode the RS block (Tsai, Shieh, Ke, & Deng, 2010).

Packet-level FEC is a method commonly used to handle losses in real-time communication (Tsai, Chilamkurti, & Shieh, 2011). Figure 4 shows the packet-level FEC concept. The concepts are de-

Figure 3. Byte-level FEC in MAC-layer concept

Mac Header		Frame Body (M blocks)								FCS
Header	Header FEC	Payload 1	FEC	Payload 2	FEC	...	Payload M	FEC	FCS
24	16	239	16	239	16	...	239	16	4

Figure 4. Packet-level FEC concept

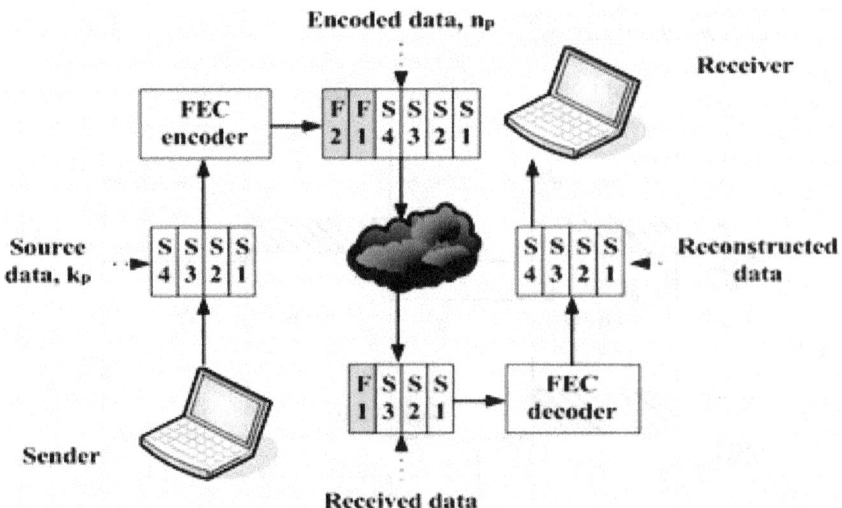

tailed in Tsai, Chilamkurti, and Shieh, (2011) and described as follow. A (n_p, k_p) block erasure code converts k_p source packets into a group of n_p coded packets so that any k_p of the n encoded packets can be used to reconstruct the original source packets (Tsai, Shieh, Hwang, & Deng, 2009). The remaining $(n_p - k_p)$ packets are referred to as parity packets. Packet-level can correct both errors and erasures in a block of n_p symbols. An error is defined as a corrupted symbol in an unknown position, while an erasure is a corrupted symbol in a known position (Tsai, Chilamkurti, Park, & Shieh, 2010). In the case of streamed media packets, loss detection is performed based on the sequence numbers in packets (Tsai, Chilamkurti, & Shieh, 2010).

2.2. Related Work

Many researchers have proposed adaptive FEC mechanisms in vehicular networks. Ahmad, Habibi, and Rahman (2008) proposed a FEC mechanism for mobile wireless communications at vehicular speeds. This paper estimates and adjusts the size of extra parity bits to suit the channel conditions which is an enhancement over standard FEC that uses a fixed block of parity bits. Bucciol, Zechinelli-Martini, and Vargas-Solar (2009) address real-time transmission optimization of loss tolerant information streams over wireless vehicular ad hoc networks. The proposed FEC mechanism combines interleaving technology to provide better recovery performance. The proposed FEC mechanism can not only dynamically adapt the inter-arrival frequency of the receiver reports, which allows a better estimate of channel conditions, but also of the FEC and interleaving parameters. Ahmad and Habibi (2008) adjust the size of byte-level FEC mechanism redundancy in response to packet retransmission requests as an enhancement to standard FEC mechanism which uses a fixed FEC redundancy. The sender side can obtain the channel condition based on the adaptive interval sampling time from the information feedback from the receiver side. Hence, the sender side can dynamically adjust the FEC redundancy based on current channel condition. However, these previously proposed mechanisms (Ahmad, Habibi, & Rahman, 2008; Bucciol, Zechinelli-Martini, & Vargas-Solar, 2009; Ahmad & Habibi, 2008) have a critical problem whereby they will not always recover the corrupt packet. The reason is that the byte-level FEC mechanism requires a redundancy of two bytes to recover one byte of error. In Ahmad, Habibi, and Rahman (2008), Bucciol, Zechinelli-Martini, and Vargas-Solar (2009), and Ahmad and Habibi (2008) for example, if the source data is 200 KB, the bit error rate computed by the equation modelling in Ahmad, Habibi, and Rahman (2008), Bucciol, Zechinelli-Martini, and Vargas-Solar (2009), and Ahmad and Habibi (2008) is 20%. For example, the mechanisms add an FEC redundancy of 80 KB (20%) into the source data packet. Hence the total size of the transmission data packet is 280 KB, which at 20% bit error rate gives a data packet with 56 KB bytes of error at the receiver. Hence, mechanism based on the methodology of Ahmad, Habibi, and Rahman (2008), Bucciol, Zechinelli-Martini, and Vargas-Solar (2009), and Ahmad and Habibi (2008) can never fully recover the data packet even when the data packet is protected by a byte-level FEC mechanism because the byte-level FEC mechanism only tolerates 40 KB of byte errors. Hence, Chilamkurti, Tsai, Ke, Park, and Kang (2010) proposed a RTM mechanism to solve previous problem. However, previous proposed FEC mechanisms only enhance the FEC recovery performance in road-side unit to vehicle transmission. They can't work when the vehicles aren't in range of roadside unit. Morillo-Pozo, Trullols, Barcelo, and Garcia-Vidal (2008) proposed a CARQ mechanism to be used in wireless vehicular networks where cars download delay-tolerant information form road-side units on the road, suffering an intermittent connectivity. Cooperation among cars is established in the dark areas where connectivity with the road-side units is lost. However, the data packet losses cannot be recovered when the data packet losses are the same in the different neighbour vehicles. Moreover, the transmission channel of road-side unit to vehicles and transmission channel of vehicles to vehicles are the same which will cause channel collision.

3. COOPERATIVE ERROR CONTROL MECHANISM COMBINING COGNITIVE TECHNOLOGY

3.1. System Overview

In this section, we will introduce the CEC mechanism combining CT that allows vehicles to work cooperatively in order to increase the transmission rate of all of them in packets received from a fixed road-side unit. In Figure 5, for example, car 1 and car 2 will come from the range of road-side unit 1 range to the range of road-side unit 2. When car 1 and car 2 are in the range of road-side unit 1 and road-side unit 2, they will receive the packet from the road-side unit using the primary channel. When packet loss happens as they leave the range of road-side unit 1 and road-side unit 2, they will be within vehicular range and will try to receive the lost packet from the other cars' free channel which is searched by CT. The concept of the CT will be described in Section 3.2.

In Figure 6, for example, the road-side unit transmits six source packets and two FEC redundant packets to car 1 and car 2 simultaneously. However, the packet loss rate in vehicular networks is higher than in wired networks. Hence, the packet 5, packet 6, and packet F1 are lost in car 1 and the packet 5, packet 6, and packet F1 are lost in car 2. In car 1 and car 2, the FEC block cannot recover source packets because the number of lost packets in larger than the number of FEC redundant packets. Hence, the previous method (Morillo-Pozo, Trullols, Barcelo, & Garcia-Vidal, 2008) cannot enhance the transmission rate because packet 5 and packet 6 are lost. Moreover, the previous method (Morillo-Pozo, Trullols, Barcelo, & Garcia-Vidal, 2008) will transmit the data packet using the same channel as the road-side unit to vehicles in vehicles to vehicles. Hence, the transmission performance will decrease. However, in the CEC mechanism combining CT, car 1 will receive the packet F1 from car 2 and car 2 will receive the packet 4 from car 1within vehicular range using the free channel which is searched by cognitive technology. Hence, not only

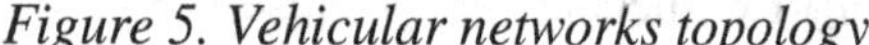

Figure 5. Vehicular networks topology

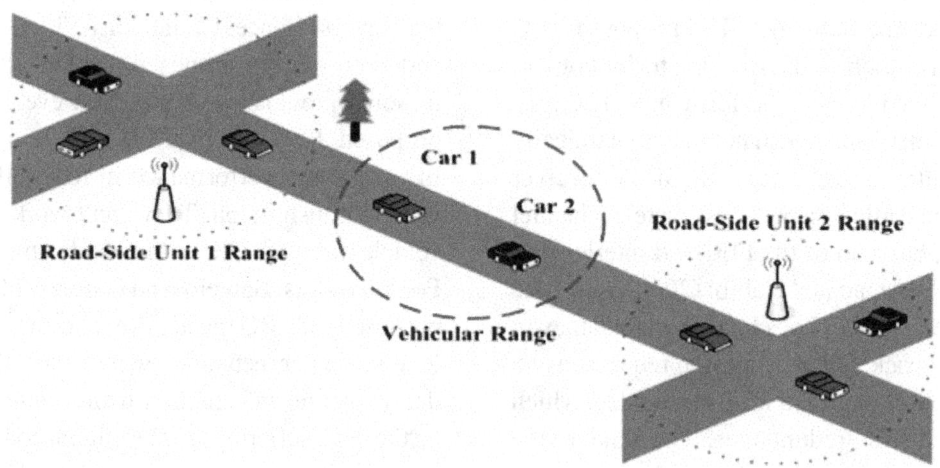

Figure 6. CEC mechanism concept

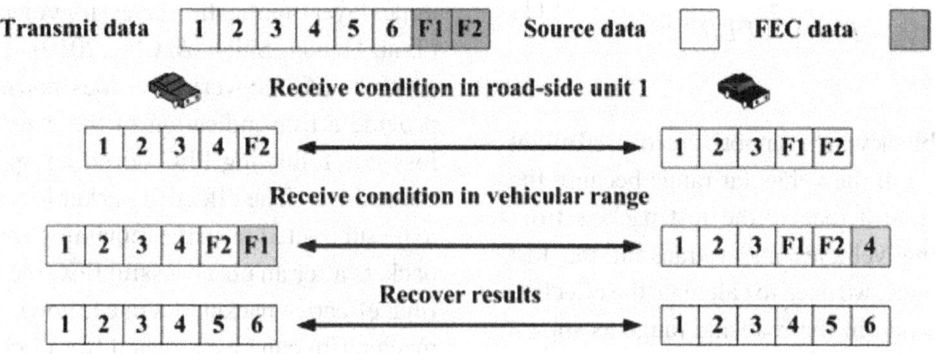

can car can recover packet 5 and packet 6 but also the car 2 can recover the packet 4 relying on the FEC mechanism in recover results. Accordingly, the lost packet also can be recovered if car 1 and car 2 lost the same packet. Moreover, CEC mechanism combining CT will not cause channel collision during the data transmission between the vehicles.

3.2. Cognitive Technology

The devices are able to use the unused frequency channel and reconfigure the current operating frequency channel when the frequency channel is full or subject to heavy competition in cognitive technology (Shen, Jiang, Liu, & Zhang, 2009). Hence, the devices can share the frequency channel to increase the utilization of frequency channels relying on cognitive technology. Many related works (Chung, Yoo, & Kim, 2009; Tsai, Chilamkurti, Zeadally, & Vinel, 2011; Tsai, Huang, Ke, & Hwang, 2011; Huang, Huang, Chilamkurti, Cheng, & Shieh, 2010) have proposed many schemes for the effective usage of the frequency channel relying on cognitive technology. The frequency channel agility can let the devices change their operating frequency channel to the optimal frequency channel. For example, dynamic frequency channel selection is a method of selection which senses signals from nearby transmitters and chooses an optimal operating environment. Hence, the devices can exploit all opportunities for the usage of the frequency channel. Moreover, power control of transmission can adjust the transmit power up to full power to allow more efficient sharing of the frequency channel. Furthermore, location awareness is an important parameter for increasing frequency channel re-use. The devices are able to determine their location and the location of other devices operating at the same frequency, relying on cognitive technology. Hence, in this paper, we assume the devices of mobile node are equipped with two wireless cards. The frequency channel of one wireless card connects to the road-side unit and the other one connects to the other vehicles. The devices use cognitive technology to find the free channel for transmitting data within vehicular range.

3.3. Mathematical Model of CEC Mechanism

We use a mathematical model to represent and analyse the protection ability of the CEC mechanism. We assume that the number of total packets in one FEC block is n, the number of source packets is k and the number of FEC redundant packets is a function of $(n-k)$, i.e., h. When the packet loss rate, *PLR*, in the vehicular network is known, the probability of a FEC block that cannot be recovered by FEC redundancy in the range of road-side is given by Equation 1.

$$P_{RS} = \sum_{i=0}^{k-1} C_i^n \times (1 - PLR)^i \times PLR^{n-i} \quad (1)$$

A FEC block which cannot be recovered might be recovered in the vehicular range because the receiver can still receive the lost packets from neighbouring vehicles and retransmit the lost packets. Hence, we need to calculate the effective packet loss rate in the roadside range as shown in Equation 2.

$$P_{e\_RS} = \frac{P_{RS} \times n - \sum_{i=0}^{k-1} C_i^n \times (1 - PLR)^i \times PLR^{n-i} \times i}{n} \quad (2)$$

We assume that the total number of vehicles in the vehicular range is v, hence the effective packet loss rate, when the lost packet received relies on neighbouring vehicles in the vehicular range, can be estimated from Equation 3.

$$P_{e\_V} = P_{e\_RS} \times \left(1 - \left(1 - P_{e\_RS}\right) \times \left(1 - PLR\right)\right)^{v-1} \quad (3)$$

Moreover, the lost packets received that rely on neighbouring vehicles in the vehicular range can help the FEC mechanism to recover the lost packets. Hence, the probability of a FEC block that still cannot be recovered in the vehicular range is given by Equation 4.

$$P_{afterV} = \sum_{i=0}^{k-1} C_i^n \times \left(1 - P_{e\_V}\right)^i \times P_{e\_V}^{\ n-i} \quad (4)$$

Furthermore, the receiver can still receive (k-1) or fewer source video packets from the FEC blocks that could previously not be recovered. However, these source video packets can still be passed to the upper layer (application layer) for video decoding (Tsai, Huang, Shieh, & Chu, 2010). The unsuccessful FEC recovery rate does not necessarily provide a true indication of the average packet loss rate following FEC decoding operations at the receiver. The effective packet loss rate needs to be subtracted from the remaining source video packets after an unsuccessful FEC recovery. The final effective packet loss rate following the CEC mechanism can be estimated from Equation 5.

$$P_{effective} = P_{afterV} - \frac{\sum_{i=0}^{k-1} C_i^n \times \left(1 - P_{e\_V}\right)^i \times P_{e\_V}^{\ n-i} \times i}{n} \quad (5)$$

For a given value of the PLR, the analytical model given in Equation 5 provides a simple yet accurate means of predicting the effective packet loss rate by the CEC mechanism. The effective packet loss rate can then be extended to evaluate the quality of the delivered video by applying a video performance metric.

3.4. Minimum Recovery Overhead Model

On the one hand, the CEC mechanism has to reduce the FEC redundant packets, because minimising the recovery overhead is important to reduce the consumption of network bandwidth in wireless networks. On the other hand, the CEC mechanism should allow the vehicular network maximum packet recovery capability with more FEC redundant packets. Hence, the CEC mechanism calculates the minimum recovery overhead as described in the following paragraph. If the lost packets can be recovered from the FEC block, the recovery overhead is the number of redundant packets alone. Conversely, if they cannot be recovered, the recovery overhead includes the FEC redundant packets and the receipt packets that cannot be recovered, because they waste network bandwidth. Therefore, the recovery overhead model can be expressed by Equation 6.

$$Overhead_{CEC} = P_{afterV} \times \left(P_{effective} \times k + h\right) + \left(1 - P_{afterV}\right) \times h \quad (6)$$

According to Equation 6, the recovery overhead increases when k or h increases. On the other hand, video streaming will have a high protection and recovery overhead when h is large. In order to determine the minimum recovery overhead, the CEC mechanism assumes that the application can tolerate the average packet loss rate of video streaming up to a certain acceptable range $P_{acceptable}$, e.g., 0.5% (Tsai, Huang, Shieh, & Chu, 2010). The concepts are detailed in Tsai, Chilamkurti, Shieh, et al. (2011), Tsai, Huang, et al. (2010), Shevtekar, Stille, and Ansari (2008), Tsai, Shieh, Ke, et al. (2010), Tsai, Chilamkurti, and Shieh (2011), Tsai, Shieh, Hwang, et al. (2009), and Tsai, Chilamkurti, Park, et al. (2010) and described as follow. Accordingly, the CEC mechanism can calculate the optimal FEC parameters, i.e., k, h, v and PLR, for the minimum recovery overhead. The CEC mechanism can calculate the minimum recovery overhead with Equation 7.

$$\left(k,\ h,\ \text{v},\ PLR\right) = \arg\ \min\ \left(Overhead_{CEC}\right) \quad (7)$$
$$s.t.\ \mathrm{P}_{effective} \leq \mathrm{P}_{acceptable}$$

CEC mechanism can use the minimum recovery overhead model to find the optimal FEC parameters in order to dynamically adjust the appropriate number of FEC redundant packets according to network conditions. Accordingly, CEC mechanism not only gives video streaming appropriate protection, but also avoids unnecessary network bandwidth consumption. The FEC redundancy of CEC mechanism combining cognitive technology is encoded in the road-side unit. The road-side unit can depend on momentary network condition to create FEC redundancy. When the packet loss rate is high, the road-side unit will generate a larger FEC redundancy to protect the source data packets. On the other head, the road-side unit will generate less FEC redundancy when the packet loss rate is low, in order to reduce the error recovery overhead. Hence, the FEC redundancy of CEC mechanism combining CT can be determined more accurately on the fly. The road-side unit needs to know the packet loss rate to rely on statistically. If the road-side unit knows the packet loss rate, we can analyse the recovery performance of the FEC mechanism. Hence, CEC mechanism combining CT can choose the appropriate FEC redundancy according to the user's settings.

4. EXPERIMENTAL RESULTS

The simulation topology is shown in Figure 7. The video server sends a video stream to video receivers. The video clip is 'Foreman' which is encoded in MPEG-4 QCIF format. The 'Foreman' is codec at 960 Kbps and 30 frames per second. The trace is a transmission involving a total of 1024 video source packets. The data rate of the wireless link is following the IEEE 802.11p standard. The handoff problem is beyond the scope of this paper. The video server transmit data from 0 s to 3 s using primary channel in order to simulate the data of video streaming will transmit through the road-side unit 1 when car 1 and car 2 are in range of road-side unit 1. The video receiver can transmit data to each other from 3 s to 6 s in order to simulate the data of video streaming and will transmit using the free channel when car 1 and car 2 are in vehicular range. The video server transmits data from 6 s to 9 s using the primary channel in order to simulate the data of video streaming and will transmit through road-side unit 2 when car 1 and car 2 are in range of road-

Figure 7. Simulation topology

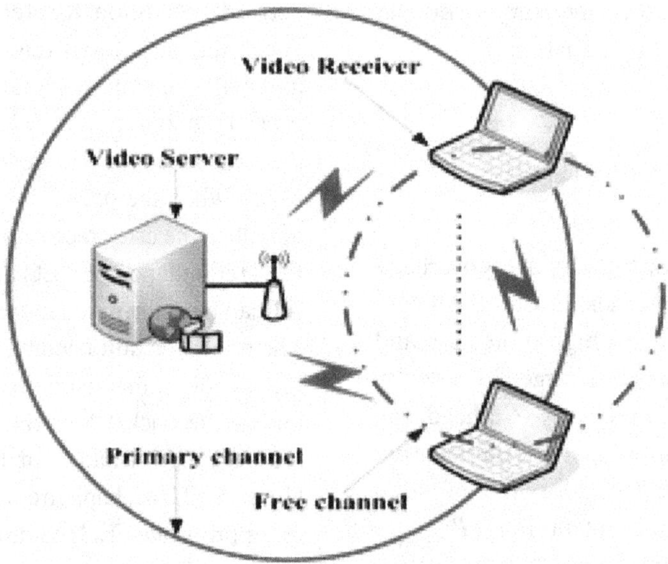

side unit 2. In this subsection, CEC mechanism combining CT of encoding and decoding FEC redundancy relies on the Reed–Solomon Code (Babu & Venkataram, 2009).

4.1. Model Evaluation

We operate the sender under network conditions with different average packet loss rates (where the average packet loss rate in the roadside and the vehicular range are the same) and observe the effective packet loss rate at the receiver, as shown in Figure 8, Figure 9, and Figure 10. Figure 8 shows the effective packet loss rate in the roadside range (without entering the vehicular range) in network conditions with different average packet loss rates. Hence, the effective packet loss rate decreases as the FEC redundant packet increases in ever vehicles. According to Figure 8, we validate the efficiency and accuracy of Equation 2 in the CEC mechanism. Figure 9 presents the effective packet loss rate when the received lost packets rely on neighbouring vehicles in the vehicular range, under network conditions with different average packet loss rates. There are no the other vehicles that can help retransmit the lost packets in the vehicular range when v is one. Hence, the effective packet loss rate decreases when the FEC redundant packets increase when v equals one, which is confirmed in Figure 8. However, there are the other vehicles that can help to retransmit the lost packet when v is greater than one. Comparing the situation where k is eight, h is zero, v is two and average packet loss rate is 10% in Figure 8 and Figure 9. In this condition, we can see the performance of the retransmission of lost packets. The effective packet loss rate in Figure 8 is 10% and the effective packet loss rate in the Figure 9 is 1.9%. Moreover, under the same conditions but when v is three and four, the effective packet loss rate in Figure 9 is 0.3% and 0.07%. Hence, the effective packet loss rate decreases when the number of vehicles is greater in the vehicular range. Furthermore, the retransmission scheme in the vehicular range can reduce the effective packet loss rate, which can help the FEC scheme to recover the lost packets in the further. Hence, for example, comparing the situation when k is eight, h is one, v is two and average packet loss rate is 10% in Figure 8 and Figure 9. The effective

Figure 8. The effective packet loss rate in the range of road-side

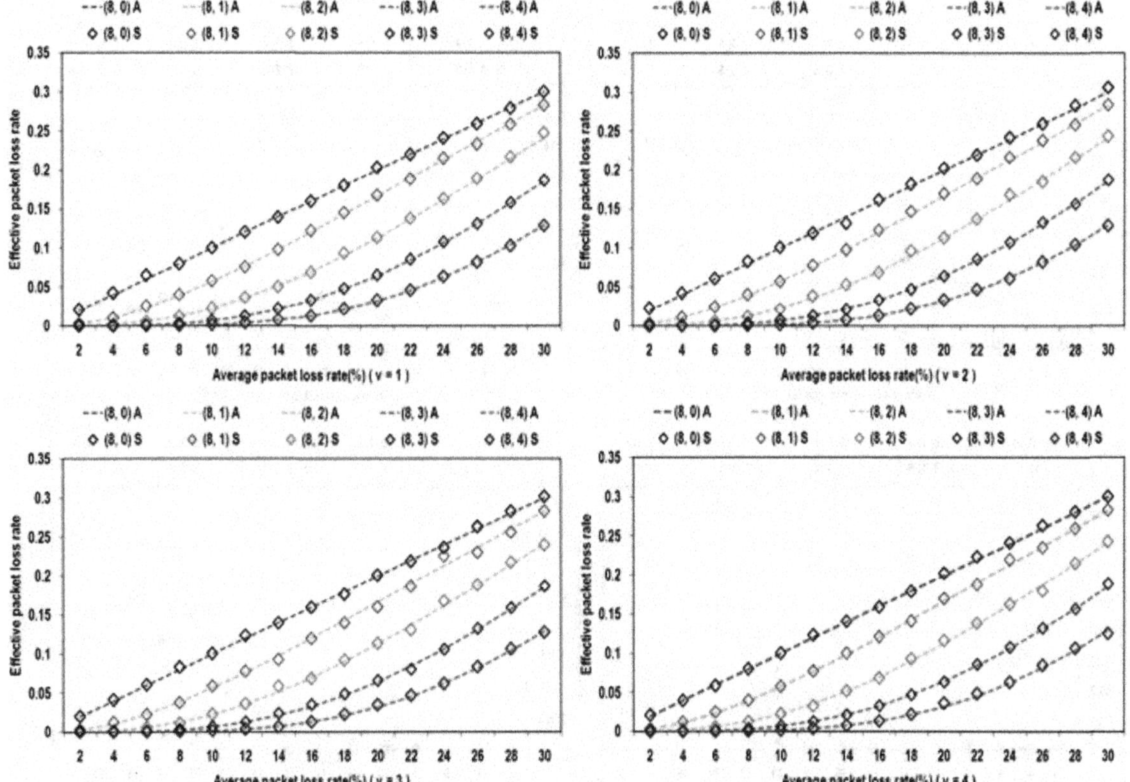

packet loss rate in Figure 8 is 5.6% and the effective packet loss rate in Figure 9 is 0.88%. Moreover, under the same conditions but when v is three and four, the effective packet loss rate in Figure 9 is 0.13% and 0.02%. We will then compare this value to those in Figure 10. According to Figure 9, we validate the efficiency and accuracy of Equation 3 by the CEC mechanism.

Figure 10 is the final effective packet loss rate following the CEC mechanism under network conditions of different average packet loss rates. The FEC redundancy can help recover performance when the retransmission scheme can work in the vehicular range which reduces the effective packet loss rate in Figure 9. For example, comparing the situation where k is eight, h is one, v is two and the average packet loss rate is 10% in Figure 8, Figure 9 and Figure 10. In this condition, we can see the performance of the retransmission of lost packets and FEC recovery. The effective packet loss rate will decrease when the retransmission scheme in the vehicular range can be works that means that v is greater than one as shown in Figure 9. The effective packet loss rate in Figure 8 is 5.6% (the effective packet loss rate is 10% when h is zero) and the effective packet loss rate in Figure 10 is 0.06% (the effective packet loss rate in Figure 9 is 0.88%). Moreover, under the same conditions but when v is three and four, the effective packet loss rate in Figure 10 is 0.001% and 0% (the effective packet loss rate in Figure 9 is 0.13% and 0.02%). Hence, the effective packet loss rate which includes FEC redundant packets is lower than the effective packet loss rate without including FEC redundant packets. Moreover, the effective packet loss rate which includes the retransmission scheme in the vehicular range is lower than the effective packet loss rate without

Figure 9. The effective packet loss rate when received lost packets rely on neighbouring vehicles in the vehicular range

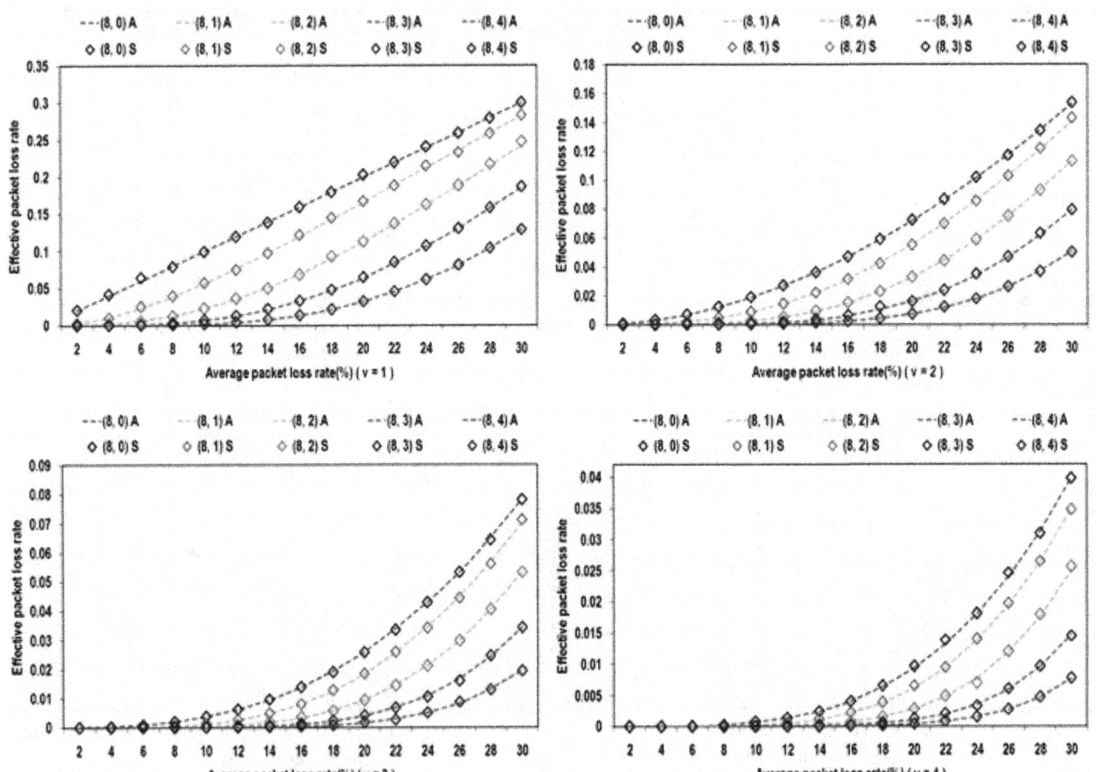

including the retransmission scheme in the vehicular range. According to Figure 10, we validate the efficiency and accuracy of Equation 5 in the CEC mechanism. Next section shows the performance of the CEC mechanism performs better than the other related mechanisms.

4.2. Comparison with the Other Related Work

The simulation topology and simulation scenario is the same as in Section 4.1. The effective packet loss rate of CEC mechanism combining CT has a lower bound less than 0.005 conditions. We use metrics which are the EPLR, DFR and PSNR to compare the CEC mechanism combining CT with RTM (Chilamkurti, Tsai, Ke, Park, & Kang, 2010) and CARQ (Morillo-Pozo, Trullols, Barcelo, & Garcia-Vidal, 2008) in performance against packet loss. The EPLR of the CEC mechanism combining CT and CARQ mechanisms for different packet loss rates are shown in Figure 11. The RTM mechanism relies on byte level FEC redundancy to protect the original source packets. However, the whole packet will be discarded when the length of error bytes is larger than the length of byte level FEC redundancy. Moreover, the RTM mechanism will not recover the lost packets relying on neighbouring vehicles in the vehicular range. Hence, the curve of the RTM are the same when v is one to four. The CARQ mechanism relies on retransmission in the vehicular range for reducing the packet losses of transmission from the roadside unit to vehicles in a platoon. However, the effective packet loss rate is highest when v is one, which means this is without any

Figure 10. The final effective packet loss rate following the CEC mechanism

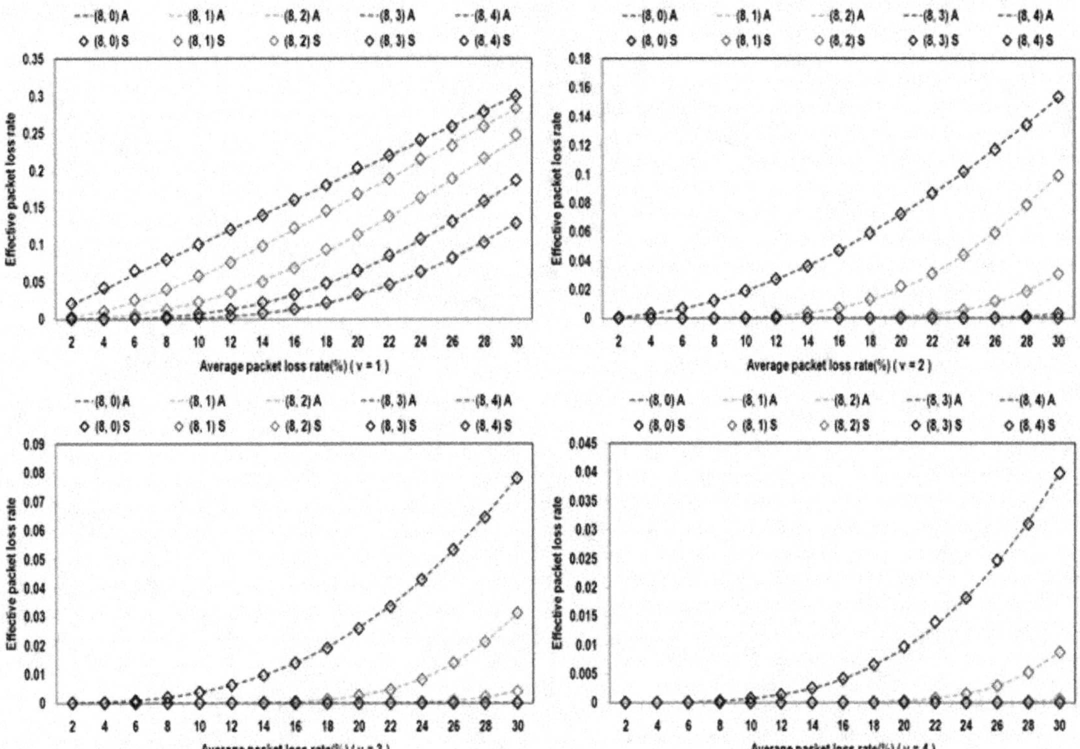

neighbouring vehicle in the vehicular range. The effective packet loss rate decreases when the number of neighbouring vehicle in the vehicular range increases. Moreover, the effective packet loss rate of the CARQ mechanism is not lowest because the data packet losses cannot be recovered when the data packet losses are the same in the different neighbouring vehicles. CEC mechanism combining CT uses cooperative FEC redundancy which can recover any lost packets relying on retransmission within vehicular range. Hence, CEC mechanism combining CT can obtain better recovery performance than the RTM mechanism and CARQ mechanism in all cases in Figure 11. Moreover, CEC mechanism combining CT uses an error recover scheme to recover the lost packets not only from the road-side unit using primary channel but also will receive redundancy from the other vehicles using the free channel. Hence, we believe the CEC mechanism combining CT will not cause channel collisions during transmission.

The DFR is an application level metric used to evaluate the quality of video streaming. Application level metrics are better for user-evaluation of the quality of video streaming than network level metrics such as the effective packet loss rate. The value of the DFR lies between zero and one. Higher values of the DFR mean better video quality. Because of the RTM mechanism, the whole packet will be discarded when the length of error bytes is larger than the length of byte level FEC redundancy and the CARQ mechanism which the data packet losses cannot be recovered when the data packet losses are the same in the different neighbouring vehicles. Moreover, the DFR curves of the CARQ mechanism and CEC mechanism improve when v increases because the lost packets can be recovered by other vehicles in the ve-

Figure 11. Performance of the effective packet loss rate

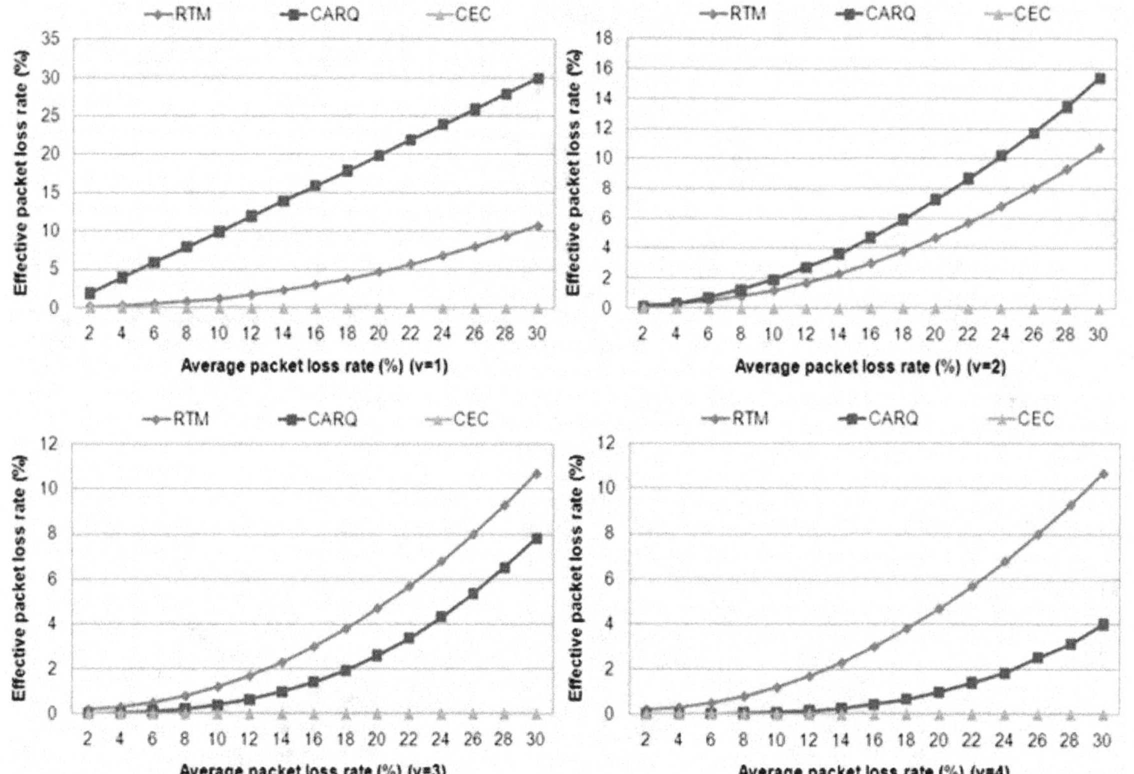

hicular range. Furthermore, the CEC mechanism relies on the recovered lost packets in the vehicular range which can enhance the recovery performance of the second recovery process of the FEC scheme. Hence, we observe that the CEC mechanism combining CT has a higher DFR than the RTM mechanism and CARQ mechanism in networks having different average loss rates in Figure 12.

The PSNR is an objective metric for measuring the quality of video streaming reconstruction. A lower value of PSNR means a poorer video quality. For the same reason as previously described, the PSNR curves of the RTM mechanism will not improve, because the RTM mechanism will not recover the lost packets from the neighbouring vehicles in the vehicular range. The RTM mechanism will discard the whole packet when the length of error bytes is larger than the length of the byte level FEC redundancy. The PSNR curves of the CARQ mechanism will improve because the CARQ mechanism will recover the lost packets from the neighbouring vehicles in the vehicular range, but the data packet losses cannot be recovered when the data packet losses are the same in the different neighbouring vehicles. Accordingly, CEC mechanisms do not have the drawback of the RTM mechanism (in that they cannot recover the lost packets by the neighbouring vehicles in the vehicular range) and the CARQ mechanism (cannot be recovered when the data packet losses are the same in the different neighbouring vehicles). Hence, we observe that the CEC mechanism combining CT has a higher PSNR than the RTM mechanism and CARQ mechanism in networks having different average

Figure 12. Performance of the decodable frame rate

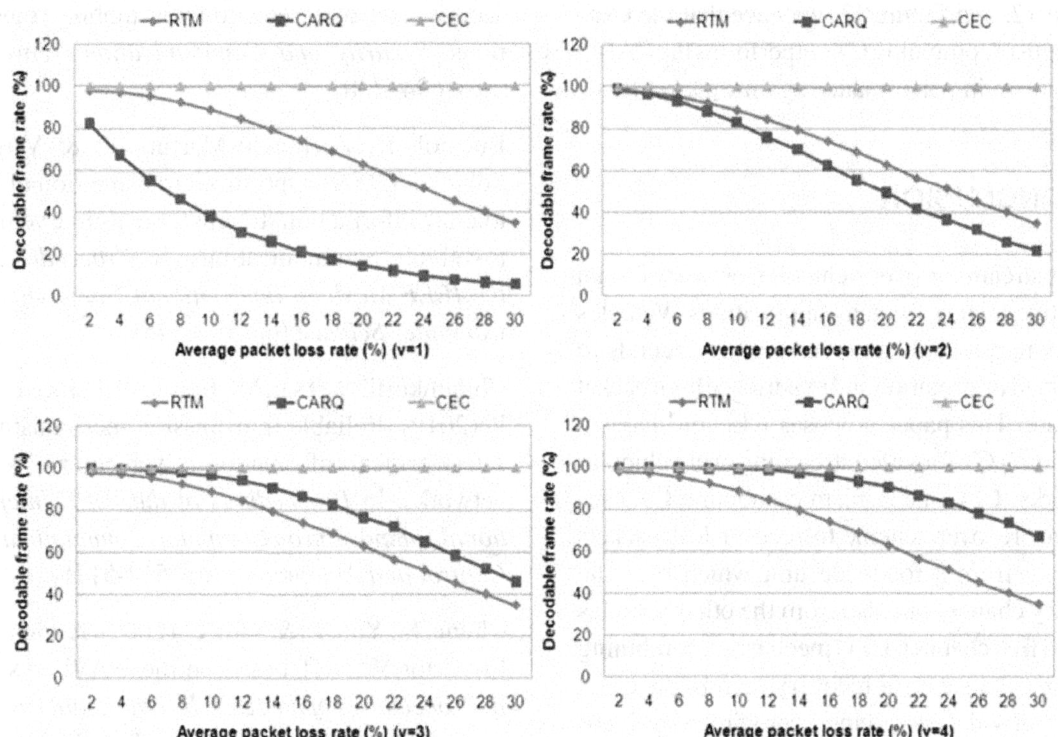

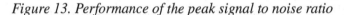

Figure 13. Performance of the peak signal to noise ratio

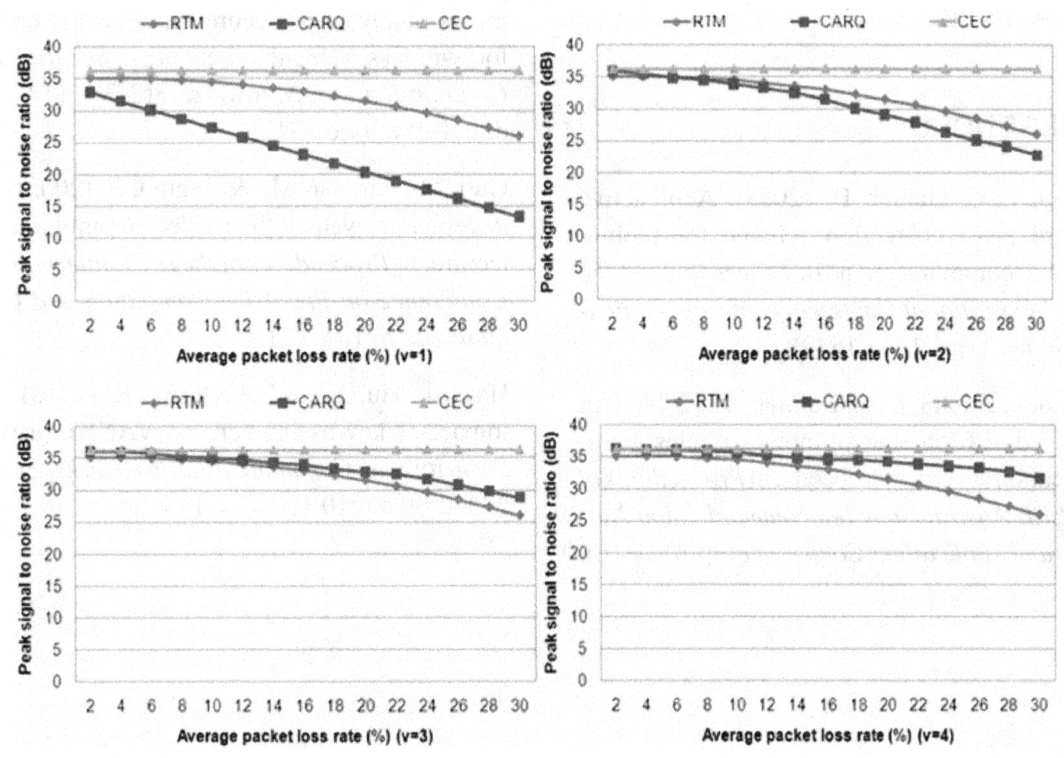

loss rates in Figure 13. According to Figure 11, Figure 12, and Figure 13, we have that the CEC mechanism combining CT outperforms the CARQ mechanism in performance against packet loss.

5. CONCLUSION

Video streaming over vehicular networks is an attractive feature to many applications. Vehicles accessing road-side units have a few seconds to download information and experience high packet loss rate. This paper proposes CEC mechanism combining CT for video streaming over vehicular networks. CEC mechanism combining CT uses an error recover scheme to recover lost packets not only from a road-side unit which uses the primary channel but also from the other vehicles using a free channel. CEC mechanism combining CT can enhance error recovery performance and quality of video streaming over wireless vehicular networks. Simulation results show the CEC mechanism combining CT performs better than RTM and CARQ mechanisms in terms of EPLR, DFR and PSNR.

REFERENCES

Ahmad, I., & Habibi, D. (2008). A proactive forward error correction scheme for mobile WiMAX communication. In *Proceedings of the IEEE International Conference on Communication Systems* (pp. 1647-1649).

Ahmad, I., Habibi, D., & Rahman, Z. (2008). An improved FEC scheme for mobile wireless communication at vehicular speeds. In *Proceedings of the IEEE Australasian Telecommunication Networks and Applications Conference* (pp. 312-316).

Babu, B., & Venkataram, P. (2009). Random security scheme selection for mobile transactions. *Security and Communication Networks*, *2*(6), 694–708.

Bucciol, P., Zechinelli-Martini, J., & Vargas-Solar, G. (2009). Optimized transmission of loss tolerant information streams for real-time vehicle-to-vehicle communications. In *Proceedings of the IEEE Mexican International Conference on Computer Science* (pp. 142-145).

Chilamkurti, N., Tsai, M., Ke, C., Park, J., & Kang, P. (2010). Reliable transmission mechanism for safety-critical information in vehicular wireless networks. In *Proceedings of the IEEE International Symposium on Computer, Communication, Control and Automation* (pp. 514-517).

Chung, S., Yoo, J., & Kim, C. (2009). A cognitive MAC for VANET based on the WAVE system. In *Proceedings of the IEEE International Conference on Advanced Communications Technology* (pp. 41-46).

Coronado, E., & Cherkaoui, S. (2010). Performance analysis of secure on-demand services for wireless vehicular networks. *Security and Communication Networks*, *3*(2-3), 114–129. doi:10.1002/sec.134.

Guo, M., Ammar, M., & Zegura, E. (2005). V3: A vehicle-to-vehicle live video streaming architecture. In *Proceedings of the IEEE International Conference on Pervasive Computing and Communications* (pp. 171-180).

Haas, J., Hu, Y., & Laberteauz, K. (2010). The impact of key assignment on VANET privacy. *Security and Communication Networks*, *3*(2-3), 233–249. doi:10.1002/sec.143.

Ho, K., Kang, P., Hsu, C., & Lin, C. (2010). Implementation of WAVE/DSRC devices for vehicular communications. In *Proceedings of the IEEE International Symposium on Computer Communication Control and Automation* (pp. 522-525).

Huang, H., Huang, T., Chilamkurti, N., Cheng, R., & Shieh, C. (2010). Adaptive forward error correction with cognitive technology mechanism for video streaming over wireless networks. In *Proceedings of the IEEE International Symposium on Computer Communication Control and Automation* (pp. 518-521).

IEEE. (2010a). IEEE P1609.1TM/D1.3: Draft standard for wireless access in vehicular environments (WAVE) – Remote management service. Piscataway, NJ: IEEE.

IEEE. (2010b). IEEE P1609.2TM/D6.0: Draft standard for wireless access in vehicular environments (WAVE) – Security services for applications and management messages. Piscataway, NJ: IEEE.

IEEE. (2010c). *IEEE Std 1609.3-2010: Standard for wireless access in vehicular environments (WAVE) – Networking services*. Piscataway, NJ: IEEE.

IEEE. (2011). *IEEE Std 1609.4-2010: Standard for wireless access in vehicular environments (WAVE) – Multi-channel operation*. Piscataway, NJ: IEEE.

Leinmueller, T., Schoch, E., Kargl, F., & Maihofer, C. (2010). Decentralized position verification in geographic ad hoc routing. *Security and Communication Networks*, *3*(4), 289–302. doi:10.1002/sec.56.

Li, H., & Lin, K. (2010). ITRI WAVE/DSRC communication unit. In *Proceedings of the IEEE Vehicular Technology Conference* (pp. 1-2).

Morillo-Pozo, J., Trullols, O., Barcelo, J., & Garcia-Vidal, J. (2008). A cooperative ARQ for delay-tolerant vehicular networks. In *Proceedings of the IEEE International Conference on Distributed Computing Systems Workshops* (pp. 192-197).

Shen, J., Jiang, T., Liu, S., & Zhang, Z. (2009). Maximum channel throughput via cooperative spectrum sensing in cognitive radio networks. *IEEE Transactions on Wireless Communications*, *8*(10), 5166–5175. doi:10.1109/TWC.2009.081110.

Shevtekar, A., Stille, J., & Ansari, N. (2008). On the impacts of low rate DoS attacks on VoIP traffic. *Security and Communication Networks*, *1*(1), 45–56. doi:10.1002/sec.7.

Tsai, M., Chilamkurti, N., Park, J., & Shieh, C. (2010). Multi-path transmission control scheme combining bandwidth aggregation and packet scheduling for real-time streaming in multi-path environment. *IET Communications*, *4*(8), 937–945. doi:10.1049/iet-com.2009.0661.

Tsai, M., Chilamkurti, N., & Shieh, C. (2010). A network adaptive forward error correction mechanism to overcome burst packet losses for video streaming over wireless networks. *Journal of Internet Technology*, *11*(4), 473–482.

Tsai, M., Chilamkurti, N., & Shieh, C. (2011). An adaptive packet and block length forward error correction for video streaming over wireless networks. *Wireless Personal Communications*, *56*(3), 435–446. doi:10.1007/s11277-010-9981-z.

Tsai, M., Chilamkurti, N., Shieh, C., & Vinel, A. (2011). MAC-level forward error correction mechanism for minimum error recovery overhead and retransmission. *Mathematical and Computer Modelling*, *53*(11-12), 2067–2077. doi:10.1016/j.mcm.2010.05.019.

Tsai, M., Chilamkurti, N., Zeadally, S., & Vinel, A. (2011). Concurrent multipath transmission combining forward error correction and path interleaving for video streaming in wireless networks. *Computer Communications*, *34*(9), 1125–1136. doi:10.1016/j.comcom.2010.02.001.

Tsai, M., Huang, T., Ke, C., & Hwang, W. (2011). Adaptive hybrid error correction model for video streaming over wireless networks. *Multimedia Systems Journal*, *17*(4), 327–340. doi:10.1007/s00530-010-0213-x.

Tsai, M., Huang, T., Shieh, C., & Chu, K. (2010). Dynamical combination of byte level and sub-packet level FEC in HARQ mechanism to reduce error recovery overhead on video streaming over wireless networks. *Computer Networks*, *54*(17), 3049–3067. doi:10.1016/j.comnet.2010.06.003.

Tsai, M., Shieh, C., Hwang, W., & Deng, D. (2009). An adaptive multi-hop FEC protection scheme for enhancing the QoS of video streaming transmission over wireless mesh networks. *International Journal of Communication Systems*, *22*(10), 1297–1318. doi:10.1002/dac.1032.

Tsai, M., Shieh, C., Ke, C., & Deng, D. (2010). Sub-packet forward error correction mechanism for video streaming over wireless networks. *Multimedia Tools and Applications*, *47*(1), 49–69. doi:10.1007/s11042-009-0406-5.

Tseng, Y., Jan, R., Chen, C., Wang, C., & Li, H. (2010). A vehicle-density-based forwarding scheme for emergency message broadcasts in VANETs. In *Proceedings of the IEEE International Conference on Mobile Adhoc and Sensor Systems* (pp. 703-708).

This work was previously published in the International Journal of Wireless Networks and Broadband Technologies, Volume 1, Issue 4, edited by Naveen Chilamkurti, pp. 22-39, copyright 2011 by IGI Publishing (an imprint of IGI Global).

Chapter 20
A Framework for External Interference-Aware Distributed Channel Assignment

Felix Juraschek
Humboldt Universität zu Berlin, Germany

Mesut Günes
Freie Universität Berlin, Germany

Bastian Blywis
Freie Universität Berlin, Germany

ABSTRACT

DES-Chan is a framework for experimentally driven research on distributed channel assignment algorithms in wireless mesh networks. DES-Chan eases the development process by providing a set of common services required by distributed channel assignment algorithms. A new challenge for channel assignment algorithms are sources of external interferences. With the increasing number of wireless devices in the unlicensed radio spectrum, co-located devices that share the same radio channel may have a severe impact on the network performance. DES-Chan provides a sensing component to detect such external devices and predict their future activity. As a proof of concept, the authors present a reference implementation of a distributed greedy channel assignment algorithm. The authors evaluate its performance in the DES-Testbed, a multi-transceiver wireless mesh network with 128 nodes at the Freie Universität Berlin.

INTRODUCTION

Channel assignment for multi-transceiver wireless mesh networks (WMNs) attempts to increase the network performance by decreasing the interference of simultaneous transmissions. Multi-transceiver mesh routers allow the communication over several wireless network interfaces at the same time. However, this can result in high interference of the wireless interfaces leading to a low network performance. With channel assignment, the reduction of interference is achieved by exploiting the availability of fully or partially non-overlapping channels. Channel assignment can be applied to all wireless networks based on technologies that

DOI: 10.4018/978-1-4666-3902-7.ch020

provide non-overlapping or orthogonal channels. Currently, wide-spread technologies are IEEE 802.11a/b/g, IEEE 802.11n, and IEEE 802.16 (WiMAX). With the low cost for IEEE 802.11 hardware, the number of deployments based on this technology is increasing and channel assignment algorithms are gaining in importance.

With the success of IEEE 802.11 technology, there is a dense distribution in urban areas of private and commercial network deployments of WLANs. These co-located networks compete for the wireless medium and can interfere with each other, thus decreasing the achievable network performance in terms of throughput and latency. Additionally, non-IEEE 802.11 devices, such as cordless phones, microwave ovens, and Bluetooth devices, operate on the unlicensed 2.4 GHz and 5 GHz frequency bands and can further decrease the network performance. It is therefore an important issue for efficient channel assignment, to also address the external interference. This task is not trivial, since the external networks and devices are not under the control of the network operator.

Although channel assignment is still a young research area, many different approaches have already been developed (Si et al., 2009). These approaches can be distinguished into centralized and distributed algorithms. Centralized algorithms rely on a central entity, usually called channel assignment server (CAS), which calculates the network-wide channel assignment and sends the result to the network nodes. In distributed approaches, each node calculates its channel assignment based on local information. Distributed approaches can react faster to topology changes due to node failures or mobility and usually introduce less protocol overhead, since communication with the CAS is not necessary. As a result, distributed approaches are more suitable once the network is operational and running. Another classification considers the frequency of channel switches on a network node. In fast channel switching approaches, channel switches may occur frequently, in the extreme for every subsequent packet a different channel is chosen. The limiting factor for such algorithms is the relative long channel switching time with commodity IEEE 802.11 hardware, which is in the order of milliseconds. Slow channel switching approaches switch the interfaces to a particular channel for a longer period, usually in the order of minutes or hours. Hybrid approaches combine both methods.

The focus of this paper is on the experimentally-driven research of distributed, slow channel switching algorithms on wireless testbeds. This process yields several challenges and pitfalls because the researcher has to deal with operating system specifics, drivers for the wireless interfaces, and the capabilities and limitations of the particular hardware. If more than one particular algorithm should be studied, the same problems and services have to be addressed multiple times because algorithms of this domain often require a set of common services. Among them are interface management, message exchange for node-to-node communication over the wireless medium, and data structures for network and conflict graphs. Additionally, a research framework for channel assignment algorithms can speed up the development process significantly by using the already available services.

The contribution of this article is DES-Chan, a framework for external interference-aware distributed channel assignment in multi-transceiver WMNs. We analyze several algorithms for distributed channel assignment and derive the required services for their implementation. We describe the architecture of DES-Chan that provides these services as well as the implementation of the greedy distributed algorithm (DGA) as a proof of concept. We present results obtained from the algorithm implementation in the DES-Testbed, a 128 multi-radio node wireless mesh testbed.

The remainder of this article is structured as follows. In the next section we present channel assignment algorithms and derive their required services. We then present the DES-Chan framework for distributed channel assignment. A reference implementation for DES-Chan is described and evaluated. The article concludes with an outlook and future work.

RELATED WORK

In this section, we present different distributed algorithms for channel assignment. The list of algorithms is not complete, for a more complete overview we refer to the survey in Si et al. (2009).

In the greedy channel assignment algorithm by Ko et al. (2007), one wireless interface of each node is switched to a common channel in order to ensure the network connectivity. For additional interfaces, a greedy algorithm selects the least interfering channel in the interference set using an interference cost function which takes the degree of overlap of two channels into account. The interference set comprises all nodes within the 3-hop neighborhood. As an additional constraint, at least one neighbor must have a radio tuned to the selected channel in order to avoid dead interfaces. Channel changes are communicated with a 3-way handshake.

The distributed greedy algorithm (DGA) assigns channels to links instead of interfaces and is therefore topology preserving, meaning that all links are sustained during the channel assignment procedure (Subramanian et al., 2008). A binary interference model, which specifies an interference range of m hops is used. A conflict graph is used to formulate the problem so that the number of edges in the conflict graph shall be minimized. In order to avoid oscillation, each vertex and channel combination can only be changed once.

The link-based channel assignment approach by Sridhar et al. (2009) is similar to DGA but additionally takes the expected traffic load on the nodes into account. The weights for the edges in the conflict graph are specified using a load-matrix that takes the expected traffic for each link into account. The channel assignment problem is then defined as minimizing the sum of the weighted edges of the conflict graph.

The Skeleton Assisted Partition Free (SAFE) algorithm uses minimal spanning trees (MSTs) to preserve the network connectivity (Shin et al., 2006). The 2-hop neighborhood is used as interference model. A conflict graph is used and the goal of the algorithm is to minimize the number of edges in the conflict graph. The channel assignment algorithm consists of two components. A random channel assignment is applied if $C < 2K$, where K is the number of wireless network cards on every node and C the number of non-overlapping channels. Due to the pigeonhole principle, two nodes will share a common channel although they are assigned randomly. The second component of the algorithm introduces the condition that all edges of a MST, the skeleton, of the network have to be preserved when $C \geq 2K$. For this, every node randomly chooses a channel set with $K-1$ channels, leaving one interface unassigned. The node broadcasts its chosen channel set, and if links to all skeleton neighbors are already established it assigns a random channel to the unassigned interface. Otherwise it tries to establish links with the not connected skeleton neighbors by assigning a channel which is in the channel set of all skeleton neighbors. If there is such a channel, it is assigned to the interface, if not, a global common channel is used.

A fast channel switching algorithm is proposed by Kyasanur et al. (2006). The set of network interfaces on each node are divided into *fixed* interfaces, which stay on a fixed channel, and *switchable* interfaces. If a node wants to communicate with a neighbor, it tunes one of the switchable interfaces to a channel of a fixed interface of the receiving node. The crucial part of this approach is the way how channels are assigned to the fixed interfaces. The authors present a simple solution, in which a well-known function calculates the channel for fixed interfaces based on the node ID. As an alternative, neighborhood information is considered for the channel selection which requires the periodic exchange of topology information messages. Each node then selects the least used channel for its fixed radio. The NET-X framework was created to implement this fast channel switching algorithm for a wireless testbed environment based on the 2.4 Linux kernel (Wireless Networking Group, 2010). The implementation of this particular algorithm requires several changes to the Linux network stack and to the driver of the wireless network interfaces which is supported by the framework.

DES-Chan

The DES-Chan framework has been developed for experimentally-driven research on distributed channel assignment in real network environments. It is currently in use at the DES-Testbed at the Freie Universität Berlin. The DES-Testbed comprises a stationary wireless mesh backbone with 128 multi-radio indoor and outdoor DES-Nodes spanning three different buildings (Günes et al., 2009). Every mesh router runs the Linux operating system and is equipped with three IEEE 802.11a/b/g wireless network interfaces. DES-Chan has been implemented as a Python framework and it is available at the website of the DES-Testbed http://www.des-testbed.net.

With DES-Chan a wide range of different algorithms can be implemented, validated, and compared in a real network environment. DES-Chan does not require any changes of the Linux kernel or the wireless network interface drivers. It is therefore easy to integrate into existing wireless mesh network testbeds. However, DES-Chan is not suitable for fast channel switching, since the channel switching time without changes to the WNIC drivers is in the order of milliseconds.

The benefits of DES-Chan are two-fold. Firstly, DES-Chan provides an abstraction layer to the low-level and operating system specific tasks. This abstraction layer enables the researcher to spend most development time on the algorithm logic instead of, for instance, memory management and handling the wireless interfaces. Secondly, DES-Chan provides basic services and data structures often required for typical tasks in channel assignment algorithms. For instance, appropriate data structures have been provided for network graphs, conflict graphs, and interference models. We start the description with the analysis of the common requirements of the presented channel assignment algorithms.

Required Services

The presented algorithms require several key services for their implementation and these should be provided by a generic channel assignment framework. The derived services are the following:

- **Interface Handling:** This basic service is mandatory for all algorithms in order to change the wireless network interface settings, for example to carry out channel switches.
- **Neighborhood Discovery:** Information about the local topology is needed as input for channel assignment decisions in the presented algorithms.
- **Message Exchange:** A message exchange service for node to node communication between the instances of an algorithm is needed to exchange topology information messages and to implement handshake mechanisms.
- **Topology Monitoring:** A topology monitoring service periodically assesses the local topology and its state. Neighbors, links, and their respective quality can be monitored which enables the algorithms to adapt to topology changes by refining the channel assignment.
- **Interference Models:** These models are used to estimate the local interference. They range from simple heuristics such as the m-hop neighborhood, where usually $m \in \{2,3\}$, to measurement-based ap-

proaches for particular network topologies (Padhye et al., 2005; Reis et al., 2006). Even though, the simple heuristics are not very realistic (Padhye et al., 2005), they are easy to implement and therefore widely used. Another common model is the conflict graph with algorithms trying to minimize its edges in order to minimize the network-wide interference (Katzela & Naghshineh, 1996). Therefore, a framework should provide appropriate data structures and operations for modeling conflict graphs.

In addition to these services, fast channel switching algorithms usually need a modified queuing system to avoid channel switching on a per-packet basis. Frames need to be stored which are sent over a channel which is currently not utilized by the wireless network card. In slow channel switching algorithms, different queues are not required since an interface operates on the same channel for minutes or hours.

Table 1 shows the described services in relation to the presented algorithms. As a conclusion, a generic framework for off-the-shelf IEEE 802.11 hardware should provide all described services but the per-channel-queuing. Due to the relative long channel switching time, slow channel switching are better suitable for these kinds of networks.

Architecture and Components

The DES-Chan framework comprises three main components as depicted in Figure 1. DES-Chan Core is a Python library that provides common functions and data structures for channel assignment algorithms. It comprises the following services: interface management, node communication, graph representation, and interference models. The Neighborhood-Discovery module provides a basic service for each node to get information about all neighboring nodes. DES-Sense is software spectrum sensing service that enables the detection of external devices that share the same radio channel. As future algorithms will require additional services, DES-Chan has been designed to be extensible by additional modules. Therefore, a modular architecture has been chosen. As next, the components are described in detail.

1. **Interface Management:** The interface management module acts as interface to the operating system and hides testbed-specific characteristics from the channel assignment algorithms. It provides various functions for configuring network interfaces and retrieving information about their state. Thus it is possible to set up and shut down interfaces, check whether an interface has been set up, and get information of unused interfaces. The module furthermore allows tuning an inter-

Table 1. The table shows the key services in relation to the algorithms. The bullet indicates that the algorithm requires the particular service, a dash means it does not. For slow channel switching algorithms, the interface management, neighbor discovery, message exchange, topology monitoring, and interference models are important.

Algorithm	Interface Management	Neighbor Discovery	Topology Monitoring	Message Exchange	Simple Interference Model	Conflict Graph	Queues per Channel
Ko	•	•	–	•	•	–	–
DGA	•	•	–	•	•	•	–
Sridhar	•	•	–	•	–	•	–
Net-X	•	• / –	–	• / –	•	–	•
SAFE	•	•	•	•	•	•	–

Figure 1. The DES-Chan framework provides common services for distributed channel assignment algorithms

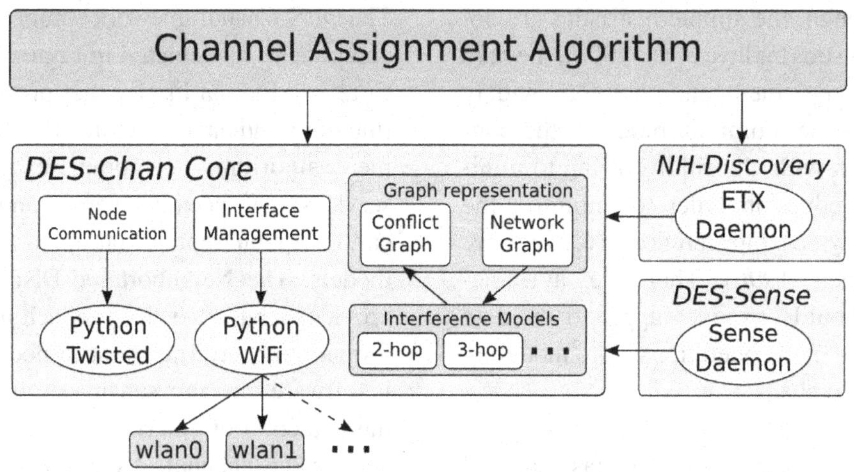

face to a specific channel and thereby implements a crucial requirement for all channel assignment algorithms. The settings of the wireless interfaces are changed with Python WiFi (Joost & Robinson, 2010), a library based on the Linux Wireless Extensions.

2. **Neighborhood-Discovery:** The Neighborhood-Discovery module determines the links and their quality on each network node based on the ETX link metric. The ETX link metric estimates how many transmissions for a packet are required until it is successfully received (De Couto et al., 2003). ETX values are calculated by each node sending broadcast probes and logging how many probes from their neighbors were successfully received. The forward and reverse delivery ratios are then used for the calculation of ETX, because for unicast communication in IEEE 802.11, an ACK frame has to be successfully received at the sender. The ETX value for a link is then calculated as

$$ETX = \frac{1}{df * dr}$$

where df is the forward delivery ratio and dr the reverse delivery ratio with $0 < df, dr \leq 1$.

The ETX implementation etxd for the DES-Chan framework has been realized as a Linux daemon. By default, a probe interval of one second and a window size of 10 seconds is used. The values can be configured via command line arguments. The link quality values are averaged over a moving window, which spans 10 probe intervals. The daemon sends UDP probe packets on the broadcast addresses of the specified network interfaces and listens for incoming probes. In order to always provide up-to-date information, etxd dynamically adapts its configuration if network interfaces have been reconfigured, shut down, or brought up.

To offer the neighborhood and link quality information to other programs, etxd provides an inter process communication (IPC) interface that can be accessed via sockets. A simple, textual protocol allows other applications such as channel assignment algorithms to get the neighbors of a node, as well as the quality and the channel of a certain link. The daemon can be queried to return only those neighbors, which are reached via reliable links, i.e., links whose quality exceeds a certain value. In addition to the link quality, etxd also returns the current channel of the WNIC that is used to reach the respective neighbor. As a result, the ETX daemon provides the neighborhood discovery and the topology monitoring service. Via the IPC interface, channel assignment algorithms can query the current state of their links and react adaptively to topology changes.

3. **Node Communication:** Different channel assignment algorithms use different communication protocols and message formats. Thus, a flexible networking library is needed, that allows implementing various protocols with few efforts. The Python Twisted (Python Twisted, 2010) library serves as the foundation for the node to node communication in DES-Chan. The library provides an asynchronous networking engine and hides technical details like creating sockets and establishing connections from the developer. The core of Twisted is a global reactor object that can be instructed to monitor sockets and to connect to servers. Based on Twisted, the researcher can quickly develop the required protocol implementation for exchanging messages among the network nodes, for instance in order to propagate changes in channel assignment or to carry out three-way handshakes prior to the actual channel switch.
4. **Graph Representation:** DES-Chan provides several data structures and functions for graph representation. A data structure for the network graph contains a vertex for every network node and an edge for every link between two network nodes. The edges can be labeled with the number of the channel that is used for the communication. The NetworkGraph class also allows storing a list of channels for each edge, thereby supporting multiple links between node pairs. The NetworkGraph class offers intuitive methods for accessing vertices and edges, and provides several utility functions. It supports printing a human-readable adjacency matrix, which can be used for debugging purposes and for saving the graph in a file. Graph objects can also be stored in the *DOT* format that can be processed to generate an image of the graph (Graphviz, http://www.graphviz.org/). A function provides the possibility to merge two NetworkGraph objects together by creating the union of both vertex sets and edge sets. This can be used for example to merge network graphs that were obtained from different neighbors.

Additionally, a corresponding ConflictGraph data structure is provided. It allows applying an interference model to a network graph, in order to calculate the interference between all link pairs in the corresponding network. The class maintains the relation between edges in the network graph and vertices in the conflict graph, and provides corresponding transformation methods. It allows manipulating the channel of the corresponding edge in the network graph and automatically updates the interference information in the conflict graph. This functionality is needed by algorithms that successively apply several channel assignments to find a configuration that minimizes interference. The ConflictGraph can be easily extended to implement other concepts, such as multi-radio conflict graph (Ramachandran et al., 2006).

5. **Interference Models:** The DES-Chan framework supports the implementation of various interference models. Currently, only the 2-hop interference model is implemented for reference, but the framework can be easily extended. The current implementation of the 2-hop heuristic supports a binary notion of

interference, i.e., two links either interfere or do not interfere, and a more accurate notion of the interference cost taking the spectral distance of two channels into account.

The following equation shows the interference cost function $I(x)$ for two center frequencies f_1 and f_2. The additional parameter ± denotes the minimum frequency difference of orthogonal channels.

$$I(f_1, f_2, \alpha) = \begin{cases} 0, \; if |f_1 - f_2| \geq \alpha \\ 1 - \frac{(|f_1 - f_2|)}{\alpha}, \; otherwise \end{cases}$$

For IEEE 802.11b/g three channels of the 2.4GHz frequency band are theoretically orthogonal. This can be modeled by setting ± to 30MHz, which results in the set of orthogonal channels $\{1,7,13\}$. However, experiments showed that this channel distance does not guarantee orthogonality in real network deployments (Fuxjager et al., 2007; Juraschek et al., 2011).

6. **DES-Sense:** DES-Sense is a software-based spectrum sensing solution which allows detecting external devices utilizing the radio channels. DES-Sense consists of the following two components:
   a. **Sensing Component:** The sensing component is a daemon that periodically retrieves statistics about the channel occupancy. This is achieved by retrieving the carrier sensing statistics of the wireless network card via the driver. Based on the statistics we are able to determine the percentage of a certain interval the medium was sensed busy and therefore could not have been used for wireless transmissions. We can efficiently determine the channel usage and predict the future activity of external networks and devices by using the statistics as input for corresponding models. The sensing component can be configured dynamically with the set of channels $C = \{c_1, c_2, \; .., c_k\}$ that will be monitored and the duration $T = \{t_1, t_2, \; .., t_k\}$ each channel is monitored.
   b. **IPC Interface:** For the integration into DES-Chan, an inter process communication (IPC) interface is provided that allows algorithms to retrieve the channel usage statistics to fuse them into their channel assignment decision. The algorithms can query the daemon via the interface to update their channel usage statistics.

Several challenges had to be addressed to efficiently implement this method. The carrier sensing statistics are retrieved via the *ath5k* driver for Atheros-based interfaces (Linux Wireless, 2011), which is one of the few drivers for the Linux system that currently provide these statistics. However, from the carrier sense statistics alone, it can only be derived that the channel has been utilized, but not by which station. It is therefore hard, to distinguish between traffic from our own network and external networks and devices. To solve this problem, the monitoring interface can be set in monitor mode and thus capture and analyze the received packets. This way, we can distinguish between internal and external traffic and can treat the channel usage of our network different than that of other networks.

In order to assess the channel usage, we need to perform periodic measurements on the available channels. With multi-radio mesh nodes, this can either be solved with a dedicated interface that permanently monitors the channel usage. The drawback of this approach is that the dedicated interface cannot be utilized for data transmissions. Another method is to perform the monitoring measurements event-based, in case a change of the link quality is observed, for example, when the throughput drops. With this method, a traffic flow must be stopped and can only be resumed after a less congested channel is found and switched to. This will lead to a higher delay for this particular flow.

Another challenge results from the fact, that a wireless interface can monitor only one channel at a time. This means, that simultaneous monitoring of all available channels is not possible with one wireless interface. The total duration T_M for monitoring all channels in C is:

$$T_M = \sum_{i=1}^{k} t_i$$

Depending on the values for t_i, the risk exists that bursty traffic might not be monitored on a particular channel. However, with periodic measurements, we aspire to gather enough data to derive realistic estimations of the future channel usage time.

For a first evaluation of the sensing component, we ensured that no frames were sent by the DES-Testbed nodes, so that we only capture the activity of external devices. We activated one wireless interface on each testbed node and started the monitoring phase of the available channels on the 2.4 GHz frequency band. The results of the channel usage measurements for one particular node are shown in Figure 2. The figure shows the channel occupancy for the channels 1 to 12 of the

Figure 2. Channel occupancy on the 2.4 GHz frequency spectrum. The channels 1, 6, 11 are already utilized up to 40%, which implies that the channels are already highly congested. Co-deployed to the DES-Testbed, the APs of the university WLAN use exactly these channels.

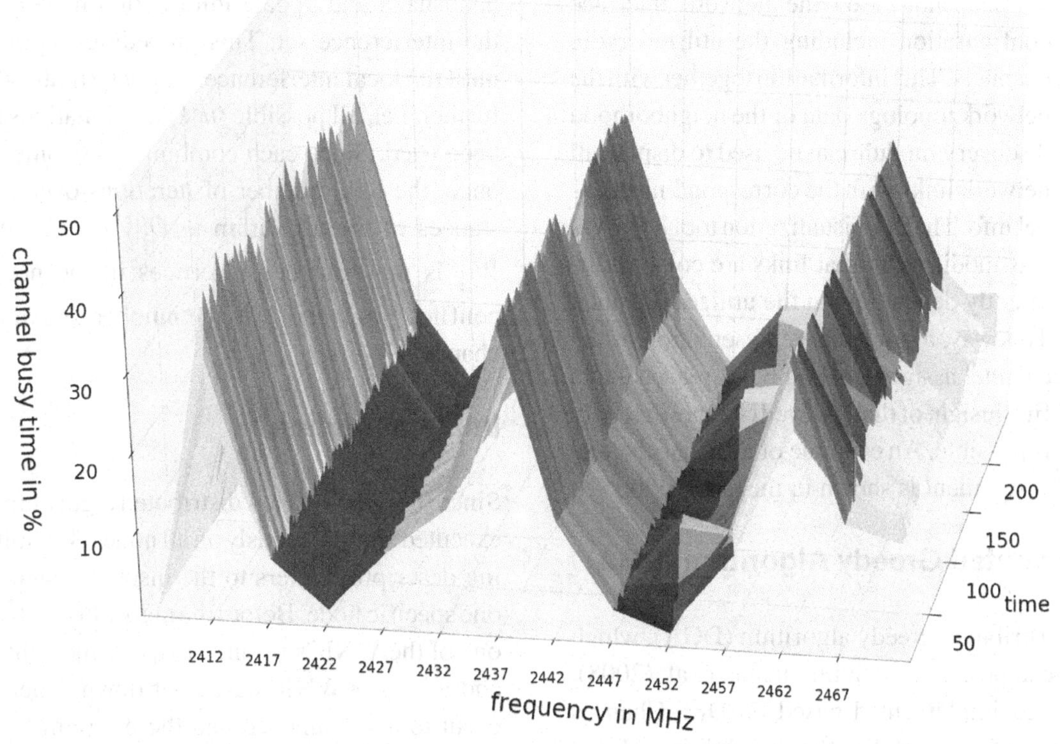

2.4 GHz band. As can be seen, the channels 1, 6, 11 are already utilized up to 40%. This complies very well with our knowledge that the co-located faculty WLANs used by students and research staff are using exactly these channels.

It is interesting to note, that the channel occupancy for all channels has only little deviation. A further analysis of the captured packets traffic revealed, that at the time of the monitoring mostly beacon frames were received, which are sent periodically from the APs of the co-located networks.

7. **Channel Assignment Visualization:** A visualization of the final channel assignment and network topology is helpful in analyzing the performance of channel assignment algorithms. In order to monitor the channel assignment, we extended the visualization of the DES-Testbed Management System (DES-TBMS) (Günes et al., 2009). The network monitoring tool DES-Mon gathers snapshot-based the network interface configuration including the utilized radio channels. This information together with the network topology data of the neighborhood discovery module can be used to display all network links with the corresponding channel info. The 3D-visualization tool DES-Vis was modified so that links are colored differently depending on the utilized channel. This way, a graphical representation of the channel assignment can be displayed and a first insight of the achieved channel diversity is possible. An example of the final channel assignment is shown in the last section.

Distributed Greedy Algorithm (DGA)

The distributed greedy algorithm (DGA), which has been proposed in Subramanian et al. (2008), has been implemented based on DES-Chan as a proof-of-concept for the capabilities of the framework. DGA assigns channels to links and is therefore topology preserving, meaning that all links are sustained during the channel assignment procedure. A conflict graph is used to formulate the problem so that the number of edges in the conflict graph shall be minimized. Each wireless link between two nodes is owned by the node with the higher node ID and only this node may assign a channel to the link.

At the network initialization, all links are assigned to the same channel. Each node then iterates over all owned links and changes the channel of the link which results in the largest decrease of interference in the local neighborhood. The largest decrease is achieved with the combination of link u and channel k that removes the highest numbers of edges in the local conflict graph. The interface constraint is respected, which means that no more channels can be assigned to a node than it has interfaces. In order to avoid oscillation, each vertex and channel combination can only be changed once.

Channel switches are carried out using a 3-way handshake and update information message for the interference set. This procedure is repeated until the local interference cannot be reduced any further, i.e., all possible (u, k) combinations have been tried. Since each combination is only tried once, the total number of iterations over all instances of the algorithm is $O(|V_c| * K)$, where $|V_c|$ is the number of vertices in the network conflict graph and K is the number of available channels.

Implementation

Since the approach is a distributed algorithm, it is executed simultaneously on all nodes. The following description refers to the instance running at one specific node. Before the algorithm is started, one of the WNICs is tuned to a common channel and all other WNICs are shut down. Then, the event loop is launched and the program flow is according to the diagram in Figure 3.

Figure 3. Flowchart of the distributed DGA implementation. Methods of the DGA class are colored gray and methods of the Messaging class are colored white. Input/Output operations are represented by parallelograms.

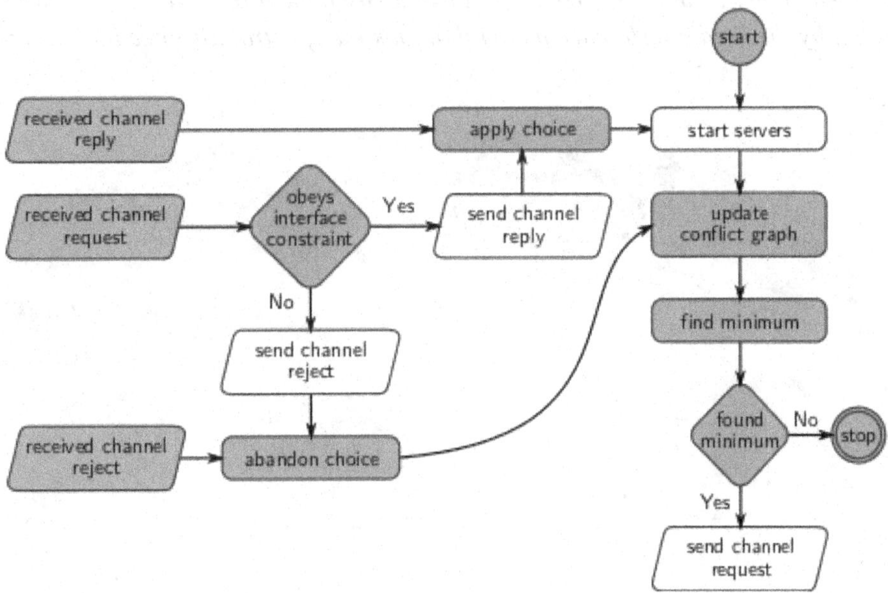

The start method uses the topology monitoring module from DES-Chan to retrieve the initial network graph. The local conflict graph, which is still empty at that point, is updated with the information of the network graph. Afterwards, the findMinimum method is called. It tries to find a vertex-channel combination that minimizes the interference in the local conflict graph. If the interference can be further reduced, the program tries to apply the combination that provides the largest decrease. Therefore, it sends a channel request message to the corresponding neighbor, asking the DGA instance on that node to change the channel accordingly.

After the channel request message has been sent, the program flow is interrupted until a reply is received. Thereby, three message types are distinguished: channel request, channel reply, and channel reject. When receiving the channel request message, the node checks whether it is able to change the channel without violating the interface constraint. If there is an unused interface available or no other links would be affected by the channel switch, the node answers with a channel reply message, indicating that the channel can be changed. After the reply has been sent, the node configures its interfaces according to the new assignment. Otherwise the request is answered with a channel reject message and the vertex-channel combination is abandoned. When the originator of the request receives a channel reply message, it applies the chosen channel.

If the channel request is answered with a channel reject message, the link-channel combination is removed from the set of possible combinations and will not be considered again. Thereby the termination of the algorithm is guaranteed. After the requested channel has been either approved or rejected, the next iteration is started.

Evaluation

The DES-Testbed was used for the evaluation of the DGA implementation based on DES-Chan. The evaluation comprises a graph-theoretic analysis of the remaining interference after channel assign-

Figure 4. Resulting channel assignment of DGA and RAND at the DES-Testbed. The images have been taken from the DES-Vis 3D-visualization tool for the DES-Testbed. First, all links are shown which are used for the channel assignment. Different link colors correspond to different channels. While the random assignment created higher channel diversity, DGA mainly used the three channels that were reported to being orthogonal by the used interference model that took the spectral distance into account.

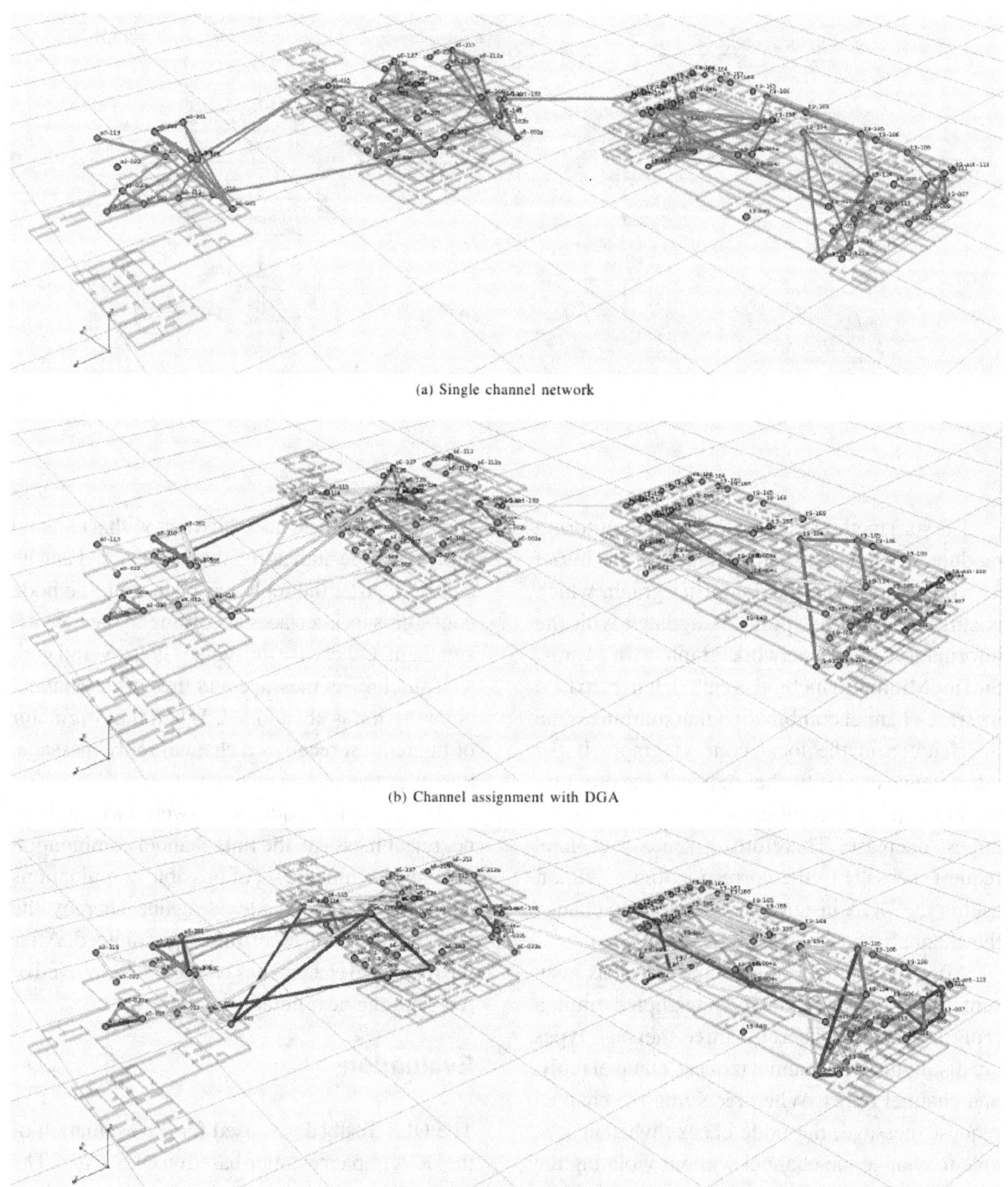

(a) Single channel network

(b) Channel assignment with DGA

(c) Channel assignment with RAND

ment as proposed by the authors of DGA. The results were compared to the single channel case, where each node uses only one of its interfaces on the same channel. The interference reduction was determined for the already presented DGA algorithm and an additional RAND algorithm which assigns channels to links randomly.

The channels 1 to 13 of the 2.4 GHz spectrum were set as the available channels for the algorithms. For DGA, an interference cost function taking the spectral distance of the channels was used as described in the previous section. The experiments were performed on 100 nodes in the DES-Testbed and replicated for 30 times. Each replication comprised three phases. First, all nodes were configured to set up only one interface on channel 1 in order to create the single channel network case. In the second phase, the DGA algorithm was executed. The same was repeated for the RAND algorithm.

A sample channel assignment that resulted from both algorithms and the single channel network is depicted in screen shots of the DES-Vis visualization tool in Figure 4. Hereby, each link color stands for a different channel. As can be observed from the link colors, the DGA algorithm mainly used the three channels that were reported to being orthogonal by the used interference model. In contrast, the random assignment resulted in higher channel diversity. The properties of these assignments are further investigated below.

As in the evaluation in Subramanian et al. (2008), the fractional network interference (FNI) was used as metric for the graph-theoretic consideration. The FNI is given by:

$$FNI = \frac{\sum_{i \in E_{CA}} weight(i)}{\sum_{i \in E_{SC}} weight(i)}$$

where E_{CA} is the set of edges of the conflict graph after channel assignment and E_{SC} is the set of edges of the conflict graph in the single channel network.

The results of 30 experiment replications are shown in Figure 5. After the channel assignment with DGA, the FNI was reduced to 0.28 in the median. Thus, the overall interference in the network

Figure 5. FNI evaluation of DGA and RAND. DGA reduces interference by 72% compared to a single channel network and by 15% compared to RAND.

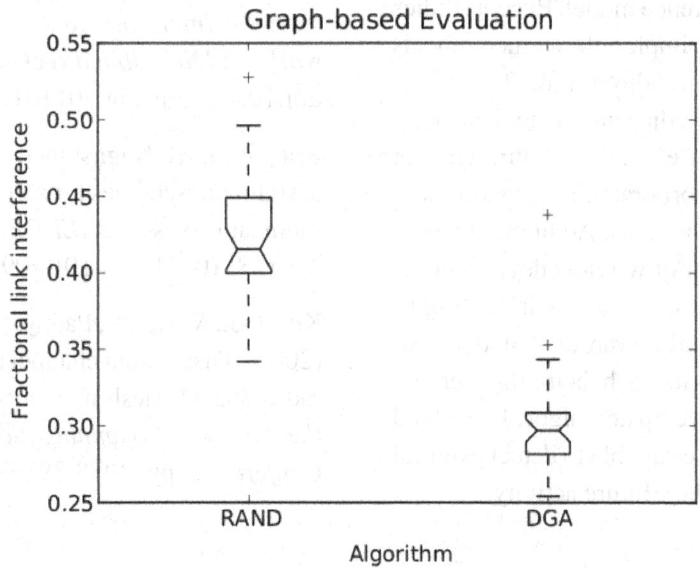

was reduced by 72% compared to the interference in a single-channel network. Furthermore, the low variation in the results shows that the algorithm achieves reproducible results. The results from RAND show a FNI of 0.43 in the median. Thus, even random channel assignment significantly reduces the interference compared to a single-channel network, but the DGA algorithm still achieves a further reduction of 15%. The presented results correspond to those that Subramanian et al. (2008) obtained for their distributed greedy algorithm on random graphs. For a setup with 50 nodes, three orthogonal channels, and three WNICs per node, the authors found that their algorithm achieves an average FNI of 0.3.

SUMMARY AND OUTLOOK

The DES-Chan framework has been developed to support the development process of distributed channel assignment algorithms in wireless mesh networks. We presented the components of DES-Chan and a reference implementation of a well-known algorithm. The evaluation showed that the obtained graph-based results are close to the results published by the authors of the algorithm. However, the results rely on the accuracy of the applied 2-hop interference model. Recent studies have shown that the simple interference models are not very accurate (Padhye et al., 2005). The results of studies on adjacent channel interference (ACI) (Fuxjager et al., 2007; Juraschek et al., 2011) will be incorporated to develop more accurate interference models. Additionally, with the increasing number of wireless devices in the unlicensed frequency spectrum, channel assignment algorithms need to be aware of such sources of external interferences to reach their full potential. Therefore, we will develop new algorithms based on DES-Sense that are capable to detect external devices and predict their future activity.

REFERENCES

De Couto, D. S. J., Aguayo, D., Bicket, J., & Morris, R. (2003). A high-throughput path metric for multi-hop wireless routing. In *Proceedings of the 9th Annual International Conference on Mobile Computing and Networking* (pp. 134-146).

Fuxjager, P., Valerio, D., & Ricciato, F. (2007). The myth of non-overlapping channels: interference measurements in IEEE 802.11. In *Proceedings of the Fourth Annual Conference on Wireless on Demand Network Systems and Services* (pp. 1-8).

Günes, M., Blywis, B., Juraschek, F., & Watteroth, O. (2009). Experimentation made easy. In *Proceedings of the First International Conference on Ad Hoc Networks* (Vol. 28).

Günes, M., Juraschek, F., Blywis, B., Mushtaq, Q., & Schiller, J. (2009). A testbed for next generation wireless networks research. *Mobile Ad-hoc Networks*, *4*, 208–212.

Joost, R., & Robinson, S. (2010). *Pythonwiϕι*. Retrieved December 24, 2010, from http://pythonwifi.wikispot.org/

Juraschek, F., Günes, M., Philipp, M., & Blywis, B. (2011). Insights from experimental research on distributed channel assignment in wireless testbeds. *International Journal of Wireless Networks and Broadband Technologies*, *1*(1), 32–49. doi:10.4018/ijwnbt.2011010103.

Katzela, I., & Naghshineh, M. (1996). Channel assignment schemes for cellular mobile telecommunication systems. *IEEE Personal Communications*, *3*, 10–31. doi:10.1109/98.511762.

Ko, B.-J., Misra, V., Padhye, J., & Rubenstein, D. (2007). Distributed channel assignment in multi-radio 802.11 mesh networks. In *Proceedings of the Wireless Communications and Networking Conference* (pp. 3978-3983).

Kyasanur, P., So, J., Chereddi, C., & Vaidya, N. H. (2006). Multichannel mesh networks: challenges and protocols. *IEEE Wireless Communications, 13*(2), 30–36. doi:10.1109/MWC.2006.1632478.

Linux Wireless. (2011). *Ath5k driver homepage*. Retrieved from http://linuxwireless.org/en/users/Drivers/ath5k

Padhye, J., Agarwal, S., Padmanabhan, V. N., Qiu, L., Rao, A., & Zill, B. (2005). Estimation of link interference in static multi-hop wireless networks. In *Proceedings of the 5th ACM SIGCOMM Conference on Internet Measurement* (pp. 28-28).

Python Twisted. (2010). *The python twisted documentation*. Retrieved December 24, 2010, from http://twistedmatrix.com/projects/core/documentation/howto/book.pdf

Ramachandran, K. N., Belding, E. M., Almeroth, K. C., & Buddhikot, M. M. (2006). Interference-aware channel assignment in multi-radio wireless mesh networks. In *Proceedings of the 25th IEEE International Conference on Computer Communications* (pp. 1-12).

Reis, C., Mahajan, R., Rodrig, M., Wetherall, D., & Zahorjan, J. (2006). Measurement-based models of delivery and interference in static wireless networks. *SIGCOMM Computer Communications, 36*(4), 51–62. doi:10.1145/1151659.1159921.

Shin, M., Lee, S., & Kim, Y. (2006). Distributed channel assignment for multi-radio wireless networks. In *Proceedings of the IEEE International Conference on Mobile Adhoc and Sensor Systems Conference* (pp. 417-426).

Si, W., Selvakennedy, S., & Zomaya, A. Y. (2009). An overview of channel assignment methods for multi-radio multi-channel wireless mesh networks. *Journal of Parallel and Distributed Computing, 70*(5).

Sridhar, S., Guo, J., & Jha, S. (2009). Channel assignment in multi-radio wireless mesh networks: A graph-theoretic approach. In *Proceedings of the First International Communication Systems and Networks and Workshops* (pp. 1-10).

Subramanian, A. P., Gupta, H., Das, S. R., & Cao, J. (2008). Minimum interference channel assignment in multi-radio wireless mesh networks. *IEEE Transactions on Mobile Computing, 7*(12), 1459–1473. doi:10.1109/TMC.2008.70.

Wireless Networking Group at UIUC. (2010). *Net-x channel assignment framework*. Retrieved December 24, 2010, from http://www.crhc.illinois.edu/wireless/netx.html

This work was previously published in the International Journal of Wireless Networks and Broadband Technologies, Volume 1, Issue 4, edited by Naveen Chilamkurti, pp. 40-54, copyright 2011 by IGI Publishing (an imprint of IGI Global).

Chapter 21
An Efficient Data Dissemination Scheme for Warning Messages in Vehicular Ad Hoc Networks

Muhammad A. Javed
The University of Newcastle, Australia

Jamil Y. Khan
The University of Newcastle, Australia

ABSTRACT

Vehicular ad hoc networks (VANETs) are expected to be used for the dissemination of emergency warning messages on the roads. The emergency warning messages such as post crash warning notification would require an efficient multi hop broadcast scheme to notify all the vehicles within a particular area about the emergency. Such emergency warning applications have low delay and transmission overhead requirements to effectively transmit the emergency notification. In this paper, an adaptive distance based backoff scheme is presented for efficient dissemination of warning messages on the road. The proposed scheme adaptively selects the furthest vehicle as the next forwarder of the emergency message based on channel conditions. The detailed performance figures of the protocol are presented in the paper using simulations in the OPNET network simulator. The proposed protocol introduces lower packet delay and broadcast overhead as compared to standard packet broadcasting protocols for vehicular networks.

INTRODUCTION

Vehicular Ad hoc Networks (VANETs) will be an integral part of the future traffic management systems to support safety and traffic management applications as well as other data transmission services. VANETs can be seen as self-organizing autonomous system formed by the moving vehicles (Fiore, Harri, Filali, & Bonnet, 2007). VANETs can distribute road traffic and emergency information along with conventional data traffic such as entertainment applications and internet access to vehicles in a timely manner. To be a part of VANETs, each vehicle must be equipped with an onboard unit (OBU) containing wireless transceiver, a GPS receiver, appropriate sensors

DOI: 10.4018/978-1-4666-3902-7.ch021

and control modules. In addition, additional fixed infrastructure in the form of Road Side Units (RSU) would help increase the network connectivity and provide connection with the centralized traffic control servers (Bernsen & Manivannan, 2009). To support numerous ITS applications, two modes of communications in such vehicular networks is envisaged: Vehicle to Vehicle (V2V) communication involves direct communication between vehicles, Vehicle to Road Side (V2R) communication involves communication between vehicles and static road side units (Morgan, 2010). VANETs have several advantages over the conventional wireless networks such as Universal Mobile Telecommunications System (UMTS), Long Term Evolution (LTE) and Worldwide Interoperability for Microwave Access (WiMAX). Main advantages are low cost of implementation and maintenance, self- organization and lower local information dissemination time (Javed & Khan; Karim, 2008). For the future implementation of an efficient VANET architecture, the IEEE802.11p protocol has been proposed (2010 IEEE Standards Association). The CSMA/CA (Carrier Sense Multiple Access with Collision Detection) mechanism which is used in the IEEE 802.11p MAC reduces the signaling overhead compared to the conventional cellular networks. However, use of the CSMA/CA scheme could reduce the reliability of a network when used to support emergency messaging services such as accident or breakdown notifications (Javed & Khan).

Many research studies have been undertaken to improve the performance of the broadcasting techniques in VANETs. However, the broadcast suppression mechanism used in current systems results in large overhead, warning notification time and packet loss. Consider a typical scenario shown in Figure 1 where a vehicle broadcasts post crash warning notification to vehicles behind it so that they can change their route and to the rescue vehicles so that they can reach the site with minimum delay. Using a simple flooding packet transmission technique in which each node on receiving a broadcast message rebroadcasts it to all other nodes within its transmission range could result in network contention and increased overhead. Using the flooding technique many nodes will rebroadcast the warning message simultaneously causing collisions hence, dissemination of urgent information such as post crash warning notification will be either delayed or interrupted. Probability based broadcast suppression technique could also result in collisions in dense networks because nodes within a certain area have a similar probability of rebroadcast (Wisitpongphan et al., 2007). Similarly in the timer based techniques, nodes with similar assigned time slots will have similar probability of rebroadcast which will cause contention (Wisitpongphan et al., 2007). In this paper we propose an efficient geocast protocol in the IEEE802.11p network to support emergency messaging services. The proposed protocol achieves the goal of minimizing the

Figure 1. Post crash warning notification message dissemination

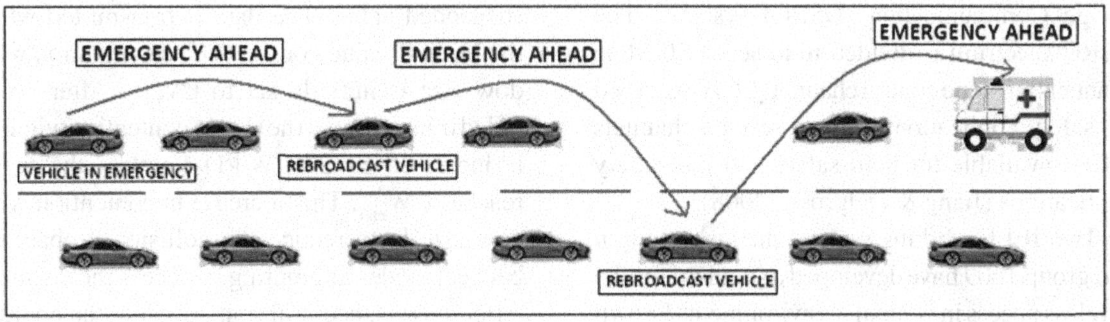

hop count for emergency message dissemination by selecting the furthest node for rebroadcast at each hop. The selection of furthest node is made possible by a new contention window design based on the distance from the sender node. The furthest node is adaptively selected each time based on the channel conditions using an adaptive distance based backoff technique. The protocol helps in broadcast suppression and reducing the contention probability of safety message distribution in a VANET which is shown by a higher packet success rate and a lower convergence time as compared to current protocols. The proposed protocol performance has been compared with the standard packet broadcasting protocols such as flooding and a timer based protocol.

The paper is structured as follows. The next section describes the IEEE 802.11p protocol architecture in detail. This is followed by review of some of the existing protocols to develop the reliability features of VANET applications particularly for car to car communication services. The next section introduces the proposed geocasting protocol for VANET applications. Simulation model is discussed in the following section. Finally, simulation results are presented followed by the conclusions.

IEEE 802.11P PROTOCOL ARCHITECTURE

Intelligent Transport Systems (ITS) have been assigned 75 MHz of licensed free spectrum centering around 5.9 GHz to support Dedicated Short Range Communication (DSRC) systems. The DSRC spectrum is divided in to seven 10 MHz channels with one control channel (CCH) reserved for safety applications and six service channels (SCH) available for both safety and non-safety applications (Jiang & Delgrossi, 2008).

Two IEEE working groups, the task group *p* and group 1609 have developed a standard for the wireless access in vehicular environments known as the WAVE. The WAVE protocol stack consists of the IEEE802.11p (2010 IEEE Standards Association) and IEEE 1609 standards (IEEE, 2006a, 2006b, 2010a, 2010b). The IEEE802.11p standard deals with the MAC and PHY layer functionalities; while the upper layer functionalities are handled by the IEEE 1609 protocol. The WAVE protocol architecture is shown in Figure 2.

The IEEE 802.11p protocol is an amended version of the IEEE 802.11 standard developed to support the WAVE architecture. It deals with the enhancements of MAC and PHY layers of the IEEE802.11 standard. The IEEE 802.11p MAC is based on the Enhanced Distributed Channel Access (EDCA) procedure defined in the IEEE 802.11e standard. The IEEE 802.11e channel access diagram is shown in Figure 3. The CSMA/CA based IEEE 802.11 MAC protocol uses the Distributed Coordination Function (DCF) to coordinate multiple channel access. A node which has data to send waits for an additional amount of time after detecting the channel as idle. This time duration is known as Arbitration Inter Frame Spacing (AIFS). If the channel is idle for AIFS amount of time, the node takes a random backoff from the values in the contention window (CW) to defer its access. The contention window maintained by each node contains a set of integer values from which random backoff value is picked. After the channel is idle for AIFS time duration, the backoff value is decremented at each time slot. If the channel becomes busy during this process, the backoff value is suspended. After the channel becomes idle again for AIFS duration of time, the backoff value resumes decrementation from the suspended value. The data is transmitted when the backoff value goes to 0. The contention window size is initially set to CW_{min}. After every failed transmission, the size of contention window is increased by 2*(CW+1)-1 unless the value reaches CW_{max}. This increase in contention window size is to reduce the collision probability between nodes attempting to access the channel. After every successful transmission, the conten-

Figure 2. The IEEE 802.11p/ WAVE protocol stack

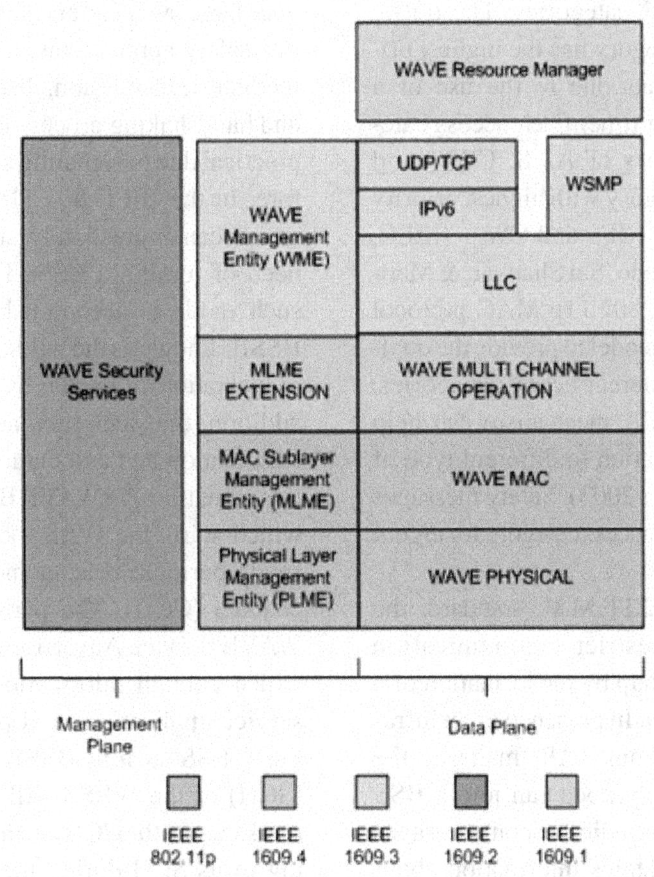

Figure 3. IEEE 802.11e EDCA scheme (IEEE, 2007)

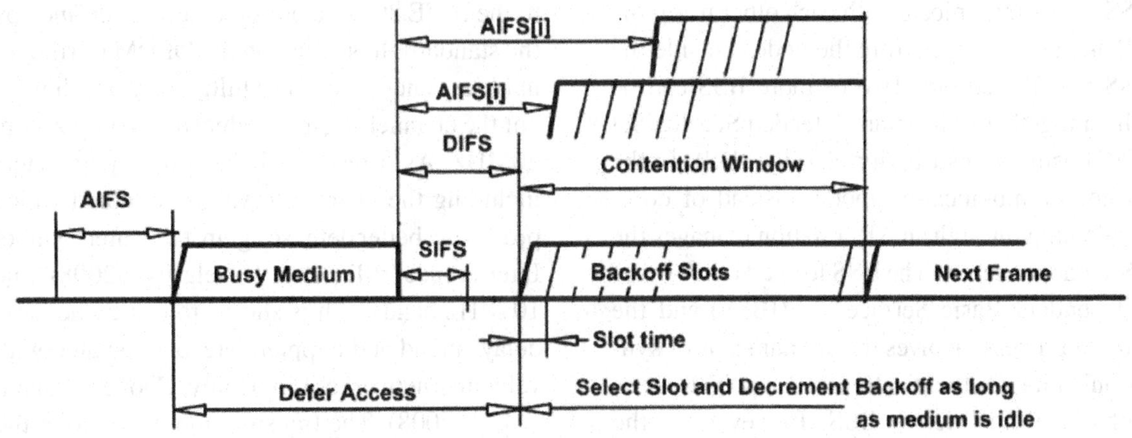

tion window size is again reduced to CW_{min}. The EDCA technique in IEEE 802.11e prioritizes the traffic in to four access categories. The traffic with highest access category has the highest priority to transmit packets due to the use of a shorter channel sensing time. Each access category has different values of AIFS, CW_{min} and CW_{max}. The access category with highest priority has the lowest value of AIFS and CW_{min} (IEEE, 2007; Sunghyun, del Prado, Sai Shankar, & Mangold, 2003). The IEEE 802.11p MAC protocol also follows the EDCA model to provide the quality of service using different access categories. In VANETs, such EDCA mechanism can help provide traffic prioritization to different type of traffics (Sunghyun et al., 2003). Safety messages can be assigned a higher access category for urgent access to the channel.

Using the IEEE 802.11 MAC standard, the wireless nodes can restrict communication within a cooperative group by the formation of a Basic Service Set (BSS). In presence of an infrastructure, the Access Point (AP) manages the formation of a BSS. Any node can join a BSS after listening to the periodic beacon messages from the AP which contains information about the Basic Service Set ID (BSSID). The process of joining involves the synchronization, authentication and association steps. All nodes within a BSS can communicate with each other using the AP and all messages from the nodes outside the BSS are filtered out. Two or more BSS can be joined together to form an Extended Service Set (ESS) using a Distribution Service (DS). In the ad hoc communication mode, instead of communicating through an AP, a station manages the BSS independently. The BSS formed is known as Independent Basic Service Set (IBSS) and the joining process involves the similar steps of synchronization, authentication and association, as with the infrastructure BSS. However for the vehicular communication environments where connection duration between vehicles is of the order of few seconds, steps such as synchronization, authentication and association would introduce large overhead and unacceptable long delays. For safety applications, a message requires immediate transmission, listening to the beacons and handshaking process for joining a BSS is not practical due to signaling and long delays. Therefore, in the IEEE 802.11p MAC standard, two nodes can immediately share data without the need of joining a BSS. The communication in such mode is accomplished through a special BSSID known as the wildcard BSSID. This mode of operation is known as the WAVE mode. In addition, the steps such as synchronization, authentication and association are not required for the formation of a WAVE BSS (WBSS). A station which starts the WBSS known as the provider sends periodic beacon messages in its control channel (CCH). The periodic beacon contains WAVE Service Advertisement (WSA) messages which contain information about the available service applications and parameters for joining the WBSS such as BSSID and service channel (SCH) of the WBSS. All nodes scan the WSA messages on the CCH and join a WBSS by switching to its SCH during the SCH interval (Campolo, Cortese, & Molinaro, 2009; Jiang & Delgrossi, 2008).

The IEEE 802.11p PHY layer is almost similar to the IEEE 802.11a physical layer as defined by the standard. It is based on the OFDM (Orthogonal Frequency Division Multiplexing) technique but the channel width is reduced to 10MHz from 20MHz. As a result, all the timing parameters including the guard interval are doubled which provides a better defense against the Inter Symbol Interference (ISI) (Jiang & Delgrossi, 2008). The 10 MHz bandwidth is shown to overcome large delay spread and doppler spread associated with vehicular networks (Lin, Henty, Cooper, Stancil, & Fan, 2008). The transmission data rate of the IEEE 802.11p standard is half of the data rate of the IEEE 802.11a standard. The management functions of MAC and PHY layers are handled by the MAC Sublayer Management Entity (MLME) and the Physical Layer Management Entity (PLME) as shown in Figure 2.

The multi channel operation for the WAVE is defined in the IEEE 1609.4 standard which sits in between the MAC layer and the IEEE 802.2 Logic link control (LLC). The WAVE architecture supports a multi channel mechanism comprising one control channel (CCH) and multiple service channels (SCH). The safety messages and the WAVE Service Advertisement (WSA) are transmitted on the CCH while service and business applications are run on the SCH. The future WAVE devices would include both single and multi channel radios. Single channel radio can either transmit or receive data on any radio channel. On the other hand, multi channel radio device can simultaneously transmit on one channel and receive on another radio channel. For single channel radio devices to work with the multi channel devices in future vehicular environments, time division multiple access is used. The time is divided in to Control Channel Interval (CCHI) and Service Channel Interval (SCHI) of 50ms each. A coordination scheme allows all WAVE stations to monitor during the control channel at the same time and switch to a desired application on a service channel during the SCH interval. This synchronisation is done through the Coordinated Universal Time (UTC) global clock signal available from GPS. A guard interval of 4ms is introduced at the start of each channel interval. The guard interval allows sufficient time to overcome the inaccuracy in timing and channel switching for the radio devices. The safety messages are shared on the CCH in the CCHI while the service messages and data exchange is done on the SCH during the SCHI. In case of emergency, vehicles can also utilize the SCHI for transmission of safety messages. The IEEE 1609.4 stack manages functions of channel routing, user priority control and channel coordination. The channel routing function manages the routing of data from LLC layer to the appropriate channel and access category. Two types of packets are defined in the WAVE standard. The IP packets are used to transmit non safety messages while the Wave Short Message (WSM) packets are used for safety messages. The WSM are short messages whose transmission parameters such as channel, data rate and transmit power can be directly controlled by the higher layers so that all the nodes receive the messages within a certain time delay. They are suited for safety applications which have higher priority and require low transmission delay. The channel router identifies the type of packet from the ethertype field in the 802.2 header and forwards the WSM packet to the CCH or to the SCH and the IP packet to the current SCH. The user priority is assigned using the EDCA access mechanism and four access categories are defined for medium access contention (1609.4) (Uzcategui & Acosta-Marum, 2009).

The channel coordination mechanism coordinates the channel intervals so that appropriate radio channel is selected at the right time for data exchange. In IEEE 1609.4, the MLME functions defined in IEEE 802.11 standard are extended to support channel synchronisation for channel coordination, channel access and transmission of management frames. The channel synchronisation makes sure that all devices are synchronised in time using the Coordinated Universal Time (UTC) for channel coordination. WAVE devices without built in timing signal obtains the timing information through other WAVE devices. The channel access in MLME controls the type of access given to CCH and SCH channels. It defines three types of access including the continuous access, alternating CCH and SCH, and immediate SCH. The MLME also maintains a management information base (MIB) which stores system and application related information. The system related information includes network and address information while the application related information includes provider and user information and channel information (1609.4) (Uzcategui & Acosta-Marum, 2009).

The IEEE 1609.3 standard defines functions and services at the LLC, network and transport layers of the WAVE protocol stack. These services are known as the Networking services which includes the data plane and Wave Management Entity (WME) as shown in Figure 2. The data plane components are specified according to the vehicular network requirements of low latency. The IEEE 1609.3 standard has defined a new transport layer protocol known as the Wave Short Message Protocol (WSMP) which enables the transmission of WSM messages for high priority safety purposes. For non safety purposes, the IPv6 has been supported along with the TCP and UDP transport layers. The parameters of IP messages are stored in a transmitter profile. The LLC layer is also implemented in every WAVE device. After the LLC receives a MAC service data unit, it forwards the packet to the IPv6 layer or the WSMP layer according to the ethertype field of the packet. At the management plane, the WME manages service requests including message includes the provider service request, user service request, WSM service request, CCH service request, etc. The provider service request could be made by any WAVE device which is interested to share a service and send WSA messages using the SCH channel. The WAVE Management Entity (WME) maintains a ProviderServiceRequestTableEntry in the MIB and considers the new service in the channel access. Similarly any user interested in the service could indicate WME which maintains a UserServiceRequestTableEntry in the MIB. The WSM service request is made by a WAVE device intending to receive WSM messages with a particular provider service identifier (PSID). The CCH service request indicates the WME that a WAVE device requires access to the CCH to monitor WSM messages or WSA during a particular channel interval. The WME also manages the changing service requests by a WAVE device and updates the MIB. The WME also monitors the WSA messages and the available link quality of the services in those messages (1609.3) (Uzcategui & Acosta-Marum, 2009).

Security services for the messages and applications are defined in IEEE 1609.2 standard. It includes algorithms and mechanisms for secure data transfer between WAVE units (IEEE, 2006b; Uzcategui & Acosta-Marum, 2009).

The IEEE 1609.1 standard specifies a WAVE application for remote resource access (IEEE, 2006a; Uzcategui & Acosta-Marum, 2009).

RELATED WORK

Many techniques have been proposed for broadcasting information in Vehicular Ad hoc networks. The most common packet broadcasting technique used is the flooding in which each node on receiving a broadcast message rebroadcasts it to all other nodes within its transmission range. This generates lot of redundant broadcast messages and increases the network traffic level consequently increasing the contention level at the MAC layer often causing the broadcast storm. To resolve the broadcast storm, many schemes have been proposed at MAC and network layers. In the Intelligent Broadcast with Implicit Acknowledgement (I-BIA) proposal, each vehicle continues to broadcast until it receives the same message from a vehicle at the back (Biswas, Tatchikou, & Dion, 2006). This implies that all the vehicles behind a transmitter can receive the broadcast through that vehicle. The introduction of implicit acknowledgement has been shown to reduce the probability of collision for cooperative collision avoidance application. Several timer and probability based broadcast schemes have been developed to suppress or to control the broadcast storm (Wisitpongphan et al., 2007). In a weighted p-persistence scheme, a node rebroadcasts at the first instance of a packet with a probability p which is proportional to the distance from the sender. Hence, the node furthest from the sender rebroadcast with the highest probability. In a slotted 1-persistence scheme, a node rebroadcasts the first instance of a packet in an assigned time slot *T* with a probability 1, else it discards the packet. Another proposed scheme known as the Distributed Vehicular Broadcast (DV-CAST) protocol considers the broadcast storm problem for different traffic scenarios ranging from sparse to dense distributions (Tonguz, Wisitpongphan, & Fan, 2010). In sparse traffic distribution scenario, where average interspacing between vehicles is high, a store carry and forward approach is used to allow the message propagate further down the road. In dense traffic scenarios, where average interspacing between vehicle is low, broadcast suppression techniques have been developed (Wisitpongphan et al., 2007). To improve the rebroadcast decision by considering network topology, a probabilistic Inter vehicle Geocast (p-IVG) protocol was introduced. Using the protocol, a vehicle waits for a specific time depending on the distance from the sender for duplicate packets before rebroadcasting a received packet with a probability *p* (Ibrahim, Weigle, & Abuelela, 2009). The probability factor introduced in the rebroadcast decision was selected based on the vehicle density. This approach helps to combat situations where many vehicles were at a similar distance from the sender and rebroadcast packets simultaneously. The Urban Multi-Hop Broadcast (UMB) partitions a highway in to small segments and sender node forwards the packet to a node in the furthest non empty segment in the direction of a broadcast (Korkmaz, Ekici, Ozguner, & Ozguner, 2004). The selected node is reserved by exchanging RTB (Request to Broadcast)/CTB (Clear to Broadcast) messages. All transmissions are also acknowledged. The protocol aims at solving the hidden node problem along with the broadcast storm problem. However, using the RTB/CTB reservation mechanism increases the end to end delay and transmission overhead. The mobicast routing protocol is an example of a geocast protocol which forwards multicast messages to vehicles in a geographical area of interest known as Zone of relevance (ZOR) at a specific time *t* using nodes with in a forwarding area called Zone of Forwarding (ZOF) (Yuh-Shyan, Yun-Wei, & Sing-Ling, 2009). The mobicast approach has been shown to achieve performance improvements in terms of packet success rate and packet delay. A Multi-Hop vehicular Broadcast (MHVB) scheme presents two different techniques for the efficient packet broadcast (Tatsuaki, Lan, & Massimiliano, 2006). In the first technique known as backfire algorithm, the node with the largest coverage area and the least distance towards the destination rebroadcasts packets with a shorter waiting time. The second technique known as the traffic congestion detection reduces the transmission of beacon messages if a node is detected to be in middle of congestion. The traffic congestion is measured by counting the number of nodes around a vehicle. However, the protocol suffers with packet loss issues due to transmission by a number of nodes. In Street Broadcast Reduction (SBR) scheme, a node rebroadcasts the emergency message only if its distance from the sender is greater than a fixed threshold (Martinez, Toh, Cano, Calafate, & Manzoni, 2011). The simulation results show a higher success rate and reduced overhead of the protocol as compared to flooding. However, the fixed rebroadcast threshold may cause large packet loss in fading scenarios where packet may not be received beyond the threshold distance.

GEOCASTING PACKET TRANSMISSION PROTOCOL

In this section we propose Geocast Packet transmission Protocol for faster delivery of emergency messages in a VANET (Javed & Khan). We assume that each vehicle is equipped with a GPS and knows its position on the road. This is a realistic assumption as the modern vehicles enjoy the facility of the GPS. If a vehicle meets an accident or experiences a fault, it broadcasts this information to the vehicles behind it that may be approaching the current position of the transmitter. The transmission range of a single vehicle is not large enough to inform all the vehicles on the road about the emergency situation. Hence we propose the use of a multi-hop packet transmission technique to distribute the emergency information to other vehicles on the road approaching the direction of an accident or breakdown. The goal of proposed protocol is

to efficiently broadcast the urgent information to all vehicles moving towards the direction of the transmitter using minimum number of rebroadcast nodes to avoid the broadcast storm problem and to reduce the message dissemination delay using the IEEE802.11p protocol. The proposed packet transmission technique uses a distance based backoff approach to achieve the above goals (Javed & Khan). The operation of the proposed geocast protocol scheme is shown in Figure 4.

In the proposed scheme, an emergency message is only received by a vehicle if that message is not a duplicate copy. After a packet is received by a vehicle, a distance based backoff is calculated such that vehicles at the farthest distance from the source have a lower contention window size. The contention window at each vehicle is chosen according to the formula.

ContentionWindow=[X,X+CW_{min}]

where

X=B×K (1)

$$B = \frac{(T_R + D_C) - D_{SD}}{D_C}$$

where

- T_R: One Hop Transmission Range (m)
- D_C: Non overlapping contention window range
- D_{SD}: Distance between source and destination (m)

B has a unique integer value for each vehicle at a different distance from the source. The value of B depends on the distance between the source and the destination. As the distance between the source and the destination is increased, the value of B is decreased. The value of B also depends on the constant D_c which is the non overlapping contention window range. The non overlapping contention window range ensures that all cars separated by a distance equal to D_c have a non overlapping contention window. The higher the value of D_c, the more is the collision probability as vehicles within a greater distance range have a similar contention window. The non overlapping contention window size is set to a minimum constant value of 10 in the simulations. K is any integer greater than CW_{min}. By choosing such value of K, the contention window size after every successful transmission which is given by [0, CW_{min}] is separated from the contention window for rebroadcast. The CW_{min} in our simulations is 15. We chose the value of K as 20 which is greater than CW_{min} for our simulations. Due to the use of the distance based backoff, the vehicles further away from the source have to wait less number of backoff slots before it can rebroadcast a packet. If the same packet is received by any vehicle during the backoff period, the duplicate packet is destroyed. If no such duplicate is received, then the packet is rebroadcasted. As a result, only the vehicle furthest away from the vehicle in emergency situation rebroadcasts the packet.

Figure 4. Flow diagram of proposed geocast scheme

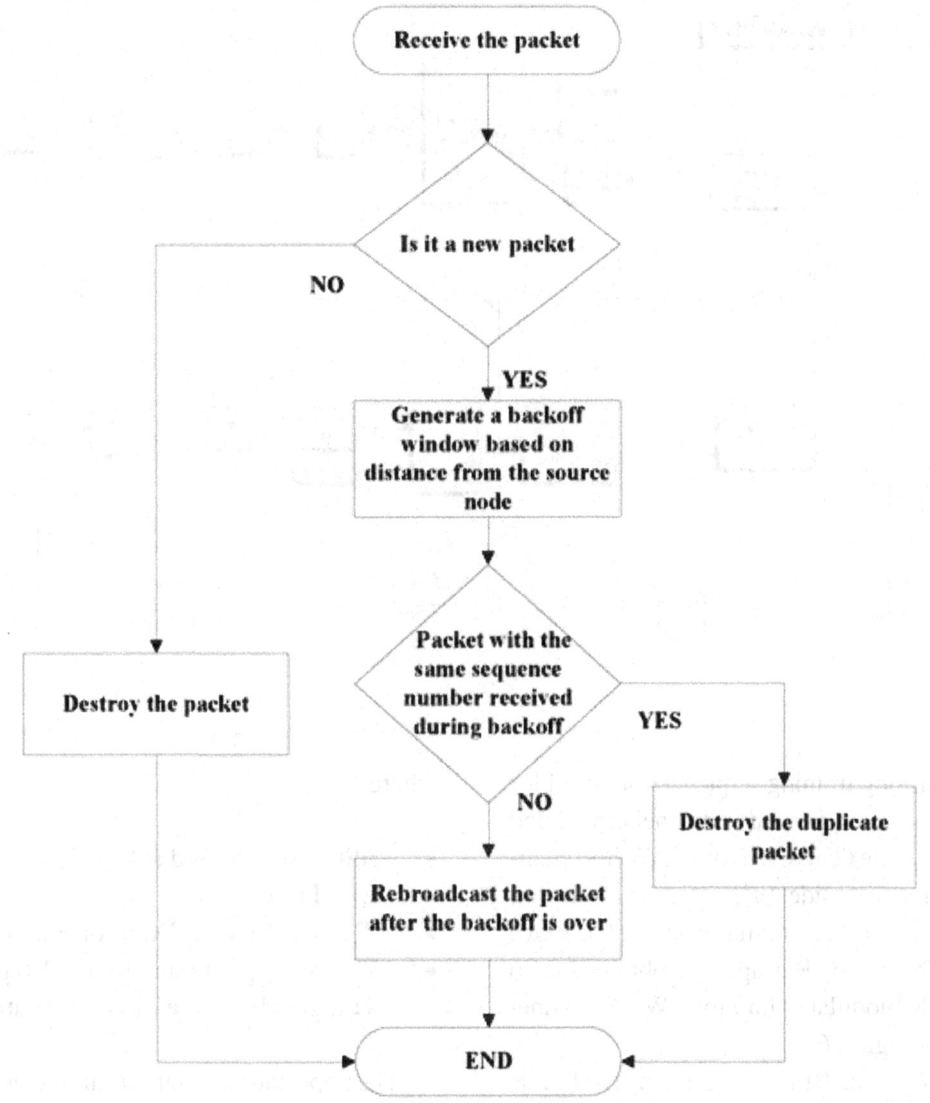

SIMULATION MODEL

The simulations are performed in OPNET Modeler 16.0. The IEEE 802.11 PHY/MAC layer model in OPNET was modified to account for the changes in IEEE 802.11p. The transmission frequency was set to 5.9 GHz and channel bandwidth was reduced to 10MHz as in the IEEE 802.11p standard. The transmission link was modeled using path loss with an exponent of 2. We used the transmission data rate of 6 Mbps using a QPSK modulated link. The BER profile of the QPSK technique was generated using the IEEE 802.11a PHY model in MATLAB as shown in Figure 5. The physical parameters of the model were changed according to IEEE 802.11p. The IEEE 802.11p model consists of a variable data source which generates message bits according to a desired data rate. The variable data is then coded, interleaved and modulated according to a modulation technique. After modulation, the OFDM transmission is done using 52 subcarriers, 4 pilot symbols, 64 point IFFT and 16 sample cyclic

Figure 5. IEEE 802.11a WLAN PHY model in MATLAB (2003)

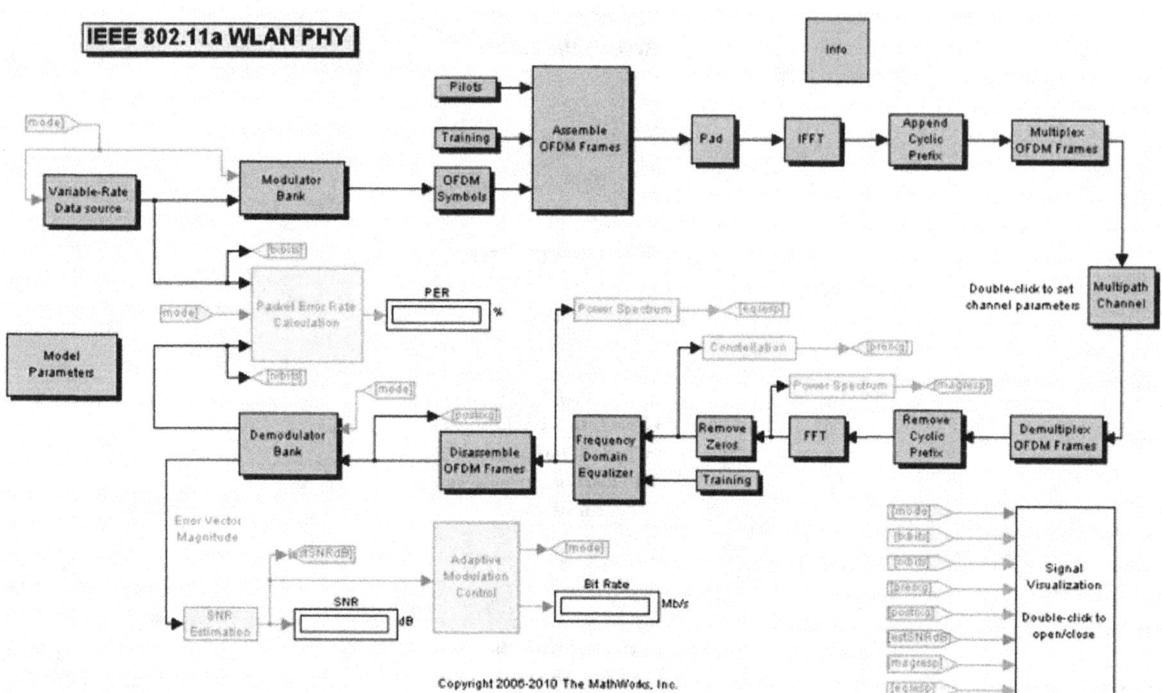

prefix. Four long training sequences are used for preamble insertion. The radio channel is modeled as Additive white Gaussian Noise (AWGN) channel. At the receiver side, FFT is applied followed by an equalizer and demodulator. Using the above model, Eb/No vs. BER graph was obtained for 6 Mbps QPSK modulated link for AWGN channel as shown in Figure 6.

The Eb/No vs. BER curve from MATLAB was then integrated in the bit error rate pipeline stage in OPNET. Eb/No was converted to SNR using the following correction formula (Kumar & Sharma, 2007).

$$SNR(dB) = \frac{Eb}{No} + 10\log_{10}\left({}^{nS_c}\!/_{nFFT} \right)$$

$$+\ 10\log_{10}\left({}^{T_d}\!/_{(T_d + T_c)} \right)$$

$$+10\log_{10}(k)$$

$$+10\log_{10}(\text{coderate})$$

where

- **nSC:** No. Of used subcarriers = 52
- **nFFT:** FFT size = 64
- T_d: Data Symbol Duration= 6.4μs
- T_c: Cyclic prefix duration = 1.6 μs
- k: $\log_2$ (M), where M = modulation order

The important simulation parameters are listed in Table 1.

PERFORMANCE ANALYSIS

We have performed simulations for a number of scenarios to evaluate the performance of the proposed protocol. The performance of the protocol is compared with the standard flooding technique. We have also compared our results with the slotted 1-persistence scheme in a multi lane network which has the best performance among timer and probability based protocols in terms

Figure 6. Eb/No vs BER curve for 6Mbps QPSK modulated link for AWGN channel

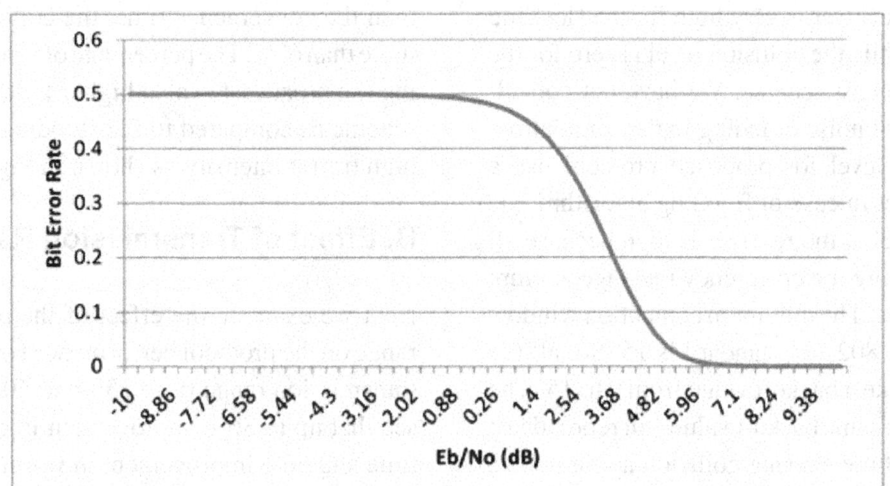

Table 1. Simulation parameters

MAC/PHY	IEEE 802.11p
Data Rate	6 Mbps
Modulation	QPSK
Vehicle Density	10-100 vehicles/km/lane
Road length	1km, 10km
Number of vehicle lanes	1-4
Transmission Range	250-1000m
Transmit Power	1.2-20mW
Packet Interarrival Time	1 sec
Packet Length	256-2048 bytes
Car Length	5m

of packet loss ratio (Wisitpongphan et al., 2007). Each simulation was run for 10,000 broadcast packets. Simulations are performed for different seed values representing different transmission scenarios. The results were averaged over a number of simulations. The performance metric used for the simulations is warning notification time (WNT), convergence time (CT), percentage of vehicles receiving the broadcast packet (VRB), the average collision level at the receiver (ACL) and the total broadcast overhead.

The warning notification time (WNT) is measured by taking the difference of a source message initiation time and the time the last vehicle on a road segment completes reception of the message. The convergence time (CT) is the difference of time instant when the first source message is transmitted and the instant when the last packet broadcast is completed. Broadcast overhead is the ratio of the number of duplicate messages to the number of vehicle nodes in the network.

A. Effect of Packet Interarrival Time

First we examine the effect of traffic intensity on the proposed protocol by varying the packet interarrival time. In this scenario, 60 stationary vehicles are placed on a straight road of 1km length. The vehicle in an emergency situation broadcasts a packet of 1024 byte with a transmission range of each vehicle set to 300m. The convergence time of the standard flooding and proposed geocast scheme for the transmission range is 43.5ms and 24.1ms respectively. The packet interarrival time is increased from 20ms to 50ms to observe the effect on the average collision level and the percentage of vehicles receiving the broadcast packet. It can be seen from the Figure 7 that the average collision level at the receiver is very high for the flooding algorithm while the collision level is zero for the proposed geocast scheme. We have not considered effects of noise or fading in this simulation, so collision level for proposed protocol has a value of zero. In case of flooding algorithm, the collision level at the receiver is high because all nodes receiving the emergency message attempt to rebroadcast. The minimum contention window size in IEEE 802.11p standard is 15. So, all the nodes will take a backoff value from 0 to 15. The vehicles with same backoff value will rebroadcast at the same time causing collision at the nodes. On the other hand, the proposed protocol assigns different contention windows to all the vehicles based on their distance from the transmitter node. As a result, only furthest vehicle rebroadcast the emergency message causing collision level to reduce to 0. The collision level of the flooding technique can go up to 90% of transmitted packets for the packet interarrival time of 20ms which is 46% of its convergence time. As the packet interarrival time increases, the collision reduces for the flooding algorithm due to low traffic intensity. However, even for a packet interarrival time longer than the convergence time, the collision level is more than 60%. The percentage of vehicles receiving the broadcast is also higher for the proposed scheme as compared to the flooding scheme for high traffic intensity as shown in Figure 7.

B. Effect of Transmission Range

Next we examine the effect of the transmission range on the protocol performance by varying the transmission range from 250m to 500m. We can see that up to 86% improvement in convergence time and 66% improvement in warning notification time at a transmission range of 500m for our proposed scheme over the flooding scheme as shown in Figure 8. In case of flooding, the increase in transmission range does not result in much improvement in warning notification time and convergence time. This is because every vehicle rebroadcasts the packet on reception, so even at higher transmission range the number of hops is not reduced. However, in the proposed protocol less number of rebroadcast hops is required to notify each vehicle with increased

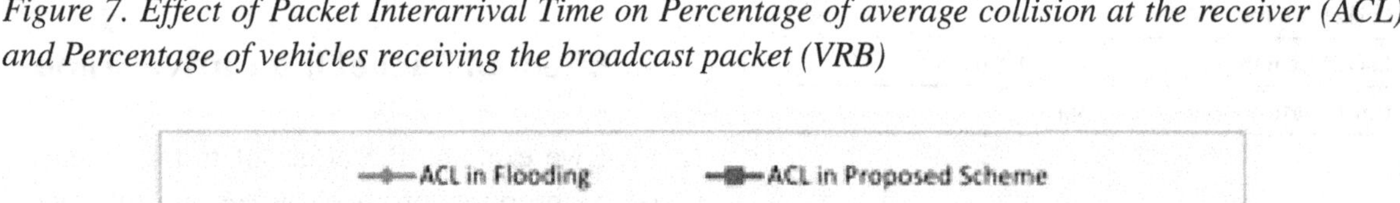

Figure 7. Effect of Packet Interarrival Time on Percentage of average collision at the receiver (ACL) and Percentage of vehicles receiving the broadcast packet (VRB)

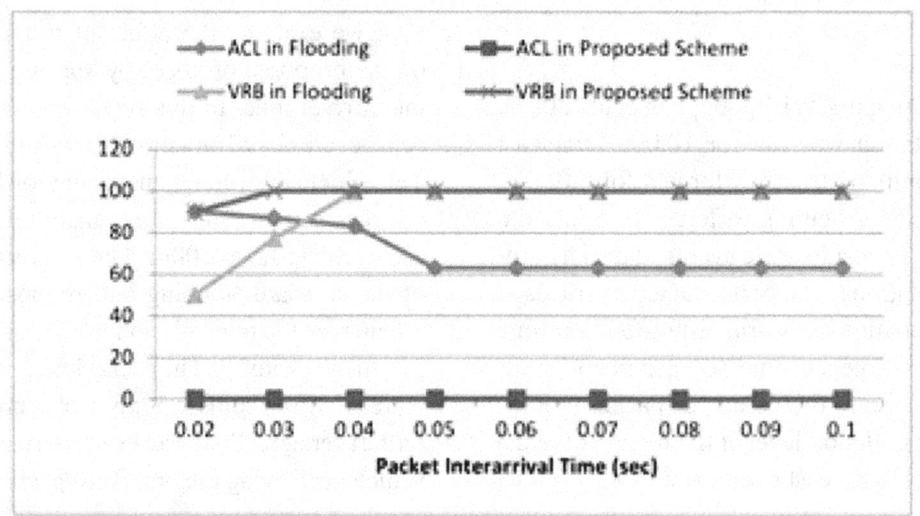

Figure 8. Effect of Transmission Range: Effective improvements achieved by the proposed protocol over the flooding protocol

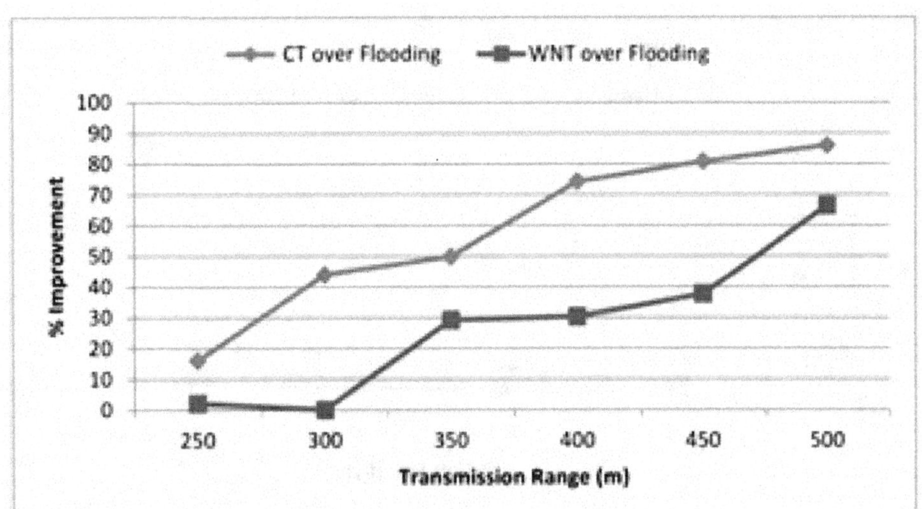

transmission range as furthest vehicle is always selected as the emergency message forwarder at each hop. Therefore, the warning notification time and convergence time decrease in the proposed protocol. Hence, the improvement of the proposed protocol over flooding increases with increased transmission range.

C. Effect of Packet Size

Next the effect of Packet size on performance is observed. The transmission range of each vehicle is set to 500m and the road length is 1000m with 60 stationary vehicles. It can be seen from the Figure 9 that the convergence and warning notification time of flooding algorithm is much higher as compared to proposed scheme. The proposed geocast scheme offers up to 87% and 69% improvement in convergence time and warning notification time over the flooding scheme for a packet size of 2048 bytes as shown in Figure 10. The performance improvement is due to less number of rebroadcast nodes in the proposed protocol as compared to flooding which reduces the collision level and hence lower the warning notification time and convergence time. However, the performance improvement for a range of packet sizes remains constant. This is due to the reason that the increase in packet size only increases the propagation delay which increases the warning notification time and convergence time. The contention probability is not effected by increasing the packet size.

D. Effect of Noise

To observe the effects of noise on performance of the proposed protocol, we use the previous scenario with a packet size of 1024 bytes. The proposed scheme performs much better than the flooding scheme in terms of warning notification time and convergence time as shown in Figure 11. The convergence time is nearly 4 times less than the flooding scheme at the bit error rate of $9.41 * 10^{-2}$. The proposed protocol performs well in a noisy channel with high error rates. As the noise is increased, the packet is received up to nodes at lesser distance. The contention window design of the proposed protocol always makes sure that the furthest vehicle from the

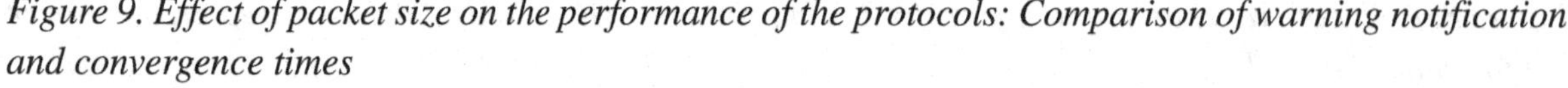

Figure 9. Effect of packet size on the performance of the protocols: Comparison of warning notification and convergence times

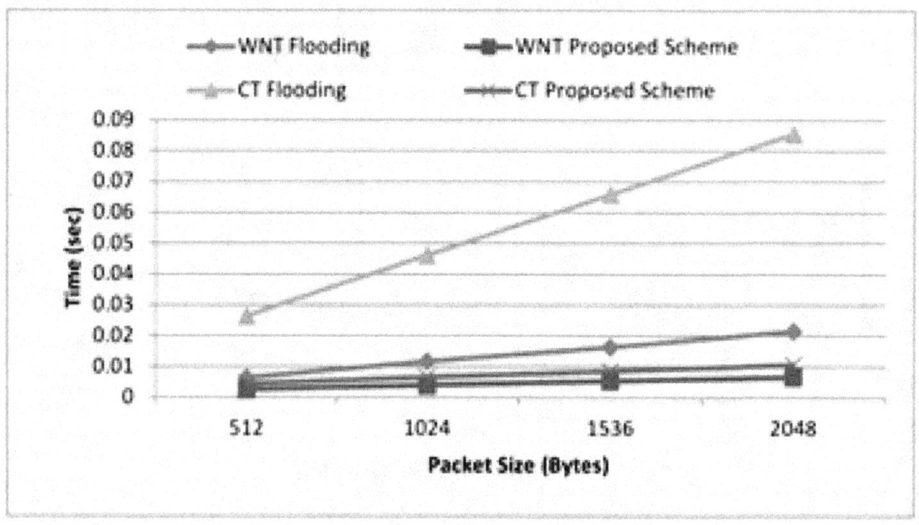

Figure 10. Effect of packet size on the performance of the protocols: Percentage of improvement of warning notification and convergence times over the flooding protocol

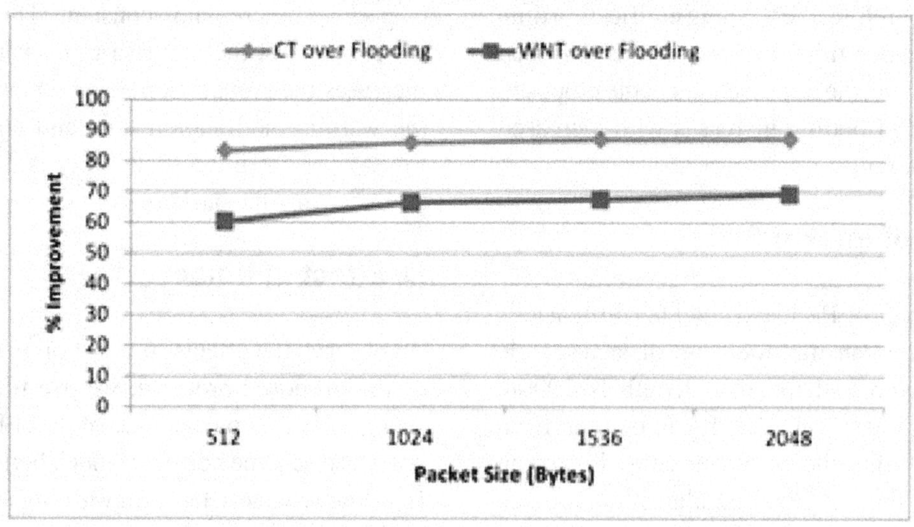

sender has the lowest contention window and hence the first one to rebroadcast. Therefore, the rebroadcast nodes are adaptively selected at maximum distance from the sender according to the channel conditions. The proposed scheme also has 8 times less broadcast overhead than the flooding scheme as shown in Figure 12. In flooding the simultaneous rebroadcast of many nodes increases the duplicate packets. As a result, the broadcast overhead is increased.

Figure 11. Effect of noise on protocols: warning notification and convergence time comparison

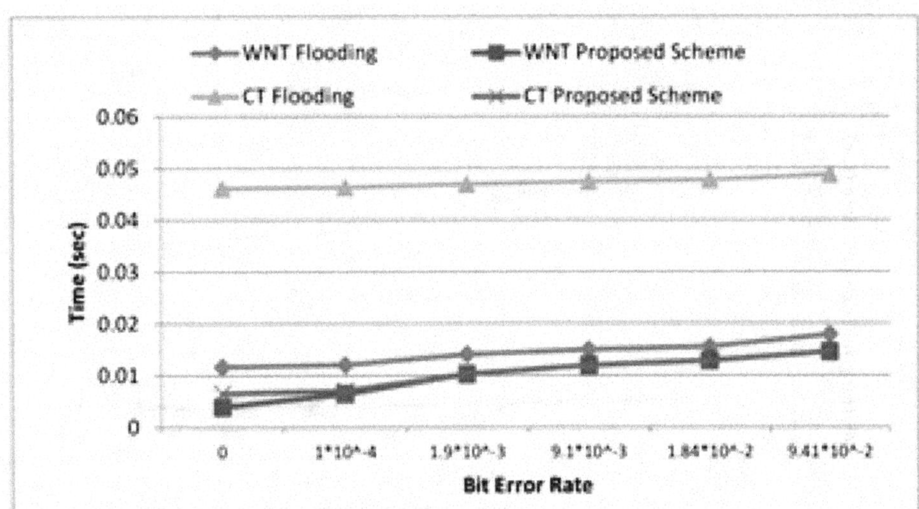

Figure 12. Effect of noise on protocols: broadcast overhead comparison

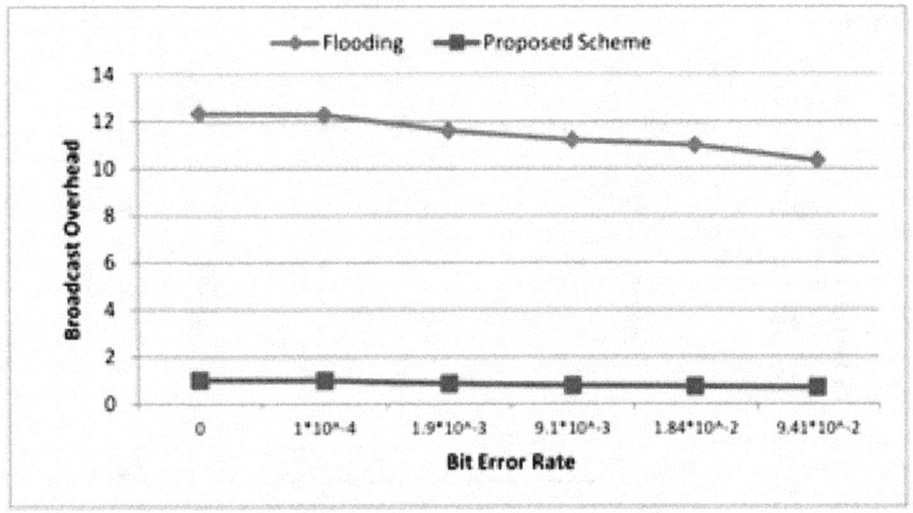

E. Effect of Node Density in Multiple Lane Road

In this scenario we observe the effects of node density on protocol performance and compare our results with slotted 1-persistence protocol (Wisitpongphan et al., 2007) which is also the suppression mechanism used in (Tonguz et al., 2010). We consider a 10km road section with 4 lanes to directly compare results with the reference (Wisitpongphan et al., 2007). The transmission range is set to 1000m and node density is varied from 10vehicles/km/lane to 100 vehicles/km/lane. Figures 13 and 14 show the warning notification time of the slotted 1-persistence protocol compared to the proposed scheme. The results show that the

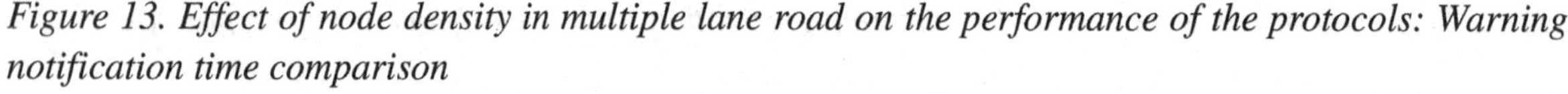

Figure 13. Effect of node density in multiple lane road on the performance of the protocols: Warning notification time comparison

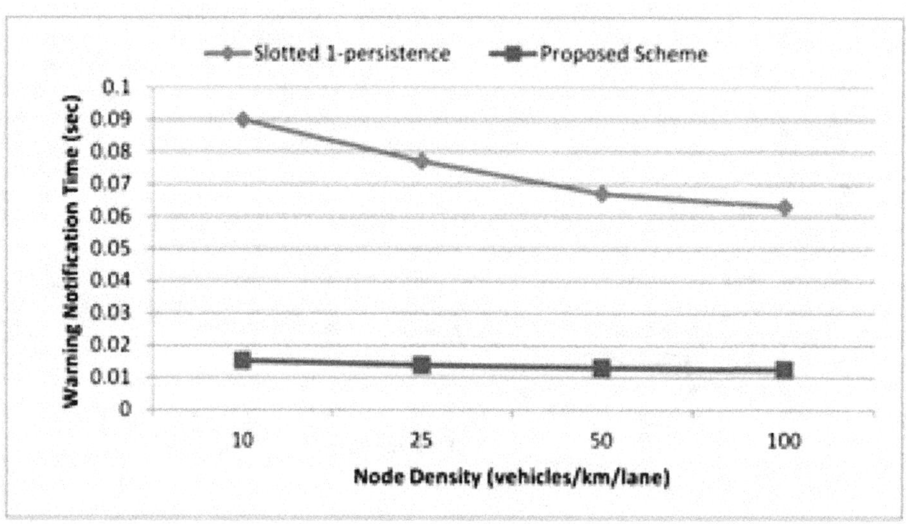

Figure 14. Effect of node density in multiple lane road on the performance of the protocols: Packet loss ratio comparison

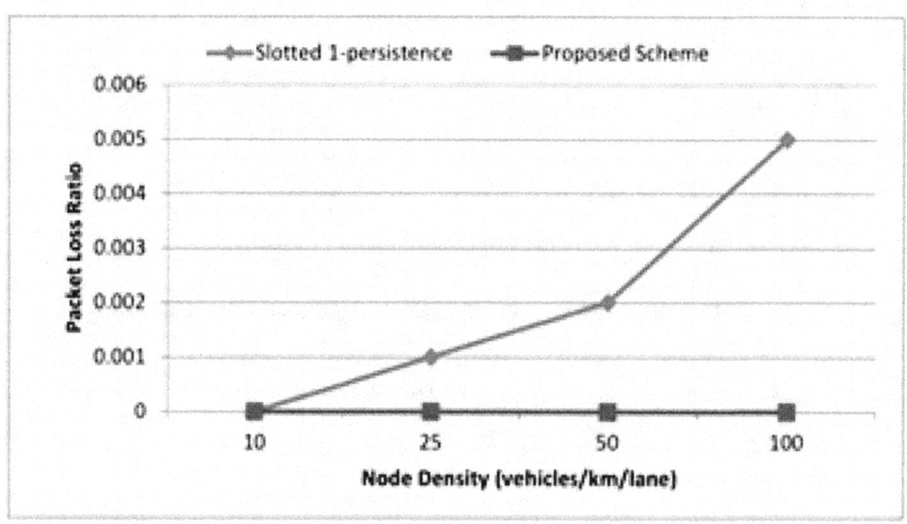

proposed geocast protocol can notify the nodes at a distance of 10km away from the emergency vehicle within 15ms as compared to 4 to 5 times larger delay in slotted 1-persistence for different node densities. Figure 12 shows that packet loss ratio is 0% in the proposed scheme as compared to up to 5% packet loss ratio in case of the slotted 1-persistence at a node density of 100 vehicles/km/lane. The reason that the proposed protocol performs better than the slotted 1-persistence protocol is that there are only 5 time slots used in slotted 1-persistence protocol. Therefore, for a transmission range of 1km, vehicles within 200m distance rebroadcast simultaneously and results in greater contention and overhead. However, the proposed scheme has a contention window design which enables only furthest vehicle within the transmission range to rebroadcast the message. This results in lower warning notification time. Similarly, packet loss ratio for the proposed scheme is 0% even at high node densities. Note that the simulation has not considered noise or fading in this scenario, which allows us to compare the broadcast suppression performance of the proposed scheme with slotted 1-persistence protocol.

CONCLUSION AND FUTURE WORK

This paper presented an efficient packet transmission scheme for emergency warning message propagation in VANETs. The main idea behind the proposed protocol is to select the furthest vehicle from the original transmitter for rebroadcast with the help of a newly proposed backoff window design which reduces the number of packet transmissions consequently lowering the contention level. Simulation results show that the proposed protocol offers very low convergence and warning notification times compared to the flooding and slotted 1-persistence protocol. Low warning notification time could significantly improve the performance of a VANET deployed for broadcasting safety information. The proposed protocol also generates lower broadcast overhead and packet loss ratio as compared to flooding and slotted 1-persistence protocol. Currently the work is looking at developing and extending the protocol to operate in high BER and fading scenarios. We are aiming at developing an efficient protocol for mobile urban and highway scenarios in future. The effect of GPS inaccuracy on the protocol will also be a subject of future study.

REFERENCES

Bernsen, J., & Manivannan, D. (2009). Unicast routing protocols for vehicular ad hoc networks: A critical comparison and classification. *Pervasive and Mobile Computing*, *5*(1), 1–18. doi:10.1016/j.pmcj.2008.09.001.

Biswas, S., Tatchikou, R., & Dion, F. (2006). Vehicle-to-vehicle wireless communication protocols for enhancing highway traffic safety. *IEEE Communications Magazine*, *44*(1), 74–82. doi:10.1109/MCOM.2006.1580935.

Campolo, C., Cortese, A., & Molinaro, A. (2009, October 12-14). CRaSCH: A cooperative scheme for service channel reservation in 802.11p/WAVE vehicular ad hoc networks. In *Proceedings of the International Conference on Ultra Modern Telecommunications & Workshops* (pp. 1-8).

Fiore, M., Harri, J., Filali, F., & Bonnet, C. (2007, March 26-28). Vehicular mobility simulation for VANETs. In *Proceedings of the 40th Annual Simulation Symposium* (pp. 301-309).

Ibrahim, K., Weigle, M. C., & Abuelela, M. (2009, April 26-29). p-IVG: Probabilistic Inter-vehicle geocast for dense vehicular networks. In *Proceedings of the 69th IEEE Vehicular Technology Conference* (pp. 1-5).

IEEE. (2006a). *IEEE Std 1609.1-2006: Trial-use standard for wireless access in vehicular environments (WAVE) - Resource manager* (pp. c1–c63). Piscataway, NJ: IEEE.

IEEE. (2006b). IEEE Std. 1609.2-2006: Trial-use standard for wireless access in vehicular environments - Security services for applications and management messages (pp. 0_1-105). Piscataway, NJ: IEEE.

IEEE. (2007). *Std 802.11-2007: Standard for information technology-telecommunications and information exchange between systems-local and metropolitan area networks-specific requirements - Part 11: Wireless LAN Medium Access Control (MAC) and Physical Layer (PHY) specifications. (Revision of IEEE Std 802.11-1999)* (pp. C1–C1184). Piscataway, NJ: IEEE.

IEEE. (2010a). *IEEE Std 1609.3-2010: Standard for wireless access in vehicular environments (WAVE) -Networking services (Revision of IEEE Std 1609.3-2007)* (pp. 1–144). Piscataway, NJ: IEEE.

IEEE. (2010b). *IEEE Std. 1609.4-2010: Standard for wireless access in vehicular environments (WAVE)--Multi-channel operation (Revision of IEEE Std 1609.4-2006)* (pp. 1–89). Piscataway, NJ: IEEE.

IEEE. (2010c). IEEE Std 802.11p-2010: (Amendment to IEEE Std 802.11-2007 as amended by IEEE Std 802.11k-2008, IEEE Std 802.11r-2008, IEEE Std 802.11y-2008, IEEE Std 802.11n-2009, and IEEE Std 802.11w-2009) (pp. 1-51). Piscataway, NJ: IEEE.

Javed, M. A., & Khan, J. Y. (2011 November 9-11). A geocasting technique in an IEEE802.11p based vehicular ad hoc network for road traffic management. In *Proceedings of the Australasian Telecommunication Networks and Applications Conference* (pp. 1-6).

Jiang, D., & Delgrossi, L. (2008, May 11-14). IEEE 802.11p: Towards an international standard for wireless access in vehicular environments. In *Proceedings of the IEEE Vehicular Technology Conference* (pp. 2036-2040).

Karim, R. (2008). *VANET: Superior system for content distribution in vehicular network applications*. Rutgers, NJ: Rutgers University.

Korkmaz, G., Ekici, E., Ozguner, F., & Ozguner, U. (2004). Urban multi-hop broadcast protocol for inter-vehicle communication systems. In *Proceedings of the 1st ACM International Workshop on Vehicular Ad Hoc Networks* (pp. 76-85).

Kumar, S., & Sharma, S. (2007). Error probability of different modulation schemes for OFDM based WLAN standard IEEE 802.11a. *International Journal of Engineering*, *4*(4), 262–320.

Lin, C., Henty, B. E., Cooper, R., Stancil, D. D., & Fan, B. (2008). A measurement study of time-scaled 802.11a waveforms over the mobile-to-mobile vehicular channel at 5.9 GHz. *IEEE Communications Magazine*, *46*(5), 84–91. doi:10.1109/MCOM.2008.4511654.

Martinez, F., Toh, C.-K., Cano, J.-C., Calafate, C., & Manzoni, P. (2011). A street broadcast reduction scheme (SBR) to mitigate the broadcast storm problem in VANETs. *Wireless Personal Communications*, *56*(3), 559–572. doi:10.1007/s11277-010-9989-4.

MATLAB. (2003). *Version 7.11.0.584 (R2010b)*. Natick, MA: The Mathworks Inc..

Morgan, Y. L. (2010). Notes on DSRC & WAVE Standards Suite: Its architecture, design, and characteristics. *IEEE Communications Surveys & Tutorials*, *12*(4), 504–518. doi:10.1109/SURV.2010.033010.00024.

Sunghyun, C., del Prado, J., Sai Shankar, N., & Mangold, S. (2003, May 11-15). IEEE 802.11 e contention-based channel access (EDCF) performance evaluation. In *Proceedings of the IEEE International Conference on Communications* (pp. 1151-1156).

Tatsuaki, O., Lan, L., & Massimiliano, L. (2006, June). Multi-Hop Vehicular Broadcast (MHVB). In *Proceedings of the IEEE 6th International Conference on ITS Telecommunications* (pp. 757-760).

Tonguz, O. K., Wisitpongphan, N., & Fan, B. (2010). DV-CAST: A distributed vehicular broadcast protocol for vehicular ad hoc networks. *IEEE Wireless Communications*, *17*(2), 47–57. doi:10.1109/MWC.2010.5450660.

Uzcategui, R., & Acosta-Marum, G. (2009). Wave: A tutorial. *IEEE Communications Magazine*, *47*(5), 126–133. doi:10.1109/MCOM.2009.4939288.

Wisitpongphan, N., Tonguz, O. K., Parikh, J. S., Mudalige, P., Bai, F., & Sadekar, V. (2007). Broadcast storm mitigation techniques in vehicular ad hoc networks. *IEEE Wireless Communications*, *14*(6), 84–94. doi:10.1109/MWC.2007.4407231.

Yuh-Shyan, C., Yun-Wei, L., & Sing-Ling, L. (2009, November 30-December 4). A Mobicast routing protocol in vehicular ad-hoc networks. In *Proceedings of the IEEE Global Telecommunications Conference* (pp. 1-6).

This work was previously published in the International Journal of Wireless Networks and Broadband Technologies, Volume 1, Issue 4, edited by Naveen Chilamkurti, pp. 55-72, copyright 2011 by IGI Publishing (an imprint of IGI Global).

Compilation of References

Aad, I., Ni, Q., Barakat, C., & Turletti, T. (2005). Enhancing IEEE 802.11 MAC in congested environments. *Computer Communications*, *28*(14), 1605–1617. doi:10.1016/j.comcom.2005.02.010.

Abboud, O., Zinner, T., Lidanski, E., Pussep, K., & Steinmetz, R. (2010, June). StreamSocial: a P2P Streaming System with Social Incentives. In *Proceedings of the IEEE International Symposium on a World of Wireless Mobile and MultiMedia Networks*, Montreal, QC, Canada.

Acquisiti, A., & Gross, R. (2005, November). Information Revelation and Privacy in Online Social Networks (the Facebook case). In *Proceedings of the ACM Workshop on Privacy in the Electronic Society*.

Acquisiti, A., & Gross, R. (2006, June). Imagined Communities: Awareness, Information Sharing, and Privacy on Facebook. In *Proceedings of the 6th Workshop on Privacy Enhancing Technologies*.

Ahmad, I., & Habibi, D. (2008). A proactive forward error correction scheme for mobile WiMAX communication. In *Proceedings of the IEEE International Conference on Communication Systems* (pp. 1647-1649).

Ahmad, I., Habibi, D., & Rahman, Z. (2008). An improved FEC scheme for mobile wireless communication at vehicular speeds. In *Proceedings of the IEEE Australasian Telecommunication Networks and Applications Conference* (pp. 312-316).

Ahmed, S., & Kawai, M. (2010). Interleaver-based subcarrier allocation schemes for BER fairness in OFDMA. In *Proceedings of the 9th International Conference on Wireless Networks* (pp. 262-266).

Ahmed, S., & Kawai, M. (2011). Effects of interleavers on BER fairness and peak to average power ratio in OFDMA System. In *Proceedings of the 6th International Symposium on Wireless and Pervasive Computing* (pp. 1-6).

Akyildiz, F., Wang, X., & Wang, W. (2005). Wireless Mesh Networks: a survey. *Journal of Computer Networks*, *47*(4), 445–487. doi:10.1016/j.comnet.2004.12.001.

Akyildiz, I. F., & Ho, J. S. M. (1995). Mobile user location update and paging mechanisms under delay constraints. *ACM-Baltzar Journal on Wireless Networks*, *1*(4), 413–425. doi:10.1007/BF01985754.

Akyildiz, I. F., Lee, W.-Y., Vuran, M. C., & Mohanty, S. (2006). Next generation/dynamic spectrum access/cognitive radio wireless networks: A survey. *International Journal of Computer and Telecommunications Networking*, *50*(13), 2127–2159.

Akyildiz, I. F., Wang, X., & Wang, W. (2005). Wireless mesh networks: a survey. *Computer Networks*, *47*, 445–487. doi:10.1016/j.comnet.2004.12.001.

Akyildiz, I., & Wang, X. (2005). A survey on wireless mesh networks. *IEEE Communications Magazine*, *43*(9), 23–30. doi:10.1109/MCOM.2005.1509968.

Alamouti, S. M. (1998). A simple diversity technique for wireless communications. *IEEE Journal on Selected Areas in Communications*, *16*, 1451–1458. doi:10.1109/49.730453.

Alexiou, A., & Haardt, M. (2004). Smart antenna technologies for future wireless systems: Trends and challenges. *IEEE Communications Magazine*, *42*(9), 90–97. doi:10.1109/MCOM.2004.1336725.

Ali, G. G., Chan, E., & Li, W. (2011). Two-step Joint Scheduling Scheme for Road Side Units (RSUs)-based Vehicular Ad Hoc Networks (VANETs). In J. Xu, G. Yu, S. Zhou, & R. Unland (Eds.), *Proceedings of the 16th International Conference on Database Systems for Advanced Applications* (LNCS 6637, pp. 453-464).

Al-Karaki, J. N., Ul-Mustafa, R., & Kamal, A. E. (2004, April). Data aggregation in wireless sensor networks - Exact and approximate algorithms. In *Proceedings of the IEEE Workshop on High Performance Switching and Routing*, Phoenix, AZ (pp. 241-245).

Al-Karaki, J. N., & Al-Mashaqbeh, G. A. (2007). Energy-centric routing in wireless sensor networks. *Microprocessors and Microsystems, 31*, 252–262. doi:10.1016/j.micpro.2007.02.008.

Al-Karaki, J. N., & Kamal, A. E. (2004). Routing techniques in wireless sensor networks: A survey. *IEEE Wireless Communications, 11*(6), 6–28. doi:10.1109/MWC.2004.1368893.

Al-Karaki, J. N., & Kamal, A. E. (2005). End-to-end support for statistical quality of service in heterogeneous mobile ad hoc networks. *Computer Communications, 28*(18), 2119–2132. doi:10.1016/j.comcom.2004.07.038.

Altman, E. (2002). Applications of Markov decision processes in communication networks. In Feinberg, A. E., & Schwartz, A. (Eds.), *Handbook of Markov decision processes* (pp. 489–536). New York, NY: Springer. doi:10.1007/978-1-4615-0805-2_16.

Andreev, S., Galinina, O., & Turlikov, A. (2010). Basic client relay model for wireless cellular networks. In *Proceedings of the International Congress on Ultra Modern Telecommunications and Control Systems and Workshops* (pp. 909-915).

Andreev, S., Galinina, O., & Vinel, A. (2010). Cross-layer channel-aware approaches for modern wireless networks. In A. Vinel, B. Bellalta, C. Sacchi, A. Lyakhov, M. Telek, & M. Oliver (Eds.), *Proceedings of the Third International Workshop on Multiple Access Communications* (LNCS 6235, pp. 163-179).

Andreev, S., Koucheryavy, Y., Himayat, N., Gonchukov, P., & Turlikov, A. (2010, December). *Active-mode power optimization in OFDMA-based wireless networks.* Paper presented at the 6th IEEE Broadband Wireless Access Workshop, Miami, FL.

Anisimov, A., Andreev, S., Galinina, O., & Turlikov, A. (2010). Comparative analysis of sleep mode control algorithms for contemporary metropolitan area wireless networks. In S. Balandin, R. Dunaytsev, & Y. Koucheryavy (Eds.), *Proceedings of the 10th International Conference on Smart Spaces and Next Generation Wired/Wireless Networking* (LNCS 6294, pp. 184-195).

Azgin, A., Altunbasak, Y., & AlRegib, G. (2005). Co-operative mac and routing protocols for wireless ad hoc networks. In. *Proceedings of the IEEE Global Telecommunications Conference, 5*, 2854–2859.

Babu, B., & Venkataram, P. (2009). Random security scheme selection for mobile transactions. *Security and Communication Networks, 2*(6), 694–708.

Bai, F., Sadagopan, N., & Helmy, A. (2003). The IMPORTANT framework for analyzing the impact of mobility on performance of routing protocols for adhoc networks. *Journal of Ad Hoc Networks, 1*, 383–403. doi:10.1016/S1570-8705(03)00040-4.

Baldo, N., & Zorzi, M. (2007, January). Fuzzy logic for cross-layer optimization in cognitive radio networks. In *Proceedings of the 4th IEEE Consumer Communications and Networking Conference* (pp. 1128-1133).

Banchs, A., Serrano, P., & Azcorra, A. (2006). End-to-end delay analysis and admission control in 802.11 DCF WLANs. *Communications of the ACM, 29*(7), 842–854.

Banchs, A., & Vollero, L. (2006). Throughput analysis and optimal configuration of 802.11e EDCA. *Computer Networks, 50*(11), 1749–1768. doi:10.1016/j.comnet.2005.07.008.

Baynat, B., Nogueira, G., Maqbool, M., & Coupechoux, M. (2009). An efficient analytical model for the dimensioning of WiMAX networks. In *Proceedings of the International Conference on Mobile Technology, Applications, and Systems* (pp. 521-534).

Bejerano, Y., & Cidon, I. (2003). An anchor chain scheme for IP mobility management. *Wireless Networks*, *9*(5), 409–420. doi:10.1023/A:1024627814601.

Benini, L., Bogliolo, A., & de Micheli, G. (2000). A survey of design techniques for system-level dynamic power management. *IEEE Transactions on Very Large Scale Integration*, *8*(3), 299–316. doi:10.1109/92.845896.

Beres, E., & Adve, R. (2008). Selection cooperation in multi-source cooperative networks. *IEEE Transactions on Wireless Communications*, *7*(1), 118–127. doi:10.1109/TWC.2008.060184.

Bernsen, J., & Manivannan, D. (2009). Unicast routing protocols for vehicular ad hoc networks: A critical comparison and classification. *Pervasive and Mobile Computing*, *5*(1), 1–18. doi:10.1016/j.pmcj.2008.09.001.

Bharghavan, V. (1994). MACAW: A media access protocol for wireless LANs. In *Proceedings of the Annual Conference on Communications, Architectures, Protocols, and Applications* (pp. 212-225). New York, NY: ACM Press.

Bianchi, G., Fratta, L., & Oliveri, M. (1996, October 15-18). Performance evaluation and enhancement of the CSMA/CA MAC protocol for 802.11 wireless LANs. In *Proceedings of the 7th IEEE International Symposium on Personal, Indoor and Mobile Radio Communications*, Taipei, Taiwan (pp. 392-396). Washington, DC: IEEE Computer Society.

Bianchi, G. (2000). Performance analysis of the IEEE 802.11 distributed coordination function. *IEEE Journal on Selected Areas in Communications*, *18*(3), 535–547. doi:10.1109/49.840210.

Bianchi, G., Borgonovo, F., Fratta, L., Musumeci, L., & Zorzi, M. (1997). C-PRMA: A centralized packet reservation multiple access for local wireless communications. *IEEE Transactions on Vehicular Technology*, *46*(2), 422–436. doi:10.1109/25.580781.

Biswas, S., Tatchikou, R., & Dion, F. (2006). Vehicle-to-vehicle wireless communication protocols for enhancing highway traffic safety. *IEEE Communications Magazine*, *44*(1), 74–82. doi:10.1109/MCOM.2006.1580935.

Bletsas, A., Khisti, A., Reed, D. P., & Lippman, A. (2006). A simple cooperative diversity method based on network path selection. *IEEE Journal on Selected Areas in Communications*, *24*(3), 659–672. doi:10.1109/JSAC.2005.862417.

Bletsas, A., Khisti, A., & Win, M. Z. (2008). Opportunistic cooperative diversity with feedback and cheap radios. *IEEE Transactions on Wireless Communications*, *7*(5), 1823–1827. doi:10.1109/TWC.2008.070193.

Bokareva, T., Hu, W., Kanhere, S., Ristic, B., Gordon, N., Bessell, T., et al. (2006). *Wireless sensor networks for battlefield surveillance.* Retrieved from http://www.cse.unsw.edu.au/~tbokareva/papers/lwc.html

Boukerche, A., El-Khatib, K., Xu, L., & Korba, L. (2004). SDAR: A secure distributed anonymous routing protocol for wireless and mobile ad hoc networks. In *Proceedings of the 29th Annual IEEE International Conference on Local Computer Networks* (pp. 618-624).

Box, G. E. P., & Jenkins, G. M. (1976). *Time series analysis: Forecasting and control.* San Francisco, CA: Holden-Day.

Brockwell, P. J., & Davis, R. A. (2002). *Introduction to time series and forecasting* (2nd ed.). New York, NY: Springer.

Bruno, R., Conti, M., & Gregori, E. (2002). Optimization of efficiency and energy consumption in p-persistent CSMA-based wireless LANs. *IEEE Transactions on Mobile Computing*, *1*(1), 10–31. doi:10.1109/TMC.2002.1011056.

Bruno, R., Conti, M., & Gregori, E. (2007). Throughput Analysis and Measurements in IEEE 802.11 WLANs with TCP and UDP Traffic Flows. *IEEE Transactions on Mobile Computing*, 171–186.

Brzezinski, A., Zussman, G., & Modiano, E. (2006). Enabling distributed throughput maximization in wireless mesh networks: a partitioning approach. In *Proceedings of the 12th Annual International Conference on Mobile Computing and Networking* (pp. 26-37).

Bucciol, P., Zechinelli-Martini, J., & Vargas-Solar, G. (2009). Optimized transmission of loss tolerant information streams for real-time vehicle-to-vehicle communications. In *Proceedings of the IEEE Mexican International Conference on Computer Science* (pp. 142-145).

Cali, F., & Conti, M. (2000). Dynamic tuning of the IEEE 802.11 protocol to achieve a theoretical throughput limit. *IEEE/ACM Transactions on Networking, 8*(6), 785–799. doi:10.1109/90.893874.

Campolo, C., Cortese, A., & Molinaro, A. (2009, October 12-14). CRaSCH: A cooperative scheme for service channel reservation in 802.11p/WAVE vehicular ad hoc networks. In *Proceedings of the International Conference on Ultra Modern Telecommunications & Workshops* (pp. 1-8).

Cao, Z., Tureli, U., & Liu, P. (2004). Optimum subcarrier assignment for OFDMA uplink. In *Proceedings of the Thirty-Seventh Asilomar Conference on Signals, Systems and Computers* (Vol.1, pp. 708-712).

Castelluccia, C. (2001). Extending mobile IP with adaptive individual paging: A performance analysis. *ACM Mobile Computing and Communication Review, 5*(2).

Castelluccia, C., & Mutaf, P. (2001). An adaptive per-host IP paging architecture. *ACM SIGCOMM Computer Communication Review*, 48-56.

Castro, M. C., Dely, P., Kassler, A. J., & Vaidya, N. H. (2009). Qos-aware channel scheduling for multi-radio/multi-channel wireless mesh networks. In *Proceedings of the 4th ACM International Workshop on Experimental Evaluation and Characterization* (pp. 11-18). New York, NY: ACM Press.

Cavalcanti, D., Das, S., Wang, J., & Challapali, K. (2008). Cognitive radio based wireless sensor networks. In *Proceedings of the 17th International Conference on Computer Communications and Networks* (pp. 1-6).

Challal, Y., Bettahar, H., & Bouabdallah, A. (2004). SAKM: A scalable and adaptive key management approach for multicast communications. *ACM Computer Communication Review, 34*(2), 55–70. doi:10.1145/997150.997157.

Chang, W. P., Li, J., & Morikawa, H. (2005). Distance-based localized mobile IP mobility management. In *Proceedings of the 8th International Symposium on Parallel Architectures, Algorithms and Networks* (pp. 154-159).

Chang, C., Cheng, R., & Shih, H., C. Y. (2007). Maximum freedom last scheduling algorithm for downlinks of DSRC networks. *IEEE Transactions on Intelligent Transportation Systems, 8*(2), 223–232. doi:10.1109/TITS.2006.889440.

Chatzimisios, P., Vitsas, V., & Boucouvalas, A. C. (2002). Throughput and delay analysis of IEEE 802.11 protocol. In *Proceedings of the 5th IEEE International Workshop on Networked Appliances University,* Liverpool, UK (pp. 168-174). Washington, DC: IEEE Computer Society.

Chen, Z. D., Kung, H., & Vlah, D. (2001). Ad hoc relay wireless networks over moving vehicles on highways. In *Proceedings of the 2nd ACM International Symposium on Mobile Ad Hoc Networking and Computing.*

Chen, H.-H., Chu, S.-W., & Guizani, M. (2008). On next generation CDMA technologies: The REAL approach for perfect orthogonal code generation. *IEEE Transactions on Vehicular Technology, 57*(5), 2822–2833. doi:10.1109/TVT.2007.914055.

Chen, J., Cao, X., Zhang, Y., Xu, W., & Sun, Y. (2009). Measuring the performance of movement-assisted certificate revocation list distribution in VANETs. *Wireless Communications and Mobile Computing, 11*(7), 888–898. doi:10.1002/wcm.858.

Chen, S., & Nahrstedt, K. (1999). Distributed quality-of-service routing in ad-hoc net- works. *IEEE Journal on Selected Areas in Communications, 17,* 1488–1505. doi:10.1109/49.780354.

Chieochan, S., Hossain, E., & Diamond, J. (2010). Channel assignment schemes for infrastructure-based 802.11 WLANs: A survey. *IEEE Communications Surveys & Tutorials, 12*(1), 124–136. doi:10.1109/SURV.2010.020110.00047.

Chilamkurti, N., Tsai, M., Ke, C., Park, J., & Kang, P. (2010). Reliable transmission mechanism for safety-critical information in vehicular wireless networks. In *Proceedings of the IEEE International Symposium on Computer, Communication, Control and Automation* (pp. 514-517).

Choi, H.-K., & Limb, J. O. (1999). A behavioral model of web traffic. In *Proceedings of the Seventh International Conference on Network Protocols* (pp. 327-334). Washington, DC: IEEE Computer Society.

Choi, T., Kim, L., Nah, J., & Song, J. (2004). Combinatorial mobile IP: A new efficient mobility management using minimized paging and local registration in mobile environments. *Wireless Networks*, *10*, 311–321. doi:10.1023/B:WINE.0000023864.73253.b7.

Chou, C.-T., Yang, J., & Wang, D. (2007). Cooperative mac protocol with automatic relay selection in distributed wireless networks. In *Proceedings of the IEEE Pervasive Computing and Communications Workshops* (pp. 526-531).

Choudhury, G., Leung, K., & Whitt, W. (1995). An algorithm to compute blocking probabilities in multi-rate multi-class multi-resource loss models. *Advances in Applied Probability*, *27*, 1104–1143. doi:10.2307/1427936.

Choudhury, S., & Gibson, J. D. (2007). Payload length and rate adaptation for multimedia communications in wireless LANs. *IEEE Journal on Selected Areas in Communications*, *25*(4), 796–807. doi:10.1109/JSAC.2007.070515.

Chowdhury, K. R., Felice, M. D., & Bononi, L. (2009). A Fading and Interference Aware Routing Protocol for Multi-Channel Multi-Radio Wireless Mesh Networks. In *Proceedings of the ACM Symposium on Performance Evaluation of Wireless Ad Hoc, Sensor, and Ubiquitous Networks* (pp. 1-8).

Chung, S., Yoo, J., & Kim, C. (2009). A cognitive MAC for VANET based on the WAVE system. In *Proceedings of the IEEE International Conference on Advanced Communications Technology* (pp. 41-46).

Cooper, R. (1981). *Introduction to queueing theory*. Amsterdam, The Netherlands: North Holland Publishing.

Coronado, E., & Cherkaoui, S. (2010). Performance analysis of secure on-demand services for wireless vehicular networks. *Security and Communication Networks*, *3*(2-3), 114–129. doi:10.1002/sec.134.

Cui, S., Goldsmith, A., & Bahai, A. (2004). Energy-efficiency of MIMO and cooperative MIMO techniques in sensor networks. *IEEE Journal on Selected Areas in Communications*, *22*, 1089–1098. doi:10.1109/JSAC.2004.830916.

David, M., Ken, D., & Doug, L. (2007). Modeling the 802.11 distributed coordination function in non-saturated heterogeneous conditions. *IEEE/ACM Transactions on Networking*, *15*(1), 159–172. doi:10.1109/TNET.2006.890136.

De Couto, D. S. J., Aguayo, D., Bicket, J., & Morris, R. (2003). A high-throughput path metric for multi-hop wireless routing. In *Proceedings of the 9th Annual International Conference on Mobile Computing and Networking* (pp. 134-146).

De Sales, L. M., Almeida, H. O., & Perkusich, A. (2008). On the performance of TCP, UDP and DCCP over 802.11 g networks. In *Proceedings of the ACM Symposium on Applied Computing*, Fortaleza, Ceara, Brazil (pp. 2074-2078). New York, NY: ACM.

de Silva, P., & Sirisena, H. (2002). A mobility management protocol for IP-based cellular networks. *IEEE Wireless Communications*, 31-37.

Deart, V., Mankov, V., & Pilugin, A. (2009). HTTP traffic measurements on access networks: Analysis of results and simulation. In S. Balandin, D. Moltchanov, & Y. Koucheryavy (Eds.), *Proceedings of the 9th International Conference on Next Generation Wired/Wireless Advanced Networking and Second Conference on Smart Spaces* (LNCS 5764, pp. 180-190).

Di Caro, G., & Dorigo, M. (1997). *AntNet: A mobile agents approach to adaptive routing* (Tech. Rep. No. IRIDIA/97-12). Brussels, Belgium: Universite Libre de Bruxelles.

Di Caro, G., & Dorigo, M. (1998). Ant colonies for adaptive routing in packet switched communications networks. In *Proceedings of the 5th International Conference on Parallel Problem Solving from Nature* (pp. 673-682).

Di Caro, G., & Dorigo, M. (1998). AntNet: Distributed stigmergetic control for communications networks. *Journal of Artificial Intelligence Research*, 317–365.

Dong, J., Ackermann, K., & Nita-Rotaru, C. (2009). Secure group communication in wireless mesh networks. *Ad Hoc Networks*, *7*(8), 1563–1576. doi:10.1016/j.adhoc.2009.03.004.

Dorigo, M., Maniezzo, V., & Colorni, A. (1991). *The Ant System: An Autocatalytic Optimizing Process* (Tech. Rep. No. 91-016). Milan, Italy: Politecnico di Milano.

Dorigo, M., & Di Caro, G. (1991). The ant colony optimization meta-heuristic. In Corne, D., Dorigo, M., Glover, F., Dasgupta, D., Moscato, P., Poli, R., & Price, K. V. (Eds.), *New Ideas in Optimization*. London, UK: McGraw Hill.

Draves, R., Padhye, J., & Zill, B. (2004). Routing in multi-radio, multi-hop wireless mesh networks. In *Proceedings of the 10th Annual International Conference on Mobile Computing and Networking* (pp. 114-128).

Dusit, N., & Ekram, H. (2006). A queuing-theoretic and optimization-based model for radio resource management in IEEE 802.16 broadband wireless networks. *IEEE Transactions on Computers*, *55*, 1473–1488. doi:10.1109/TC.2006.172.

Eklund, C., Marks, B., Stanwood, K. L., & Wang, S. (2002). IEEE Standard 802.16: A technical overview of the WirelessMAN air interface for broadband wireless access. *IEEE Communications Magazine*, *40*(6), 98–107. doi:10.1109/MCOM.2002.1007415.

El Rhazi, A., & Pierre, S. (2007). A data collection algorithm using energy maps in sensor networks. In *Proceedings of the Third IEEE International Conference on Wireless and Mobile Computing, Networking and Communications* (p. 64).

El Rhazi, A., & Pierre, S. (2009). A tabu search algorithm for cluster building in wireless sensor networks. *IEEE Transactions on Mobile Computing*, *8*(4), 433–444. doi:10.1109/TMC.2008.125.

Elayoubi, S., & Fourestié, B. (2008). Performance evaluation of admission control and adaptive modulation in OFDMA WiMAX systems. *IEEE/ACM Transactions on Networking*, *16*, 1200–1211. doi:10.1109/TNET.2007.911426.

Emma, D., Loreto, S., Pescape, A., & Ventre, G. (2006). Measuring SCTP Throughput and Jitter over Heterogeneous Networks. In *Proceedings of the 20th International Conference on Advanced Information Networking and Applications* (Vol. 2, pp. 395-399). Washington, DC: IEEE Computer Society.

Ergen, M., & Varaiya, P. (2008). Formulation of distributed coordination function of IEEE 802.11 for asynchronous networks: Mixed data rate and packet size. *IEEE Transactions on Vehicular Technology*, *57*(1), 436–447. doi:10.1109/TVT.2007.901887.

Escrig, B. (2010). On-demand cooperation mac protocols with optimal diversity-multiplexing tradeo-off. In *Proceedings of the IEEE International Conference on Wireless Communications and Networking* (pp. 576-581).

Espina, J., Falck, T., Muehlsteff, J., Yilin, J., Adan, M. A., & Aubert, X. (2008). Wearable body sensor network towards continuous cuff-less blood pressure monitoring. In *Proceedings of the 5th International Summer School and Symposium on Medical Devices and Biosensors* (pp. 28-32).

Fall, K., & Varadham, K. (2001). *The ns Manual: The VINT project documents*. Retrieved from http://www.cs.cornell.edu/people/egs/615/ns2-manual.pdf

Fall, K., & Varadhan, K. (2000). *NS notes and documentation: The VINT project*. Retrieved from http://www.isi.edu/nsnam/ns/

Fehske, A., Gaeddert, J., & Reed, J. H. (2005). A new approach to signal classification using spectral correlation and neural networks. In *Proceedings of the First IEEE International Symposium on New Frontiers in Dynamic Spectrum Access Networks* (pp. 144-150).

Ferre, P., Doufexi, A., Nix, A., & Bull, D. (2004). Throughput analysis of IEEE 802.11 and IEEE 802.11e MAC. In[). Washington, DC: IEEE Computer Society.]. *Proceedings of the IEEE Wireless Communications and Networking Conference*, *2*, 783–788.

Fette, B. (2006). *Cognitive radio technology*. Oxford, UK: Newnes.

Finn, G. (1987). *Routing and addressing problems in large metropolitan-scale internetworks* (Tech. Rep. No. ISI/RR-87-180). Los Angeles, CA: University of Southern California.

Fiore, M., Harri, J., Filali, F., & Bonnet, C. (2007, March 26-28). Vehicular mobility simulation for VANETs. In *Proceedings of the 40th Annual Simulation Symposium* (pp. 301-309).

Frei, C. (2000). *Abstraction techniques for resource allocation in communication networks.* Unpublished doctoral dissertation, Swiss Federal Institute of Technology, Lausanne, Switzerland.

Fuxjager, P., Valerio, D., & Ricciato, F. (2007). The myth of non-overlapping channels: interference measurements in IEEE 802.11. In *Proceedings of the Fourth Annual Conference on Wireless on Demand Network Systems and Services* (pp. 1-8).

Gambiroza, V., Sadeghi, B., & Knightly, E. W. (2004). End-to-end performance and fairness in multihop wireless backhaul networks. In *Proceedings of the ACM SIGCOMM Mobile Computing and Networking* (pp. 287-301).

Ganguly, S., Navda, V., Kim, K., Kashyap, A., Niculescu, D., & Izmailov, R. et al. (2006). Performance optimizations for deploying VoIP services in mesh networks. *IEEE Journal on Selected Areas in Communications, 24*(11), 2147–2158. doi:10.1109/JSAC.2006.881594.

Gilboa, I., & Schmeidler, D. (2001). *A theory of case-based decisions.* Cambridge, UK: Cambridge University Press. doi:10.1017/CBO9780511493539.

Goldsmith, A. (2005). Path loss and shadowing. In Goldsmith, A. (Ed.), *Wireless communications* (pp. 27–41). Cambridge, UK: Cambridge University Press.

Gomez, J., Alonso-Zarate, J., Verikoukis, C., Perez-Neira, A., & Alonso, L. (2007). Cooperation on demand protocols for wireless networks. In *Proceedings of the IEEE International Symposium on Personal, Indoor and Mobile Radio Communications* (pp. 1-5).

Goussevskaia, O., Machado, M. D. V., Mini, R. A. F., Lourerio, A. A. F., Mateus, G. R., & Nogueira, J. M. (2005). Data dissemination based on the energy map. *IEEE Communications Magazine, 43*(7), 134–143. doi:10.1109/MCOM.2005.1470845.

Graffi, K., Gross, C., Stingl, D., Hartung, D., Kovacevic, A., & Steinmetz, R. (2011, January). LifeSocial.KOM: A secure and P2P-based solution for online social networks. In *Proceedings of the IEEE Consumer Communications and Networking Conference* (pp. 554-558).

Günes, M., Blywis, B., & Juraschek, F. (2008). *Concept and design of the hybrid distributed embedded systems testbed* (Tech. Rep. No. TR-B-08-10). Berlin, Germany: Freie Universität Berlin.

Günes, M., Blywis, B., Juraschek, F., & Schmidt, P. (2008). *Practical issues of implementing a hybrid multi-NIC wireless mesh-network* (Tech. Rep. No. TR-B-08-11). Berlin, Germany: Freie Universität Berlin.

Günes, M., Blywis, B., Juraschek, F., & Watteroth, O. (2009). Experimentation made easy. In J. Zheng, S. Mao, S. Midkiff, & H. Zhu (Eds.), Ad Hoc Networks (LNCS 28, pp. 493-505).

Günes, M., Juraschek, F., Blywis, B., Mushtaq, Q., & Schiller, J. (2009). A testbed for next generation wireless networks research. *Mobile Ad-hoc Networks, 4*, 208–212.

Guo, M., Ammar, M., & Zegura, E. (2005). V3: A vehicle-to-vehicle live video streaming architecture. In *Proceedings of the IEEE International Conference on Pervasive Computing and Communications* (pp. 171-180).

Gupta, P., & Kumar, P. (2000). The capacity of wireless networks. *IEEE Transactions on Information Theory*, 388–404. doi:10.1109/18.825799.

Gustafsson, E., Jonsson, A., & Perkins, C. (2005). *Mobile IPv4 regional registration.* Retrieved from http://tools.ietf.org/html/rfc4857

Haas, J., Hu, Y., & Laberteauz, K. (2010). The impact of key assignment on VANET privacy. *Security and Communication Networks, 3*(2-3), 233–249. doi:10.1002/sec.143.

Haenggi, M., & Puccinelli, D. (2005). Routing in ad hoc networks: A case for long hops. *IEEE Communications Magazine, 43*, 112–119. doi:10.1109/MCOM.2005.1522131.

Han, S., & Chan, E. (2005). Continuous residual energy monitoring in wireless sensor networks. In J. Cao, L. T. Yang, M. Guo, & F. Lau (Eds.), *Proceedings of the Second International Symposium on Parallel and Distributed Processing and Applications* (LNCS 3358, pp. 169-177).

Han, Y., & Gui, X. (2009). Adaptive secure multicast in wireless networks. *International Journal of Communication Systems, 22*(9), 1213–1239. doi:10.1002/dac.1023.

Han, Y., Gui, X., Wu, X., & Yang, X. (2011). Proxy encryption based secure multicast in wireless mesh networks. *Journal of Network and Computer Applications, 34*(2), 469–477. doi:10.1016/j.jnca.2010.05.002.

Harrold, T. J., Faris, P. C., & Beach, M. A. (2008, September 18). Distributed spectrum detection algorithms for cognitive radio. In *Proceedings of the IET Seminar on Cognitive Radio and Software Defined Radios: Technologies and Techniques* (pp. 1-5).

Hart, J. K., & Martinez, K. (2006). Environmental sensor networks: A revolution in the earth system science. *Earth-Science Reviews*, 177–191. doi:10.1016/j.earscirev.2006.05.001.

Haverinen, H. (2000). *Mobile IP regional paging.* Retrieved from http://tools.ietf.org/html/draft-haverinen-mobileip-reg-paging-00

Hayajneh, M., & Abdallah, C. T. (2004). Distributed joint rate and power control game- theoretic algorithms for wireless data. *IEEE Communications Letters*, 8.

He, X., Xiao, Y., & Li, S. (2008). PAPR reduction of random interleaved uplink OFDMA system. In *Proceedings of the 3rd International Conference on Communications and Networking in China* (pp. 478-481).

Heinzelman, W. B., Chandrakasan, A. P., & Balakrishnan, H. (2002). Application specific protocol architecture for wireless microsensor networks. *IEEE Transactions on Wireless Networking, 1*(4), 660–670. doi:10.1109/TWC.2002.804190.

Heissenbuettel, M. (2005). *Routing and Broadcasting in Ad Hoc Networks* (Unpublished doctoral dissertation). University Bern, Bern, Switzerland.

Heusse, M., Rousseau, F., Berger-Sabbatel, G., & Duda, A. (2003). Performance anomaly of 802.11b. In *Proceedings of the IEEE International Conference INFOCOM* (Vol. 2, pp. 836-843).

Ho, K., Kang, P., Hsu, C., & Lin, C. (2010). Implementation of WAVE/DSRC devices for vehicular communications. In *Proceedings of the IEEE International Symposium on Computer Communication Control and Automation* (pp. 522-525).

Holland, G., & Vaidya, N. (1999, August). Analysis of TCP performance over mobile ad hoc networks. In *Proceedings of the 5th Annual ACM/IEEE International Conference on Mobile Computing and Networking* (pp. 219-230). New York, NY: ACM.

Holland, G., Vaidya, N. H., & Bahl, P. (2001, July). A Rate-Adaptive MAC Protocol for Multi-Hop Wireless Networks. In *Proceedings of the ACM International Conference on Mobile Computing and Networking.*

Holland, G., Vaidya, N. H., & Bahl, P. (2000). *A Rate-Adaptive MAC Protocol for Multi-Hop Wireless Networks (Tech. Rep. No. UMI TR00-019).* College Station, TX: Texas A & M University.

Hou, F. P., Ho, H., & Shen, X. (2006, November). Performance analysis of a reservation-based connection admission scheme in 802.16 networks. In *Proceedings of the IEEE Global Telecommunications Conference* (pp. 1-5).

Huang, H., Huang, T., Chilamkurti, N., Cheng, R., & Shieh, C. (2010). Adaptive forward error correction with cognitive technology mechanism for video streaming over wireless networks. In *Proceedings of the IEEE International Symposium on Computer Communication Control and Automation* (pp. 518-521).

Huang, D., & Medhi, D. (2004). A key-chain based keying scheme for many-to-many secure group communication. *ACM Transactions on Information and System Security, 7*(4), 1–30. doi:10.1145/1042031.1042033.

Hunter, T. E., & Nosratinia, A. (2006). Diversity through coded cooperation. *IEEE Transactions on Wireless Communications, 5*(2), 283–289. doi:10.1109/TWC.2006.1611050.

Hussain, S., Schaffner, S., & Moseychuck, D. (2009). Applications of wireless sensor networks and RFID in a smart home environment. In *Proceedings of the Seventh Annual Communication Networks and Services Research Conference* (pp. 153-157).

Hwang, I., Hwang, B., & Su, R. (2011). Maximizing downlink bandwidth allocation method based on SVC in mobile WiMAX networks for generic broadband services. *International Scholarly Research Network Communications and Networking, 2011*, 5.

Ibrahim, K., Weigle, M. C., & Abuelela, M. (2009, April 26-29). p-IVG: Probabilistic Inter-vehicle geocast for dense vehicular networks. In *Proceedings of the 69th IEEE Vehicular Technology Conference* (pp. 1-5).

IEEE Standards Association. (1997). *IEEE 802.11 WG Standard for wireless LAN: Medium access control (MAC) and physical layer (PHY) specifications, IEEE 802.11 Standard.* Retrieved from http://standards.ieee.org/getieee802/download/802.11-2007.pdf

IEEE Standards Association. (1999). *IEEE 802.11 Standard for Wireless LAN: Medium Access Control (MAC) and Physical Layer (PHY) Specification.* Retrieved from http://standards.ieee.org/about/get/802/802.11.html

IEEE Standards Association. (1999). *IEEE 802.11b WG, Part II: Wireless LAN medium access control (MAC) and physical layer (PHY) specifications: High-speed physical layer extension in the 2.4 GHz band, IEEE 802.11b Standard.* Retrieved from http://standards.ieee.org/about/get/802/802.11.html

IEEE Standards Association. (2003). *IEEE Standard 802.15.4.* Washington, DC: Author.

IEEE Standards Association. (2005). *IEEE P802.11e Amendment to IEEE Std 802.11, Part II: Wireless LAN Medium Access Control (MAC) and Physical Layer (PHY) Specifications: MAC Quality of Server (QoS) Enhancements.* Retrieved from http://standards.ieee.org/about/get/802/802.11.html

IEEE Standards Association. (2007). *IEEE 802.11: Wireless LAN Medium Access Control (MAC) and Physical Layer (PHY) Specifications* (Rev. ed.). Washington, DC: Author.

IEEE. (2006). *IEEE Std 1609.1-2006: Trial-use standard for wireless access in vehicular environments (WAVE) - Resource manager* (pp. c1–c63). Piscataway, NJ: IEEE.

IEEE. (2007). *Std 802.11-2007: Standard for information technology-telecommunications and information exchange between systems-local and metropolitan area networks-specific requirements - Part 11: Wireless LAN Medium Access Control (MAC) and Physical Layer (PHY) specifications. (Revision of IEEE Std 802.11-1999)* (pp. C1–C1184). Piscataway, NJ: IEEE.

IEEE. (2010). IEEE P1609.1TM/D1.3: Draft standard for wireless access in vehicular environments (WAVE) – Remote management service. Piscataway, NJ: IEEE.

IEEE. (2010). *IEEE Std 1609.3-2010: Standard for wireless access in vehicular environments (WAVE) -Networking services (Revision of IEEE Std 1609.3-2007)* (pp. 1–144). Piscataway, NJ: IEEE.

IEEE. (2010). IEEE P1609.2TM/D6.0: Draft standard for wireless access in vehicular environments (WAVE) – Security services for applications and management messages. Piscataway, NJ: IEEE.

IEEE. (2010). *IEEE Std. 1609.4-2010: Standard for wireless access in vehicular environments (WAVE)--Multi-channel operation (Revision of IEEE Std 1609.4-2006)* (pp. 1–89). Piscataway, NJ: IEEE.

IEEE. (2010). *IEEE Std 1609.3-2010: Standard for wireless access in vehicular environments (WAVE) – Networking services.* Piscataway, NJ: IEEE.

IEEE. (2010). IEEE Std 802.11p-2010: (Amendment to IEEE Std 802.11-2007 as amended by IEEE Std 802.11k-2008, IEEE Std 802.11r-2008, IEEE Std 802.11y-2008, IEEE Std 802.11n-2009, and IEEE Std 802.11w-2009) (pp. 1-51). Piscataway, NJ: IEEE.

IEEE. (2011). *IEEE Std 1609.4-2010: Standard for wireless access in vehicular environments (WAVE) – Multi-channel operation.* Piscataway, NJ: IEEE.

IEEE. Computer Society. (2005). IEEE Standard 802.16e-2005: Amendment to IEEE standard for local and metropolitan area networks - Part 16: Air interface for fixed broadband wireless access systems- Physical and medium access control layers for combined fixed and mobile operation in licensed bands. Washington, DC: IEEE Computer Society.

IEEE. Computer Society. (2007). *Part 11: Wireless lan medium access control (mac) and physical layer (phy) specifications* (Tech. Rep. No. 802.11-2007). Washington, DC: Author.

IEEE. Computer Society. (2010). IEEE 802.11: Amendment 6: Wireless access in vehicular environments. Washington, DC: IEEE Computer Society.

Information Sciences Institute. (n. d.). *NS-2-the network simulation.* Retrieved from http://www.isi.edu/nsnam/ns/index.html

Jacquet, P., Muhlethaler, P., & Qayyum, A. (2003). *Optimized link state routing (OLSR) protocol.* Retrieved from http://www.ietf.org/rfc/rfc3626.txt

Jacquet, P., Muhlethaler, P., Clausen, T., Laouiti, A., Qayyum, A., & Viennot, L. (2001). Optimized link state routing protocol for ad hoc networks. In *Proceedings of the 5th IEEE Multi Topic Conference.*

Jagadeesan, S., Manoj, B. S., & Murthy, C. S. R. (2003). Interleaved carrier sense multiple access: An efficient MAC protocol for ad hoc wireless networks. In *Proceedings of the IEEE International Conference on Communications*, Anchorage, AK (Vol. 2, pp. 1124-1128). Washington, DC: IEEE Computer Society.

Jain, R. (1991). *The Art of Computer Systems Performance Analysis: Techniques for Experimental Design, Measurement, Simulation, and Modeling.* New York, NY: John Wiley & Sons.

Jaiswal, N. (1968). *Priority queues.* New York, NY: Academic Press.

Jalloul, L. M. A., & Holtzman, J. M. (1994). Performance analysis of DS/CDMA with noncoherent M-ary orthogonal modulation in multipath fading channels. *IEEE Journal on Selected Areas in Communications, 12*(5), 862–870. doi:10.1109/49.298060.

Javed, M. A., & Khan, J. Y. (2011 November 9-11). A geocasting technique in an IEEE802.11p based vehicular ad hoc network for road traffic management. In *Proceedings of the Australasian Telecommunication Networks and Applications Conference* (pp. 1-6).

Jayaweera, S. K. (2004). An energy-efficient virtual MIMO architecture based on V-BLAST processing for distributed wireless sensor networks. In *Proceedings of the First Annual IEEE Communications Conference on Sensor and Ad Hoc Communications and Networks* (pp. 299-308).

Jiang, D., & Delgrossi, L. (2008, May 11-14). IEEE 802.11p: Towards an international standard for wireless access in vehicular environments. In *Proceedings of the IEEE Vehicular Technology Conference* (pp. 2036-2040).

Jiang, L. B., & Liew, S. C. (2008). Improving throughput and fairness by reducing exposed and hidden nodes in 802.11 networks. *IEEE Transactions on Mobile Computing, 7*(1), 34–49. doi:10.1109/TMC.2007.1070.

Jiang, T., & Wu, Y. (2008). Overview: Peak-to-average power ratio reduction techniques for OFDM signals. *IEEE Transactions on Broadcasting, 54*(2), 257–268. doi:10.1109/TBC.2008.915770.

Johnson, D. B. (1994). Routing in Ad Hoc Networks of Mobile Hosts. In *Proceedings of the Workshop on Mobile Computing Systems and Applications*, Santa Cruz, NM (pp. 158-163).

Johnson, D., & Maltz, D. (1994). Dynamic Source Routing in Ad Hoc Wireless Networks. In *Proceedings of the Conference on Mobile Computing Systems and Applications*, Santa Cruz, NM (pp. 158-163).

Johnson, D. B., & Maltz, D. A. (1996). Dynamic source routing in ad hoc wireless networks. In Imielinski, T., & Korth, H. (Eds.), *Mobile computing* (p. 353). Boston, MA: Kluwer Academic. doi:10.1007/978-0-585-29603-6_5.

Johnson, D. B., Maltz, D. A., & Broch, J. (2001). DSR: The Dynamic Source Routing Protocol for Multi-Hop Wireless Ad Hoc Net- works. In Johnson, D. B. (Ed.), *Ad Hoc Networking* (pp. 139–172). Reading, MA: Addison-Wesley.

Joost, R., & Robinson, S. (2010). *Python WiFi.* Retrieved from http://pythonwifi.wikispot.org/

Joost, R., & Robinson, S. (2010). *Pythonwiφι.* Retrieved December 24, 2010, from http://pythonwifi.wikispot.org/

Junbeom, H., & Yoon, H. (2010). A Multi-service Group Key Management Scheme for Stateless Receivers in Wireless Mesh Networks. *Mobile Networks and Applications, 15*(5), 680–692. doi:10.1007/s11036-009-0191-4.

Juraschek, F., Günes, M., Philipp, M., & Blywis, B. (2011). Insights from experimental research on distributed channel assignment in wireless testbeds. *International Journal of Wireless Networks and Broadband Technologies, 1*(1), 32–49. doi:10.4018/ijwnbt.2011010103.

Kalpakis, K., Dasgupta, K., & Namjoshi, P. (2003). Efficient algorithms for maximum lifetime data gathering and aggregation in wireless sensor networks. *Computer Networks, 42*(6), 697–716. doi:10.1016/S1389-1286(03)00212-3.

Kanal, L. N., & Sastry, A. R. K. (1978). Models for channels with memory and their applications to error control. *Proceedings of the IEEE, 66*(7), 724–744. doi:10.1109/PROC.1978.11013.

Karim, R. (2008). *VANET: Superior system for content distribution in vehicular network applications*. Rutgers, NJ: Rutgers University.

Katzela, I., & Naghshineh, M. (1996). Channel assignment schemes for cellular mobile telecommunication systems. *IEEE Personal Communications, 3*, 10–31. doi:10.1109/98.511762.

Kempf, J. (2001). *Dormant host alerting ("IP paging") problem statement. Retrieved* from http://tools.ietf.org/html/rfc3132

Kempf, J., Castelluccia, C., Mutaf, P., Nakajima, N., Ohba, Y., Ramjee, R., et al. (2001). *Requirements and functional architecture for an IP host alerting protocol.* Retrieved from http://tools.ietf.org/html/rfc3154

Khemiri, S., Boussetta, K., Achir, N., & Pujolle, G. (2007). Wimax bandwidth provisioning service to residential customers. In *Proceedings of the International Conference on Mobile Wireless Communications Networks* (pp. 116-120).

Khemiri, S., Boussetta, K., Achir, N., & Pujolle, G. (2007). Optimal call admission control for an IEEE802.16 wireless metropolitan area network. In *Proceedings of the International Conference on Network Control and Optimization.*

Kim, H. (2009). *Exploring tradeoffs in wireless networks under flow-level traffic: Energy, capacity and QoS.* Unpublished doctoral dissertation, University of Texas, Austin.

Kim, K.-H., & Shin, K. G. (2007). Self-healing multi-radio wireless mesh networks. In *Proceedings of the 13th Annual ACM International Conference on Mobile Computing and Networking* (pp. 326-329).

Kim, H., Hou, J. C., Hu, C., & Ge, Y. (2006). QoS provisioning in IEEE 802.11-compliant networks: Past, present, and future. *Computer Networks. The International Journal of Computer and Telecommunications Networking, 51*(8).

Kim, H., Lee, J., Choi, Y., Chung, Y., & Lee, H. (2011). Dynamic bandwidth provisioning using ARIMA-based traffic forecasting for mobile WiMAX. *Computer Communications, 34*(1), 99–106. doi:10.1016/j.comcom.2010.08.008.

Kim, K. J., Kwon, S. Y., Hong, E. K., & Whang, K. C. (2000). Effect of tap spacing on the performance of direct-sequence spread-spectrum RAKE receiver. *IEEE Transactions on Communications, 48*(6), 1029–1036. doi:10.1109/26.848565.

Kleinrock, L. (1975). Queueing systems: *Vol. 1. Theory.* New York, NY: John Wiley & Sons.

Ko, B.-J., Misra, V., Padhye, J., & Rubenstein, D. (2007). Distributed channel assignment in multi-radio 802.11 mesh networks. In *Proceedings of the IEEE Wireless Communications and Networking Conference* (pp. 3978-3983).

Kodialam, M., & Nandagopal, T. (2005). Characterizing the capacity region in multi-radio multi-channel wireless mesh networks. In *Proceedings of the Annual ACM International Conference on Mobile Computing and Networking* (pp. 73-87).

Kohler, E., Handley, M., & Floyd, S. (2006). Designing DCCP: congestion control without reliability. In *Proceedings of the Conference on Applications, Technologies, Architectures, and Protocols for Computer Communication* (pp. 27-38). New York, NY: ACM.

Kong, J., & Hong, X. (2003). ANODR: Anonymous on demand routing with untraceable routes for mobile ad-hoc networks. In *Proceedings of the 4th ACM International Symposium on Mobile and Ad Hoc Networking and Computing* (pp. 291-302).

Korkmaz, G., Ekici, E., Ozguner, F., & Ozguner, U. (2004). Urban multi-hop broadcast protocol for inter-vehicle communication systems. In *Proceedings of the 1st ACM International Workshop on Vehicular Ad Hoc Networks* (pp. 76-85).

Kos, A. (2006). *Management of Selforganizing Wireless Sensor Networks* (Unpublished master's thesis). Danube University Krems, University of Applied Sciences St. Poelten, Krems, Austria.

Kubo, H., Shinkuma, R., & Takahashi, T. (2010, December). Mobile P2P Multicast based on Social Network Reducing Psychological Forwarding Cost. In *Proceedings of the IEEE Conference on Global Communications* (pp. 1-5).

Kulkarni, G., Adlakha, S., & Srivastava, M. (2005). Subcarrier allocation and bit loading algorithms for OFDMA-based wireless networks. *IEEE Transactions on Mobile Computing*, *4*(6), 652–662. doi:10.1109/TMC.2005.90.

Kumar, S., & Sharma, S. (2007). Error probability of different modulation schemes for OFDM based WLAN standard IEEE 802.11a. *International Journal of Engineering*, *4*(4), 262–320.

Kuo, W.-K. (2007). Energy efficiency modeling for IEEE 802.11 DCF system without retry limits. *Computer Communications*, *30*(4), 856–862. doi:10.1016/j.comcom.2006.10.005.

Kusume, K., & Bauch, G. (2008). Simple constructruction of multiple interleavers: Cyclically shifting a single interleaver. *IEEE Transactions on Communications*, *56*(9), 1394–1397. doi:10.1109/TCOMM.2008.060420.

Kwon, Y., Fang, Y., & Latchman, H. (2003). A novel MAC protocol with fast collision resolution for wireless LANs. In *Proceedings of the 22nd IEEE Computer and Communications Annual Joint Conference of INFOCOM* (Vol. 2, pp. 853-862). Washington, DC: IEEE Computer Society.

Kwon, S.-J., Nam, S. Y., Hwang, H. Y., & Sung, D. K. (2004). Analysis of a mobility management scheme considering battery power conservation in IP-based mobile networks. *IEEE Transactions on Vehicular Technology*, *53*(6), 1882–1890. doi:10.1109/TVT.2004.836964.

Kyasanur, P., So, J., Chereddi, C., & Vaidya, N. H. (2006). Multichannel mesh networks: challenges and protocols. *IEEE Wireless Communications*, *13*(2), 30–36. doi:10.1109/MWC.2006.1632478.

Kyasanur, P., & Vaidya, N. H. (2006). Routing and link-layer protocols for multi-channel multi-interface ad hoc wireless networks. *Mobile Computing and Communications*, *10*(1), 31–43. doi:10.1145/1119759.1119762.

Lahiri, K., Raghunathan, A., Dey, S., & Panigrahi, D. (2002). Battery-driven system design: A new frontier in low power design. In *Proceedings of the 15th IEEE International Conference on Design Automation and the 7th International Conference on Very Large Scale Integration Design* (pp. 261-267).

Laneman, J. N., Tse, D. N. C., & Wornell, G. W. (2004). Cooperative diversity in wireless networks: Efficient protocols and outage behavior. *IEEE Transactions on Information Theory*, *50*(12), 3062–3080. doi:10.1109/TIT.2004.838089.

Laneman, J. N., & Wornell, G. W. (2000). Energy-efficient antenna sharing and relaying for wireless networks. In. *Proceedings of the IEEE International Conference on Wireless Communications and Networking*, *1*, 7–12.

Laneman, J. N., & Wornell, G. W. (2003). Distributed space-time-coded protocols for exploiting cooperative diversity in wireless networks. *IEEE Transactions on Information Theory*, *49*(10), 2415–2425. doi:10.1109/TIT.2003.817829.

Le, B., Rondeau, T. W., Maldonado, D., Scaperoth, D., & Bostian, C. W. (2006). Signal recognition for cognitive radios. In *Proceedings of the Software Defined Radio Forum Technical Conference.*

Lee, J. W., Lee, H.-J., & Cho, D.-H. (2005). Effect of dormant registration on performance of mobility management based on IP paging in wireless data networks. *International Journal of Electronics and Communications, 59*, 319–323. doi:10.1016/j.aeue.2004.11.009.

Lee, J., Ernst, T., & Chilamkurti, N. (2011). Performance analysis of PMIPv6 based network mobility for intelligent transportation systems. *IEEE Transactions on Vehicular Technology, 61*(1), 74–85. doi:10.1109/TVT.2011.2157949.

Lee, V., Wu, X., & Ng, J. K. (2006). Scheduling real-time requests in on-demand data broadcast environments. *Real-Time Systems, 34*(2), 83–99. doi:10.1007/s11241-006-7982-5.

Leinmueller, T., Schoch, E., Kargl, F., & Maihofer, C. (2010). Decentralized position verification in geographic ad hoc routing. *Security and Communication Networks, 3*(4), 289–302. doi:10.1002/sec.56.

Leskovec, J., Backstrom, L., Kumar, R., & Tomkins, A. (2008). Microscopic Evolution of Social Networks. In *Proceedings of the ACM SIGKDD International Conference on Knowledge Discovery and Data Mining*, Las Vegas, NV (pp. 462-470).

Li, H., & Lin, K. (2010). ITRI WAVE/DSRC communication unit. In *Proceedings of the IEEE Vehicular Technology Conference* (pp. 1-2).

Li, Y., Cao, B., Wang, C., You, X., Daneshmand, A., Zhuang, H., et al. (2009). Dynamical cooperative mac based on optimal selection of multiple helpers. In *Proceedings of the IEEE Global Telecommunications Conference* (pp. 3044-3049).

Li, B., & Battiti, R. (2007). Achieving optimal performance in IEEE 802.11 wireless LANs with the combination of link adaptation and adaptive backoff. *Computer Networks: The International Journal of Computer and Telecommunications Networking, 51*(6), 1574–1600.

Li, M., & Liu, Y. (2010). Iso-Map: Energy-efficient contour mapping in wireless sensor networks. *IEEE Transactions on Knowledge and Data Engineering, 22*(5), 699–710. doi:10.1109/TKDE.2009.157.

Lin, Z., Sammour, M., Sfar, S., Charlton, G., Chitrapu, P., & Reznik, A. (2009, April). MAC v. PHY: How to relay in cellular networks. In *Proceedings of the IEEE Conference on Wireless Communications and Networking*, Budapest, Hungary (pp. 1-6).

Lin, C., Henty, B. E., Cooper, R., Stancil, D. D., & Fan, B. (2008). A measurement study of time-scaled 802.11a waveforms over the mobile-to-mobile vehicular channel at 5.9 GHz. *IEEE Communications Magazine, 46*(5), 84–91. doi:10.1109/MCOM.2008.4511654.

Lindsey, S., Raghavendra, C., & Sivalingam, K. M. (2002). Data gathering algorithms in sensor networks using energy metrics. *IEEE Transactions on Parallel and Distributed Systems, 13*(9), 924–935. doi:10.1109/TPDS.2002.1036066.

Lin, R., & Jan, J. (2007). A tree-based scheme for security of many to many communications. *Journal of High Speed Networking, 16*(1), 69–79.

Linux Foundation. (2009, November 19). *DCCP Testing*. Retrieved August 8, 2010, from http://www.linux-foundation.org/ collaborate/workgroups/networking/dccp_testing

Linux Wireless. (2011). *Ath5k driver homepage*. Retrieved from http://linuxwireless.org/en/users/Drivers/ath5k

Lin, W.-Y., & Wu, J.-S. (2007). Modified EDCF to improve the performance of IEEE 802.11e WLAN. *Computer Communications, 30*(4), 841–848. doi:10.1016/j.comcom.2006.10.013.

Liping, W., Fuqiang, L., Yusheng, J., & Nararat, R. (2007). Admission control for non-preprovisioned service flow in wireless metropolitan area networks. In *Proceedings of the Fourth European Conference on Universal Multiservice Networks* (pp. 243-249).

Liu, K., & Lee, V. (2010). RSU-based real-time data access in dynamic vehicular networks. In *Proceedings of the IEEE Annual Conference on Intelligent Transportation Systems* (pp. 1051-1056).

Liu, L., & Feng, G. (2005). Swarm Intelligence Based Node disjoint Multipath Routing Protocol for Mobile Ad Hoc Networks. In *Proceedings of the 5th International Conference on Information, Communications and Signal Processing*, Bangkok, Thailand.

Liu, J.-S., & Lin, C.-H. R. (2005). Energy-efficiency clustering protocol in wireless sensor networks. *Ad Hoc Networks, 3*(3), 371–388. doi:10.1016/j.adhoc.2003.09.012.

Liu, K., & Lee, V. (2010). Performance analysis of data scheduling algorithms for multi-item requests in multi-channel broadcast environments. *Journal of Communications Systems, 23*(4), 529–542.

Liu, P., Tao, Z., Narayanan, S., Korakis, T., & Panwar, S. S. (2007). Coopmac: A cooperative mac for wireless lans. *IEEE Journal on Selected Areas in Communications, 25*(2), 340–354. doi:10.1109/JSAC.2007.070210.

Liu, Z., Shen, Y., Ross, K. W., Panwar, S. S., & Wang, Y. (2009). LayerP2P: Using Layered Video Chunks in P2P Live Streaming. *IEEE Transactions on Multimedia, 11*(7), 1340–1352. doi:10.1109/TMM.2009.2030656.

Li, Y., & Chen, I. (2010). Design and performance analysis of mobility management schemes based on pointer forwarding for wireless mesh networks. *IEEE Transactions on Mobile Computing, 10*(3), 349–361. doi:10.1109/TMC.2010.166.

Lochert, C., Scheuermann, B., Caliskan, M., & Mauve, M. (2007). The feasibility of information dissemination vehicular ad-hoc networks. In *Proceedings of the 4th Annual Conference on Wireless On-demand Network Systems and Services.*

Lu, M.-H., Steenkiste, P., & Chen, T. (2009). Design, Implementation and Evaluation of an Efficient Opportunistic Retransmission Protocol. In *Proceedings of the ACM International Conference on Mobile Computing and Networking* (pp. 73-84).

Lu, S., Sun, Y., Ge, Y., Dutkiewicz, E., & Zhou, J. (2010). Joint power and rate control in ad hoc networks using a supermodular game approach. In *Proceedings of the IEEE Conference on Wireless Communications and Networking* (p. 1-6).

Luo, J., Rosenberg, C., & Girard, A. (2010). Engineering Wireless Mesh Networks: Joint Scheduling, Routing, Power Control and Rate Adaptation. *IEEE/ACM Transactions on Networking, 18*(5), 1387–1400. doi:10.1109/TNET.2010.2041788.

Madden, S. R., Franklin, M. J., Hellerstein, J. M., & Hong, W. (2002, December). TAG: Tiny aggregation service for ad-hoc sensor networks. In *Proceedings of the Fifth Symposium on Operating Systems Design and implementation*, Boston, MA.

Mah, B. (1997). An empirical model of HTTP network traffic. In *Proceedings of the Sixteenth Annual Joint Conference of the IEEE Computer and Communications Societies* (p. 592). Washington, DC: IEEE Computer Society.

Mahmoud, Q. (2007). *Cognitive networks: Towards self-aware networks.* New York, NY: Wiley-Interscience.

Mak, T., Laberteaux, K., & Sengupta, R. (2005). A multi-channel VANET providing concurrent safety and commercial services. In *Proceedings of the 2nd ACM International Workshop on Vehicular Ad Hoc Networks*, Cologne, Germany (pp. 1-9).

Marks, R. B., Nikolich, P., & Snyder, R. (in press). *IEEE Std 802.16m, Amendment to IEEE standard for local and metropolitan area networks – Part 16: Air interface for broadband wireless access systems – Advanced air interface.* Retrieved from http://ieee802.org/16/pubs/80216m.html

Marks, B. R., Stanwood, K., & Chang, D. (2004). *ANSI/IEEE std. 802.16-2004: Local and metropolitan area networks part 16: Air interface for fixed broadband wireless access systems (Revision of IEEE Std. 802.16-2001).* Washington, DC: IEEE Computer Society.

Martinez, F., Toh, C.-K., Cano, J.-C., Calafate, C., & Manzoni, P. (2011). A street broadcast reduction scheme (SBR) to mitigate the broadcast storm problem in VANETs. *Wireless Personal Communications, 56*(3), 559–572. doi:10.1007/s11277-010-9989-4.

MATLAB. (2003). *Version 7.11.0.584 (R2010).* Natick, MA: The Mathworks Inc..

Ma, W., & Fang, Y. (2004). Dynamic hierarchical mobility management strategy for mobile IP networks. *IEEE Journal on Selected Areas in Communications, 22*(4), 664–676. doi:10.1109/JSAC.2004.825968.

Miao, G. (2008). *Cross-layer optimization for spectral and energy efficiency*. Unpublished doctoral dissertation, Georgia Institute of Technology, School of Electrical and Computer Engineering, Atlanta.

Min, H., Yi, S., & Hong, J. (2008). Energy-efficient data aggregation protocol for location-aware wireless sensor networks. In *Proceedings of the International Symposium on Parallel and Distributed Processing with Applications* (pp. 751-756).

Mini, R. A. F., Loureiro, A. A. F., & Nath, B. (2004, October). Energy map construction for wireless sensor network under a finite energy budget. In *Proceedings of the 7th ACM International Symposium on Modeling, Analysis, and Simulation of Wireless and Mobile Systems* (pp. 165-169).

Mini, R. A. F., Loureiro, A. A. F., & Nath, B. (2004). The distinctive design characteristic of a wireless sensor network: The energy map. *Computer Communications, 27*, 935–945. doi:10.1016/j.comcom.2004.01.004.

Mini, R. A. F., Loureiro, A. A. F., & Nath, B. (2005). Prediction-based energy map for wireless sensor networks. *Ad Hoc Networks Journal, 3*, 235–253. doi:10.1016/j.adhoc.2004.07.008.

Morgan, Y. L. (2010). Notes on DSRC & WAVE Standards Suite: Its architecture, design, and characteristics. *IEEE Communications Surveys & Tutorials, 12*(4), 504–518. doi:10.1109/SURV.2010.033010.00024.

Morillo-Pozo, J., Trullols, O., Barcelo, J., & Garcia-Vidal, J. (2008). A cooperative ARQ for delay-tolerant vehicular networks. In *Proceedings of the IEEE International Conference on Distributed Computing Systems Workshops* (pp. 192-197).

Murthy, S., & Garcia-Luna-Aceves, J. (1996). An Efficient Routing Protocol for Wireless Networks. *ACM Mobile Networks and Applications, 1*(2), 183–197. doi:10.1007/BF01193336.

Nadeem, T., Shankar, P., & Iftode, L. (2006). A comparative study of data dissemination models for VANETs. In *Proceedings of the 3rd Annual International Conference on Mobile and Ubiquitous Systems: Networking & Services* (pp. 1-10).

Nagle, F., & Singh, L. (2009, July 20-22). Can Friends Be Trusted? Exploring Privacy in Online Social Networks. In *Proceedings of the International Conference on Advances in Social Network Analysis and Mining* (pp. 312-315).

Naguib, A. F., & Paulraj, A. (1996). Performance of wireless CDMA with M -ary orthogonal modulation and cell site antenna arrays. *IEEE Journal on Selected Areas in Communications, 14*(9), 1770–1783. doi:10.1109/49.545700.

Naguib, A. F., Paulraj, A., & Kailath, T. (1994). Capacity improvement with base-station antenna arrays in cellular CDMA. *IEEE Transactions on Vehicular Technology, 43*(3), 691–698. doi:10.1109/25.312780.

Nakamura, T. (in press). *LTE release 10 & beyond (LTE-Advanced)*. Retrieved from http://www.3gpp.org/article/lte-advanced

Natkaniec, M., & Pach, A. R. (2000, July 3-6). An analysis of backoff mechanism used in IEEE 802.11 networks. In *Proceedings of the 5th IEEE Symposium on Computers and Communications*, Antibes-Juan les Pins, France (pp. 444-449). Washington, DC: IEEE Computer Society.

Natkaniec, M., & Pach, A. R. (2002, October 27-30). PUMA - a new channel access protocol for wireless LANs. In *Proceedings of the 5th International Symposium on Wireless Personal Multimedia Communications* (Vol. 3, pp. 1351-1355).

Navaratnam, P., Akhtar, N., & Tafazolli, R. (2006). On the performance of DCCP in wireless mesh networks. In *Proceedings of the 4th ACM international Workshop on Mobility Management and Wireless Access*, Terromolinos, Spain (pp. 144-147). New York, NY: ACM.

Nelakuditi, S., Lee, S., Yu, Y., Wang, J., Zhong, Z., Lu, G. H., et al. (2005). Blacklist-aided forwarding in static multihop wireless networks. In *Proceedings of the Second Annual Conference on Sensor and Ad Hoc Communications and Networks* (pp. 252-262).

Niazi, M., & Hussain, A. (2009). Agent-based tools for modeling and simulation of self-organization in peer-to-peer, ad hoc, and other complex networks. *IEEE Communications Magazine*, *47*(3), 166–173. doi:10.1109/MCOM.2009.4804403.

Niculescu, D., & Nath, B. (2002). *Trajectory-based forwarding and its applications* (Tech. Rep. No. DCS-TR-488). New Brunswick, NJ: Rutgers University.

Open SS7. (2008, October 31). *Iperf 2.0.8 Released.* Retrieved August 8, 2010, from http://www.openss7.org/rel20081029_1.html

Ord, J. K. (1972). *Families of Frequency Distributions.* New York, NY: Hafner.

Ozugur, T., Naghshineh, M., Kermani, P., & Copeland, J. A. (1999, December 5-9). Fair media access for wireless LANs. In *Proceedings of the IEEE Global Telecommunications Conference*, Rio de Janeireo, Brazil (Vol. 1, pp. 570-579). Los Alamitos, CA: IEEE Press.

Padhye, J., Agarwal, S., Padmanabhan, V. N., Qiu, L., Rao, A., & Zill, B. (2005). Estimation of link interference in static multi-hop wireless networks. In *Proceedings of the 5th ACM SIGCOMM Conference on Internet Measurement* (pp. 28-28).

Padhye, J., Firoiu, V., Towsley, D., & Kurose, J. (1998). Modeling TCP throughput: A simple model and its empirical validation. *Communications of the ACM*, *28*(4), 303–314. doi:10.1145/285243.285291.

Park, V., & Corson, M. (1997). A Highly Adaptive Distributed Routing Algorithm for Mobile Wireless Networks. In *Proceedings of the IEEE International Conference INFOCOM.*

Pathak, P., & Dutta, R. (2011). A survey of network design problems and joint de- sign approaches in wireless mesh networks. *IEEE Communications Surveys and Tutorials, 13*(3).

Perkins, C. (2002). *IP mobility support for IPv4.* Retrieved from http://www.ietf.org/rfc/rfc3220.txt

Perkins, C. E., & Bhagwat, P. (1994). Highly dynamic destination sequenced distance-vector routing (DSDV) for mobile computers. In *Proceedings of the ACM SIGCOMM Conference on Communications Architectures, Protocols and Applications.*

Perkins, C. E., Belding-Royer, E., & Das, S. (2003). *RFC 3561: Ad hoc On-Demand Distance Vector (AODV) Routing.* Retrieved from http://www.ietf.org/rfc/rfc3561.txt

Perkins, C., & Bhagwat, P. (1994). Highly dynamic Destination-Sequenced Distance-Vector routing (DSDV) for mobile computers. In *Proceedings of the SIGCOM Conference on Communications Architecture, Protocols and Applications* (pp. 234-244).

Perkins, C., & Royer, E. (1999). Ad-Hoc On-Demand Distance Vector Routing. In *Proceedings of the 2nd IEEE Workshop Mobile Computing Systems and Applications*, New Orleans, LA.

Perkins, C. (1997). Mobile IP. *IEEE Communications Magazine*, 84–99. doi:10.1109/35.592101.

Ping, G., Zheng, J., Kai, N., Lu, T., & Wu, W. (2009, April). A timing synchronization scheme for space-time cooperative relay OFDM system. In *Proceedings of the Conference on Wireless and Optical Communication Networks*, Cairo, Egypt (pp. 1-5).

Pouwelse, J. A., Garbacki, P., Wang, J., Bakker, A., Yang, J., & Iosup, A. et al. (2006). Tribler: A social-based Peer-to-Peer System. *Concurrency and Computation: Recent Advances in Peer-to-Peer Systems and Security, 20*(2), 127–138.

Prasad, R. (2004). Basics of OFDM and synchronization. In Prasad, R. (Ed.), *OFDM for wireless communications systems* (pp. 117–147). Boston, MA: Artech House.

Pyattaev, A., Andreev, S., Koucheryavy, Y., & Moltchanov, D. (2010, December). *Some modeling approaches for client relay networks.* Paper presented at the 15th IEEE International Workshop on Computer Aided Modeling Analysis and Design of Communication Links and Networks, Miami, FL.

Pyattaev, A., Andreev, S., Vinel, A., & Sokolov, B. (2010). *Client relay simulation model for centralized wireless networks.* Paper presented at the Federation of European Simulation Societies Congress, Prague, Czech Republic.

Python Twisted. (2010). *The python twisted documentation.* Retrieved December 24, 2010, from http://twistedmatrix.com/projects/core/documentation/howto/book.pdf

Qiao, D., Choi, S., & Shin, K. G. (2002). Good- put Analysis and Link Adaptation for IEEE 802.11a Wireless LANs. *IEEE Transactions on Mobile Computing*, 278–292. doi:10.1109/TMC.2002.1175541.

Qureshi, B., Min, G., & Kouvatsos, D. (2010). A Framework for Building Trust Based Communities in P2P Mobile Social Networks. In *Proceedings of the IEEE Conference on Computer and Information Technology* (pp. 567-574).

Rabaey, J., Ammer, J., da Silva, J., Jr., & Patel, D. (2000). PicoRadio: Ad-hoc wireless networking of ubiquitous low-energy sensor/monitor nodes. In *Proceedings of the IEEE Computer Society Annual Workshop on Very Large Scale Integration* (pp. 9-12). Washington, DC: IEEE Computer Society.

Rafaeli, S., & Hutchison, D. A. (2003). Survey of key management for secure group communication. *ACM Computing Surveys*, *35*(3), 309–329. doi:10.1145/937503.937506.

Raghavendra, C. S. (2007). *Wireless sensor networks*. New York, NY: Springer.

Rajagopalan, R., & Varshney, P. K. (2006). Data-aggregation techniques in sensor networks: A survey. *IEEE Communications Surveys & Tutorials*, *8*(4), 48–63. doi:10.1109/COMST.2006.283821.

Ramachandran, K. N., Belding, E. M., Almeroth, K. C., & Buddhikot, M. M. (2006). Interference-aware channel assignment in multi-radio wireless mesh networks. In *Proceedings of the 25th IEEE International Conference on Computer Communications* (pp. 1-12).

Raman, V., & Vaidya, N. H. (2009). *Adjacent channel interference reduction in multichannel wireless networks using intelligent channel allocation* (Tech. Rep. No. 06-27074). Urbana, IL: University of Illinois at Urbana-Champaign.

Ramjee, R., Li, L., La Porta, T., & Kasera, S. (2002). IP paging service for mobile hosts. *Wireless Networks*, *8*, 427–441. doi:10.1023/A:1016534027402.

Ramjee, R., Varadhan, K., Salgarelli, L., Thuel, S. R., Wang, S.-Y., & La Porta, T. (2002). HAWAII: A domain-based approach for supporting mobility in wide-area wireless networks. *IEEE/ACM Transactions on Networking*, *10*(3), 396–410. doi:10.1109/TNET.2002.1012370.

Rappaport, T. S. (2002). *Wireless Communications: Principles and Practice* (2nd ed.). Upper Saddle River, NJ: Prentice Hall.

Reddy, A., Estrin, D., & Govindan, R. (2000). Large scale fault isolation. *IEEE Journal on Selected Areas in Communications*, *18*(5), 733–743. doi:10.1109/49.842989.

Redlich, O., Ezri, D., & Wulich, D. (2009). Snr estimation in maximum likelihood decoded spatial multiplexing. Retrieved from http://arxiv.org/abs/0909.1209

Reis, C., Mahajan, R., Rodrig, M., Wetherall, D., & Zahorjan, J. (2006). Measurement-based models of delivery and interference in static wireless networks. *Communications of the ACM*, *36*(4), 51–62. doi:10.1145/1151659.1159921.

Resta, G., Santi, P., & Simon, J. (2007, September). Message propagation in vehicular ad hoc networks. In *Proceedings of the ACM International Symposium on Mobile Ad Hoc Networking and Computing*, Montreal, QC, Canada (pp. 140-149).

Rhee, I., & Xu, L. (2007). Limitations of equation-based congestion control. *IEEE/ACM Transactions on Networking*, *15*(4), 852–865. doi:10.1109/TNET.2007.893883.

Robinson, J., Papagiannaki, K., Diota, C., Guo, X., & Krishnamurthy, L. (2005). *Experimenting with a multi-radio mesh networking testbed*. Paper presented at the First Workshop on Wireless Network Measurements.

Rondeau, T. W., Le, B., Rieser, C. J., & Bostian, C. W. (2004). Cognitive radios with genetic algorithms: Intelligent control of software defined radios. In *Proceedings of the Software Defined Radio Forum Technical Conference* (pp. 3-8).

Rondeau, T. W., & Bostian, C. W. (2009). *Artificial intelligence in wireless communications*. Boston, MA: Artech House.

Rong, B., & Ephremides, A. (2009). On opportunistic cooperation for improving the stability region with multipacket reception. In *Proceedings of the 3rd Euro-NF Conference on Network Control and Optimization* (pp. 45-59).

Rong, B., Qian, Y., Lu, K., Hsiao-Hwa, C., & Mohsen, G. (2008). Call admission control optimization in WiMAX networks. *IEEE Transactions on Vehicular Technology*, *57*, 621–632.

Rozner, E., Seshadri, J., Mehta, Y. A., & Qiu, L. (2009). SOAR: Simple Opportunistic Adaptive Routing Protocol for Wireless Mesh Networks. *IEEE Transactions on Mobile Computing*, *8*(2), 1622–1635. doi:10.1109/TMC.2009.82.

Sarkar, N. I. (2006, January 2-4). Fairness studies of IEEE 802.11b DCF under heavy traffic conditions. In *Proceedings of the First IEEE International Conference on Next-Generation Wireless Systems* (pp. 11-16). Washington, DC: IEEE Computer Society.

Sarkar, N. I., & Sowerby, K. W. (2006, November 27-30). Wi-Fi performance measurements in the crowded office environment: a case study. In *Proceedings of the 10th IEEE International Conference on Communication Technology*, Guilin, China (pp. 37-40). Washington, DC: IEEE Computer Society.

Schilling, D. L., Pickholtz, R. L., & Milstein, L. B. (1990). Spread spectrum goes commercial. *IEEE Spectrum*, *27*(8), 40–45. doi:10.1109/6.58433.

Schoch, E., Kargl, F. M., & Leinmüller, T. (2008). Communication patterns in VANETs. *IEEE Communications Magazine*, *46*(11), 119–125. doi:10.1109/MCOM.2008.4689254.

Schurgers, C. (2002). *Energy-aware wireless communications*. Unpublished doctoral dissertation, University of California, Los Angeles.

Schwetman, H. (2001). CSIM19: A powerful tool for building system models. In *Proceedings of the 33th IEEE Winter Simulation Conference*, Arlington, VA (pp. 250-255).

Sendonaris, A., Erkip, E., & Aazhang, B. (2003). User cooperation diversitypart I: System description. *IEEE Transactions on Communications*, *51*(11), 1927–1938. doi:10.1109/TCOMM.2003.818096.

Shafaq, B., & Ratan, K. (2007). Adaptive connection admission control and packet scheduling for QoS provisioning in mobile WiMAX. In *Proceedings of the IEEE International Conference on Signal Processing and Communications*, Dubai, UAE (p. 1355).

Shah, R. C., & Rabaey, J. (2002, March). Energy aware routing for low energy ad hoc sensor networks. In *Proceedings of the IEEE Wireless Communications and Networking Conference*, Orlando, FL (pp. 350-355).

Shah, V., Mehta, N. B., & Yim, R. (2009). Relay selection and data transmission throughput tradeoff in cooperative systems. In *Proceedings of the IEEE Global Telecommunications Conference* (pp. 230-235).

Shah, V., Mehta, N. B., & Yim, R. (2010). Splitting algorithms for fast relay selection: Generalizations, analysis, and a unifed view. *IEEE Transactions on Wireless Communications*, *9*(4), 1525–1535. doi:10.1109/TWC.2010.04.091364.

Shen, J., Jiang, T., Liu, S., & Zhang, Z. (2009). Maximum channel throughput via cooperative spectrum sensing in cognitive radio networks. *IEEE Transactions on Wireless Communications*, *8*(10), 5166–5175. doi:10.1109/TWC.2009.081110.

Shevtekar, A., Stille, J., & Ansari, N. (2008). On the impacts of low rate DoS attacks on VoIP traffic. *Security and Communication Networks*, *1*(1), 45–56. doi:10.1002/sec.7.

Shin, M., Lee, S., & ah Kim, Y. (2006). Distributed channel assignment for multi-radio wireless networks. In *Proceedings of the IEEE International Conference on Mobile Adhoc and Sensor Systems* (pp. 417-426).

Shin, M., Lee, S., & Kim, Y. (2006). Distributed channel assignment for multi-radio wireless networks. In *Proceedings of the IEEE International Conference on Mobile Adhoc and Sensor Systems Conference* (pp. 417-426).

Shokri, R., Yazdani, N., & Khonsari, A. (2007). Chain-based anonymous routing for wireless ad hoc networks. In *Proceedings of the 4th IEEE Consumer Communications and Networking Conference* (pp. 297-302).

Shore, W. (2010). *Wireless Transport Layer Congestion Control Evaluation* (Unpublished master's thesis). University of North Florida, Jacksonville, FL.

Shuai, L., Xie, G., & Yang, J. (2008). Characterization of HTTP behavior on access networks in WEB2.0. In *Proceedings of the IEEE International Thermoelectric Society* (pp. 1-6).

Simanapalli, S. (1994). Adaptive array methods for mobile communications. In *Proceedings of the 44th International Conference on Vehicular Technology*, Stockholm, Sweden (Vol. 3, pp. 1503-1506).

Singh, R., Singh, H., & Kaler, R. S. (2010). An Adaptive Energy Saving and Reliable Routing Protocol for Limited Power Sensor Networks. In *Proceedings of the International Conference on Advances in Computer Engineering* (pp. 79-85).

Siringoringo, W., & Sarkar, N. I. (2009). Teaching and learning Wi-Fi networking fundamentals using limited resources. In Gutierrez, J. (Ed.), *Selected readings on telecommunications and networking* (pp. 22–40). Hershey, PA: IGI Global.

Si, W., Selvakennedy, S., & Zomaya, A. Y. (2009). An overview of channel assignment methods for multi-radio multi-channel wireless mesh networks. *Journal of Parallel and Distributed Computing*, *70*(5).

Sohraby, K., Minoli, D., & Znati, T. F. (2007). *Wireless sensor networks: Technology, protocols, and applications*. New York, NY: Wiley-Interscience. doi:10.1002/047011276X.

Song, G. (2005). *Cross-layer optimization for spectral and energy efficiency*. Unpublished doctoral dissertation, Georgia Institute of Technology, School of Electrical and Computer Engineering, Atlanta.

Song, C., & Guizani, M. (2006). Energy map: Mining wireless sensor network data. In. *Proceedings of the IEEE International Conference on Communications*, *8*, 3525–3529.

Song, C., Guizani, M., & Sharif, H. (2007). Adaptive clustering in wireless sensor networks by mining sensor energy data. *Computer Communications*, *30*, 2968–2975. doi:10.1016/j.comcom.2007.05.027.

Song, G., & Li, Y. (2006). Asymptotic throughput analysis for channel-aware scheduling. *IEEE Transactions on Communications*, *54*(10), 1827–1834. doi:10.1109/TCOMM.2006.881254.

Spagnolini, U. (2004). A simplified model to evaluate the probability of error in DS-CDMA systems with adaptive antenna arrays. *IEEE Transactions on Wireless Communications*, *3*(2), 578–587. doi:10.1109/TWC.2003.819020.

Sridhar, S., Guo, J., & Jha, S. (2009). Channel assignment in multi-radio wireless mesh networks: A graph-theoretic approach. In *Proceedings of the First International Communication Systems and Networks and Workshops* (pp. 1-10).

Stallings, W. (2005). *Wireless Communications and Networks* (2nd ed.). Upper Saddle River, NJ: Prentice Hall.

Stuber, G. (2001). *Principles of mobile communication*. Boston, MA: Kluwer Academic Publishers.

Subramanian, A. P., Gupta, H., Das, S. R., & Cao, J. (2008). Minimum interference channel assignment in multi-radio wireless mesh networks. *IEEE Transactions on Mobile Computing*, *7*(12), 1459–1473. doi:10.1109/TMC.2008.70.

Sunghyun, C., del Prado, J., Sai Shankar, N., & Mangold, S. (2003, May 11-15). IEEE 802.11 e contention-based channel access (EDCF) performance evaluation. In *Proceedings of the IEEE International Conference on Communications* (pp. 1151-1156).

Sun, J., Zhang, C., & Zhang, Y. (2011). SAT: A security Architecture Achieving Anonymity and Traceability in Wireless Mesh Networks. *IEEE Transactions on Dependable and Secure Computing*, *8*(2), 295–307. doi:10.1109/TDSC.2009.50.

Sun, Y., & Liu, K. J. R. (2007). Hierarchical group access control for secure multicast communications. *IEEE/ACM Transactions on Networking*, *15*(6), 1514–1526. doi:10.1109/TNET.2007.897955.

Swales, S. C., Beach, M. A., Edwards, D. J., & McGeehn, J. P. (1990). The performance enhancement of multibeam adaptive base station antennas for cellular land mobile radio systems. *IEEE Transactions on Vehicular Technology*, *39*(1), 56–67. doi:10.1109/25.54956.

Sy, D., Chen, R., & Bao, L. (2006). ODAR: On-demand anonymous routing in ad hoc networks. In *Proceedings of the IEEE Conference on Mobile Adhoc and Sensor Systems* (pp. 267-276).

Takeuchi, S., Koga, H., Iida, K., Kadobayashi, Y., & Yamaguchi, S. (2005). Performance Evaluations of DCCP for Bursty Traffic in Real-Time Applications. In *Proceedings of the IEEE/IPSJ International Symposium on Applications and the Internet* (pp. 142-149). Washington, DC: IEEE Computer Society.

Tarhini, C., & Chahed, T. (2008). Density-based admission control in IEEE802.16e mobile WiMAX. In *Proceedings of the 1st IFIP Wireless Days Conference* (pp. 1-5).

Tarokh, V., Seshadri, N., & Calderbank, A. R. (1998). Space-time codes for high data rate wireless communicatoin: Performance analysis and code construction. *IEEE Transactions on Information Theory*, *44*, 744–765. doi:10.1109/18.661517.

Tatsuaki, O., Lan, L., & Massimiliano, L. (2006, June). Multi-Hop Vehicular Broadcast (MHVB). In *Proceedings of the IEEE 6th International Conference on ITS Telecommunications* (pp. 757-760).

Tenenbaum, J. B., Silva, V. D., & Langford, J. C. (2000). A global geometric framework for nonlinear dimensionality reduction. *Science*, *290*, 2319–2323. doi:10.1126/science.290.5500.2319.

Texas Instruments. (2006). *Datasheet CC2420, 2.4 GHz IEEE 802.15.4 / ZigBee-ready RF Transceiver*. Dallas, TX: Author.

Tisue, S., & Wilensky, U. (2004). NetLogo: A Simple Environment for Modeling Complexity. In *Proceedings of the International Conference on Complex Systems*, Boston, MA.

Toktas, E., Biyikoglu, E.-U., & Yilmaz, A. O. (2009). Subcarrier allocation in OFDMA with time varying channel and packet arrivals. In *Proceedings of the European Wireless Conference* (pp. 178-183).

Tonguz, O. K., Wisitpongphan, N., & Fan, B. (2010). DV-CAST: A distributed vehicular broadcast protocol for vehicular ad hoc networks. *IEEE Wireless Communications*, *17*(2), 47–57. doi:10.1109/MWC.2010.5450660.

Torrent-Moreno, M., Mittag, J., Santi, P., & Hartenstein, H. (2009). Vehicle-to-vehicle communication: Fair transmit power control for safety-critical information. *IEEE Transactions on Vehicular Technology*, *58*(7), 3684–3703. doi:10.1109/TVT.2009.2017545.

Tsai, M., Chilamkurti, N., Park, J., & Shieh, C. (2010). Multi-path transmission control scheme combining bandwidth aggregation and packet scheduling for real-time streaming in multi-path environment. *IET Communications*, *4*(8), 937–945. doi:10.1049/iet-com.2009.0661.

Tsai, M., Chilamkurti, N., & Shieh, C. (2010). A network adaptive forward error correction mechanism to overcome burst packet losses for video streaming over wireless networks. *Journal of Internet Technology*, *11*(4), 473–482.

Tsai, M., Chilamkurti, N., & Shieh, C. (2011). An adaptive packet and block length forward error correction for video streaming over wireless networks. *Wireless Personal Communications*, *56*(3), 435–446. doi:10.1007/s11277-010-9981-z.

Tsai, M., Chilamkurti, N., Shieh, C., & Vinel, A. (2011). MAC-level forward error correction mechanism for minimum error recovery overhead and retransmission. *Mathematical and Computer Modelling*, *53*(11-12), 2067–2077. doi:10.1016/j.mcm.2010.05.019.

Tsai, M., Chilamkurti, N., Zeadally, S., & Vinel, A. (2011). Concurrent multipath transmission combining forward error correction and path interleaving for video streaming in wireless networks. *Computer Communications*, *34*(9), 1125–1136. doi:10.1016/j.comcom.2010.02.001.

Tsai, M., Huang, T., Ke, C., & Hwang, W. (2011). Adaptive hybrid error correction model for video streaming over wireless networks. *Multimedia Systems Journal*, *17*(4), 327–340. doi:10.1007/s00530-010-0213-x.

Tsai, M., Huang, T., Shieh, C., & Chu, K. (2010). Dynamical combination of byte level and sub-packet level FEC in HARQ mechanism to reduce error recovery overhead on video streaming over wireless networks. *Computer Networks, 54*(17), 3049–3067. doi:10.1016/j.comnet.2010.06.003.

Tsai, M., Shieh, C., Hwang, W., & Deng, D. (2009). An adaptive multi-hop FEC protection scheme for enhancing the QoS of video streaming transmission over wireless mesh networks. *International Journal of Communication Systems, 22*(10), 1297–1318. doi:10.1002/dac.1032.

Tsai, M., Shieh, C., Ke, C., & Deng, D. (2010). Sub-packet forward error correction mechanism for video streaming over wireless networks. *Multimedia Tools and Applications, 47*(1), 49–69. doi:10.1007/s11042-009-0406-5.

Tseng, Y., Jan, R., Chen, C., Wang, C., & Li, H. (2010). A vehicle-density-based forwarding scheme for emergency message broadcasts in VANETs. In *Proceedings of the IEEE International Conference on Mobile Adhoc and Sensor Systems* (pp. 703-708).

Tung, H. Y., Tsang, K., Lee, L., & Ko, K. (2008). QoS for mobile wimax networks: Call admission control and bandwidth allocation. In *Proceedings of the IEEE Consumer Communications and Networking Conference* (pp. 576-580).

Turyn, R. (1963). Ambiguity functions of complementary sequences. *IEEE Transactions on Information Theory, 9*(1), 46–47. doi:10.1109/TIT.1963.1057807.

Upadhyay, P. C., & Tiwari, S. (in press). Distributed and fixed mobility management strategy for IP-based mobile networks. *International Journal of Business Data Communications and Networking*.

Uzcategui, R., & Acosta-Marum, G. (2009). Wave: A tutorial. *IEEE Communications Magazine, 47*(5), 126–133. doi:10.1109/MCOM.2009.4939288.

Valko, A. G. (1999). Cellular IP: A new approach to internet host mobility. *Computer Communication Review, 29*(1), 50–65.

Vinel, A., Dudin, A., Andreev, S., & Xia, F. (2010, September). Performance modeling methodology of emergency dissemination algorithms for vehicular ad-hoc networks. In *Proceedings of the 7th International Symposium on Communication Systems, Networks and Digital Signal Processing*, Newcastle, UK (pp. 397-400).

Viterbi, A. M., & Viterbi, A. J. (1993). Erlang capacity of a power controlled CDMA system. *IEEE Journal on Selected Areas in Communications, 11*(6), 892–900. doi:10.1109/49.232298.

Wang, H., & Li, W. (2005). Dynamic admission control and QoS for 802.16 wireless man. In *Proceedings of the Wireless Telecommunications Symposium* (pp. 60-66).

Wang, Y., & Garcia-Luna-Aceves, J. J. (2003, March 16-20). Throughput and fairness in a hybrid channel access scheme for ad hoc networks. In *Proceedings of the IEEE Wireless Communications and Networking Conference* (pp. 988-993). Washington, DC: IEEE Computer Society.

Wang, H., & Chen, B. (2004). Asymptotic distributions and peak power analysis for uplink OFDMA signals. In. *Proceedings of the IEEE International Conference on Acoustics, Speech, and Signal Processing, 4*, 1085–1088.

Wang, W., Liew, S. C., & Li, V. O. K. (2005). Solutions to performance problems in VoIP over a 802.11 wireless LAN. *IEEE Transactions on Vehicular Technology, 54*(1), 366–384. doi:10.1109/TVT.2004.838890.

Wan, Z., Ren, K., Zhu, B., Preneel, B., & Ming, G. (2010). Anonymous User Communication for Privacy Protection in Wireless Metropolitan Mesh Networks. *IEEE Transactions on Vehicular Technology, 59*(2), 519–532. doi:10.1109/TVT.2009.2028892.

Wiegle, M. (2007). *PackMime-HTTP: Web traffic generation in NS2*. Retrieved from http://www.isi.edu/nsnam/ns/doc/node552.html.

Wilensky, U., & Stroup, W. (1999). Learning through Participatory Simulations: Network based Design for Systems Learning in Classrooms. In *Proceedings of the Computer Supported Collaborative Learning Conference*, Stanford, CA.

Wireless Networking Group at UIUC. (2010). *Net-x channel assignment framework*. Retrieved December 24, 2010, from http://www.crhc.illinois.edu/wireless/netx.html

Wisitpongphan, N., Tonguz, O. K., Parikh, J. S., Mudalige, P., Bai, F., & Sadekar, V. (2007). Broadcast storm mitigation techniques in vehicular ad hoc networks. *IEEE Wireless Communications, 14*(6), 84–94. doi:10.1109/MWC.2007.4407231.

Woo, A., Tong, T., & Culler, D. (2003). Taming the Underlying Challenges of Reliable Multihop Routing in Sensor Networks. In *Proceedings of the 1st International Conference on Embedded Networked Sensor Systems* (pp. 14-27).

Wu, D., Liu, Y., & Ross, K. W. (2009, April). Queuing Network Models for Multi-Channel P2P Live Streaming Systems. In *Proceedings of the IEEE Conference on INFOCOM* (pp.73-81).

Wu, J., Fujimoto, R., Guensler, R., & Hunte, M. (2004). MDDV: A mobility-centric data dissemination algorithm for vehicular networks. In *Proceedings of the 1st ACM International Workshop on Vehicular Ad Hoc Networks* (pp. 47-56).

Wu, Q., Mu, Y., Susilo, W., Qin, B., & Domingo-Ferrer, J. (2009). Asymmetric group key agreement. In A. Joux (Ed.), *Proceedings of the 28th International Conference on the Theory and Applications of Cryptographic Techniques* (LNCS 5479, pp. 153-170).

Wu, T., Xue, Y., & Cui, Y. (2006). Preserving traffic privacy in wireless mesh networks. In *Proceedings of the International Symposium on a World of Wireless, Mobile and Multimedia Networks* (pp. 459-461).

Wu, X., & Li, N. (2006). Achieving privacy in mesh networks. In *Proceedings of the 4th ACM Workshop on Security of Ad Hoc and Sensor Networks* (pp. 13-22).

Wu, Y., & Cao, G. (2001). Stretch-optimal scheduling for on-demand data broadcasts. In *Proceedings of the Tenth International Conference on Computer Communications and Networks* (pp. 500-504).

Xiangping, R. Q. B. (2004). Opportunistic splitting algorithms for wireless networks. In *Proceedings of the Twenty-Third Annual Joint Conference of the IEEE Computer and Communications Societies* (Vol. 3, pp. 1662).

Xiao, Y. (2004, March 21-25). Concatenation and piggyback mechanisms for the IEEE 802.11 MAC. In *Proceedings of the IEEE Wireless Communications and Networking Conference* (pp. 1642-1647). Washington, DC: IEEE Computer Society.

Xiao, Y., Peng, Y., & Li, S. (2007). PAPR reduction for interleaved OFDMA with low complexity. In *Proceedings of the 6th International Conference on Information, Communications and Signal Processing* (pp. 1-4).

Xie, J. (2006). User independent paging scheme for mobile IP. *Wireless Networks, 12*, 145–158. doi:10.1007/s11276-005-5262-2.

Xie, J., & Akylidiz, I. F. (2002). A novel distributed dynamic location management scheme for minimizing signaling costs in mobile IP. *IEEE Transactions on Mobile Computing, 1*(3), 163–175. doi:10.1109/TMC.2002.1081753.

Xu, J., Hu, Q., Lee, W., & Lee, D. L. (2004). Performance evaluation of an optimal cache replacement policy for wireless data dissemination. *IEEE Transactions on Knowledge and Data Engineering, 16*(1), 125–139. doi:10.1109/TKDE.2004.1264827.

Xu, J., Tang, X., & Lee, W. (2006). Time-critical on-demand data broadcast algorithms, analysis and performance evaluation. *IEEE Transactions on Parallel and Distributed Systems, 17*(1), 3–14. doi:10.1109/TPDS.2006.14.

Yan, H., Xu, Y., & Gidlund, M. (2009, January). Experimental e-health applications in wireless sensor networks. In *Proceedings of the WRI International Conference on Communications and Mobile Computing* (pp. 563-567).

Ye, G., Saadawi, T., & Lee, M. (2002). SCTP congestion control performance in wireless multi-hop networks. In *Proceedings of the Conference on Military Communications* (Vol. 2, pp. 934-939). Washington, DC: IEEE Computer Society.

Ye, W., Heidemann, J., & Estrin, D. (2002). An energy-efficient MAC protocol for wireless sensor networks. In *Proceedings of the 21st IEEE Conference on Computer and Communications Societies* (Vol. 3, pp. 1567-1576).

Yi, L. Z., Bin, L., Tong, Z., & Wei, Y. (2008). On scheduling of data dissemination in vehicular networks with mesh backhaul. In *Proceedings of the IEEE International Conference on Communications Workshops* (pp. 385-392).

Yin, J., Wang, X., & Agrawal, D. P. (2005). Modeling and optimization of wireless local area network. *Computer Communications*, *28*(10), 1204–1213. doi:10.1016/j.comcom.2004.07.027.

Yoo, T., & Goldsmith, A. (2006). On the optimality of multi-antenna broadcast scheduling using zero-forcing beamforming. *IEEE Journal on Selected Areas in Communications*, *24*(3), 528–541. doi:10.1109/JSAC.2005.862421.

You, T., Yeh, C.-H., & Hassanein, H. (2003). CSMA/IC: A new class of collision-free MAC protocols for ad hoc wireless networks. In *Proceedings of the Eighth IEEE International Symposium on Computers and Communication* (pp. 843-848). Washington, DC: IEEE Computer Society.

Yuh-Shyan, C., Yun-Wei, L., & Sing-Ling, L. (2009, November 30-December 4). A Mobicast routing protocol in vehicular ad-hoc networks. In *Proceedings of the IEEE Global Telecommunications Conference* (pp. 1-6).

Yun, C. W., Sung, D. K., & Aghvami, A. H. (2003). Steady state analysis of P-MIP mobility management. *IEEE Communications Letters*, *7*(6), 278–280. doi:10.1109/LCOMM.2003.813797.

Yun, L., Ke-Ping, L., Wei-Liang, Z., & Qian-Bin, C. (2006). A novel random backoff algorithm to enhance the performance of IEEE 802.11 DCF. *Wireless Personal Communications*, *36*(1), 29–44. doi:10.1007/s11277-006-6176-8.

Zhang, J., & Varadharajan, V. (2006). A scalable multi-service group key management scheme. In *Proceedings of the Advanced International Conference on Telecommunications and International Conference on Internet and Web Applications and Services*.

Zhang, W., Jaehnert, J., & Dolzer, K. (2003, April 22-25). Design and evaluation of a handover decision strategy for 4th generation mobile networks. In *Proceedings of the 57th IEEE Semiannual Vehicular Technology Conference* (Vol. 3, pp. 1969-1973).

Zhang, X., Gomez Castellanos, J., & Campbell, A. T. (2002). P-MIP: Paging extensions for mobile IP. *Mobile Networks and Applications*, *7*(2), 127–141. doi:10.1023/A:1013774805067.

Zhang, Y., & Fang, Y. (2006). ARSA: An Attack-Resilient Security Architecture for Multihop Wireless Mesh Networks. *IEEE Journal on Selected Areas in Communications*, *24*(10), 1916–1928. doi:10.1109/JSAC.2006.877223.

Zhang, Y., Zhao, J., & Cao, G. (2010). Service scheduling of vehicle-roadside data access. *Mobile Networks and Applications*, *15*(1), 83–96. doi:10.1007/s11036-009-0170-9.

Zhao, Y. J., Govindan, R., & Estrin, D. (2002, March). Residual energy scans for monitoring wireless sensor networks. In *Proceedings of the IEEE Wireless Communications and Networking Conference* (pp. 356-362).

Zhao, Y. J., Govindan, R., & Estrin, D. (2003). *Computing aggregates for monitoring wireless sensor networks* (Tech. Rep. No. 02-773). Los Angeles, CA: University of Southern California.

Zhao, J., & Cao, G. (2008). VADD: Vehicle-assisted data delivery in vehicular ad hoc networks. *IEEE Transactions on Vehicular Technology*, *57*(3), 1910–1922. doi:10.1109/TVT.2007.901869.

Zhao, J., Zhang, Y., & Cao, G. (2007). Data pouring and buffering on the road: A new data dissemination paradigm for vehicular ad hoc networks. *IEEE Transactions on Vehicular Technology*, *56*(6), 3266–3277. doi:10.1109/TVT.2007.906412.

Zhao, L., Al-Dubai, A. Y., & Min, G. (2010). GLBM: A new QoS aware multicast scheme for wireless mesh networks. *Journal of Systems and Software*, *83*(8), 1318–1326. doi:10.1016/j.jss.2010.01.044.

Zheng, J., & Ma, M. (2009). A utility-based joint power and rate adaptive algorithm in wireless ad hoc networks. *IEEE Transactions on Communications*, *57*(1), 134–140. doi:10.1109/TCOMM.2009.0901.060524.

Zheng, L., & Tse, D. N. C. (2002). Diversity and multiplexing: A fundamental tradeoff in multiple antenna channels. *IEEE Transactions on Information Theory*, *49*(5), 1073–1096. doi:10.1109/TIT.2003.810646.

Zhu, H., Lin, X., Lu, R., Ho, P.-H., & Shen, X. (2008). SLAB: A Secure Localized Authentication and Billing Scheme for Wireless Mesh Networks. *IEEE Transactions on Wireless Communications*, *7*(10), 3858–3868. doi:10.1109/T-WC.2008.07418.

About the Contributors

Naveen Chilamkurti teaches at La Trobe University (Melbourne, Australia). He holds a PhD and Master of Computer Science from La Trobe University. He is a senior member of IEEE Society and active in IEEE Communications and Computer Society technical committees. Dr. Naveen organized and chaired many international conferences in the areas of wireless computing, pervasive computing, and next generation wireless networks. He serves as technical program committee, steering committee, organizing committee, and track chair in many international conferences. Dr. Naveen has authored around seventy-five scientific publications, book chapter, and journals. He serves as an associate editor for six international journals such as *Wiley Journal of Communication Systems* and others. He also serves as a guest editor for various international journals. Dr. Naveen's research interests are primarily in the areas of wireless communications, multimedia, pervasive computing, 4G communications, wireless sensor networks, green networking, WiMAX, and RFID technologies.

* * *

Sabbir Ahmed received his BE degree (1999) in Electrical and Electronic from Bangladesh University of Engineering and Technology (BUET), Bangladesh and M.E. degree (2005) in Information Science and System Engineering from Ritsumeikan University, Japan. He is currently a PhD student in Ritsumeikan University, Japan. His research interests include mobile and wireless communication, multicarrier systems, multiple access methods etc. He received the Best Student Paper Award at the 25th AIAA International Communication Satellite System Conference held in Seoul, South Korea, 2007.

Sanjay P. Ahuja has a MS and PhD in computer science and engineering from the University of Louisville. He is a full professor and the Fidelity National Financial Distinguished Professor in computer and information sciences in the school of computing at the University of North Florida. He is a senior member of the IEEE. His research interests include the performance evaluation, modeling, and simulation of computer networks and distributed systems, network security, and cloud computing.

G. G. Md. Nawaz Ali is currently a PhD student in the Department of Computer Science, City University of Hong Kong. He received his BSc degree in Computer Science and Engineering from Khulna University of Engineering & Technology, Bangladesh. He is a student member of IEEE. His current research interests include computer networking, mobile computing and ad hoc networking with a focus on vehicular ad hoc networking.

Keith Blow joined BT in 1981 and worked on the theory of non-linear optical propagation effects in fibres, principally solitons which are special pulses of light where the intensity dependent refractive index can cancel the frequency dependence of the refractive index and lead to pulses propagating without distortion. In 1990 he set up a new research group working on quantum optical properties and non-linear spatial optics. In 1996 he began working on the application of bit serial processing techniques to all-optical networks and optical computing. The bit serial approach led to the demonstration of all-optical pseudo random number generators, parity counters, binary adders and binary counters. In 1999 he was appointed professor of Photonic Networks at Aston University, UK where he is continuing to work on many of these areas. In 2000 he began working on the applications of serial techniques and reconfigurable hardware to the Internet physical layer and is now interested in the applications of ad-hoc sensor networks.

Bastian Blywis received a Diploma degree in Computer Science from Freie Universität Berlin, Germany in 2007. He started as a research associate with teaching duties in 2007 at Freie Universität Berlin. Since April 2011 he is employed for a third party funded project and finishing his PhD.

Edward Chan received his BSc ad MSc degrees in Electrical Engineering from Stanford University, and his PhD in Computer Science from Sunderland University. He worked in the Silicon Valley for a number of years in the design and implementation of computer networks and real-time control systems before joining City University of Hong Kong where he is now an Associate Professor. His current research interests include performance evaluation of high speed networks, mobile data management, power-aware computing, and network management.

Marc Eberhard received his Dipl.-Phys. from Technische Universita¨t Minchen, Germany, in 1996 and his Dr. rer. nat. from Heinrich-Heine-Universita¨t D sseldorf, Germany, in 2000. He then joined the Photonics Research Group at Aston University in Birmingham, UK, as a contract research fellow and was promoted to a Lecturer position in 2001. He has worked on different topics in theoretical physics, ranging from semiconductor devices over plasma physics to optical communication systems. His recent work at Aston is on modelling of optical communication systems with an emphasis on polarisation mode dispersion.

Scott Fowler received a BS (1998) from Minot State University, Minot, ND, U.S., MS (2001) from the University of North Dakota, Grand Forks, N.D., U.S. and PhD (2006) from Wayne State University, Detroit, MI, U.S., all degrees are in computer science. In 2006 - 2010 he was a research fellow in the Adaptive Communications Networks Research Group at Aston University, UK. Since 2010 he has been an associate professor at Linköping University, Sweden. Dr. Fowler has served on several IEEE conferences/workshops as TPC to Co-Chair. His research has been funded and supported by European Union Framework 7, Excellence Center at Linköping - Lund in information Technology (ELLIIT), Ericsson and Ascom. Dr. Fowler's research interests include Quality of Service (QoS) support over heterogeneous networks, Computer networks (wired, wireless), Mobile Computing, Performance Evaluation of Networks and Security. Dr. Fowler was a host for a Fulbright Specialist from the USA in 2011. He is Member of IEEE and ACM.

Mesut Günes received a Diploma degree in Computer Science from RWTH Aachen University, Germany in 1998 and a PhD in 2004. He was a research and teaching assistant at the Department of Computer Science, Informatik 4, RWTH Aachen University, from 2004 - 2005 he was a research fellow at the International Computer Science Institute (ICSI), Berkeley, USA. Since 2007 he is Professor, Head of Distributed Embedded Systems, Institute of Computer Science at Freie Universität Berlin, Germany.

Muhammad A. Javed received his BSc in Electrical Engineering from University of Engineering and Technology Lahore, Pakistan in 2008. Currently, he is working towards completion of his Masters of Philosophy degree in the School of Electrical Engineering & Computer Science at University of Newcastle, Australia. His research interests are vehicular ad hoc networks, communication architecture and protocol design for emerging wireless networks and wireless sensor networks.

Felix Juraschek received a Diploma degree in Computer Science from Freie Universität Berlin, Germany in 2008. In the same year, he is joined the faculty of Mathematics and Computer Science at the same university and was responsible for the EU OPNEX project. He joined the METRIK graduate school in 2011 and is working towards his PhD in the field of distributed channel assignment.

Makoto Kawai was born in 1949. He received BS, MS and Dr. Eng. degrees in electrical engineering from Kyoto University, Kyoto, Japan, in 1972, 1974 and 1987, respectively. During 1974 and 1999 he worked mainly for research and development of wireless communication systems at Nippon Telegraph and Telephone Corporation (NTT) and also worked at Advanced Telecommunications Research Institute International (ATR) from 1996 to 1999. He was an associate professor at Graduate School of Informatics, Kyoto University from 1999 to 2003. He is now a professor of Graduate School of Science and Engineering and College of Information Science and Engineering, Ritsumeikan University, Shiga, Japan. He is a member of IEEE, AIAA, IEICE, IPSJ and IEEJ.

Jamil Y. Khan received his PhD from the Department of Electronic and Electrical Engineering specializing in Communications Engineering from the University of Strathclyde, Glasgow, UK in 1991. Since his Ph.D he has worked as a research assistant in the University of Strathclyde 1991-92 in the European research program RACE then worked as a Lecturer/Senior in the Massey University, New Zealand 1992-1999. In 1999 he joined the University of Newcastle as a Senior Lecturer in Telecommunications engineering. He established the Telecommunications Engineering teaching and research programs in the university. Currently he is an Associate Professor in the School of Electrical Engineering & Computer Science. His main research interest areas are cognitive and cooperative wireless networks, smart grid communications, wireless network architecture, wireless sensor and body area networks. Jamil Khan is a senior member of the IEEE Communications Society and a member of the ACM.

Sumit Kumar (MS, CRC, IIIT Hyderabad) has been working as a research assistant in the project "Mobile and Static Cognitive Wireless Sensor Network" since April 2010 in International Institute of Information Technology, Hyderabad. He completed his B-tech in Electronics and Communication Engineering from Gurukula Kangri University in 2008. His research focuses on wireless communication systems and networks. His current research interests include cognitive networks, spectrum sensing algorithms, reconfigurable hardware for software radio and artificial intelligence.

G. Rama Murthy (Associate Professor, CRC, IIIT Hyderabad) obtained his B. Tech degree in Electronics and Communication Engineering from Sri Venka-teswara University, Tirupati, India, M.S. in Electrical Engg. from the Louisiana State University, Baton Rouge U.S.A and Ph.D. degree in computer engineering from Purdue University, West Lafayette, U.S.A. His research interests include Wireless Sensor Networks, Ad-hoc Wireless Networks, Computer Architecture, Multi Dimensional Neural Networks, Performance Evaluation, Cybernetics and Optimal Filtering. He is currently an associate professor at the International Institute of Information Technology (IIIT), Hyderabad, India. He is a member of Eta Kappa Nu, Phi Kappa Phi, IEEE and senior member of ACM. He has about 140 referred research publications and about 40 Journal publications to his credit. He is currently the editor of International Journal of Systemics, Cybernetics and Informatics and Research Journal of Telecommunication and Information Technology (INSINET publications).

Ahmed Shaikh received a MS and PhD from Aston University (2009). In 2006 - 2009 he was a research assistant in the Adaptive Communications Networks Research Group at Aston University, UK. He was also awarded the Overseas Research Student Academic Scholarship from the Higher Education Funding Council of England.

Feng She received the BS and MS from Central South University and Tongji University, respectively, and received the PhD from Shanghai Jiaotong University, Shanghai, China in 2008. He was with Huawei Technologies as a researcher from 2008 to 2009. Then he joined Shanghai University of Engineering Sciences, China, as an assistant professor in the Department of Electronic Engineering. In 2010, he held visiting research appointment at Cheng kung University (NCKU) in Tai Wan. In 2010, he joined Alcatel Lucent Shanghai Bell as a research scientist. His research work is on advanced MIMO technique in 3GPP LTE and LTE-A now. He is currently serving as an editor of Wireless Communications and Mobile and Computing (WCMC). He is a Co-Chair of IEEE Infocom2011 workshop of C&C Networks. Dr She's research interests include wireless communication and networking for multiple antenna system, with current focus on MIMO communication systems, multi-user communication theory, cross-layer design and optimization, cellular systems, and communication theory.

William R. Shore has a bachelor of science degree from the University of North Florida in electrical engineering, December 2007, and a master of science in computer and information sciences from the University of North Florida, December 2010. Shore is currently employed as a software engineer I at Beeline.com, a member of Adecco International. He is also an adjunct professor for the school of computing at the University of North Florida. He has on-going interests in networking technology, wireless communication systems, pervasive computing, and has extensive experience with embedded system development. He has broad programming experience in a variety of languages at many levels of abstraction and in many application environments. This is his first article in the networking and computing areas.

Deepti Singhal (Pursuing PhD, CRC, IIIT Hyderabad) has been working as a research assistant in the project "Mobile and Static Cognitive Wireless Sensor Network" since August 2010 in International Institute of Information Technology, Hyderabad and also pursuing her PhD. She has completed her M. Tech in Information and Communication Technology (ICT) from Dhirubhai Ambani Institute of Information and Communication Technology (DAIICT), Gandhinagar, India, in 2007. Her current interests include Wireless Sensor Networks, Ad Hoc Networks, Cognitive Radio, switching, routing, sub-netting, 3G & 4G networks and QoS provisioning in wireless networks.

Index

D

E

F

G

H

I

L

M

N

O

P

Q

R

S

T

U

V

W

Z